AF566662

ENDOTHELINS
in Biology and Medicine

Edited by

JOHN P. HUGGINS, PH.D.
Pfizer, Ltd.
Sandwich, Kent, England

JOHN T. PELTON, PH.D.
Hoechst Marion Roussel, Inc.
Cincinnati, Ohio

CRC Press
Boca Raton New York London Tokyo

Acquiring Editor: Marsha Baker
Project Editor: Renee Taub
Marketing Manager: Susie Carlisle
Direct Marketing Manager: Becky McEldowney
Cover design: Denise Craig
PrePress: Kevin Luong
Manufacturing: Sheri Schwartz

Library of Congress Cataloging-in-Publication Data

Endothelins in biology and medicine / edited by John P. Huggins, John T. Pelton.
p. cm.
Includes bibliographical references and index.
ISBN 0-8493-6975-4 (alk. paper)
1. Endothelins. I. Huggins, John P. II. Pelton, John T.
QP552.E54E543 1997
615′.36--dc20 96-27766
CIP

International Standard Book Number 0-8493-6975-4
Library of Congress Card Number 96-27766
Printed in the United States of America 1 2 3 4 5 6 7 8 9 0
Printed on acid-free paper

PREFACE

The publication of this book, *Endothelins in Biology and Medicine,* comes at an apt time in the pharmacological exploitation of these peptides for clinical benefit. Endothelin-1 (ET-1) was discovered in 1988 as a 21 residue peptide with two internal disulfide bonds, that was secreted from endothelial cells. It was characterized as being one of the most potent contractors of vascular smooth muscle, a property that is still true 7 years later. Its discovery has prompted enormous scientific interest and the attention of all major pharmaceutical companies.

In the years immediately following 1988, research into ETs led to a number of discoveries.

First, three peptides compose the ET family, these being ET-1, -2 and -3. (A mouse peptide, vasointestinal contractor or VIC, was subsequently recognized to be a murine and rat form of ET-2.) The varying relative potency of these peptides in functional assays and, more conclusively, their varying affinities in different cells and tissues led to the concept of ET receptor subtypes and is reviewed in the first two chapters of this book.

Secondly, it was rapidly shown that ETs and their receptors are widely distributed, have a variety of actions in varied tissues, and can be secreted from many cells and tissues, making the term "endothelin" somewhat a misnomer. These actions, as they have emerged in detail, comprise a significant portion of this volume. The distribution of ET receptors is reviewed by Anthony Davenport in Chapter 3, while the interactions between ETs and their receptors and the subsequent activation of intracellular signaling pathways is detailed in Chapters 4 and 5. Section III is devoted to the pathophysiology of the endothelins. In addition to the extensive cardiovascular responses (reviewed in Chapter 8) expected from the effects of ETs on vascular smooth muscle as described in the original report by Yanagisawa et al. in 1988, ETs effect and affect contractions of other smooth muscles. In Chapter 10, the role of ETs in the pulmonary system is described, while Chapter 12 is devoted to the urogenital actions of the endothelins. Endothelin also has powerful and complex effects in kidney (Chapter 9) and in nerve, including in the CNS (Chapter 11); other effects, such as ocular responses (Chapter 13) that may have relevance for early clinical testing of antagonists, are also elicited.

The third avenue of research that immediately blossomed after the initial discovery of ET-1, was the mechanism of ET synthesis (Chapter 6) and catabolism (Chapter 7). ETs are all synthesized as prepropeptides, and the final ETs are produced by the actions of endothelin converting enzymes (ECE). These have also been the subject of intense research, stimulating pharmaceutical companies and others to discover inhibitors that, by analogy to the angiotensin converting enzyme inhibitors, are expected to have major therapeutic benefit, especially in the absence of convincing evidence that ETs are stored.

Molecular biology has dramatically influenced the emerging picture of the biology of the ETs, providing the first clear evidence of receptor subtypes, demonstrating regions of receptors that are involved in binding and signal transduction, and proving the role of ETs in disease by the construction and characterization of transgenic animals (Chapter 2).

Although initial thoughts suggested that ET-1 was directly responsible for the opening of Ca^{2+} channels, its receptors have universally been shown to be of the type that couple with G-proteins and are linked to the turnover of phosphatidylinositol phosphates. A major theme that has emerged from studying effects elicited in various systems by ETs is that longer-term effects, such as mitogenesis, are induced as well as relatively acute responses such as smooth muscle contraction. A major challenge is to understand how signal transduction couples these acute and longer-term effects (Chapter 5). However, even acute responses of ETs are often characterized by slow onset and wash-out, and the binding characteristics of ETs (Chapter 3) are most unusual for G-protein coupled receptors, being pseudo-irreversible under many conditions. This has led Denis Guedin and colleagues to challenge and enlarge traditional assumptions and theories based on reversible ligand-receptor interactions (Chapter 4).

Much of the driving force of the above studies has been the discovery of agonists and antagonists of ET receptors, or their subtypes, that can be used for therapeutic benefit. Studies of structure-activity relationships (SAR) began using peptides derived from the ETs (Chapter 15); such an approach led to the discovery of receptor subtype-specific peptides (as well as ET-3, itself) that were subsequently useful to classify biological responses. The study of peptide conformation (Chapter 14) was an essential part of these SAR studies. Although there is now general agreement regarding many of the structural/conformational features present in these remarkable peptides, controversies continue over the nature of the C-terminal hexapeptide region of the molecule. Recognizing the central role of ET receptor antagonists in understanding biology and in future clinical practice, two chapters deal with these agents: Chapter 16 by Doherty and Patt describes the design and synthesis of novel receptor antagonists (both selective and nonselective, peptidic and non-peptidic) while Chapter 1 describes the pharmacology of these agents. Finally, the long-awaited step of the clinical pharmacology of newly emerging agents is described in Chapter 17, which also paves possible future paths for clinical usage of compounds now in the development pipelines of a number of pharmaceutical companies.

ETs are peptides with some unique characteristics: their potency, their acute and longer-lasting effects, and their binding characteristics, when coupled with their distribution, the location of their receptors, their evident autocrine or paracrine effects, and their altered levels in pathological situations, indicate that the therapeutic potential of agents acting either at ET receptors or ECE is likely to be substantial.

Note: The following abbreviations are used throughout the book: $Ca^{2+}{}_i$ or $[Ca^{2+}]_i$, intracellular free Ca^{2+} concentration; cAMP, cyclic AMP; cGMP, cyclic GMP; CNS, central nervous system; c.s.f., cerebrospinal fluid; ECE, endothelin converting enzyme; ET, endothelin; i.a., intraarterial; i.c.v., intracerebroventricular; InsP, inositol phosphate; i.p., intraperitoneal; i.v., intravenous; STX, sarafotoxin S6.

The Editors

John P. Huggins, Ph.D., is a Principal Scientist and Team Leader at Central Research, Pfizer in Sandwich, U.K. Dr. Huggins received his Ph.D. in 1983 from the University of Bristol, U.K., where he was subsequently supported by the British Heart Foundation. From 1985, he was a post-doctoral fellow in the newly formed Cellular Pharmacology section of Smith, Kline and French, and then in 1988 he moved to Merrell Dow Research Centre in Strasbourg, France. More recently, he returned to the U.K. to take up his current position of managing Pfizer's Heart Failure Discovery project in Sandwich. He has also had shorter research projects in Berlin and Philadelphia.

Dr. Huggins is a member of The Biochemical Society, The British Pharmacological Society, The International Society for Heart Research and The British Society for Cardiovascular Research.

Dr. Huggins' major research interest has been the fusion of molecular and structural biology with classic pharmacology, and he has pursued this goal in his study of phospholamban, endothelin receptors, and cyclic GMP-dependent protein kinase. He currently plays a key role in understanding how information from human genome sequencing can be harnessed to generate future pharmaceutical targets, in relation to Pfizer's collaboration with Incyte Pharmaceuticals. He has some 35 publications to his name.

John T. Pelton, Ph.D., is Research Scientist and Head of Structural, Theoretical, and Enzyme Chemistry at Hoechst Marion Roussel, Inc.'s Cincinnati Research Center. Dr. Pelton received his Ph.D. degree in 1981 from the University of Nebraska-Lincoln where he examined the structure-activity relationships of enkephalins and hypothalamic peptides. From 1981–1984, as a recipient of an NIH post doctoral fellowship, he studied with Prof. Victor Hruby at the University of Arizona, Tucson. He joined Merrell Dow in 1985.

Dr. Pelton is a member of the American Chemical Society, the American Society for Biochemistry and Molecular Biology, the American Peptide Society, the European Peptide Society, the American Association for the Advancement of Science and the Fédération Européenne des Sociétées de Biochimie. Dr. Pelton was a member of the organizing committee for a recent IUPHAR Satellite Symposium on Chemistry and Biology of Endothelin and a guest editor of *Neurochemistry International.*

Dr. Pelton has a long-standing interest in the application of structural and modeling studies to the design of novel therapeutic agents. He has authored several chapters and review articles and published over 70 original research papers.

Contributors

Bruno Battistini
The William Harvey Research Institute
St. Bartholomew's Hospital
London, United Kingdom
(current address)
Department of Pharmacology
Faculty of Medicine
University of Sherbrooke
Sherbrooke (PQ), Canada

Marc Bigaud
Preclinical Research
Sandoz Pharma
Basel, Switzerland

David P. Brooks
Department of Renal Pharmacology
SmithKline Beecham Pharmaceuticals
King of Prussia, Pennsylvania

Jaw-Kang Chang
Tulane University Medical Center
Department of Surgery
New Orleans, Louisiana

Anthony Davenport
University of Cambridge
Clinical Pharmacology Unit
Addenbrooke's Hospital
Cambridge, United Kingdom

Annette Doherty
Cardiovascular/Cancer/Peptide Chemistry
Parke-Davis Pharmaceutical Research Division
Warner-Lambert Company
Ann Arbor, Michigan

Hugues D'Orchymont
Department of Chemistry
Strasbourg Research Center
Hoechst Marion Roussel
Strasbourg Cedex, France

Michael J. Dunn
Medical College of Wisconsin
Milwaukee, Wisconsin

Christian Frelin
Institut de Pharmacologie Moléculaire et Cellulaire du CNRS
Université de Buce-Sophia Antipolis
Valbonne, France

Denis Guedin
Biochimie Cardiovasculaire
Roussel-UCLAF
Romainville Cedex, France

Bulen Gumusel
Tulane University Medical Center
Department of Surgery
New Orleans, Louisiana

Qingzhong Hao
Tulane University Medical Center
Department of Surgery
New Orleans, Louisiana

William G. Haynes
University of Iowa
College of Medicine
Department of Internal Medicine
Cardiovascular Diseases Division
Iowa City, Iowa

C. Robin Hiley
University of Cambridge
Department of Pharmacology
Cambridge, United Kingdom

John P. Huggins
Marion Merrell Dow
Strasbourg Research Center
Strasbourg Cedex, France
(current address)
Cardiovascular Biology
Central Research
Pfizer Ltd.
Sandwich, Kent
United Kingdom

Albert Hyman
Tulane University Medical Center
Department of Surgery
New Orleans, Louisiana

Semiko Kaw
Max-Planck-Institute for Experimental Medicine
Göttingen, Germany

Howard Lippton
Tulane University Medical Center
Department of Surgery
New Orleans, Louisiana

Ponnal Nambi
Department of Renal Pharmacology
SmithKline Beecham Pharmaceuticals
King of Prussia, Pennsylvania

Neville N. Osborne
Nuffield Laboratory of Ophthalmology
University of Oxford
Oxford, United Kingdom

William C. Patt
Parke-Davis Pharmaceutical Research Division
Warner-Lambert Company
Ann Arbor, Michigan

John T. Pelton
Theoretical and Enzyme Chemistry
Cincinnati Research Center
Hoechst Marion Roussel
Cincinnati, Ohio

Kathleen S. Sang
University of Cambridge
Department of Pharmacology
Cambridge, United Kingdom

Herbert Schramek
Institute of Physiology
University of Innsbruck
Innsbruck, Austria

Maarten van den Buuse
Baker Medical Research Institute
Prahran, Victoria, Australia

Paul Vigne
Institut de Pharmacologie Moléculaire et Cellulaire du CNRS
Université de Buce-Sophia Antipolis
Valbonne, France

Timothy D. Warner
The William Harvey Research Institute
St. Bartholomew's Hospital
London, United Kingdom

David J. Webb
Clinical Research Center
Department of Medicine
University of Edinburgh
Western General Hospital
Edinburgh, United Kingdom

Masashi Yanagisawa
Howard Hughes Medical Institute
Research Laboratories
University of Texas
Southwestern Medical Center at Dallas
Dallas, Texas

Table of Contents

Part IV. Ligand Structure and Activity Relationships

Part V. Clinical Studies

Part I

Endothelin Receptors

Chapter 1

Endothelin Receptors: Pharmacological Subtyping, Agonists, and Antagonists

C. Robin Hiley

CONTENTS

0-8493-6975-4/97/$0.00+$.50

I. THE NATURE OF ENDOTHELIN RECEPTORS

On the basis of structural analogies between the peptide which they had newly isolated and peptide toxins which interacted with voltage-sensitive ion channels, Yanagisawa et al. (1988) suggested that ET-1 might be an agonist at the dihydropyridine-sensitive Ca^{2+} channel. A large number of subsequent studies confirmed that many responses to ETs may be reduced by blockade of L-type Ca^{2+} channels. However, Gu et al. (1989) showed no interaction between ET-1 and radiolabelled Ca^{2+} channel blockers while Silberberg et al. (1989) used cell-attached patch-clamp recording to show that ET-1 interacted with L-type Ca^{2+} channels by means of a second messenger. An important early observation was that ET-1 had high sequence and structural analogy with the STX, a family of peptides which may be isolated from the venom of the burrowing asp *Atractaspis engaddensis* (Lee and Chiappinelli, 1988; Takasaki et al., 1988). These toxins interact with the same receptors as ET-1 and activate phosphoinositide hydrolysis in rat heart and brain (Kloog et al., 1988). The majority of other studies on ET receptor mechanisms suggested that ETs acted on novel G-protein coupled receptors to increase intracellular free Ca^{2+} levels and bring about other changes in cellular activity and, by 1990, this was widely accepted (Simonson and Dunn, 1990).

II. SUBTYPES OF ENDOTHELIN RECEPTORS

A. Operational and Biochemical Evidence

Early functional studies were generally consistent with activation of a single receptor type by the ETs, that is, concentration/effect curves usually had Hill slopes close to unity. Nevertheless, there were a number of observations which suggested more than one type of receptor might be involved in some vascular responses. For example, [$Ala^{1,15}$]ET-1, in which one of the two disulfide bridges characterizing the ETs has been removed by substitution of two alanine residues for two cysteines, was found to have dose/effect relations in the blood-perfused superior mesenteric arterial bed of the rat which were best explained by it interacting with two receptor types (Hiley et al., 1989; Randall et al., 1989). Also, [$Ala^{1,3,11,15}$]ET-1, in which both disulfide bridges are absent, produced a much smaller maximal response than ET-1 in rat blood-perfused hindquarters but was unlikely to be a partial agonist at a single site shared with ET-1 since it did not antagonize the parent peptide (Randall, 1989, 1991). It was also found that ET-3 was a more powerful vasodilator than it was a vasoconstrictor in the rat isolated perfused mesentery (Warner et al., 1989) and that, in contrast to their relative potency in causing contraction of this preparation (ET-1 was 4.4 times more potent than ET-3), ET-3 was apparently more potent as a vasodilator than ET-1, giving the maximum response at 1 pmol as opposed to 50 pmol (Douglas and Hiley, 1990).

The possibility of receptor heterogeneity was supported by cross-linking experiments showing multiple bands in gel electrophoresis. Masuda et al. (1989) reported preferential labeling in rat heart of a 44-kDa band by [^{125}I]ET-1 and a 32-kDa band by [^{125}I]ET-3. In chick heart, Watanabe et al. (1989) found a single band of 53 kDa for [^{125}I]ET-1 and [^{125}I]ET-2 cross-linked by disuccidinyl tartrate, but three bands (two major at 34 and 46 kDa and one minor at 53 kDa) for [^{125}I]ET-3. Two bands for [^{125}I]ET-3 binding, of 38 and 53 kDa, were discovered in rat brain membranes, whereas [^{125}I]ET-1 only labeled a single band of 53 kDa (Ambar et al., 1990). Thus, as in the functional studies, ET-3 appeared to have different properties relative to ET-1 which would be consistent with the existence of multiple receptors for these two peptides.

Some competition experiments with radioligand binding also supported the possibility of more than one receptor subtype. On the basis of the inhibition of [^{125}I]sarafotoxin S6b binding by ET-1, ET-3, sarafotoxins S6b (STX-b), and S6c (STX-c), Kloog et al. (1990) proposed that there might be three receptor subtypes for the ET/STX family; one with high affinity for ET-1 and STX-b in smooth muscle, another with high affinity for STX-c found in the cerebellum, and a third with no selectivity between the peptides found in the caudate nucleus. Jeng et al. (1990) reported that there was a 1000-fold difference in affinity for ET-3 between brain and lung when determined using [^{125}I]ET-1, and the ratio of the IC_{50} values for ET-1 and ET-3 varied from 2 in the brain, kidney, and liver to more than 100 in heart and spleen, suggesting that ET-3 could differentiate between binding sites.

B. ET_A and ET_B Receptors

1. *Cloning of ET_A and ET_B Receptors*

The strands of evidence for multiple ET receptors were drawn together by the simultaneous publication at the end of 1990 of the structures and characteristics of two receptor clones. Arai et al. (1990) obtained from a bovine lung cDNA library the sequence of a receptor which had K_i values of 0.92, 7.2, 52, and 900 nM for ET-1, ET-2, STX-b, and ET-3, respectively; this was named the ET_A receptor. In contrast, when the receptor cloned from a rat lung cDNA library by Sakurai et al. (1990) was transfected into COS-7 cells, the concentration/effect curves for increases in free intracellular Ca^{2+} by the ET isopeptides were superimposable, with an EC_{50} = 20 pM; furthermore, this receptor, named ET_B, was also sensitive to both STX-b and STX-c. Human forms of both receptors were soon reported and the selectivity of approximately 1000-fold of the ET_A receptor for ET-1 relative to ET-3 was confirmed (Hosoda et al., 1991) as was the lack of selectivity of the ET_B receptor (Nakamuta et al., 1991; Sakamoto et al., 1991).

Thus, operational effects (functional responses and radioligand binding) of the ETs came to be divided into those mediated by ET_A receptors, where ET-1 was more potent than ET-3, and those due to ET_B receptors, where ET-1 and ET-3 were approximately equipotent. This approach did not, however, take adequate account of the fact that the cloned ET_A receptor showed a 1000-fold selectivity for ET-1 over ET-3 as compared to the variable potency ratios, rarely exceeding 100-fold, found in most functional studies. This discrepancy may be due to the tissues concerned containing a mixture of the two receptor subtypes (see discussion by Huggins et al., 1993).

2. *Further Subtypes of ET_A and ET_B Receptors*

Mixtures of ET_A and ET_B receptors, defined respectively as being very selective for ET-1 relative to ET-3 and nonselective between these two isopeptides, did not explain the results obtained by Bertelsen et al. (1992) using [^{125}I]ET-1 in the CNS of Fischer 344 rats. Their conclusions were based on changes in the affinity ratio of ET-1 to ET-2. The cloned receptors show that ET-2 is eight times less potent than ET-1 at the ET_A receptor but equipotent at the ET_B receptor. Both ET-1 and ET-2 had 25 to 100 times higher affinity than ET-3 for binding sites in cerebral cortex and spinal cord, but the relative potencies of ET-1 and ET-2 differed in these two regions of the CNS. In cerebral cortex, ET-1 was six times more potent than ET-2, whereas since ET-2 had a higher affinity in the spinal cord, the ratio of affinities in the cord was only 1.5-fold. Since the affinity of ET-3 was not significantly different between the two regions, it was proposed that there were two subtypes of ET_A receptor: ET_{A1} in the cerebral cortex and ET_{A2} in the spinal cord.

Another proposed subdivision of mammalian ET_A receptors into ET_{A1} and ET_{A2} subtypes was made by Sudjarwo et al. (1994). They showed that the rabbit saphenous vein contracted in response to the ET_B-selective agonists STX-c and IRL 1,620 (see Section III.B.1) as well as to ET-1 and ET-3. The ET_B receptor-mediated response could be readily desensitized by pretreatment with STX-c, leaving an ET-3 response which was abolished by BQ 123, a selective ET_A receptor antagonist (see Section IV.A.1), and hence ET_A receptor mediated. However, when the ET_B receptor had been desensitized, the response to ET-1 was only partly sensitive to BQ 123, leading the authors to conclude that this tissue contained both a BQ 123-sensitive ET_{A1} receptor and a BQ 123-insensitive ET_{A2} receptor subtype. BQ 123-insensitive receptors, apparently of the ET_A subtype as defined by the order of agonist potency, have also been reported by Bodelsson and Stjernquist (1993) in human umbilical artery.

The possibility that there might be two subtypes of ET_A receptor in the goat cerebral artery, distinguished by differing affinity for BQ 123, was raised by Salom et al. (1993). These observations were similar to those of Sumner et al. (1992) who found that BQ 123 had an apparently greater potency ($pA_2 = 8.3$) against ET-3-induced contraction of rat aorta than it did against ET-1 ($pA_2 = 6.9$) but, as noted below (Section IV.A.1.b), there are problems in interpreting the antagonism by BQ 123 of the effectively irreversible agonist ET-1 as compared to less-tightly bound agonists like STX-b and ET-3.

Kumar et al. (1993) also proposed subtypes of the ET_A receptor, but in a non-mammalian species. They showed that follicular oocytes from *Xenopus laevis* contained native binding sites for [^{125}I]ET-1 where ET-1 was about 1000-fold more potent than ET-3 at inhibiting binding and 1 μM STX-c was inactive. On this basis they suggested that this binding site was an ET_A receptor. However, 1 μM BQ 123 only inhibited binding by 25% even though BQ 123 had an IC_{50} of 20 nM against human ET_A receptors expressed in defolliculated oocytes. They named this receptor the ET_{AX} receptor.

It has also been proposed that there are subtypes of the ET_B receptor, with Sokolovsky et al. (1992) making the division into ET_{B1} and ET_{B2} on the basis of being able to detect both high ($K_d = 0.5$ to 1.6 nM) and "super-high" affinity ($K_d = 30$ to 65 pM) sites for [^{125}I]ET-1, [^{125}I]ET-2, and [^{125}I]ET-3 in several brain regions and in the atrium of the rat. Functional evidence for more than one ET_B receptor subtype was presented by Battistini et al. (1993) who compared the contractile effects in the guinea-pig trachea of six analogues of ET-1 and compared their results with those obtained by other groups in various vascular tissues and the cerebellum. In particular, they noted an absence of activity of [Ala^{1,3,11,15}]ET-1 (an ET_B receptor-selective agonist, see Section III.B.1.a) in the trachea, which otherwise appeared to contain predominantly ET_B receptors. Additionally, Warner et al. (1993) have provided strong evidence that endothelium-dependent relaxation in response to ET-1 and STX-c is mediated by different PD 142,893-sensitive (see Section IV.A.1) ET_B receptors, relative to the ET_B receptors mediating ET-1-induced contraction of the smooth muscle of rat stomach or rabbit pulmonary artery which were very much less sensitive to PD 142,893.

The lack of full definition of the characteristics of those functional responses, which do not seem to comply with standard definitions of ET_A and ET_B receptors, and the lack of overlap in the use of techniques and ligands, resulted in no sufficiently clear pattern emerging to allow unequivocal acceptance of subtypes within the ET_A and ET_B receptor classification by the IUPHAR Nomenclature Committee reporting in 1994 (Masaki et al., 1994). Nevertheless, a consensus with regard to the ET_B receptor is emerging as a result of the work of Warner, Clozel, and others (Douglas et al., 1994). Thus, in their paper characterizing a new nonpeptide antagonist, bosentan, Clozel et al. (1994) used ET_{B1} to refer to endothelial receptors and ET_{B2} to refer to those causing vasoconstriction, but no cloning study has yet confirmed the existence of these subtypes. Development of new ligands and reinvestigation of some tissues, such as those discussed above, with identical protocols and ligands will no doubt lead to unequivocal evidence for the existence of further subdivisions within the receptors for the ET peptides.

The question of "atypical" sites for ET action, that is, those which do not fit simply into the ET_A and ET_B classification as now defined by the availability of selective agonists and antagonists, has been discussed in detail by Bax and Saxena (1994).

C. Other Endothelin Receptor Subtypes

1. *Operational Studies*

The most positive indications of a third group of responses have arisen from determinations of the ET-1:ET-3 potency ratio which have revealed responses where ET-3 is much more potent than ET-1, or where ET-1 has no operational effect at all. Such a site has been named the ET_C receptor. Thus, Emori et al. (1990) reported binding of [^{125}I]ET-3 to bovine carotid artery endothelial cells which was inhibited by ET-3 at subnanomolar concentrations but not by ET-1 at concentrations below 1 nM. ET-1 had no effect up to 100 nM on free intracellular Ca^{2+} concentration although the EC_{50} for ET-3 on the Ca^{2+} response was ~3 nM; this receptor may be coupled to nitric oxide release since ET-3 has an EC_{50} of 5 nM for this effect (Emori et al., 1991). ET-3 is also more potent than ET-1 at causing endothelium-dependent relaxation in the rat mesenteric bed (Douglas and Hiley, 1990) and releases nitric oxide from bovine aorta while ET-1 does not (Warner et al., 1992). In another study of endothelial cells, ET-3, STX-c, and [Ala^{1,3,11,15}]ET-1, but not ET-1 up to 100 nM, accelerated wound healing in cultured human umbilical vein endothelial cell monolayers (Wren et al., 1993).

Samson et al. (1990) found an EC_{50} of 5 nM for the inhibition by ET-3 of prolactin release from anterior pituicytes of male or ovarectomized female donors, although ET-1 was inactive up to 1 μM. In later studies of these cells from intact female rats (Samson and Skala, 1992; Samson, 1992), prolactin release was found to be sensitive to ET-1 acting through the ET_A receptor since, in these cells, ET-1, ET-2, and vasoactive intestinal contractor (VIC) were ~100 times more potent than ET-3 and the response was blocked by the ET_A receptor antagonist BQ 123.

Thus it seems that both endothelial cells and pituicytes from male or ovariectomized female rats may contain a receptor that is selective for ET-3 and might be termed the ET_C receptor although a receptor with these properties has only been cloned from *Xenopus laevis* and not from mammalian sources.

There have been a number of reports showing patterns of tissue activation by agonists and of antagonism by the various compounds now available (described in Section IV), which suggest that, even with the addition of an ET-3-selective receptor, the canon of ET receptors is not yet complete. Some of the observations might be explained by the existence of subtypes of ET_A and ET_B receptors, such as the group of responses discussed above, which appear to be mediated by ET_A receptors but are insensitive to the ET_A-selective antagonist BQ 123. There are also the observations of Eglezos et al. (1993) who found that the enhancement by ET-1 of the electrically evoked twitch response of rat vas deferens was apparently both non-ET_B receptor mediated and BQ 123 insensitive, and of Battistini et al. (1994) who found that there was a contractile response to ET-1 with similar properties in the guinea-pig gall bladder.

2. *Cloned Receptors*

The existence of a genomically encoded ET receptor which is selective for ET-3 was confirmed by Karne et al. (1993) who cloned a G-protein-coupled receptor they termed ET_C from *Xenopus laevis* dermal melanophores. This receptor had K_i values for ET-3 and ET-1 of 45.5 and 114 nM, respectively, although the EC_{50} of melanophore activation by ET-3 was 24 nM as compared to over 10 μM for ET-1 and ET-2. Since no equivalent receptor has been cloned from a mammalian source, it remains to be seen whether or not this amphibian ET_C receptor is closely related to the ET-3 selective receptor described in endothelial and pituitary cells.

Another receptor has been cloned from a cDNA library constructed from the heart of *Xenopus laevis* (Kumar et al., 1994). This receptor, when expressed in COS cells, showed a selectivity of more than 100-fold for ET-1 relative to ET-3, both for inhibition of binding of $[^{125}I]$ET-1 and stimulation of phosphatidylinositol metabolism; STX-c had no effect on either binding or inositol phosphate accumulation nor did BQ 123 block either effect. In view of an overall 74% homology between this clone and the human ET_A receptor, and on other structural features, it was suggested that this receptor was the ET_{AX} receptor described by the same group (see Section II.B.2) as occurring naturally in *Xenopus laevis* oocytes (Kumar et al., 1993). It remains to be resolved whether this receptor is unique to the amphibian, and is the amphibian ET_A receptor, or also has counterparts in mammalian systems.

III. AGONISTS AT ENDOTHELIN RECEPTORS

A. Naturally Occurring Peptides

1. *Endothelins*

The human ET family consists of three peptides encoded in the genome (Inoue et al., 1989) which have high sequence homology and are characterized by 21 amino acid residues (linked by disulfide bonds between Cys^1-Cys^{15} and Cys^3-Cys^{11}), a free amino terminus, and a C-terminal carboxylic acid.

Human ET-1 is identical with the peptide first isolated from the supernatant of cultured porcine aortic endothelial cells by Yanagisawa et al. (1988) and is taken to be the parent compound of the family. With the exception of a very few sites (see Section II.C and below), ET-1 is active wherever any of the other peptides has activity and has similar or greater potency than the other peptides at those sites. A very wide range of potencies have been reported, though EC_{50} values generally fall into the range 0.1 to 5 nM.

Human ET-2, [Trp^6, Leu^7]ET-1, is generally found to have a potency slightly less than that of ET-1 at binding sites and in functional studies; it is not able to distinguish between ET_A and ET_B receptors. This is also true of [Asn^4, Trp^6, Leu^7]ET-1 which was originally identified in the mouse genome and called vasoactive intestinal constrictor (VIC) but is now believed to be the rodent form of ET-2 (Bloch et al., 1991). However, Morton and Davenport (1992) found that some neurons in cultures of embryonic rat brain responded selectively to ET-2 with an increase in intracellular Ca^{2+} concentration, suggesting the existence of ET-2-selective receptors which can be detected at the level of the single cell.

The third member of the human ET family, ET-3 ([Thr^2, Phe^4, Thr^5, Tyr^6, Lys^7, Tyr^{14}]ET-1), has six substitutions relative to ET-1 and, as discussed in Section II, has a distinct pharmacology which has led it to be considered a selective ET_B receptor agonist and has made it a central tool for identification of ET receptors. ET-3 has a low affinity in cloned ET_A receptors in an expression system (e.g., K_d value of ~1 μM in cloned human ET_A receptors; Hosoda et al., 1991), but it often has a potency in the low- to mid-nanomolar concentration range at what might be determined to be ET_A receptor sites in functional experiments on the basis of the ET-1:ET-3 potency ratio. Thus, in the rat pulmonary artery, the respective EC_{50} values for ET-1 and ET-3 are 0.68 and 40.6 nM (Nakajima et al., 1989): a ratio of 60 which has been interpreted as indicating a response mediated by ET_A receptors. However, both ET_A and ET_B receptor mRNA are expressed in this vessel, though the ET_B receptor mRNA might be confined to the endothelial cells and the receptor mediates vasodilatation. Interestingly, it is the levels of ET_B receptor mRNA which increase in animals exposed to chronic hypoxia (Li et al., 1994). Hence, the ET-1:ET-3 potency ratio alone should not be taken as evidence that only ET_A receptors are present. More substantial evidence, such as from the use of a more selective ET_B receptor agonist or of selective antagonists must also be obtained.

2. *Sarafotoxins*

This peptide family, found in the venom of a burrowing asp, has expanded over the last few years and shows heterogeneity in both the N- and C-terminal portions (Figure 1). However, of the ten naturally occurring ET/STX peptides described by the end of 1994, nine amino acid residues are conserved throughout; namely, the four Cys residues, Asp^8, Glu^{10}, His^{16}, and the terminal dipeptide Ile^{20}-Trp^{21}. Readers are referred to Chapter 15 and the review by Huggins et al. (1993) for a detailed examination of the structure/activity relations for both ETs and STXs since they address the importance of the various amino acid residues in determining the biological activity of these peptides.

The two most widely studied members of the STX family are STX-b, which has similar potency to ET-1 in both binding and functional experiments, and STX-c, which is a highly selective ligand for ET_B receptors. STX-c was shown to be a weak inhibitor of [^{125}I]ET-1 binding to rat ventricular membranes (Nayler et al., 1989), but interest in this peptide increased when Williams et al. (1991) found that it had K_i values for the predominantly ET_A binding sites for [^{125}I]ET-1 in rat aorta and atrium

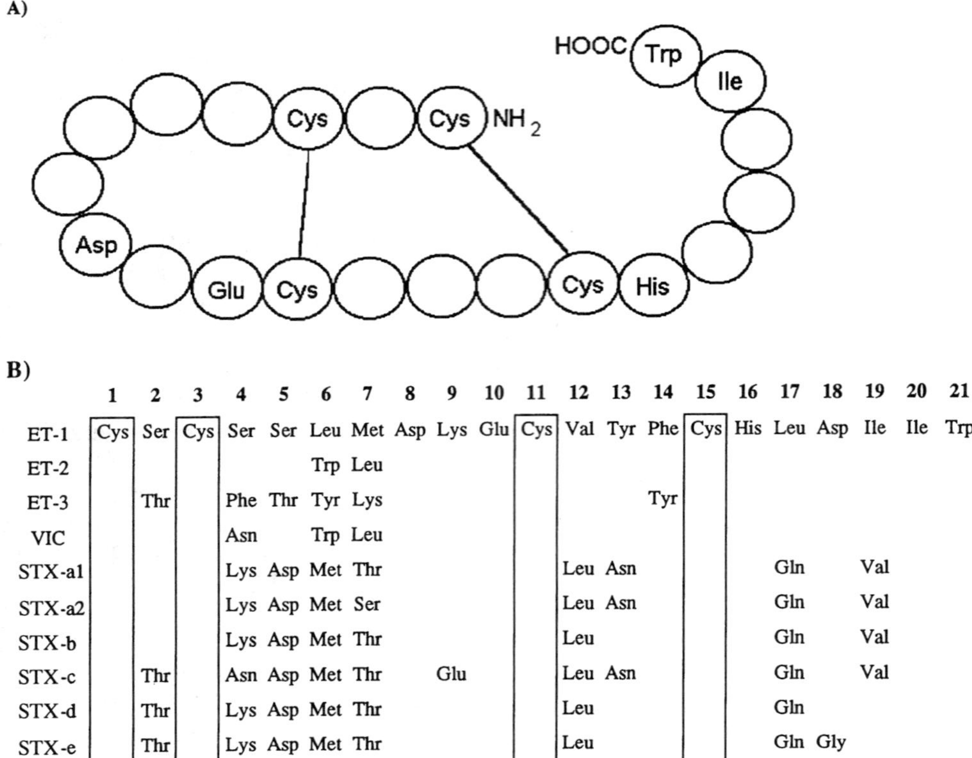

	1	2	3	4	5	6	7	8	9	10	11	12	13	14	15	16	17	18	19	20	21
ET-1	Cys	Ser	Cys	Ser	Ser	Leu	Met	Asp	Lys	Glu	Cys	Val	Tyr	Phe	Cys	His	Leu	Asp	Ile	Ile	Trp
ET-2						Trp	Leu														
ET-3		Thr		Phe	Thr	Tyr	Lys							Tyr							
VIC				Asn		Trp	Leu														
STX-a1				Lys	Asp	Met	Thr					Leu	Asn				Gln		Val		
STX-a2				Lys	Asp	Met	Ser					Leu	Asn				Gln		Val		
STX-b				Lys	Asp	Met	Thr					Leu					Gln		Val		
STX-c		Thr		Asn	Asp	Met	Thr		Glu			Leu	Asn				Gln		Val		
STX-d		Thr		Lys	Asp	Met	Thr					Leu					Gln				
STX-e		Thr		Lys	Asp	Met	Thr					Leu					Gln	Gly			

FIGURE 1
Sequences of the ten known members of the ET/STX family of peptides. (A) The nine conserved amino acid residues in the ten peptides, the positions of the disulfide bridges, and the free carboxy terminus. (B) The full sequence of ET-1 and the substitutions relative to this peptide found in the other members of the family so far described. The boxes show the positions of the Cys residues. N.B. The original STX-d sequence was revised by Lamthanh et al. (1994) and STX-d is now believed to be identical to STX-e as reported by Ducancel et al. (1993). Thus, Asp^{18} is also conserved throughout nine peptides.

of >5 μM and 4.2 μM, respectively, but bound with high affinity (K_i = 16 nM) to the ET_B receptor-rich rat cerebellum. STX-c was more than 50,000 times less potent than ET-1 in the aorta and atrium, suggesting very low levels of ET_B receptors in these tissues. However, the ET-1:ET-3 potency ratios in these two tissues were only 22 and 47, respectively, and Li et al. (1994) showed that detectable levels of mRNA for ET_B receptors were expressed in rat atria and ventricles. In addition to contracting some vessels containing ET_B receptors on their vascular smooth muscle, STX-c also causes endothelium-dependent relaxation of preconstricted blood vessels (Warner et al., 1993).

STX-a is ten times less potent than STX-b in contracting endothelium-denuded rat aorta; STX-b was three times less potent than ET-1 in this preparation (Kitazumi et al., 1990). STX-d has no activity in the guinea-pig aorta, a predominantly ET_A receptor preparation, and was at least 133-fold less potent than STX-c (which in turn was 8 times more potent than ET-1) in the pig coronary artery where the contractions are presumably mediated by the ET_B receptor (Schoeffter and Randriantsoa, 1993). No biological data are currently available on STX-a2.

B. Analogues of Endothelin and Sarafotoxin Peptides

Many full-length analogues of the ET and STX peptides have been synthesized (see review by Huggins et al., 1993) but few have proven to have properties of lasting interest. Furthermore, not even all of those which have shown useful pharmacology have gone on to find extensive use in ET receptor studies. For example, Takayanagi et al. (1991) reported that [Glu^9]STX-b (Glu^9 being an important substitution in STX-c relative to STX-b) was a selective agonist at ET_B receptors which was approximately equipotent with STX-c in binding studies of [^{125}I]ET-3 in intact rat aorta.

The reason that [Glu^9]STX-b has not found extensive use is that it is one of several ET_B receptor-selective agonists, and others, to be described below, have now become established in the literature. An extensive range of agonists and antagonists is now available and some, like [Glu^9]STX-b, have been derived from the structures of the ET/STX peptides. However, the current list of ET receptor ligands is missing a selective ET_A receptor agonist.

1. ET_B Receptor-Selective Agonists

a. [Ala^{1,3,11,15}]ET-1

One full-length ET analogue that has been widely investigated is the tetra-alanyl-substituted linear peptide, [Ala^{1,3,11,15}]ET-1 (4-Ala-ET-1), in which the characteristic Cys-Cys bridges of the ETs are absent. 4-Ala-ET-1 is approximately equipotent with ET-1 at inhibiting [^{125}I]ET-1 binding to rat cerebellar membranes (Hiley et al., 1990) but has no contractile activity in the rat aorta (Topouzis et al., 1991). Saeki et al. (1991) went on to show that 4-Ala-ET-1 is a selective ET_B receptor antagonist and that several truncated analogues, especially 4-Ala-ET-1 (6-21) and 4-Ala-ET-1 (8-21), were potent ET_B receptor agonists (IC_{50} values of 0.32 and 1.2 nM as compared to 0.11 nM for 4-Ala-ET-1) with selectivity ratios in binding experiments of 1000 and 1200, respectively, as compared to 1700 for [Ala^{1,3,11,15}]ET-1. Both had similar potency to 4-Ala-ET-1 in causing relaxation of precontracted porcine pulmonary artery.

b. BQ 3,020 and IRL 1,620

The work of Saeki et al. (1991) led to the introduction of BQ 3,020 {N-acetyl-[Ala^{11,15}]ET-1 (6-21)} as a selective ET_B receptor agonist (Ihara et al., 1992). In binding studies with [^{125}I]ET-1, BQ 3,020 had an ET_B:ET_A selectivity ratio of 4700 and it contracted rabbit pulmonary artery with an EC_{50} of 0.57 nM (as compared to 0.21 and 0.081 nM for ET-1 and ET-3, respectively), a value that was unaffected by the presence of the selective ET_A receptor antagonist, BQ 123 (1 μM). [^{125}I]BQ 3,020 has been characterized as a useful ligand for ET_B receptors (Ihara et al., 1992; Molenaar et al., 1992) as it has a lower nonspecific binding than [^{125}I]ET-3 (Kobayashi et al., 1993).

Another truncated analogue of ET-1, IRL 1,620 {N-succinyl-[Glu^9, Ala^{11,15}]-ET-1 (8-21)} is also a potent selective agonist at ET_B receptors with a potency ratio of 120,000 in binding experiments in porcine lung (K_i = 1.9 μM at ET_A receptors and 16 pM at ET_B receptors), and it had half the potency of ET-3 as a contractile agonist in the guinea-pig trachea where the ET-1:ET-3 potency ratio is 4.2 (Takai et al., 1992). This suggests that this tissue apparently has a mixed population of ET_A and ET_B receptors, a conclusion confirmed by the observation that the selective ET_A receptor antagonist FR 139,317 does not block the contractions elicited by ET-3 but does antagonize those evoked by ET-1 and ET-2 (Cardell et al., 1993).

2. C-Terminal Hexapeptides

The C-terminus after the final Cys is conserved in all four ETs, giving the hexapeptide His-Leu-Asp-Ile-Ile-Trp which is biologically active and was the first stimulus for division of ET receptors into two types. Maggi et al. (1989) found that ET-1 (16-21) was an agonist in guinea-pig bronchus with an EC_{50} of 228 nM as compared to 6.9 nM for ET-1. However, the hexapeptide was inactive as an agonist on the deendothelialized aorta at concentrations up to 10 μM. They proposed that the aortic contractions were mediated by an ET_A receptor and those in the bronchus by an ET_B receptor — this terminology is not related to that later used for the cloned receptors. ET-1 (16-21) also caused endothelium-dependent relaxation of the perfused mesenteric arterial bed of the rat (Douglas and Hiley, 1991) which would be consistent with it being an ET_B receptor agonist, but this has not been confirmed by use of a selective antagonist. Rovero et al. (1990) reported a structure/activity study for ET-1 (16-21) which showed that removal of the N-terminal His, to give ET-1 (17-21), reduced its potency 100-fold in the guinea-pig bronchus; several substitutions in ET-1 (16-21) were reported, but the only one that was tolerated was replacement of Ile^{19} with Val.

With the exception of STX-d/e, the STXs also have a common C-terminal hexapeptide which shows two substitutions, Gln^{17} and Val^{19}, relative to that of the ETs. The C-terminal hexapeptide derived from STX-a, STX-b, and STX-c, His-Gln-Asp-Val-Ile-Trp, and that from STX-d, His-Gln-Asp-Ile-Ile-Trp, bind weakly to $[^{125}I]$ET-1 sites in rabbit aorta, pulmonary artery, and rat heart ventricle, but none showed agonist or antagonist (against ET-1) effects at concentrations of up to 30 μM in rabbit pulmonary artery of rat left atrium (Doherty et al., 1991).

IV. ANTAGONISTS AT ENDOTHELIN RECEPTORS

A. Peptide Antagonists

The first three ET antagonists were reported in 1991, and all were peptides. One, $[Dpr^1-Asp^{15}]$ET-1, was a full-length analogue which has been the subject of only a limited number of studies since its discovery. A second, BE 18,275B, was a cyclic pentapeptide and played an important role in ET pharmacology by serving as a lead compound which led to the discovery of what has become the standard ET_A receptor antagonist, BQ 123. Finally, ET (11-21) was described as a selective antagonist of ET-3 contraction of rat aorta; a peptide thought to be of this structure was later given the code IRL 1,038 and characterized as a selective ET_B receptor antagonist, but there are now doubts as to its identity and effectiveness (see Section IV.A.2.a below).

1. ET_A Receptor-Selective Antagonists

a. $[Dpr^1-Asp^{15}]$ET-1

By replacing Cys^1 with diaminoproprionic acid, Cys^{15} with Asp, and forming an amide link between these substitutions, Spinella et al. (1991) obtained an antagonist of ET-1-mediated constriction of the guinea-pig isolated perfused lung; a concentration of 0.1 μM giving an approximately five-fold shift in the dose/response curve (equivalent to a pA_2 of 7.6). Since $[Dpr^1-Asp^{15}]$ET-1 was inactive against ET-3 it was concluded that it was an ET_A-selective antagonist. In a binding assay, $[Dpr^1-Asp^{15}]$ET-1 (IC_{50} = 2 nM) was 40 times less potent than ET-1 (50 pM) at inhibiting binding of 15 pM $[^{125}I]$ET-1 to cultured rat pulmonary artery smooth muscle cells. A derivative, [Dpr^1, Asp^{15}]ET-1, which had not been cyclized between positions 1 and 15, was only slightly less potent at inhibiting binding of [^{125}I]ET-1 (IC_{50} = 7 nM) but was an agonist 300 to 500 times less potent than ET-1 in the perfused lung preparation. Later both [Dpr^1-Asp^{15}]ET-1 and [Dpr^1-Asp^{15}]ET-3 were found to block ET_A receptor-induced bronchoconstriction in conscious sheep, though the ET-3 analogue appeared to be a little less potent (Abraham et al., 1993).

b. BE 18,275B, BQ 123, BQ 153, and BQ 610

A number of companies chose to screen either fermentation products or chemical banks in order to obtain lead compounds in the search for high potency ET antagonists. Ihara et al. (1991a) obtained a cyclic pentapeptide, BE 18,257B (cyclo[-D-Glu-L-Ala-allo-D-Ile-L-Leu-D-Trp-]), from a fermentation broth of *Streptomyces misakiensis* which inhibited [^{125}I]ET-1 (10 pM) binding to cultured porcine aortic vascular smooth muscle cells (IC_{50} = 0.5 μM) and to ventricular membranes (IC_{50} = 0.8 μM), but not to cerebellar membranes (IC_{50} > 100 μM) from the pig. BE 18,257B was thus an ET_A-selective agent and in rabbit iliac artery the pA_2 was 5.9 against ET-1-induced contraction.

Ihara and colleagues at Banyu Pharmaceuticals used BE 18,257B as a lead compound which led them to very much more potent compounds, one of which, BQ 123 (cyclo[-D-Asp-L-Pro-D-Val-L-Leu-D-Trp-]) has become a standard agent for investigation of ET_A receptor-mediated effects. Ihara et al. (1991b) showed that both BQ 123 and BQ 153 (cyclo[-D-sulfoalanine-L-Ala-L-Pro-D-Val-L-Leu-D-Trp-]) were very potent inhibitors of [^{125}I]ET-1 (10 pM) binding to cultured porcine aortic smooth muscle cells (BQ 123, IC_{50} = 7.3 nM; BQ 153, IC_{50} = 8.6 nM). They were relatively inactive at the ET_B binding sites in pig cerebellum with IC_{50}s of 18 μM (BQ 123) and 54 μM (BQ 153). Both compounds had identical pA_2 values of 7.4 (Schild slope = 0.96) against ET-1-induced contraction of pig coronary artery. *In vivo*, BQ 153 antagonized the pressor but not the depressor response to ET-1 (1 nmol kg^{-1}, i.v.) in conscious Wistar-Kyoto rats over the dose range 0.1–10 mg kg^{-1}), but a bolus dose of 10 mg kg^{-1} BQ 153 had no effect on the basal blood pressure; this led to the suggestion that ET peptides, acting at ET_A receptors, had no physiological role in the maintenance of blood pressure. However, the relatively nonselective antagonist SB 209,670 (see below) did decrease blood pressure in spontaneously hypertensive rats (Ohlstein et al., 1994).

In cultured SK-N-MC cells, BQ 123 appeared to act as a noncompetitive antagonist of the increase in cytosolic free Ca^{2+} concentration mediated by ET-1 (Hiley et al., 1992) since the maximum response to ET-1 was reduced in magnitude and there was no shift in the EC_{50}. Subsequently, Vigne et al. (1993) showed that parallel shifts in the response curve were observed if the antagonist were added at the same time as the agonist and concluded that the essentially irreversible combination of ET-1 with its receptors was responsible for the noncompetitive behavior seen in the previous study.

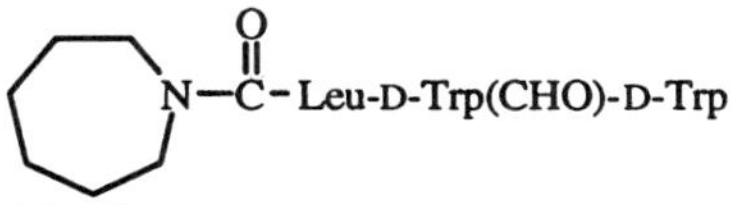

N—C(=O)-Leu-D-Trp(CHO)-D-Trp

BQ-610

More recently, the group at Banyu Pharmaceuticals has introduced BQ 610 as a selective ET_A receptor antagonist which antagonizes the cerebral vasospasm produced by subarachnoid hemorrhage (Zuccarello et al., 1994).

c. *FR 139,317*

A linear tripeptide, FR 139,317 is an ET_A receptor antagonist of similar potency to BQ 123. It has a selectivity ratio for ET_A binding sites of approximately 90-fold over ET_B sites, an IC_{50} of 53 nM for the specific binding of [^{125}I]ET-1 to porcine aortic microsomes, and a pA_2 of 7.2 against ET-1-induced contraction in rabbit aorta (Sogabe et al., 1993). It, too, reduces cerebral vasospasm caused by subarachnoid hemorrhage (Nirei et al., 1993) when given as a dose of 0.1 mg intracisternally, reduces the renal damage associated with the chronic renal failure induced by reduction of kidney mass (Benigni et al., 1993), and at a dose of 15 mg kg^{-1} decreases the area of infarction caused by occlusion of the left coronary artery in the rat (Lee et al., 1994). *In vivo,* however, it had a shorter duration of action (<60 min on the blood pressure responses to 1 nmol kg^{-1} ET-1) in conscious rats than PD 142,893 and PD 145,065 (the effects of which lasted over 2 h) when all the antagonists were used at a dose of 10 μmol kg^{-1} (Doherty et al., 1993).

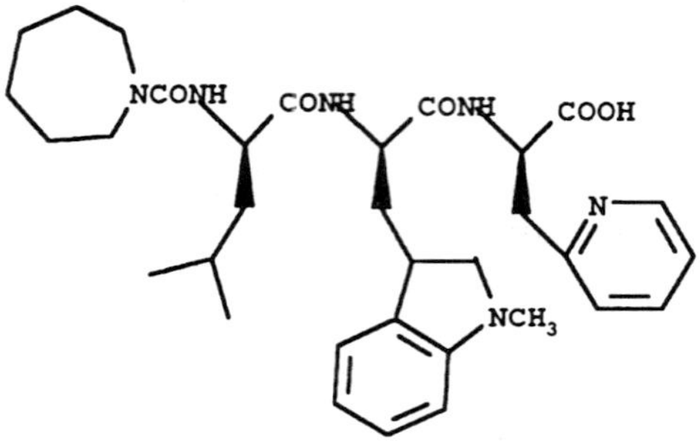

NCONH CONH CONH COOH N NCH_3

d. *Other Selective ET_A Receptor Peptoid Antagonists*

The tetrapeptide, PD 151,242 {N-[(hexahydro-1-azepinyl)carbonyl]-L-Leu(1-methyl)-D-Trp-D-Tyr}, is a selective ET_A receptor antagonist with a pA_2 value of 5.92 in isolated, deendothelialized, human coronary arteries (Davenport et al., 1994). In radioiodinated form it was used as a selective ligand for ET_A receptors and had, in human aorta, a K_d of 0.76 nM. ET_A-selective ligands inhibited [^{125}I]PD 151242 binding in aorta with high affinities (BQ 123, K_i = 0.41 nM; FR 139,317, K_i = 0.55 nM) whereas the ET_B receptor-selective agent, BQ 3,020, competed with a K_i of 1.36 μM (Davenport et al., 1994).

TAK 044 is a cyclic hexapeptide derived by Kikuchi et al. (1994), working at Takeda Chemical Industries, from BE 18,275B as a lead compound. It is a highly selective ET_A ligand when used to inhibit [^{125}I]ET-1 binding with IC_{50} values of 82 pM and 120 nM, respectively, for binding sites in pig cardiac ventricle and bovine brain membranes — values which indicated considerably higher affinity for both sites than BQ 123. Despite this apparent selectivity, functionally it appears to be a nonselective antagonist with 100 nM inhibiting both ET-1-mediated contraction of porcine coronary artery and STX-c-induced contraction of pig coronary vein (Kikuchi et al., 1994). It was suggested that the higher potency at the ET_B sites in the functional assay might be due to different subtypes of this receptor in the two preparations. *In vivo,* TAK 044 gave dose-related inhibition of both the transient depressor and the longer-lasting pressor responses to i.v. ET-1, with significant decreases in the responses occurring at 0.1 mg kg^{-1} for the pressor response and 1 mg kg^{-1} for the depressor response. In comparison with BQ 123, it was approximately equipotent on the pressor response and ten times more potent on the depressor effect (Ikeda et al., 1994). TAK 044 also inhibited the STX-c-induced renal vasoconstriction at a dose of 3 mg kg^{-1} whereas the same dose of BQ 123 had no effect (Ikeda et al., 1994).

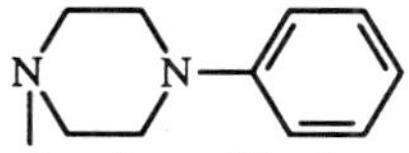

cyclo(D-Asp-Asp-Asp-D-Thg-Leu-D-Trp)
D-Thg = 2-(2-thienyl)glycine
TAK-044

2. ET_B Receptor-Selective Antagonists

a. IRL 1,038

A truncated, cyclized, sequence from ET-1, [Cys^{11}-Cys^{15}]ET-1 (11-21) (Figure 2) was introduced as a selective ET_B receptor antagonist by a group working at CIBA-Geigy, Japan and given the code number IRL 1,038 (Urade et al., 1992). It was reported to have a selectivity ratio for the ET_B receptor of at least 60 and had K_i values of 6 and 11 nM at ET_B binding sites in rat and guinea-pig cerebellum as compared to 700 and 630 nM at ET_A binding sites in rat aorta and guinea-pig heart. It was less potent as an antagonist at ET_B receptors in guinea-pig trachea since a concentration of 3 μM gave ~3-fold shift in the ET-3 log concentration/response curve (Urade et al., 1992). Difficulties in obtaining consistent batches of this compound were reported, with possible molecular aggregation, such that it appears unsuited for routine use (Urade et al., 1994).

b. BQ 788

Ishikawa et al. (1994) described BQ 788 as being a potent antagonist (pA_2 = 8.4) of BQ 3,020-induced contraction of rabbit pulmonary artery. A dose of 1 mg kg^{-1} BQ 788 abolished the transient depressor response, and enhanced the peak pressor response, to ET-1 (0.3 nmol kg^{-1}, i.v.). BQ 788 (50 μg kg^{-1}, i.v.) inhibited the broncho-constriction induced by i.v. ET-1 by 50%. In binding studies, BQ 788 had an IC_{50} for inhibition of [^{125}I]ET-1 binding to the ET_B receptors on human Girardi heart cells of 1.2 nM. In a later study, BQ 788 was found to have an IC_{50} of 3 μM for antagonism of ET-3-induced relaxation of preconstricted rat aorta (Karaki et al., 1994a).

c. RES 701-1

This is another selective ET_B receptor antagonist which was first isolated from *Streptomyces* sp. RE-701 (Karaki et al., 1994b) and which antagonized the relaxant response to ET-3 of the precontracted rat aorta with concentration ratios of approximately 2 and 16 at 0.3 and 3 μM, respectively (equivalent to a pA_2 value of 6.5 to 6.7). RES 701-1 had no effect on the contractile response of deendothelialized rat aorta to either ET-1 or ET-3, confirming the conclusion, which would be reached from the ET-1:ET-3 potency ratio of 220, that contraction of this preparation is only mediated by ET_A receptors. Subsequent use of RES 701-1 has shown that ET_B receptors mediate the ET-induced release of dopamine from rat striatal slices (Kohzuma et al., 1994).

3. Nonselective Antagonists

The structure/activity relations of peptides derived from the C-terminal hexapeptide of the ETs have been extensively studied by Doherty and co-workers working at Parke Davis in the U.S. Although, as noted above, they were unable to

ET_B receptor antagonists

IRL 1038

Cys-Val-Tyr-Phe-Cys-His-Leu-Asp-Ile-Ile-Trp

RES-701-1

Gly-Asn-Trp-His-Gly-Thr-Ala-Pro-Asp-Trp-Phe-Phe-Asn-Tyr-Tyr-Trp

BQ-788

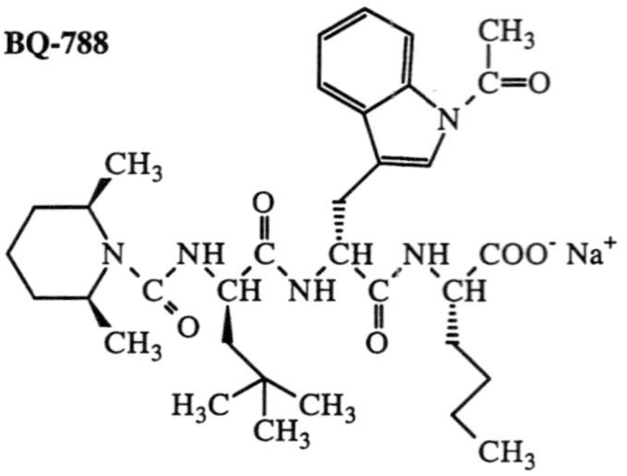

Non-selective ET_A and ET_B receptor peptide antagonists

PD 142893

Acetyl-D-Dip-Leu-Asp-Ile-Ile-Trp

D-Dip = 3,3-D-diphenylalanine

PD 145065

Acetyl-D-Bhg-Leu-Asp-Ile-Ile-Trp

D-Bhg = 5H-bibenzym[a,d]cycloheptene-10,11-dihydroglycine

FIGURE 2
ET_B receptor antagonists.

show that any of the hexapeptides derived from the ET or STX peptides had agonist activity at concentrations up to 30 μM in the rabbit pulmonary artery or the paced left atrium of the rat (Doherty et al., 1991), they have produced a number of antagonists, including the nonselective agents PD 142,893 and PD 145,065. PD 142,893 (as the disodium salt) had IC_{50} values for inhibition of [^{125}I]ET-1 binding of 31 and 54 nM at ET_A (rabbit renal artery vascular smooth muscle cells) and ET_B (rat cerebellum) binding sites, respectively, with pA_2 values of 6.6 for ET-1-mediated contraction of rabbit femoral artery (ET_A) and of 6.3 for STX-c-evoked contraction of rabbit pulmonary artery (ET_B) (Doherty et al., 1993). PD 145,065 (also the disodium salt) was slightly more potent than PD 142,893 in these assays with pA_2 values of 6.9 and 7.1 in the femoral and pulmonary arteries, respectively, and with IC_{50} values of 4 and 15 nM in the ET_A and ET_B receptor binding assays. At a dose of 10 μmol kg^{-1}, PD 145,065 antagonized both the depressor and pressor responses to ET-1 (1 nmol kg^{-1}, i.v.) in anesthetized, ganglion-blocked rats, but PD 142,893 (10 μmol kg^{-1}) only inhibited the depressor response.

Warner et al. (1993) showed that PD 142,893, at a concentration of 1 μM, was inactive against ET-1-induced, but ET_B receptor-mediated, contractions of rabbit pulmonary artery rings, though it readily antagonized the relaxation in response to STX-c of the rat isolated perfused mesenteric bed. This suggests that PD 142,893 is selective for the ET_{B1} receptors on the endothelium.

B. Nonpeptide Antagonists

The first nonpeptide antagonists appeared in 1992, a year after their peptide counterparts, and, with few exceptions, they are characterized by being to some extent selective at ET_A receptors. Indeed, of those agents reported so far, only Ro 46-2005 has not yet been found to have any selectivity in the assays used.

1. 50-235 and 97-139

27-O-caffeoyl myricerone (50-235), the first nonpeptide ET antagonist, was isolated from bayberry and is selective for ET_A receptors. It had a K_i value of 130 nM against [^{125}I]ET-1 binding in rat A7r5 vascular smooth muscle cells, but was inactive against ET-1-induced increases in free intracellular Ca^{2+} concentration in Girardi heart cells which are a source of ET_B receptors (Fujimoto et al., 1992). Structure/activity studies showed that the 3-keto, 17-carboxyl, and 27-caffeoyl groups were important for conferring affinity on the molecule; for example, replacement of the caffeoyl moiety with cinnamoyl, which lacks the catechol hydroxyls of 50-235, reduced the affinity threefold while replacement of the 3-keto group with an α hydrogen and a β hydroxyl reduced affinity by 60-fold (Mihara et al., 1993).

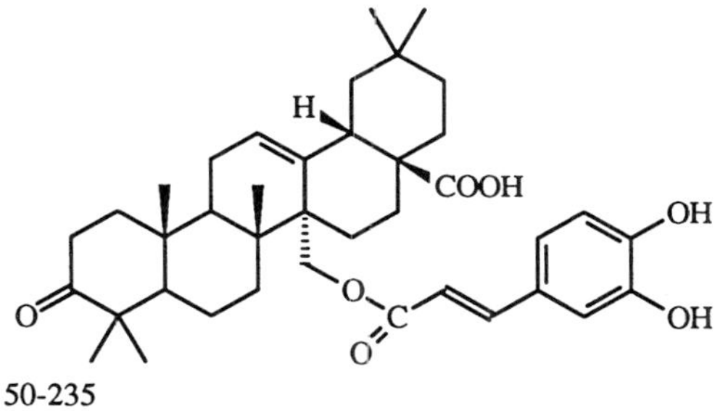

H COOH OH OH O O C O

50-235

Further development of 50-235 led to a 50-fold improvement in affinity in the related compound 97-139, which had K_i values of 1 nM and 1 μM against [^{125}I]ET-1 binding to rat A7r5 vascular smooth muscle cells and Girardi heart cells, respectively (Mihara et al., 1994). When tested against ET-1-induced incorporation of [^{3}H]thymidine into A7r5 cells, 97-139 gave a parallel shift of the concentration/effect curve equivalent to a pA_2 of 9.4 and in rat aortic rings the pA_2 was 8.8. In pithed rats, an i.v. dose of 0.1 mg kg^{-1} significantly inhibited the pressor response to ET-1 (0.1 nmol kg^{-1}, i.v.). The potency *in vivo* of 97-139 was approximately the same as that of BQ 123 which was less than would have been expected from their potency ratios *in vitro* (BQ 123 had a pA_2 = 7.6 in rat aorta, suggesting an activity 15 times less than that of 97-139); this was attributed to binding of the antagonist by plasma proteins.

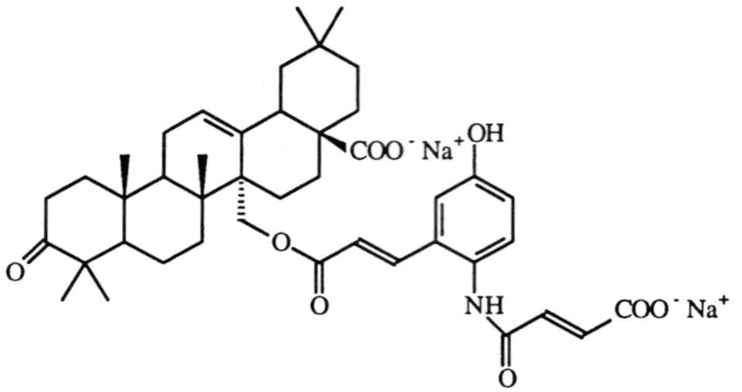

97-139

2. *Ro 46-2005 and Bosentan (Ro 47-0203)*

Following the report of the first nonpeptide ET antagonist, the next major step came with the description of the first orally active antagonist at ET receptors by Clozel et al. (1993). Ro 46-2005 was derived from a pyrimidinyl sulfonamide lead compound (synthesized as part of an antidiabetic program) which had been discovered by screening a chemical library against [^{125}I]ET-1 binding to human placental membranes; these membranes contain a single type of binding site (Fischli et al., 1989) of the ET_B type (Kilpatrick et al., 1993). Ro 46-2005 is equipotent at ET_A and ET_B receptors with pA_2 values against ET-1-induced contraction of rat aorta (ET_A) and STX-c-induced relaxation of rat small mesenteric arteries (ET_B) of 6.50 and 6.47, respectively. It has IC_{50} values against [^{125}I]ET-1 binding at both receptors of 200 to 500 nM (Breu et al., 1993). The oral availability was estimated as 30% (Clozel et al., 1993).

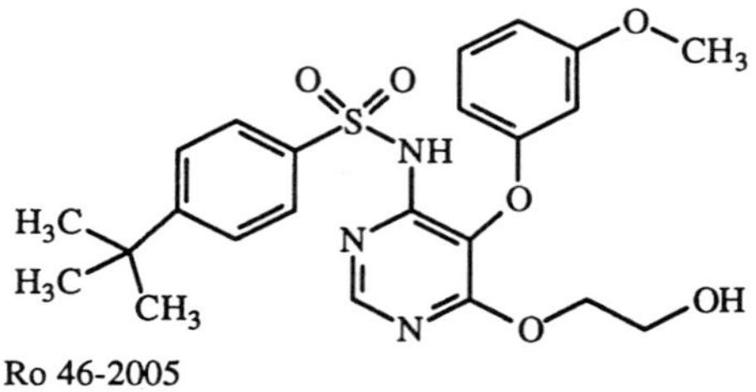

Ro 46-2005

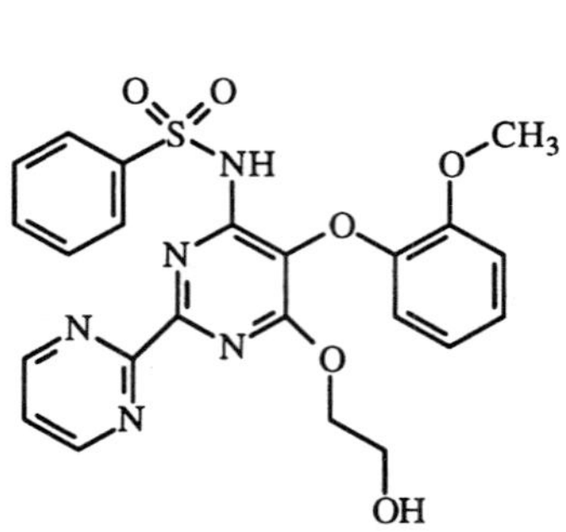

Bosentan (Ro 47-0203)

Further development by Roche led to the introduction of bosentan (Ro 47-0203) as a nonselective antagonist (Clozel et al., 1994). Inhibition of the binding of [^{125}I]ET-1 to recombinant human receptors expressed in CHO cells suggested that it has selectivity (20- to 50-fold) for ET_A over the ET_B receptors designated ET_{B1} by this group. In functional studies bosentan was somewhat less selective, with pA_2 values of 7.28, 6.72, and 5.94, respectively, against ET-1 contractions in deendothelialized rat aorta (ET_A), STX-c-induced relaxation of rabbit mesenteric artery (ET_{B1}), and STX-c-induced contraction of rat trachea (ET_{B2}). *In vivo*, in pithed rats, i.v. bosentan significantly inhibited both the depressor (at 10 mg kg^{-1}) and pressor (at 3 mg kg^{-1}) responses to i.v. ET-1, which are thought to be mediated by ET_B and ET_A receptors, respectively. A lower dose of bosentan (1 mg kg^{-1}, i.v.) reduced the depressor response to the selective ET_B agonist STX-c, but 30 mg kg^{-1} (i.v.) was required to antagonize the STX-c pressor response. Orally, doses of 1 to 100 mg kg^{-1} bosentan antagonized the pressor responses to proET-1 (0.3 nmol kg^{-1}, i.v.); 30 mg kg^{-1} still had an effect 6 h after administration (65% inhibition as compared to 61% at 45 min after giving the antagonist).

3. *SB 209,670*

This antagonist, described by Ohlstein et al. (1994), is more potent than bosentan at ET_A receptors. A lead compound, SK & F 66,861, was found by screening a chemical bank as it was fairly potent with K_i values of 7.3 and >30 μM, respectively, for ET_A and ET_B receptors (Elliott et al., 1994). The structure was subjected to molecular modeling which led to the hypothesis that the biphenyl indene structure might mimic binding by the residues Tyr^{14}, Phe^{15}, and Trp^{21} of ET, with the carboxylic acid mimicking either Asp^{18} or the free carboxyl of Trp^{21}. Further refinement led to synthesis of an indane, SB 209,670, which retained the free carboxyl group which had been postulated to be an important determinant of the affinity of SK & F 66,861 as the methyl ester had a K_i of >30 μM (Elliott et al., 1994).

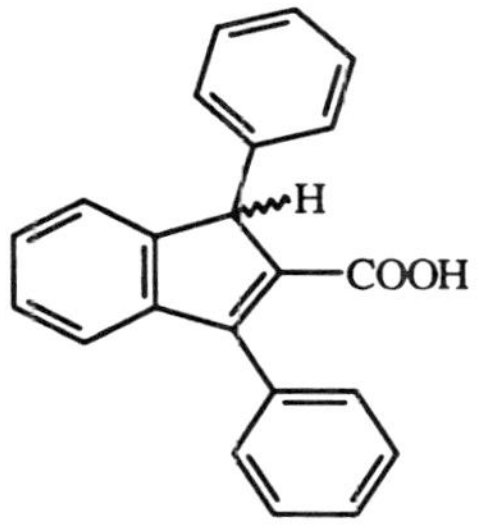

H
COOH

SK&F 66861

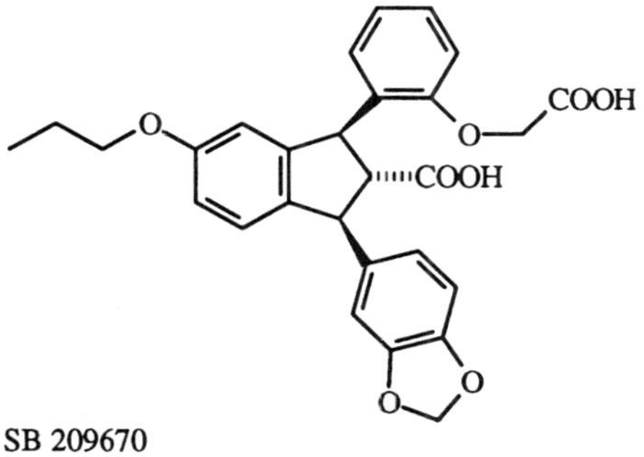

O
COOH
O
COOH
O
O

SB 209670

SB 209,670 has K_i values for inhibition of [^{125}I]ET-1 binding to cloned human ET receptors, stably expressed in CHO cells, of 0.2 nM (ET_A) and 18 nM (ET_B). In rat aorta, SB 209,670 antagonized contractions induced by ET-1 with a pA_2 of 9.39 and a Schild slope of 0.94, without affecting the maximum response. Using the rabbit pulmonary artery for assessing its effects on the ET_B receptor, the pA_2 was determined as 6.70 (Ohlstein et al., 1994). These values suggest that it is at least 90-fold selective for ET_A receptors, which makes it rather more selective than bosentan.

Although no isolated tissue values were quoted for ET_B receptor-mediated vasodilatation, *in vivo* SB 209,670 (18 μmol kg^{-1}) antagonized both the pressor and depressor responses and the transient increase in carotid artery conductance (perhaps mediated by ET_{B1} receptors) to ET-1 (0.3 nmol kg^{-1}, i.v.). The antagonism of the ET-1 pressor response (0.2 nmol kg^{-1}, i.v. every 30 min) in conscious rats by SB 209,670 (18 μmol kg^{-1}) lasted for more than 3.5 h.

4. *BMS 182,874*

Stein et al. (1994) described the development of BMS 182,874 which, like the Roche compounds, is a sulfonamide. The lead compound, sulfathiazole, was discovered to have an IC_{50} of 69 μM during the screening of a chemical library against [^{125}I]ET-1 binding to the rat A10 vascular smooth muscle cell line. BMS 182,874 had a K_i value of 150 nM in the binding assay but was less active against the ET_A receptor-mediated contraction of rabbit carotid artery induced by ET-1 (pA_2 = 6.3). Stein et al. (1994) reported that BMS 182,874 appeared to be very selective for ET_A receptors since it had a K_i >200 μM when tested as an inhibitor of [^{125}I]ET-1 binding to rat cerebellar membranes, a rich source of ET_B receptors.

Buchan et al. (1994) compared the activity of BMS 182,874 with that of Ro 46-2005 and SB 209,670 in a binding assay using [^{125}I]ET-1 and human ET_A and ET_B receptors stably expressed in CHO cells (Table 1). This confirms the selectivity for ET_A receptors shown by SB 209,670 and BMS 182,874 but suggests that, even for this receptor subtype, BMS 182,874 is rather a weak ligand and is not likely to be as useful in studies of ET receptors or their functions as SB 209,670, bosentan, or the selective ET_A and ET_B peptide ligands such as BQ 123, FR 139,317 and BQ 788.

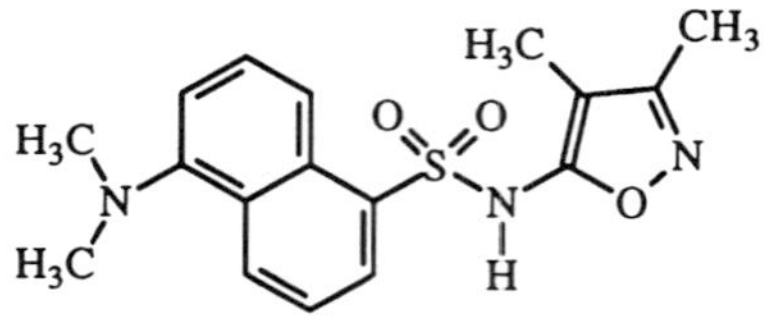

H_3C CH_3 O O H_3C N S N N O H_3C H

BMS 182874

TABLE 1

pIC_{50} Values for Inhibition of [^{125}I]ET-1 Binding to Cloned Human Lung ET_A and ET_B Receptors

Compound	ET_A Receptors	ET_B Receptors
Ro 46-2005	6.7	6.8
SB 209,670	8.7	7.5
BMS 182,874	5.8	<5

5. *Pheophorbide, WS009A and Sulfisoxazole*

Two of these are natural products discovered in screening programs. Pheophorbide was extracted with chloroform from the ears of *Artemesia capillaris* in a program investigating substances used in traditional Chinese medicine — in this case as a treatment for hepatitis. The substance obtained from the plant by chloroform extraction is pheophorbide a, a degradation product of chlorophyll a (Ohshima et al., 1994). It had IC_{50} values for inhibition of [^{125}I]ET-1 to porcine aortic membranes and rat brain membranes of 80 and 210 nM, respectively. A comparison with related compounds showed that a free carboxyl group was essential for binding activity at ET binding sites. No data concerning its actions as an antagonist have yet been published.

WS009A is an anthroquinone which was isolated from a culture of *Streptomyces* sp. No. 89009 and reported to be a selective ET_A receptor antagonist, but with low affinity (Miyata et al., 1992). It had IC_{50} values of 5.8 μM and <800 μM against [^{125}I]ET-1 binding in pig aorta and brain, respectively. WS009A 100 μM only inhibited the contractile response of rabbit aorta to ET-1 by 40%, but a dose of 10 mg kg^{-1} inhibited the peak pressor response to ET-1 (3.2 μg kg^{-1}, i.v.) by ~70% (Miyata et al., 1992).

In a different approach to new drug discovery, Chan et al. (1994) reported that a rational drug design procedure based on the crystal structure of ET-1 led to the identification of a possible pharmacophore. Operational experiments confirmed that the antibacterial sulfanilamide, sulfisoxazole, was a weak but selective ET_A antagonist (IC_{50} = 0.6 μM for inhibition of binding of [^{125}I]ET-1 to cloned human ET_A receptors in TE 671 cells and K_b = 1.2 μM for inhibition of ET-1-stimulated phosphatidylinositol hydrolysis in the same cells).

V. CONCLUDING WORDS

In the few years from the first international report of the discovery and characterization of ET-1, determination of the pharmacology of these peptides and their receptors has made enormous progress. For much of the time the pharmacology trailed the molecular biology, first in the identification of the human ET peptide family, then in the unequivocal characterization of the two principal receptor subtypes, ET_A and ET_B. The production of the current range of agonists and antagonists, both selective and otherwise, has taken time, but now presents a powerful group of tools for further characterization of ET receptors and for elucidation of the role that this peptide family plays in both health and disease.

REFERENCES

Abraham, W.M., Ahmed, A., Cortes, A., Spinella, M.J., Malik, A.B. and Andersen, T.T., A specific endothelin-1 antagonist blocks inhaled endothelin-1-induced bronchoconstriction in sheep, *J. Appl. Physiol.*, 74, 2537-2542. 1993.

Ambar, I., Kloog, Y. and Sokolovsky, M., Cross-linking of endothelin 1 and endothelin 3 to rat brain membranes: identification of the putative receptor(s), *Biochemistry*, 29, 6415-6418. 1990.

Arai, H., Hori, S., Aramori, I., Ohkubo, H. and Nakanishi, S., Cloning and expression of a cDNA encoding an endothelin receptor, *Nature*, 348, 730-732. 1990.

Battistini, B., Germain, M., Fournier, A. and Sirois, P., Structure-activity relationships of ET-1 and selected analogues in the isolated guinea-pig trachea. Evidence for the existence of different ET_B receptor subtypes, *J. Cardiovasc. Pharmacol.*, 22 (Suppl. 8), S219-S224. 1993.

Battistini, B., O'Donnell, L.J.D., Warner, T.D., Fournier, A., Farthing, M.J.G. and Vane, J.R., Characterization of endothelin receptors in the isolated gall bladder of the guinea pig. Evidence for an additional ET receptor subtype, *Br. J. Pharmacol.*, 112, 1244-1250. 1994.

Bax, W.A. and Saxena, P.R., The current endothelin receptor classification. Time for reconsideration?, *Trends Pharmacol. Sci.*, 15, 379-386. 1994.

Benigni, A., Zoja, C., Corna, D., Orisio, S., Longaretti, L., Bertani, T. and Remuzzi, G., A specific endothelin subtype A receptor antagonist protects against injury in renal disease progression, *Kidney Int.*, 44, 440-444. 1993.

Bertelsen, G.A., Rebello, S. and Gulati, A., Characteristics of endothelin receptors in the cerebral cortex and spinal cord of aged rats, *Neurobiol. Aging*, 13, 513-519. 1992.

Bloch, K.D., Hong, C.C., Eddy, R.L., Shows, T.B. and Quertermous, T., cDNA cloning and chromosomal assignment of the endothelin 2 gene: vasoactive intestinal contractor peptide is rat endothelin 2, *Genomics*, 10, 236-242. 1991.

Bodelsson, G. and Stjernquist, M., Characterization of endothelin receptors and localization of ^{125}I-endothelin-1 binding sites in human umbilical artery, *Eur. J. Pharmacol.*, 249, 299-305. 1993.

Breu, V., Löffler, B.-M. and Clozel, M., *In vitro* characterization of Ro 46-2005, a novel synthetic nonpeptide endothelin antagonist of ET_A and ET_B receptors, *FEBS Lett.*, 334, 210-214. 1993.

Cardell, L.O., Uddman, R. and Edvinsson, L., A novel ET_A-receptor antagonist, FR 139317, inhibits endothelin-induced contractions of guinea-pig pulmonary arteries, but not trachea, *Br. J. Pharmacol.*, 108, 448-452. 1993.

Chan, M.F., Okun, I., Stavros, F.L., Hwang, E., Wolff, M.E. and Balaji, V.N., Identification of a new class of ET_A selective endothelin antagonists by pharmacophore directed screening, *Biochem. Biophys. Res. Commun.*, 201, 228-234. 1994.

Clozel, M., Breu, V., Burri, K., Cassal, J.-M., Fischli, W., Gray, G.A., Hirth, G., Löffler, B.-M., Müller, M., Neidhart, W. and Ramuz, H., Pathophysiological role of endothelin revealed by the first orally active endothelin receptor antagonist, *Nature*, 365, 759-761. 1993.

Clozel, M., Breu, V., Gray, G.A., Kalina, B., Löffler, B.-M., Burri, K., Cassal, J.-M., Hirth, G., Müller, M., Neidhart, W. and Ramuz, H., Pharmacological characterization of bosentan, a new potent orally active nonpeptide endothelin receptor antagonist, *J. Pharmacol. Exp. Ther.*, 270, 228-235. 1994.

Davenport, A.P., Kuc, R.E., Fitzgerald, F., Maguire, J.J., Berryman, K. and Doherty, A.M., [^{125}I]-PD151242: a selective radioligand for human ET_A receptors, *Br. J. Pharmacol.* 111, 4-6. 1994.

Doherty, A.M., Cody, W.L., Leitz, N.L., DePue, P.L., Taylor, M.D., Rapundalo, S.T., Hingorani, G.P., Major, T.C., Panek, R.L. and Taylor, D.G., Structure-activity studies of the C-terminal region of the endothelins and the sarafotoxins, *J. Cardiovasc. Pharmacol.*, 17 (Suppl. 7), S59-S61. 1991.

Doherty, A.M., Cody, W.L., He, J.X., DePue, P.L., Cheng, X.-M., Welch, K.M., Flynn, M.A., Reynolds, E.E., LaDouceur, D.M., Davis, L.S., Keiser, J.A. and Haleen, S.J., *In vitro* and *in vivo* studies with a series of hexapeptide endothelin antagonists, *J. Cardiovasc. Pharmacol.*, 22 (Suppl. 8), S98-S102. 1993.

Douglas, S.A. and Hiley, C.R., Endothelium-dependent vascular activities of some endothelin like peptides in the isolated superior mesenteric arterial bed of the rat, *Br. J. Pharmacol.*, 101, 81-88. 1990.

Douglas, S.A. and Hiley, C.R., Endothelium-dependent mesenteric vasorelaxant effects and systemic actions of endothelin (16-21) and other endothelin-related peptides in the rat, *Br. J. Pharmacol.*, 104, 311-320. 1991.

Douglas, S.A., Meek, T.D. and Ohlstein, E.H., Novel receptor antagonists welcome a new era in endothelin biology, *Trends Pharmacol. Sci.*, 15, 313-316. 1994.

Ducancel, F., Matre, V., Dupont, C., Lajeunesse, E., Wollberg, Z., Bdolah, A., Kochva, E., Boulain, J.-C. and Ménez, A., Cloning and sequence analysis of cDNAs encoding precursors of sarafotoxins. Evidence for an unusual "rosary-type" organization, *J. Biol. Chem.*, 268, 3052-3055. 1993.

Eglezos, A., Cucchi, P., Patacchini, R., Quartara, L., Maggi, C.A. and Mizrahi, J., Differential effects of BQ-123 against endothelin-1 and endothelin-3 on the rat vas deferens. Evidence for an atypical endothelin receptor, *Br. J. Pharmacol.*, 109, 736-738. 1993.

Elliott, J.D., Lago, M.A., Cousins, R.D., Gao, A., Leber, J.D., Erhard, K.F., Nambi, P., Elshourbagy, N.A., Kumar, C., Lee, J.A., Bean, J.W., DeBrosse, C.W., Eggleston, D.S., Brooks, D.P., Feuerstein, G., Ruffolo, R.R., Weinstock, J., Gleason, J.G., Peishoff, C.E. and Ohlstein, E.H., 1, 3-Diarylindan-2-carboxylic acids, potent and selective nonpeptide endothelin receptor antagonists, *J. Med. Chem.*, 37, 1553-1557. 1994.

Emori, T., Hirata, Y. and Marumo, F., Specific receptors for endothelin-3 in cultured bovine endothelial cells and its cellular mechanism of action, *FEBS Lett.*, 263, 261-264. 1990.

Emori, T., Hirata, Y., Kanno, K., Ohta, K., Eguchi, S., Imai T., Shichiri, M. and Marumo, F., Endothelin-3 stimulates production of endothelium-derived nitric oxide via phosphoinositide breakdown, *Biochem. Biophys. Res. Commun.*, 174, 228-235. 1991.

Fischli, W., Clozel, M. and Guilly, C., Specific receptors for endothelin on membranes from human placenta. Characterization and use in a binding assay, *Life Sci.*, 44, 1429-1436. 1989.

Fujimoto, M., Mihara, S.-I., Ueda, M., Nakamaru, M. and Sakurai, K.-S., A novel nonpeptide endothelin antagonist isolated from bayberry, *Myrica cerifera*, *FEBS Lett.*, 305, 41-44. 1992.

Gu, X.H., Liu, J.J., Dillon, J.S. and Nayler, W.G., The failure of endothelin to displace bound, radioactively-labeled, calcium antagonists (PN200/110, D888 and diltiazem), *Br. J. Pharmacol.*, 96, 262-264. 1989.

Hiley, C.R., Douglas, S.A. and Randall, M.D., Pressor effects of endothelin and some analogs in the perfused superior mesenteric arterial bed of the rat, *J. Cardiovasc. Pharmacol.*, 13 (Suppl. 5), S197-S199. 1989.

Hiley, C.R., Jones C.R., Pelton J.T. and Miller R.C., Binding of [^{125}I]-endothelin-1 to rat cerebellar homogenates and its interactions with some analogues, *Br. J. Pharmacol.*, 101, 319-324. 1990.

Hiley, C.R., Cowley, D.J., Pelton, J.T. and Hargreaves, A.C., BQ-123, cyclo(-D-Trp-D-Asp-Pro-D-Val-Leu), is a non-competitive antagonist of the actions of endothelin-1 in SK-N-MC human neuroblastoma cells, *Biochem. Biophys. Res. Commun.*, 184, 504-510. 1992.

Hosoda, K., Nakao, K., Arai, H., Suga, S.-I., Ogawa, Y., Mukoyama, M., Shirakama, G., Saito, Y., Nakanishi, S. and Imura, H., Cloning and expression of human endothelin-1 receptor cDNA, *FEBS Lett.*, 287, 23-26. 1991.

Huggins, J.P., Pelton, J.T. and Miller, R.C., The structure and specificity of endothelin receptors: their importance in physiology and medicine, *Pharmacol. Ther.*, 55-123. 1993.

Ihara, M., Fukuroda, T., Saeki, T., Nishikibe, M., Kojiri, K., Suda, H. and Yano, M., An endothelin receptor (ET_A) antagonist isolated from *Streptomyces misakiensis*, *Biochem. Biophys. Res. Commun.*, 178, 132-137. 1991a.

Ihara, M., Noguchi, K., Saeki, T., Fukuroda, T., Tsuchida, S., Kimura, S., Fukami, T., Ishikawa, K., Nishikibe, M. and Yano, M., Biological profiles of highly potent novel endothelin antagonists selective for the ET_A receptor, *Life Sci.*, 50, 247-255. 1991b.

Ihara, M., Saeki, T., Fukuroda, T., Kimura, S., Ozaki, S., Patel, A.C. and Yano, M., A novel radioligand [^{125}I]BQ-3020 selective for endothelin (ET_B) receptors, *Life Sci.*, 51, PL47-PL52. 1992.

Ikeda, S., Awane, Y., Kusumoto, K., Wakimasu, M., Watanabe, T. and Fujino, M., A new endothelin receptor antagonist, TAK-044, shows long-lasting inhibition of both ET_A- and ET_B-mediated blood pressure responses in rats, *J. Pharmacol. Exp. Ther.* 270, 728-733. 1994.

Ishikawa, K., Ihara, M., Noguchi, K., Mase, T., Mino, N., Saeki, T., Fukuroda, T., Fukami, T., Ozaki, S., Nagase, T., Nishikibe, M. and Yano, M., Biochemical and pharmacological profile of a potent and selective endothelin B-receptor antagonist, BQ-788, *Proc. Natl. Acad. Sci. U.S.A.*, 91, 4892-4896. 1994.

Jeng, A.Y., Savage, P., Soriano, A. and Balwierczakm, J.L., Different affinities and selectivities of endothelin-1 and endothelin-3 binding to various rat tissues, *Biochem. Int.*, 22, 669-676. 1990.

Karaki, H., Sudjarwo, S.A. and Hori, M., Novel antagonist of endothelin ET_{B1} and ET_{B2} receptors, BQ-788. Effects on blood vessel and small intestine, *Biochem. Biophys. Res. Commun.*, 205, 168-173. 1994a.

Karaki, H., Sudjarwo, S.A., Hori, M., Tanaka, T. and Matsuda, Y., Endothelin ET_B receptor antagonist, RES-701-1: Effects on isolated blood vessels and small intestine, *Eur. J. Pharmacol.*, 262, 255-259. 1994b.

Karne, S., Jayawickreme, C.K. and Lerner, M.R., Cloning and characterization of an endothelin-3 specific receptor (ET_C receptor) from *Xenopus laevis* dermal melanophores, *J. Biol. Chem.*, 268, 19126-19133. 1993.

Kilpatrick, S.J., Roberts, J.M., Lykins, D.L. and Taylor, R.N., Characterization and ontogeny of endothelin receptors in human placenta, *Am. J. Physiol.*, 264, E367-E372. 1993.

Kikuchi, T., Ohtaki, T., Kawata, A., Imada, T., Asami, T., Masuda, Y., Sugo, T., Kusumoto, K., Kubo, K., Watanabe, T., Wakimasu, M. and Fujino, M., Cyclic hexapeptide endothelin receptor antagonists highly potent for both receptor subtypes ET_A and ET_B, *Biochem. Biophys. Res. Commun.*, 200, 1708-1712. 1994.

Kitazumi, K., Shiba, T., Nishiki, K., Furukawa, Y., Takasaki, C. and Tasaka, K., Structure-activity relationship in vasoconstrictor effects of sarafotoxins and endothelin-1, *FEBS Lett.*, 260, 269-272. 1990.

Kloog, Y., Ambar, I., Sokolovsky, M., Kochva, E., Wollberg, Z. and Bdolah, A., Sarafotoxin, a novel vasoconstrictor peptide: phosphoinositide hydrolysis in rat heart and brain, *Science*, 242, 268-270. 1988.

Kloog, Y., Bousso-Mittler, D., Bdolah, A. and Sokolovsky, M., Three apparent receptor subtypes for the endothelin/sarafotoxin family, *FEBS Lett.*, 253, 199-202. 1989.

Kobayashi, M., Ihara, M., Sato, N., Saeki, T., Ozaki, S., Ikemoto, F. and Yano, M., A novel ligand, [^{125}I]BQ-3020, reveals the localization of endothelin ET_B receptors, *Eur. J. Pharmacol.*, 235, 95-100. 1993.

Kohzuma, M., Kataoka, Y., Koizumi, S., Shibaguchi, H., Nakashima, M., Yamashita, K., Niwa, M. and Taniyama, K., ET_B receptor involvement in stimulatory and neurotoxic action of endothelin on dopamine neurons, *Neuroreport*, 5, 2653-2656. 1994.

Kumar, C. S., Nuthulaganti, P., Pullen, M. and Nambi, P., Novel endothelin receptors in the follicular membranes of *Xenopus laevis* oocytes mediate calcium responses by signal transduction through gap junctions, *Mol. Pharmacol.*, 44, 153-157. 1993.

Kumar, C., Mwangi, V., Nuthulaganti, P., Wu, H.-L., Pullen, M., Brun, K., Aiyar, H., Morris, R.A., Naughton, R. and Nambi, P., Cloning and characterization of a novel endothelin receptor from *Xenopus* heart, *J. Biol. Chem.*, 269, 13414-13420. 1994.

Lamthanh, H., Bdolah, A., Creminon, C., Grassi, J., Menez, A., Wollberg, Z. and Kochva, E., Biological activities of [Thr^2]sarafotoxin-b, a synthetic analogue of sarafotoxin-b, *Toxicon*, 32, 1105-1114. 1994.

Lee, C.Y. and Chiappinelli, V.A., Similarity of endothelin to snake venom toxin, *Nature*, 335, 303. 1988.

Lee, J.Y., Warner, R.B., Adler, A.L. and Opgenorth, T.J., Endothelin ET_A receptor antagonist reduces myocardial infarction induced by coronary artery occlusion and reperfusion in the rat, *Pharmacology*, 49, 319-324. 1994.

Li, H., Elton, T.S., Chen, Y.F. and Oparil, S., Increased endothelin receptor gene expression in hypoxic rat lung, *Am. J. Physiol.*, 266, L553-L560. 1994.

Maggi, C.A., Giuliani, S., Patacchini, R., Santicoli, P., Rovero, P., Giachetti, A. and Meli, A., The C-terminal hexapeptide, endothelin-(16-21), discriminates between different endothelin receptors, *Eur. J. Pharmacol.*, 166, 121-122. 1989.

Masaki, T., Vane, J.R. and Vanhoutte, P.M., V. International Union of Pharmacology nomenclature of endothelin receptors, *Pharmacol. Rev.*, 46, 137-142. 1994.

Mihara, S., Sakurai, K., Nakamura, M., Konoike, T. and Fujimoto, M., Structure-activity relationships of an endothelin ET_A receptor antagonist, 50-235, and its derivatives, *Eur. J. Pharmacol.*, 247, 219-221. 1993.

Mihara, S.-I., Nakajima, S., Matumura, S., Kohnoike, T. and Fujimoto, M., Pharmacological characterization of a potent nonpeptide endothelin receptor antagonist, 97-139, *J. Pharmacol. Exp. Ther.*, 268, 1122-1128. 1994.

Miyata, S., Hashimoto, M., Fujie, K., Shouho, M., Sogabe, K., Kiyoto, S., Okuhara, M. and Kohsaka, M., WS009-A and WS009-B, new endothelin receptor antagonists isolated from *Streptomyces* sp. No. 89009. 2. Biological characterization and pharmacological characterization of WS009-A and WS009-B, *J. Antibiot.*, 45, 1041-1046. 1992.

Molenaar, P., Kuc, R.E. and Davenport, A.P., Characterization of two new ET_B selective radioligands, [^{125}I]-BQ3020 and [^{125}I]-[Ala^{1,3,11,15}]ET-1 in human heart, *Br. J. Pharmacol.*, 107, 637-639. 1992.

Morton, A.J. and Davenport, A.P., Cerebellar neurons and glia respond differentially to endothelins and sarafotoxin-S6b, *Brain Res.*, 581, 299-306. 1992.

Nakamuta, M., Takayanagi., R., Sakai, Y., Sakamoto, S., Hagiwara, H., Mizuno, T., Saito, Y., Hirose, S., Yamamoto, M. and Nawata, H., Cloning and sequence analysis of a cDNA encoding human nonselective type of endothelin receptor, *Biochem. Biophys. Res. Commun.*, 177, 34-39. 1991.

Nakajima, K., Kumagaye, S.-I., Nishio, H., Kuoda, H., Watanabe, T.X., Kobayashi, Y., Tamaoki, H., Kimura, T. and Sakakibara, S., Synthesis of endothelin-1 analogs, endothelin-3 and sarafotoxin S6b; structure-activity relationships, *J. Cardiovasc. Pharmacol.*, 13 (Suppl. 5), S8-S12. 1989.

Nayler, W.G., Gu, X.H. and Casley, D.J., Sarafotoxin S6c is a relatively weak displacer of specifically bound ^{125}I-endothelin, *Biochem. Biophys. Res. Commun.*, 161, 89-94. 1989.

Nirei, H., Hamada, K., Shoubo, M., Sogabe, K., Notsu, Y. and Ono, T., An endothelin ET_A receptor antagonist, FR139317, ameliorates cerebral vasospasm in dogs, *Life Sci.*, 52, 1869-1874. 1993.

Ohlstein, E.H., Nambi, P., Douglas, S.A., Edwards, R.M., Gellai, M., Lago, A., Leber, J.D., Cousins, R.D., Gao, A., Frazee, J.S., Peishoff, C.E., Bean, J.W., Eggleston, D.S., Elshourbagy, N.A., Kumar, C., Lee, J.A., Yue, T.-L., Louden, C., Brooks, D.P., Weinstock, J., Feuerstein, G., Poste, G., Ruffolo, R.R. and Gleason, J.G., SB 209670, a rationally designed potent nonpeptide endothelin receptor antagonist, *Proc. Natl. Acad. Sci. U.S.A.*, 91, 8052-8056. 1994.

Ohshima, T., Hirata, M., Oda, T., Sasaki, A. and Shiratsuchi, M., Pheophorbide a, a potent endothelin receptor antagonist for both ET_A and ET_B subtypes, *Chem. Pharm. Bull. (Tokyo)*, 42, 2174-2176. 1994.

Randall, M.D., Vascular Activities of Endothelium-Derived Vasoactive Factors in Resistance Beds of the Rat, Ph.D. Thesis, University of Cambridge. 1989.

Randall, M.D., Vascular activities of the endothelins, *Pharmacol. Ther.*, 50, 73-93. 1991.

Randall, M.D., Douglas, S.A. and Hiley, C.R., Vascular activities of endothelin-1 and some alanyl substituted analogues in resistance beds of the rat, *Br. J. Pharmacol.*, 98, 685-699. 1989.

Rovero, P., Patacchini, R. and Maggi, C.A., Structure-activity studies on endothelin (16-21), the C-terminal hexapeptide of the endothelins, in the guinea-pig bronchus, *Br. J. Pharmacol.*, 101, 232-234. 1990.

Saeki, T., Ihara, M., Fukuroda, T., Yamagiwa, M. and Yano, M., [Ala^{1,3,11,15}]endothelin-1 analogs with ET_B agonistic activity, *Biochem. Biophys. Res. Commun.*, 179, 286-292. 1991.

Salom, J.B., Torregrosa, G., Barber, M.D., Jover, T. and Alborch, E., Endothelin receptors mediating contraction in goat cerebral arteries, *Br. J. Pharmacol.*, 109, 826-830. 1993.

Sakamoto, A., Yanagisawa, M., Sakurai, T., Takuwa, Y., Yanagisawa, H. and Masaki, T., Cloning and functional expression of human cDNA for the ET_B endothelin receptor, *Biochem. Biophys. Res. Commun.*, 178, 656-663. 1991.

Sakurai, T., Yanagisawa, M., Takuwa, Y., Miyazaki, H., Kimura, S., Goto, K. and Masaki, T., Cloning of a cDNA encoding a non-isopeptide-selective subtype of the endothelin receptor, *Nature*, 348, 732-735. 1990.

Samson, W.K., The endothelin-A receptor subtype transduces the effects of the endothelins in the anterior pituitary gland, *Biochem. Biophys. Res. Commun.*, 187, 590-595. 1992.

Samson, W.K. and Skala, K.D., Comparison of the pituitary effects of the mammalian endothelins: vasoactive intestinal contractor (endothelin-β, rat endothelin-2) is a potent inhibitor of prolactin secretion, *Endocrinology*, 130, 2964-2970. 1992.

Samson, W.K., Skala, K.D., Alexander, B.D. and Huang, F.-L.S., Pituitary site of action of endothelin: selective inhibition of prolactin release *in vitro*, *Biochem. Biophys. Res. Commun.*, 169, 737-743. 1990.

Schoeffter, P. and Randriantsoa, A., Differences between endothelin receptors mediating contraction of guinea-pig aorta and pig coronary artery, *Eur. J. Pharmacol.*, 249, 199-206. 1993.

Silberberg, S.D., Poder, T.C. and Lacerda, A.E., Endothelin increases single-channel calcium currents in coronary arterial cells, *FEBS Lett.*, 247, 68-72. 1989.

Simonson, M.S. and Dunn, M.J., Cellular signaling by peptides of the endothelin gene family, *FASEB J.*, 4, 2989-3000. 1990.

Sokolovsky, M., Ambar, I. and Galron, R., A novel subtype of endothelin receptors, *J. Biol. Chem.*, 267, 20551-20554. 1992.

Sogabe, K., Nirei, H., Shoubo, M., Nomoto, A., Ao, S., Notsu, Y. and Ono, T., Pharmacological profile of FR139317, a novel, potent endothelin ET_A receptor antagonist, *J. Pharmacol. Exp. Ther.*, 264, 1040-1046. 1993.

Spinella, M.J., Malik, A.B., Everitt, J. and Andersen, T.T., Design and synthesis of a specific endothelin 1 antagonist: effects on pulmonary vasoconstriction, *Proc. Natl. Acad. Sci. U.S.A.*, 88, 7443-7446. 1991.

Stein, P.D., Hunt, J.T., Floyd, D.M., Moreland, S., Dickinson, K.E.J., Mitchell, C., Liu, E.C.-K., Webb, M.L., Murugesan, N., Dickey, J., McMullen, D., Zhang, R., Lee, V.G., Serafino, R., Delaney, C., Schaeffer, T.R. and Kozlowski, M., The discovery of sulfonamide endothelin antagonists and the development of the orally active ET_A antagonist 5-(dimethylamino)-*N*-(3,4-dimethyl-5-isoxazolyl)-1-naphthalenesulfonamide, *J. Med. Chem.*, 37, 329-331. 1994.

Sudjarwo, S.A, Hori, M., Tanaka, T., Matsuda, Y., Okada, T. and Karaki, H., Subtypes of endothelin ET_A and ET_B receptors mediating venous smooth muscle contraction, *Biochem. Biophys. Res. Commun.*, 200, 627-633. 1994.

Sumner, M.J., Cannon, T.R., Mundin, J.W., White, D.G. and Watts, I.S., Endothelin ET_A and ET_B receptors mediate vascular smooth muscle contraction, *Br. J. Pharmacol.*, 107, 858-860. 1992.

Takai, M., Umemura, I., Yamasaki, K., Watakabe, T., Fujitani, Y., Oda, K., Urade, Y., Inui, T., Yamamura, T. and Okada, T., A potent and specific agonist, Suc-[Glu^9, $Ala^{11,15}$]-endothelin-1(8-21), IRL 1620, for the ET_B receptor, *Biochem. Biophys. Res. Commun.*, 184, 953-959. 1992.

Takasaki, C., Yanagisawa, M., Kimura, S., Goto, K. and Masaki, T., Similarity of endothelin to snake venom toxin, *Nature*, 335, 303. 1988.

Takayanagi, R., Kitazumi, K., Takasaki, C., Ohnaka, K., Aimoto, S., Tasaka, K., Ohashi, M. and Nawata, H., Presence of nonselective type of endothelin receptor on vascular endothelium and its linkage to vasodilation, *FEBS Lett.*, 282, 103-106. 1991.

Topouzis S., Huggins J.P., Pelton J.T and Miller R.C., Modulation by endothelium of the responses induced by endothelin-1 and by some of its analogues in rat isolated aorta, *Br. J. Pharmacol.*, 102, 545-549. 1991.

Urade, Y., Fujitani, Y., Oda, K., Watakabe, T., Umemura, I., Takai, M., Okada, T., Sakata, K. and Karaki, H., An endothelin B receptor antagonist: IRL 1038, [Cys^{11}-Cys^{15}]-endothelin-1 (11-21), *FEBS Lett.*, 311, 12-16. 1992.

Urade, Y., Fujitani, Y., Oda, K., Watakabe, T., Umemura, I., Takai, M., Okada, T., Sakata, K. and Karaki, H., Retraction. An endothelin B receptor-selective antagonist: IRL 1038, [Cys^{11}-Cys^{15}]-endothelin-1 (11-21), *FEBS Lett.*, 342, 103. 1994.

Vigne, P., Breittmayer, J.P. and Frelin, C., Competitive and non competitive interactions of BQ-123 with endothelin ET_A receptors, *Eur. J. Pharmacol.*, 245, 229-232. 1993.

Warner, T.D., Schmidt, H.H.H.W. and Murad, F., Interactions of endothelins and EDRF in bovine native endothelial cells: selective effects of endothelin-3, *Am. J. Physiol.*, 262, H1600-H1605. 1992.

Warner, T.D., Allcock, G.H., Corder, R. and Vane, J.R., Use of the endothelin antagonists BQ-123 and PD 142893 to reveal three endothelin receptors mediating smooth muscle contraction and the release of EDRF, *Br. J. Pharmacol.*, 110, 777-782. 1993.

Watanabe, H., Miyazaki, H., Kondoh, M., Masuda, Y., Kimura, S., Yanagisawa, M., Masaki, T. and Murakami, K., Two distinct types of endothelin receptors are present on chick cardiac membranes, *Biochem. Biophys. Res. Commun.*, 161, 1252-1259. 1989.

Williams, D.L., Jones, K.L., Pettibone, D.J., Lis, E.V. and Clineschmidt, B.V., Sarafotoxin S6c: an agonist which distinguishes between endothelin receptor subtypes, *Biochem. Biophys. Res. Commun.*, 175, 556-561. 1991.

Yanagisawa, M., Kurihara, H., Kimura, S., Tomobe, Y., Kobayashi, M., Mitsui, Y., Yazaki, Y., Goto, K., and Masaki, T., A novel potent vasoconstrictor peptide produced by vascular endothelial cells, *Nature*, 332, 411-415. 1988.

Zuccarello, M., Lewis, A.I. and Rapoport, R.M., Endothelin ET_A and ET_B receptors in subarachnoid hemorrhage-induced cerebral vasospasm, *Eur. J. Pharmacol.* 259, R1-R2. 1994.

Chapter **2**

Endothelin Receptor Molecular Biology

John P. Huggins

CONTENTS

0-8493-6975-4/97/$0.00+$.50

I. RECEPTOR SUBTYPE BIOCHEMISTRY

The short time between the discovery (and cloning) of the first ET (Yanagisawa et al., 1988) and the cloning of two receptor subtypes with high affinity for this peptide (Arai et al., 1990; Sakurai et al., 1990) was a remarkable achievement in molecular pharmacology, especially given that receptor subtype-selective antagonists had not been discovered before these receptors had been cloned. A general classification of ET receptors has emerged from the pharmacological nature of the cloned receptors (Masaki et al., 1994). The recent report of the construction of transgenic animals in which the gene for ET-1 had been deleted (Kurihara et al., 1994) was equally remarkable both in terms of the physiology of the resultant animals, that has challenged much of the current thinking about the roles of the ET peptides, and the rapidity of the achievement.

Nevertheless, there was some pharmacological and biochemical evidence available before the reports of receptor cloning that suggested that ET receptors were not uniform.

A. ETs and Analogues

Prior to 1990, the absence of general agreement in the literature concerning rank potency of different ETs in various pharmacological preparations suggested receptor heterogeneity: the majority of observations can now be rationalized using the ET_A/ET_B scheme (see Chapter 1) that emerged from initial receptor cloning. In addition, certain peptidergic analogues of ET-1 had different pharmacological properties in different systems. In particular, $[Ala^{1,3,11,15}]$ET-1, a linear analogue of ET-1 in which all the cysteines were replaced by alanine residues, was found to have agonist properties in preparations of guinea-pig cerebellum and trachea and rat mesentery and hindquarters (Crawford et al., 1990; Pelton and Miller, 1991; Randall et al., 1989), but to be neither an agonist nor antagonist of ET-1 in isolated rat aortic rings (Topouzis et al., 1991). This is notable because $[Ala^{1,3,11,15}]$ET-1 was subsequently found to be an ET_B-receptor ligand (Saeki et al., 1991).

B. Cross-Linking

Evidence was provided for receptor subtypes from multiple bands on autoradiographs of SDS polyacrylamide gels after cross-linking $[^{125}I]$ETs or sarafotoxins to their binding proteins in membranes prepared from various tissues; the M_r of the proteins labeled varies considerably between studies, but most reports indicate that "high" (73 to 44 kDa) and "low" (50 to 32) M_r bands were labeled (Masuda et al., 1989; Sugiura et al., 1989; Watanabe et al., 1989; Ambar et al., 1990; Schvartz et al., 1990; Hagiwara et al., 1990; Galron et al., 1991). In some cases, particular bands were preferentially labeled by different $[^{125}I]$ETs (Masuda et al., 1989; Watanabe et al., 1989; Ambar et al., 1990), the labeling of particular bands was inhibited by a receptor isotype-selective sarafotoxin (STXN-c) (Schvartz et al., 1991b) or cross-linking gave different results in different tissues (Martin et al., 1990; Galron et al., 1991).

The labeling of more than one band does not seem to be an evidence of ET receptor subtypes because the ET receptor cloned from bovine cerebellum and expressed in COS-7 cells gave a two-band pattern of labeling (Sakurai et al., 1990) that was reminiscent of that reported in many other studies and identical to that seen for the native receptor in bovine cerebellum (Elshourbagy et al., 1992). In addition, the "high" and "low" M_r proteins gave essentially identical radioactive protease

V8 maps, and inclusion of EDTA in the labeling incubation reduced labeling of the "low" M_r band (Schvartz et al., 1990, 1991a). Hence, artifactual proteolysis during receptor labeling seems to explain the multiple bands seen in cross-linking studies. This might well be of importance in those binding studies where EDTA is not present in the incubation, as the affinity of the proteolyzed receptor form(s) has not been studied. In addition, the presence of proteolyzed receptors that may have ligand selectivities that are different to the intact receptor may explain some of the labeling differences seen in cross-linking studies.

However, one feature of interest that emerged from this type of study, that is pronounced for ET_B-receptors, is the extraordinary stability of radioligand-ET receptor complexes at neutral pH, even in the absence of a cross-linker and presence of SDS (Takasuka et al., 1991, 1992). This stability, which is the cause of the very slow off-rate of dissociation of ETs from their receptors (Chapter 4), is a characteristic feature of ET receptors that has implications for the pathophysiology of the peptides (Huggins et al., 1993a) and has been explored by site-directed mutagenesis of ET receptors (below).

C. Binding Curves

Before the advent of recombinant ET receptors, receptor heterogeneity was evident from binding studies that indicated that Hill numbers for some ligand saturation binding curves were less than unity and that inhibition by ET-3 of [^{125}I]ET-1 binding was clearly biphasic in some tissues (Bousso-Mittler et al., 1989; Galron et al., 1989; Roubert et al., 1989; Jones et al., 1991; Maggi et al., 1991).

II. RECEPTOR CLONING

The simultaneous reports of the cloning and expression of receptors with or without selectivity for ET-1 compared to ET-3 immediately clarified much of the literature and led to the current ET_A and ET_B classification (Masaki et al., 1994; Chapter 1). It is important to note that because of uncertainties regarding agonist efficacies in any particular tissues, the classification is based on ET-1 and ET-3 *affinities*, and not potencies; these are evidently difficult to determine from functional studies alone (Huggins et al., 1993a). The consequent uncertainties inherent in functional experiments make the probing of the distribution of cloned ET receptors by *in situ* hybridization of great interest (Durham et al., 1993; Hori et al., 1992; see also Chapter 3), especially when performed at a level that enables the identification of cell types expressing particular receptors; however, the relationship between the presence of receptor mRNA and physiology is not always obvious. (For example, the distribution of m5 muscarinic receptor mRNA has been known since 1990, but a function for the encoded receptor has yet to be demonstrated; Vilaro et al., 1990.) In principle, the RT-PCR (reverse transcriptase polymerase chain reaction) technique can allow analysis of receptor expression in tiny quantities of tissue or single cells (see, e.g., Terada et al., 1992) that have been used for functional (e.g., electrophysiological) studies. Despite their strong potential for linking function with receptor distribution, these techniques have not been exploited to date by many workers in the study of ET receptors.

ET receptors have now been cloned from human placenta (Adachi et al., 1991; Cyr et al., 1991; Hosoda et al., 1991; Hayzner et al., 1992) and lung (Haendler et al., 1992), bovine lung (Arai et al., 1990) and rat A10 vascular smooth muscle cells (Lin et al., 1991) in the case of ET_A, and human placenta (Ogawa et al., 1991; Hayzer et al., 1992), liver (Nakamuta et al., 1991), jejunum (Sakamoto et al., 1991) and lung (Haendler et al., 1992), bovine lung (Saito et al., 1991), rat lung (Sakurai et al., 1990) and pig cerebellum (Elshourbagy et al., 1992) in the case of ET_B-receptors (In addition, there is a preliminary communication of a clone from retinal pericytes [King et al., 1992].) The characterization of these recombinant receptors has been performed by binding and functional studies following transient expression, usually in COS cells, that seem not to possess a native ET receptor (Arai et al., 1990). Cell lines expressing ET_A- and ET_B-receptors after transfection with a stable expression vector have been produced (Aramori and Nakanishi, 1992; Hechler et al., 1993a,b) and, in some cases, recombinant ET receptors have been used to detect (e.g., Stavros et al., 1993) or characterize (e.g., FR 139,317: Aramori et al., 1993) novel ET receptor antagonists. In addition, Haendler et al. (1993a) have produced transformants of *E. coli* that express ET_B-receptors with K_d values for ET-1, -2, and -3 similar to those found for these receptors when expressed in mammalian cells. As will become apparent from the discussion that follows, "the lack of information concerning the three-dimensional structure of G-protein-coupled receptors is a major barrier to understanding the structural requirements for ligand binding and of mechanisms of signal transduction, and biophysical studies of certain members of this family of proteins should be a major aim of molecular pharmacologists over the next decade or so" (Huggins et al., 1993a), and hence the successful expression of these receptors in large quantities by bacteria is a noteworthy advance towards an understanding of how ETs bind to the ligand-binding sites of their receptors.

In all cases, the primary structures of the recombinant receptors have indicated similarities with other G-protein-coupled receptors in that they contain seven hydrophobic, putative membrane-spanning domains and conserved residues at key positions in their sequences (see below). In this chapter the first putative transmembrane helix of a G-protein-coupled receptor is abbreviated h1, and so on. Analysis of the gene structure of the bovine ET_B-receptor indicates that it possesses 6 introns and spans 36 kb of DNA (Mizuno et al., 1992). There is no evidence for alternative splicing as occurs, for example, with dopamine receptors (Giros et al., 1989), although other, more subtle forms of posttranscriptional RNA editing (Burnashev et al., 1992) that have been observed to change the specificity of other receptors have not been studied and cannot yet be ruled out for ET receptors.

Following the original cloning strategies used (cloning by expression in *Xenopus* oocytes or by screening of mammalian cell expression libraries), most subsequent strategies have biased the rediscovery of ET_A- and ET_B-receptors by seeking sequences similar to the original reports, albeit from different species in many cases. However, not all functional evidence is compatible with the existence of only two ET receptors; the existence of responses elicited by ET-3 more potently than ET-1, as is the case for the inhibition of prolactin secretion from cultured anterior pituitary cells (Samson et al., 1990), clearly fall into this category and have been classified as being produced via ET_C-receptors (Masaki et al., 1994). The existence of these and other "unusual" responses to ETs (see Chapter 1) has prompted the question of whether more gene products exist than the subtypes hitherto described. Northern blots using RNA prepared from various tissues have not provided any clear evidence of further ET receptors (see Huggins et al., 1993a) and Southern blots of digested genomic DNA hybridized with radiolabelled DNA probes for ET_A- or ET_B-receptors indicate that if other ET receptors are encoded by mammals their structure must be less similar to the known mammalian recombinant ET receptors than these receptors are to each other (Sakamoto et al., 1991; Haendler et al., 1992).

TABLE 1

ET Receptor Sequence Identity

	ET_A			ET_{AX}	ET_B				ET_C	GRP
	Bovine	Human	Rat	*Xenopus*	Bovine	Human	Pig	Rat	*Xenopus*	Mouse
ET_A										
Bovine	100									
Human	95	100								
Rat	91	96	100							
ET_{AX} *Xenopus*	71	73	72	100						
ET_B										
Bovine	57	57	57	58	100					
Human	58	60	57	61	89	100				
Pig	58	59	59	57	90	90	100			
Rat	58	57	57	59	91	88	86	100		
ET_C *Xenopus*	49	49	48	50	50	52	50	53	100	
GRP Mouse	31	32	32	32	32	32	32	32	29	100

Note: The percentage identity between aligned sequences of cloned ET receptors is shown. The bombesin (GRP) receptor is also included for comparison. See text for references and methods.

Karne et al. (1993) claim to have cloned an ET_C-receptor from melanophores of *Xenopus laevis* that produce a pigmentation response that is marginally selective for ET-3 compared to ET-1 or -2 in terms of peptide concentration, but this recombinant receptor has not been sufficiently characterized to ascertain whether it is of the ET_C class or is the *Xenopus* form of the ET_B-receptor. In addition, Kumar et al. (1994) have cloned another ET receptor from *Xenopus,* in this case from heart. This was termed an ET_{AX}-receptor on the basis that COS cells transiently transfected with a plasmid containing the receptor sequence expressed binding sites for ET-1 which were blocked by ET-1 at 1000- to 10,000-fold lower concentrations than were necessary for ET-3, but which were unaffected by BQ-123 at concentrations up to 1 μM. These binding results were confirmed with functional data.

III. SEQUENCE ANALYSIS

The protein sequence identity of cloned ET receptors, as calculated using the BESTFIT option of the GCG package (based on the method of Smith and Waterman, 1981), is presented in Table 1 and compared to the sequence of the somewhat similar bombesin receptor (Spindel et al., 1990) as a control. This type of comparison has three drawbacks in that (1) only the putative transmembrane helices of G-protein-coupled receptors show sequence similarity with each other significantly greater than with a random string of residues, (2) the boundary residues of these helices are unknown and, most importantly, (3) the figures obtained rely heavily on the ability of a particular computer program to align sequences of low overall similarity but which contain domains of relatively high identity.

Nevertheless, a number of rather clearly defined identity levels are apparent from the table that are surprisingly similar within particular receptor subtypes, indicating that this simple approach has utility in receptor classification. ET_A- and ET_B-receptors of different species show between 86 and 96% identity with themselves and between 57 to 60% identity with members of the other receptor subclass. This compares with a maximum homology of 32% between any ET receptor and the receptor for bombesin. Similarly, the putative *Xenopus* ET_C-receptor shows no more than 53% identity to any other ET receptor, giving weight to the proposition that this does indeed represent a separate ET receptor class, especially as the other ET receptor cloned from *Xenopus* ("ET_{AX}") has much higher identity, this being between 71 and 73% to ET_A- and between 58 and 61% to ET_B-receptors, respectively. Comparisons of sequence similarity give the same pattern, although the differences obtained between receptor classes are less marked (data not shown).

When protein sequences are known for receptors from several species, more detailed (and useful) comparisons can be made to search for structures in the receptors that are responsible for ligand binding. Such a comparison, at a two-dimensional level, is reported in Plate 1,* in which amino acid residues have been color-coded that have been found in ET_A- or ET_B-receptors of all species to be acidic (Asp or Glu), basic (Lys, Arg, or His), hydrophobic (Gly, Tyr, Trp, Val, Ala, Met, Phe, Ile, or Leu), or hydrophilic (Ser, Thr, Gln or Asn) according to Cid et al. (1982). Cystines and prolines are coded individually because of the special nature of their properties (disulfide bond and α-helix breaking, respectively). Even a cursory inspection of Plate 1 shows the high degree of similarity within and even between subclasses of ET receptors. Some of the residues conserved between ET receptors are also found in all members of the G-protein-coupled receptor superfamily. These are α-helix breakers, such as Pro residues in putative helices 2, 4, 6, and 7, a negatively charged residue in h2, cystines in h3 and h6, aromatic residues in h5 and h6, and aliphatic hydrophobic residues in all helices apart from h5 and h6. Other residues which are generally conserved in other G-protein-coupled receptors but are variant in ET receptors are good candidates for residues that are involved in ligand binding (Huggins et al., 1993a,b).

The assumption used in this type of comparison is that residues that are conserved in all G-protein-coupled receptors are those involved in providing a skeleton that orients those residues important for the three functions possessed by these proteins, namely, binding ligands, binding G-proteins, and transducing information about the concentration of the former across the membrane into an activation state of the latter. The extracellular N-terminal domain was originally proposed (Arai et al., 1990) to be that part of ET receptors that conferred the ligand affinity to the ET_A- and ET_B-receptor subclasses. However, these sequences are not preserved between species and, therefore, would be predicted to play a minimal role in ligand recognition. For a peptide that is large compared to traditional ligands such as catecholamines, it is easy to imagine that the exceptionally high affinity found, for example, for the interactions of endothelins with their receptors (Chapter 4), is achieved by the corporate action of a number of binding sites on the extracellular loops or within extracellular clefts on the molecular surface formed by the putative transmembrane helices. Comparisons of the type shown in Plate 1 identify potential residues that enable ET_A- and ET_B-receptors to distinguish between subtype-selective ligands such as endothelin-3. For example, ET_B-receptors possess an acidic residue (always Asp) near the top of h4, which in ET_A-receptors is a hydrophobic residue (Val). Currently, a limited number of tools are available to molecular pharmacologists to unravel the mystery of how selectivity is achieved. These are described in the next section.

IV. MODELING AND MUTAGENESIS

In the absence of biophysical information regarding how the structure of any G-protein-coupled receptor enables high-affinity selective interactions with its

---

* Plate 1 follows page 70.

ligand, there are currently three methodologies available to provide hints on how this is achieved: chemical modification of receptors, modeling, and site-directed mutagenesis of cloned receptors. While greatly enhancing our knowledge about receptor structure in the last decade, all have drawbacks.

A. Protein Modification

Protein modification is an outmoded but still useful tool to study the biochemistry of receptors. Using diethylpyrocarbonate (Huggins et al., 1993b) and tetranitromethane (Bousso et al., 1992) to modify ET receptors, the involvement of receptor His and Tyr residues in the ligand binding site has been proposed, although in the latter case it was not clear whether the modified residue was present in the receptor or the ligand. In addition, sulfydryl-specific alkylating agents reduced the ability of [^{125}I]ET-1 to bind to its receptors (Spinella et al., 1993). This effect was inhibited by pretreatment of receptors with an antagonist that binds more reversibly than ET-1, [Drp1-Asp^{15}]ET-1, suggesting that the Cys concerned was in or near the ligand binding site.

Although the obvious caveat of these studies is the lack of knowledge about which residues are involved, such studies have the advantage that, in most cases, large changes of conformation are unlikely to be induced in the receptor by these modifications, whereas the effects of changes induced by site-directed mutagenesis (below) may be more severe as protein folding is more likely to be modified than the conformation of an already folded, existing protein.

B. Modeling

Three-dimensional receptor models have been constructed using the positions of the transmembrane helices described by cryomicroscopy for bacteriorhodopsin, orienting the helices with hydrophobic residues facing outwards, and performing energy minimization calculations on the seven-helix bundle (Hibert et al., 1991; Trumpp-Kallmeyer et al., 1992). The ET peptides are too large to enter the cleft that is proposed to contain the binding site of G-protein-coupled receptors for small ligands (Hibert et al., 1991). However, one of the earliest features to emerge from structure-activity relationships was that removal of the C-terminal Trp (that exists as a free carboxyl group, unlike in most biologically active peptides, where it is aminated) destroys biological activity and affinity for the ET receptor (Chapter 15). Huggins et al. (1993b) used the similarity of this residue to the indole structure of 5-hydroxytryptophan (5-HT) to model the possibility of interactions between this Trp and a binding site proposed to be within the dihedral cleft defined by the tightly packed seven-helix bundle, similar to that found in G-protein-coupled receptors that bind small cationic neurotransmitters. The resulting model predicted interactions from residues specifically conserved in ET receptors found in a pocket composed of putative helices 3, 4, 5, 6, and 7. A somewhat similar approach was taken by Krystek et al. (1994), whose model was used, successfully in one case, to predict point mutations in ET_A-receptors that would affect activity (below).

The disadvantage of the modeling approach is that large assumptions are necessarily made about the basic structure and conformational similarity between G-protein-coupled receptors and, especially, between these receptors and the structurally very different protein bacteriorhodopsin. This inevitably leads to uncertainty and plasticity in the final structure and hence resolution is too low for clear structural predictions in the design of new agonists and antagonists. This is particularly the case for peptidergic receptors, where only part of the binding interaction is modeled. However, the models provide a framework for performing intelligent mutagenesis experiments to test and refine structural hypotheses. The history of trying to interpret structure-activity relationships with peptides (Chapter 15) teaches us that considerations of the tertiary structure are essential (Chapter 14), and this conclusion will be even more striking for complex proteins than for relatively tiny peptides.

C. Mutagenesis

Site-directed mutagenesis and the construction of chimeric receptors has provided a means of probing the structure of receptors to make structure-activity relationships "in reverse".

Two approaches have been used to try to determine whether a mutation reduces affinity by destroying functional viability via gross changes in conformation. The chimeric receptor approach assumes that receptor similarity is sufficient so that where parts of receptors are interchanged this will have little gross conformational effect. This is certainly likely to be the case where the chimeras are ET receptor subtypes (below). The second approach is to check that agonists still give a functional response in cells transfected with a mutated receptor with reduced affinity. The supposition here is that the transduction of information by a receptor is a guide of conformational viability, but evidently mutations leading to local conformational changes in the binding domain can drastically affect ligand affinity. Hence, changes in affinity alone cannot be used to conclude the involvement of a particular amino acid in binding even where signal transduction to agonists is maintained. In some cases, multiple mutations are made at one site to try to tease apart affinity changes resulting from changes in conformation and affinity changes due to direct modification of ligand-receptor interactions. To put this discussion of *gross* conformational changes into context, removal of a single hydrogen bond between ligand and receptor could easily explain a tenfold difference in ligand affinity.

Haendler et al. (1993a,b) succeeded in expressing human ET_B-receptors in *E. coli* that bound [^{125}I]ET-1 with a K_d similar to that observed in eukaryotic cells. Replacement by Ala of each of the three Cys residues found in regions thought to correspond to extracellular loops (residues 174, 255, and 358 in extracellular loops 1, 2, and 3, respectively) caused a dramatic reduction in binding, which was especially marked with the former two Cys residues. By analogy with models of other G-protein-coupled receptors, it was concluded that Cys residues 174 and 255 form a disulfide bridge that is necessary for active receptor conformation. However, the mutations affected both affinity and receptor number (Haendler et al., 1993a) and no other attempts to examine conformation were performed.

Point mutations have also been made in ET receptors. Molecular modeling and sequence alignments demonstrated the possible role of Tyr 129 in h2 of the human ET_A-receptor in selectively binding ligands (Krystek et al., 1994; Lee et al., 1994). This residue is a His in the ET_B-subtype. Replacement of the Tyr by several other residues (Ala, Glu, Asn, His, Lys, Ser, or Phe) gave receptors with the same affinity for ET-1 and ET-2, but *enhanced* affinity for the ET_B-receptor-selective agonists ET-3 and STX-c and (except for the mutant Y129F) reduced affinity for the ET_A-receptor-selective antagonist BQ 123 (Lee et al., 1994). This confirmed similar results that had been obtained with a Y129A ET_A-mutant by Krystek et al. (1994). While these elegant results suggest a role for this residue in modulating ET_A-receptor selectivity, the picture is more complex as the H150Y mutant of ET_B-receptors did not affect binding of ligands compared to the wild-type ET_B-receptor (Lee et al., 1994). Hence, these results support the concept of complex multifaceted binding sites in peptidergic receptors and, interestingly, suggest that even if BQ 123 is not competitive with ETs as an antagonist at ET_A-receptors (Hiley et al., 1993), the sites of binding of ET-1 and BQ 123 overlap.

A second residue that has been implicated in ligand binding in ET_A-receptors is Lys 181. When this residue was replaced by Asp, the receptor's affinity was reduced by more than 1000-fold for STX-c (Zhu et al., 1992), by 600-fold for ET-3, but only marginally for ET-1 and -2 (Mauzy et al., 1992; Zhu et al., 1992). Both wild-type and mutated receptors were able to transduce binding to functional responses (InsP turnover and intracellular Ca^{2+} [Mauzy et al., 1992; Wu et al., 1992; Zhu et al., 1992]). This result evidently does not explain the difference in selectivities between ET_A- and ET_B-receptor subtypes as the Lys concerned is invariant between these subtypes. Hence, the Lys is most probably important in maintaining the physicochemical environment (perhaps conformation) of the STX-c and ET-3 binding sites, without being directly involved in ligand interactions. In addition, Lys 140 (ET_A) or Lys 161 (ET_B) have been studied by point substitution (see below).

ET receptors contain two potential glycosylation sites in their N-termini, and enzymatic deglycosylation with neuraminidase has been found to decrease [^{125}I]ET-1 binding to membranes, at least when prepared from some tissues (Bousso-Mittler et al., 1991). Truncation studies using ET_A-receptors have reported that removal of residues 25–76 (of the extracellular tail), which eliminates both sites, but not residues 25–49, which eliminates only the N-terminal site, abolishes [^{125}I]ET-1 binding, as does truncation at the C-terminus by 55 residues (Hashido et al., 1992). In all cases, Western blotting showed that the truncations did not prevent expression per se, although their effect on gross conformation or protein folding is likely. Whether the inhibition of binding by the large N-terminal deletion was due to removal of the glycosyl complex at the site nearer (in the primary sequence) the transmembrane region is not known. The work with intracellular deletions has been extended and a tail of more than 7 residues was found to be necessary for binding while more than 13 residues were needed for transduction, as indicated by increases in intracellular Ca^{2+} in *COS*-7 cells (Hashido et al., 1993). Because it was found that the binding effects were not caused by changes in receptor expression level, conformational perturbation was probably induced by truncation, emphasizing the problems inherent in this type of study.

Most of the mutagenesis studies on ET receptors have concerned ET_A/ET_B-receptor chimeras and this approach has been pioneered by the extraordinarily productive group of Miyamoto in Kamakura, who have made a series of human ET_A-receptors containing portions of ET_B sequences, and vice versa (Adachi et al., 1992). Initially, it was found that transferring a region from the ET_B-receptor, consisting of the extracellular ends of h2 and h3, and the loop bridging these, to the ET_A-subtype (to give the "ET_A-B" mutant) reduced the ability of BQ 123 to inhibit the binding of [^{125}I]ET-1 and, conversely, a similar transfer of ET_A-receptor h2, h3, and joining loop (to give "ET_B-B" mutant receptors) conferred the inhibitory effects of BQ 123 to ET_B-receptors, albeit at much higher concentrations than for wild-type ET_A-receptors (Adachi et al., 1992). Subsequently, a hepta-residue sequence within this "B" portion (Lys-Leu-Leu-Ala-Gly [KLLAG], occurring near the external surface of h2) was defined as responsible for [^{125}I]ET-1 binding, from experiments in which the equivalent sequence from the β_2-adrenoceptor was inserted into the ET_A-receptor (Adachi et al., 1993,1994a). Mutating each of these residues individually to the equivalent residue in the β-adrenoceptor indicated a very marked role of Lys 140 in the binding of ET-1 (Adachi et al., 1994a). Unfortunately, other substitutions were not made and so the precise role of the Lys is still open to speculation. Also, this Lys is absent in the ET_C-receptor cloned from *Xenopus* (Karne et al., 1993).

The site that confers selectivity between receptor subtypes in terms of ET-3 binding was more difficult to precisely define using the chimeric receptor approach (Adachi et al., 1994a,b). A chimeric receptor ("ET_A-CD") consisting of approximately one half (N-terminal) ET_A and the other half (C-terminal) ET_B was unable to distinguish between ET-1 and ET-3 in binding experiments. The best attempt to further localize the site which enabled this binding of ET-3 was the "ET_A-CD_3" mutant, in which extracellular loops 2 and 3, h6, and half of h5 were all replaced by ET_B-type sequences. This receptor had a K_i of 0.8 nM for ET-3 compared to 0.13 nM for ET-1 (Adachi et al., 1994a). In addition, in receptors ET_A-CD and ET_A-CD3, [^{125}I]ET-1 binding was inhibited by BQ 3020, as for ET_B-receptors (Adachi et al., 1994a). Equivalent experiments have also been performed with ET_B-receptors (Adachi et al., 1994b), but in this case, as well as h5 and the second extracellular loop, h2, h3, and the first extracellular loop were required for ET-3 binding. It is not easy to resolve these differences between the $ET_{A/B}$ and the $ET_{B/A}$ chimeras. As with the binding of [^{125}I]ET-1 to ET_A-receptors, the binding of [^{125}I]ET-3 to ET_B-receptors was heavily dependent on the Lys within the KLLAG sequence (Lys 161 in the human ET_B-receptor); however, in this case the binding of ET-1 was unaffected (Adachi et al., 1994b), clearly suggesting that while this residue is important in influencing the characteristics of the binding site, it is not essential for interaction with ET peptides.

The same chimeric constructs have been used to identify the region involved in binding an ET_B-receptor-selective antagonist, bosentan (Ro 47-0203); as with ET-1 the KLLAG sequence of h2 is required for binding, although this was not actually proven as the mutant containing the equivalent sequence from the β_2-adrenoceptor does not bind radioligand and as rigorous tests for allosteric binding were not employed (Adachi et al., 1994c).

Sakamoto et al., (Sakamoto, Yanagisawa, Nakao et al., 1993 and Sakamoto, Yanagisawa, Sawamura et al., 1993) have constructed a comprehensive set of $ET_{A/B}$- and $ET_{B/A}$-receptor chimeras in which putative transmembrane and hydrophilic domains have been serially transferred. These studies show the requirement of the domains from the second intracellular loop to the end of h6 of the ET_B-receptor in binding ET-3 and of the sequence from h1 to h3 as well as the extracellular moiety of h7 of the ET_A-receptor for high-affinity BQ 123 binding. They have developed a model of ligand binding already suggested for other peptides (Schwyzer, 1977) in which it is proposed that the C-terminal end of ETs binds to a site located in h1, h2, h3, and h7 that in ET_A-receptors also binds BQ 123 and in ET_B-receptors also binds BQ 3020 and IRL 1620. The N-terminal end of ET-1, termed the address domain, is proposed to bind to a site only present (or available) in h4, h5, and h6 of ET_A-receptors. It would be interesting to see how their model could be developed by studying the effects of ET peptide analogues on their chimeric receptors, thereby varying both sides of the structure-activity relationship.

Finally, a similar approach of combining chimeric and single residue replacement mutants has been used to define the region responsible for the remarkable stability of [^{125}I]ET-1:ET_B-receptor complexes in the presence of SDS (Takasuka et al., 1994). This study, that was aided by the finding that the stability of this complex is species dependent, concluded that Asp 75 and Pro 93 of human ET_B-receptors were essential, although neither the exact function of these residues in complex formation nor the implications of the complex, that would presumably be entirely stable under physiological conditions, are currently evident.

D. Hirschprung's Disease

A surprising finding to have recently emerged in the story of the biology of ET receptors is that a missense mutation in the ET_B-receptor predisposes individuals towards Hirschsprung's disease, also known as aganglionic megacolon, in which the congenital absence of ganglia in both the myenteric and submucosal plexuses induces severe constipation and intestinal blockage that usually presents in the neonatal period (Puffenberger et al., 1994). Sequencing within a recessive susceptibility locus (*HSCR2*), previously mapped to the same human chromosome region as the ET_B-receptor gene (*EDNRB*), showed the presence of a W276C mutation in h5. A higher concentration of STXN-c was required to give the same intracellular Ca^{2+} response in CHO cells expressing mutated ET_B-receptor compared to those expressing the wild-type, despite similar transfection efficiencies and apparently similar binding of the two receptors, although more detailed studies need to be performed to further characterize the nature of the defect involved. In addition, transgenic mice lacking a functional *EDNRB* gene resembled a strain of mice (piebald-lethal S') that has been characterized for its defects in the development of epidermal melanocytes and myenteric ganglions (Hosoda et al., 1994). This strain of mice was found to have a deletion of the entire *EDRNB* gene. In a complementary animal construct, the *EDN3* gene, which encodes ET-3, was disrupted (Baynash et al., 1994). These animals were also found to produce megacolons and spotting of fur color; a natural recessive mutation that induces a similar mouse phenotype (lethal spotting, *ls*) was located to a point mutation in the *EDN3* gene that prevents processing of big ET-3 by ECE.

Apart from the medical interest of these findings, they also have implications for the biology of ETs. A general colocalization of ET-3 and ET_B-receptors has suggested that ET-3 is the natural ligand for ET_B-receptors (see Huggins et al., 1993a). The fact that ET-1 does not replace the need for ET-3 in *EDN3*$^-$ mice shows that this is true, at least in some tissues, and also demonstrates that circulating ETs are not able to modify receptor function, confirming the paracrine function proposed for ETs by Huggins et al. (1993a).

E. Miscellaneous

The 5'-flanking region of the ET_A-receptor within the human genome has been sequenced and its promoter region partially characterized (Yang et al., 1993). When a 534-bp section of this region was inserted into a plasmid upstream of a luciferase gene and transfected into CHO cells, a promoter activity was demonstrated. The gene was localized to human chromosome 4. In addition to direct transcriptional control of ET receptors, control may also be exerted by regulating mRNA stability. This is suggested by the work of Sakurai et al. (1992), who found that although low levels of ET-1 or ET-3 elicit downregulation of ET_B-receptors at the mRNA level, this mRNA had a long intracellular life span. Such mechanisms have been proposed for other proteins (Zaidi and Malter, 1994).

Since the initial idea that ETs were endogenous Ca^{2+} channel agonists was disproved (see Huggins et al., 1993a), the effects of ions on ET receptors has been little studied. However, divalent cations markedly increase the affinity of ligand binding in both cell membranes (Huggins et al., 1994) and solubilized receptors (Kalina and Loffler, 1992); these effects seem to be mediated neither through changes in ligand conformation nor through the effects of G-proteins. Mn^{2+} has been found to change the Stoke's radius of solubilized ET receptor complexes (Kalina and Loffler, 1992).

V. MOLECULAR BIOLOGY OF THE ISOPEPTIDES

A. Gene Structure and Transcription

The discovery of what is now known to be ET-1 in media conditioned by endothelial cell culture led directly to the cloning of its cDNA (Yanagisawa et al., 1988). The cDNA sequence indicated that ET-1 is synthesized as a much larger prepropeptide that is converted by stepwise proteolysis into ET-1 (Chapter 6). Low stringency Southern analysis of human genomic DNA revealed the existence of two other ET sequences (Inoue et al., 1989). In addition, a fourth sequence was discovered by genomic cloning in mice (Saida et al., 1989). This is the vasointestinal contractor (VIC) or endothelin-β, which is often described as the murine and rat form of endothelin-2 (Rubanyi and Botelho, 1991) on the basis that the mRNA encoding its precursor is of identical length to that of preproET-2 and markedly different to preproET-1 and -3 (Bloch et al., 1991). The human gene for preproET-1 (EDN1) resides on chromosome arm 6p distal to HLA and has close linkage with D6S89 (Hoehe et al., 1993).

A large variety of agents stimulate the secretion of ETs from cell lines or tissues (Huggins et al., 1993a; Chapter 7), and in some cases this is associated with an increase in the level of preproET mRNA (Benatti et al., 1994). Conditions known to enhance transcription of preproETs include hypoxia of cells (Kourembanas et al., 1991) or whole animals (Elton et al., 1992; Donahue et al., 1994), fluid shear stress (Malek et al., 1993), and exposure of cells to various agents including endotoxin and proinflammatory cytokines (Nakano et al., 1994), angiotensin II (Chua et al., 1993) and factors associated with vascular injury, such as TGFβ1 and interleukin-1 (Maemura et al., 1992). In human aortic endothelial cells, it was found that gene expression also varied with the cell donor's age and the phase of cell growth (Kumazaki et al., 1994). The *cis*-acting DNA elements and *trans*-acting factors that control preproET-1 gene expression have recently been reviewed (Benatti et al., 1994) and will be described only briefly here; two promoter regions (A and B) are involved, both of which are required for full gene activation. Region A is probably the human equivalent of the GATA-2 gene, previously described in chickens; region B includes a functional AP1 consensus sequence.

A second mRNA sequence has been identified from a human placental cDNA library that encodes an identical preproET-1 molecule, but has a different 5′-untranslated region (Benatti et al., 1993), suggesting that transcriptional regulation may vary between the two sequences. Also, five alternately spliced variants of ET-2 mRNA have been detected, all of which encode identical ET-2 (O'Reilly et al., 1993). Thus control of ET-2 splicing might also affect expression of preproETs.

B. Transgenic Animals

The report of Kurihara et al. (1994) of the effects of oblation of the ET-1 gene has emphasized how much is still to be learned in the ET field. Homozygous ET-1$^-$/ET-1$^-$ mice died at birth due to respiratory failure that may have been caused partially by pronounced craniomandibular malformations and possibly also by failure of respiratory control mechanisms. Most surprising of all was that ET-1$^+$/ET-1$^-$ mice suffered mild but significant hypertension compared to wild-type mice. The response to the NO synthase inhibitor, L-*N*-monomethylarginine, was normal in the heterozygous mice, suggesting that abnormal endothelium-derived NO secretion did not contribute to the observed hypertension. In addition, the effect of disrupting the ET-3 gene was described above (Section IV. D).

One disturbing feature of the above results is that all ET receptor antagonists and ECE inhibitors developed for human therapy will have to be tested very carefully for teratogenic effects if used during pregnancy. However, the physiological and pathological conclusions are more difficult to extrapolate. It is, however, clear that much more research is needed into the effects of ETs on tissue growth and differentiation *in vitro*.

VI. CONCLUSION

This chapter has described the molecular pharmacology of ETs and their receptors. Although enormous progress has been made in the field, much remains to be done. Areas of notable advance are in the understanding of ET receptors by mutagenesis and modeling; however, the agreement between the two methods is not always good (which means that the models require modification and higher resolution) and results of mutagenesis experiments are often difficult to interpret. Similarly, while the transgenic mice lacking expression of ETs and ET receptors have enormous potential for understanding the biology of the ETs, they pose at least as many questions as answers at the current time. Hence, the dissection of the developmental role of ETs and direct biophysical study of ET receptors could be singled out as being of paramount interest in future attempts to analyze the ET peptides and receptors by molecular pharmacological means.

REFERENCES

Adachi, M., Furuichi, Y. and Miyamoto, C., Identification of a ligand-binding site of the human endothelin-A receptor and specific regions required for ligand selectivity, *Eur. J. Biochem.*, 220, 37-43. 1994a.

Adachi, M., Furuichi, Y. and Miyamoto, C., Identification of specific regions of the human endothelin-B receptor required for high affinity binding with endothelin-3, *Biochim. Biophys. Acta*, 1223, 202-208. 1994b.

Adachi, M., Furuichi, Y. and Miyamoto, C., Identification of a region of the human endothelin ET_A receptor required for interaction with bosentan, *Eur. J. Pharmacol.*, 269, 225-234. 1994c.

Adachi, M., Hashido, K., Trzeciak, A., Watanabe, T., Furuichi, Y. and Miyamoto, C., Functional domains of human endothelin receptor, *J. Cardiovasc. Pharmacol.*, 22 (Suppl. 8), S121-S124. 1993.

Adachi, M., Yang, Y.-Y., Furuichi, Y. and Miyamoto, C., Cloning and characterization of cDNA encoding human A-type endothelin receptor, *Biochem. Biophys. Res. Commun.*, 180, 1265-1272. 1991.

Adachi, M., Yang, Y.-Y., Trzeciak, A., Furuichi, Y. and Miyamoto, C., Identification of a domain of ET_A receptor required for ligand binding, *FEBS Lett.*, 311, 179-183. 1992.

Ambar, I., Kloog, Y. and Sokolovsky, M., Cross-linking of endothelin 1 and endothelin 3 to rat brain membranes: identification of the putative receptor(s), *Biochemistry*, 29, 6415-6418. 1990.

Arai, H., Hori, S., Aramori, I., Ohkubo, H. and Nakanishi, S., Cloning and expression of a cDNA encoding an endothelin receptor, *Nature*, 348, 730-732. 1990.

Aramori, I. and Nakanishi, S., Coupling of two endothelin receptor subtypes to differing signal transduction in transfected Chinese hamster ovary cells, *J. Biol. Chem.*, 267, 12468-12474. 1992.

Aramori, I., Nirei, H., Shoubo, M., Sogabe, K., Nakamura, K., Kojo, H., Notsu, Y., Ono, T. and Nakanishi, S., Subtype selectivity of a novel endothelin antagonist, FR139317, for the endothelin receptors in transfected Chinese hamster ovary cells, *Mol. Pharmacol.*, 43, 127-131. 1993.

Baynash, A. G., Hosoda, K., Giaid, A., Richardson, J. A., Emoto, N., Hammer, R. E. and Yanagisawa, M., Interaction of endothelin-3 with endothelin-B receptor is essential for development of epidermal melanocytes and enteric neurons, *Cell*, 79, 1277-1285. 1994.

Benatti, L., Bonecchi, L., Cozza, L. and Sarientos, P., Two preproendothelin 1 mRNAs transcribed by alternative promoters, *J. Clin. Invest.*, 91, 1149-1156. 1993.

Benatti, L., Fabbrini, M. S. and Patrono, C., Regulation of endothelin-1 biosynthesis, *Anal. N.Y. Acad. Sci.*, 714, 109-121. 1994.

Bloch, K. D., Hong, C. C., Eddy, R. L., Shows, T. B. and Quertermous, T., cDNA cloning and chromosomal assignment of the rat endothelin 2 gene: vasoactive intestinal contractor peptide is rat endothelin 2, *Genomics*, 10, 236-242. 1991.

Bousso, D., Bdolah, A. and Sokolovsky, M., Involvement of tyrosyl residue(s) in binding of endothelin and sarafotoxin to their receptors in rat brain and heart, *Neurosci. Lett.*, 140, 247-250. 1992.

Bousso-Mittler, D., Kloog, Y., Wollberg, Z., Bdolah. A., Kochva, E. and Sokolovsky, M., Functional endothelin/sarafotoxin receptors in the rat uterus, *Biochem. Biophys. Res. Commun.*, 162, 952-957. 1989.

Bousso-Mittler, D., Galron, R. and Sokolovsky, M., Endothelin/sarafotoxin receptor heterogeneity: evidence for different glycosylation in receptors from different tissues, *Biochem. Biophys. Res. Commun.*, 178, 921-926. 1991.

Burnashev, N., Monyer, H., Seeburg, P. H. and Sakmann, B., Divalent ion permeability of AMPA receptor channels is dominated by the edited form of a single subunit, *Neuron*, 8, 189-198. 1992.

Chua, B. H., Chua, C. C., Diglio, C. A. and Siu, B. B., Regulation of endothelin-1 mRNA by angiotensin II in rat heart endothelial cells, *Biochim. Biophys. Acta*, 1178, 201-206. 1993.

Cid, H., Bunster, M., Arriagada, E. and Campos, M., Prediction of secondary structure of proteins by means of hydrophobicity profiles, *FEBS Lett.*, 150, 247-254. 1982.

Crawford, M. L. A., Hiley, C. R. and Young, J. M., Characteristics of endothelin-1 and endothelin-3 stimulation of phosphoinositide breakdown differ between regions of guinea pig and rat brain, *Naunyn-Schmiedeberg's Arch. Pharmacol.*, 341, 268-271. 1990.

Cyr, C., Huebner, K., Druck, T. and Kris, R., Cloning and chromosomal localization of a human endothelin ET_A receptor, *Biochem. Biophys. Res. Commun.*, 181, 184-190. 1991.

Donahue, D. M., Lee, M. E., Suen, H. C., Quertermous, T. and Wain, J. C., Pulmonary hypoxia increases endothelin-1 gene expression in sheep, *J. Surg. Res.*, 57, 280-283. 1994.

Durham, S. K., Goller, N. L., Lynch, J. S., Fisher, S. M. and Rose, P. M., Endothelin receptor B expression in the rat and rabbit lung as determined by *in situ* hybridization using nonisotopic probes, *J. Cardiovasc. Pharmacol.*, 22 (Suppl. 8), S1-S3. 1993.

Elshourbagy, N. A., Lee, J. A., Korman, D. R., Nuthalaganti, P., Sylvester, D. R., Dilella, A. G., Sutiphong, J. A. and Kumar, C. S., Molecular cloning and characterization of the major endothelin receptor subtype in porcine cerebellum, *Mol. Pharmacol.*, 41, 465-473. 1992.

Elton, T. S., Oparil, S., Taylor, G. R., Hicks, P. H., Yang, R. H., Jin, H. and Chen, Y. F., Normobaric hypoxia stimulates endothelin-1 gene expression in the rat, *Am. J. Physiol.*, 263, 1260-1264. 1992.

Galron, R., Kloog, Y., Bdolah, A. and Sokolovsky, M., Functional endothelin/sarafotoxin receptors in rat heart myocytes: structure-activity relationships and receptor subtypes, *Biochem. Biophys. Res. Commun.*, 163, 936-943. 1989.

Galron, R., Bdolah, A., Kochva, E., Wollberg, Z., Kloog, Y. and Sokolovsky, M., Kinetic and cross-linking studies indicate different receptors for endothelins and sarafotoxins in the ileum and cerebellum, *FEBS Lett.*, 283, 11-14. 1991.

Giros, B., Sokoloff, P., Martres, M.-P., Riou, J.-F., Emorine, L. J. and Schwartz, J.-C., Alternative splicing directs the expression of two D2 dopamine receptor isoforms, *Nature*, 342, 923-926. 1989.

Haendler, B., Hechler, U. and Schleuning, W.-D., Molecular cloning of human endothelin (ET) receptors ET_A and ET_B, *J. Cardiovasc. Pharmacol.*, 22(Suppl. 8), S1-S4, 1992.

Haendler, B., Hechler, U., Becker, A., and Schleuning, W. D., Expression of human endothelin receptor ET_B by *Escherichia coli* transformants, *Biochem. Biophys. Res. Commun.*, 191, 633-638. 1993a.

Haendler, B., Hechler, U., Becker, A., and Schleuning, W. D., Extracellular cysteine residues 174 and 255 are essential for active expression of human endothelin receptor ET_B in *Escherichia coli*, *J. Cardiovasc. Pharmacol.*, 22 (Suppl. 8), S4-S6. 1993b.

Hagiwara, H., Kozuka, M., Eguchi, S., Shibabe, S., Ito, T. and Hirose, S., Solubilization of endothelin receptors from bovine lung plasma membranes in a non-aggregated state and estimation of their minimal functional sizes, *Biochem. Biophys. Res. Commun.*, 172, 576-581. 1990.

Hashido, K., Adachi, M., Gamou, T., Watanabe, T., Furuichi, Y. and Miyamoto, C., Identification of specific intracellular domains of the human ET_A receptor required for ligand binding and signal transduction, *Cell Mol. Biol. Res.*, 39, 3-12. 1993.

Hashido, K., Gamou, T., Adachi, M., Tabuchi, H., Watanabe, T., Furuichi, Y. and Miyamoto, C., Truncation of N-terminal extracellular or C-terminal intracellular domains of human ET_A receptor abrogated the binding activity of ET-1, *Biochem. Biophys. Res. Commun.*, 187, 1241-1248. 1992.

Hayzer, D. J., Rose, P. M., Lynch, J. S., Webb, M. L., Kienzle, B. K., Liu, E. C., Bogosian, E. A., Brinson, E. and Runge, M. S., Cloning and expression of a human endothelin receptor: subtype A, *Am. J. Med. Sci.*, 304, 231-8. 1992.

Hechler, U., Becker, A., Haendler, B. and Schleuning, W.-D., Binding characteristics of recombinant human endothelin receptor ET_A and ET_B expressed in baby hamster kidney cells, *J. Cardiovasc. Pharmacol.*, 22 (Suppl. 8), S15-S17. 1993a.

Hechler, U., Becker, A., Haendler, B. and Schleuning, W.-D., Binding characteristics of recombinant human endothelin receptor ET_A and ET_B expressed in baby hamster kidney cells, *J. Biochem. Biophys. Res. Commun.*, 194, 1305-1310. 1993b.

Hibert, M. F., Trumpp-Kallmeyer, S., Bruinvels, A. and Hoflack, J., First three-dimensional models of neurotransmitter G-binding protein-coupled receptors, *Mol. Pharmacol.*, 40, 8-15. 1991.

Hiley, C. R., Cowley, D. J., Pelton, J. T. and Hargreaves, A. C., BQ-123, cyclo(-D-Trp-D-Asp-Pro-D-Val-Leu), is a non-competitive antagonist of the actions of endothelin-1 in SK-N-MC human neuroblastoma cells, *Biochem. Biophys. Res. Commun.*, 184, 504-510. 1993.

Hoehe, M. R., Ehrenreich, H., Otterud, B., Caennazo, L., Plaetke, R., Zander, H. and Leppert, M., The human endothelin-1 gene (EDN1) encoding a peptide with potent vasoactive properties maps distal to HLA on chromosome arm 6p in close linkage to D6S89, *Cytogenet. Cell Genet.*, 62, 131-135. 1993.

Hori, S., Komatsu, Y., Shigemoto, R., Mizuno, N. and Nakanishi, S., Distinct tissue distribution and cellular localization of two messenger ribonucleic acids encoding different subtypes of rat endothelin receptors, *Endocrinology*, 130, 1885-1895. 1992.

Hosoda, K., Nakao, K., Arai, H., Suga, S.-I., Ogawa, Y., Mukoyama, M., Shirakama, G., Saito, Y., Nakanishi, S. and Imura, H., Cloning and expression of human endothelin-1 receptor cDNA, *FEBS Lett.*, 287, 23-26. 1991.

Hosoda, K., Hammer, R. E., Richardson, J. A., Baynash, A. G., Cheung, J. C., Giaid, A. and Yanagisawa, M., Targeted and natural (piebald-lethal) mutations of endothelin-B receptor gene produce megacolon associated with spotted coat color in mice, *Cell*, 79, 1267-1276. 1994.

Huggins, J. P., Pelton, J. T. and van Giersbergen, P. L. M., The receptors for endothelins and their analogues in SK-N-MC neuroblastoma cells, *Peptides*, 15, 529-536. 1994.

Huggins, J. P., Pelton, J. T. and Miller, R. J., The structure and specificity of endothelin receptors: their importance in physiology and medicine, *Pharmacol. Ther.*, 59, 55-123. 1993a.

Huggins, J. P., Trumpp-Kallmeyer, S., Hibert, M. F., Hoflack, J. M., Fanger, B. O. and Jones, C. R., Modeling and modification of the binding site of endothelin and other receptors, *Eur. J. Pharmacol.*, 245, 203-214. 1993b.

Inoue, A., Yanagisawa, M., Kimura, S., Kasuya, Y., Miyauchi, T., Goto, K. and Masaki, T., The human endothelin family: three structurally and pharmacologically distinct isopeptides predicted by three separate genes, *Proc. Natl. Acad. Sci. U.S.A.*, 86, 2863-2867. 1989.

Jones, C. R., Hiley, C. R., Pelton, J. T. and Miller, R. C., Endothelin receptor heterogeneity; structure activity, autoradiographical and functional studies, *J. Receptor Res.*, 11, 299-310. 1991.

Kalina, B. and Loffler, B. M., Crosslinking analysis of an endothelin receptor protein from human placenta, *Biochem. Int.*, 27, 735-744. 1992.

Karne, S., Jayawickreme, C. K. and Lerner, M. R., Cloning and characterization of an endothelin-3 specific receptor (ET_C receptor) from *Xenopus laevis* dermal melanophores, *J. Biol. Chem.*, 268, 19126-19133. 1993.

King G. L., Shiba, T. and Goldstein B. J., Cloning and expression of endothelin-1 (ET-1) receptor from retinal pericytes, *Invest. Ophthalmol. Visual Sci.*, 33, 816. 1992.

Kourembanas, S., Marsden, P. A., Mcquillan, L. P., and Faller, D. V., Hypoxia induces endothelin gene expression and secretion in cultured human endothelium, *J. Clin. Invest.*, 88, 1054-1057. 1991.

Krystek, S. R., Patel, P. S., Rose, P. M., Fisher, S. M., Kienzle, B. K., Lach, D. A., Liu, E. C. K., Lynch, J. S., Novotny, J. and Webb, M. L., Mutation of peptide binding site in transmembrane region of a G-protein-coupled receptor accounts for endothelin receptor subtype selectivity, *J. Biol. Chem.*, 269, 12383-12386. 1994.

Kumar, C., Mwangi, V., Nuthulaganti, P., Wu, H.-L., Pullen, M., Brun, K., Aiyar, H., Morris, R. A., Naughton, R. and Nambi, P., Cloning and characterization of a novel endothelin receptor from *Xenopus* heart, *J. Biol. Chem.*, 269, 13414-13420. 1994.

Kumazaki, T., Fujii, T., Kobayshi, M. and Mitsui, Y., Aging- and growth-dependent modulation of endothelin-1 gene expression in human vascular endothelial cells, *Exp. Cell Res.*, 211, 6-11. 1994.

Kurihara, Y., Kurihara, H., Suzuki, H., Kodama, T., Maemura, K., Nagai, R., Oda, H., Kuwaki, T., Cao, W.-H., Kamada, N., Jishage, K., Ouchi, Y., Azuma, S., Toyoda, Y., Ishikawa, T., Kumada, M. and Yazaki, Y., Elevated blood pressure and craniofacial abnormalities in mice deficient in endothelin-1, *Nature*, 368, 703-710. 1994.

Lee, J. A., Elliot, J. D., Sutiphong, J. A., Friesen, W. J., Ohlstein, E. H., Stadel, J. M., Gleason, J. G. and Peishoff, C. E., Tyr-129 is important to the peptide ligand affinity and selectivity of human endothelin type A receptor, *Proc. Natl. Acad. Sci. U.S.A.*, 91, 7164-7168. 1994.

Lin, H. Y., Kaji, E. H., Winkel, G. K., Ives, H. E. and Lodish, H. F., Cloning and functional expression of a vascular smooth muscle endothelin 1 receptor, *Proc. Natl. Acad. Sci. U.S.A.*, 88, 3185-3189. 1991.

Maemura, K., Kurihara, H., Morita, T., Oh-hashi, Y. and Yazaki, Y., Production of endothelin-1 in vascular endothelial cells is regulated by factors associated with vascular injury, *Gerontology*, 38 (Suppl. 1), 29-35. 1992.

Maggi, M., Vannelli, G. B., Peri, A., Brandi, M. L., Fantoni, G., Giannini, S., Torrisi, C., Guardabasso, V., Barni, T., Toscano, V., Massi, G. and Serio, M., Immunolocalization, binding, and biological activity of endothelin in rabbit uterus: effect of ovarian steroids, *Am. J. Physiol.*, 260, E292-E305. 1991.

Malek, A. M., Greene, A. L. and Izumo, S., Regulation of endothelin 1 gene by fluid shear stress is transcriptionally mediated and independent of protein kinase C and cAMP, *Proc. Natl. Acad. Sci. U.S. A.*, 90, 5999-6003. 1993.

Martin, E. R., Brenner, B. M. and Ballermann, B. J., Heterogeneity of cell surface endothelin receptors, *J. Biol. Chem.*, 265, 14044-14049, 1990.

Masaki, T., Vane, J. R. and Vanhoutte, P. M., V. International union of pharmacology nomenclature of endothelin receptors, *Pharmacol. Rev.*, 46, 137-142. 1994.

Masuda, Y., Miyazaki, H., Kondoh, M., Watanabe, H., Yanagisawa, M., Masaki, T. and Murakami, K., Two different forms of endothelin receptors in rat lung, *FEBS Lett.*, 257, 208-210. 1989.

Mauzy, C., Wu, L.-H., Egloff, A. M., Mirzadegan, T. and Chung, F.-Z., Replacement of lysine-181 by aspartic acid in the third transmembrane region of the endothelin (ET) type B receptor selectively reduces its high affinity binding with ET-3 peptide, *J. Cardiovasc. Pharmacol.*, 22 (Suppl. 8), S5-S7. 1992.

Mizuno, T., Imai, T., Itakura, M., Hirose, S., Hirata, Y., Marumo, F. and Hagiwara, H., Structure of the bovine endothelin-B receptor gene and its tissue-specific expression revealed by northern analysis, *J. Cardiovasc. Pharmacol.*, 22 (Suppl. 8), S8-S10. 1992.

Nakamuta, M., Takayanagi., R., Sakai, Y., Sakamoto, S., Hagiwara, H., Mizuno, T., Saito, Y., Hirose, S., Yamamoto, M. and Nawata, H., Cloning and sequence analysis of a cDNA encoding human nonselective type of endothelin receptor, *Biochem. Biophys. Res. Commun.*, 177, 34-39. 1991.

Nakano, J., Takizawa, H., Ohtoshi, T., Shoji, S., Yamaguchi, M., Ishii, A., Yanagisawa, M. and Ito, K., Endotoxin and proinflammatory cytokines stimulate endothelin-1 expression and release by airway epithelial cells, *Clin. Exp. Allergy*, 24, 330-336. 1994.

Ogawa, Y., Nakao, K., Arai, H., Nakagawa, O., Hosoda, K., Suga, S.-I., Nakanishi, S. and Imura, H., Molecular cloning of a non-isopeptide-selective human endothelin receptor, *Biochem. Biophys. Res. Commun.*, 178, 248-255. 1991.

O'Reilly, G., Charnock-Jones, D. S., Morrison, J. J., Cameron, I. T., Davenport, A. P. and Smith, S. K., Alternatively spliced mRNAs for human endothelin-2 and their tissue distribution, *Biochem. Biophys. Res. Commun.*, 193, 834-840. 1993.

Pelton, J. T. and Miller, R. C., The role of disulfide bonds in endothelin-1, *J. Pharm. Pharmacol.*, 43, 43-45. 1991.

Puffenberger, E. G., Hosoda, K., Washington, S. S., Nakao, K., deWit, D., Yanagisawa, M. and Chakravarti, A., A missense mutation of the endothelin-B receptor gene in multigenic Hirschsprung's disease, *Cell*, 79, 1257-1266. 1994.

Randall, M. D., Douglas, S. A. and Hiley, C. R., Vascular activities of endothelin-1 and some alanyl substituted analogues in resistance beds of the rat, *Br. J. Pharmacol.*, 98, 685-699. 1989.

Roubert, P., Gillard, V., Plas, P., Guillon, J.-M., Chabrier, P.-E. and Braquet, P., Angiotensin II and phorbol-esters potently downregulate endothelin (ET-I) binding sites in vascular smooth muscle cells, *Biochem. Biophys. Res. Commun.*, 164, 809-815. 1989.

Rubanyi, G. M. and Botelho, L. H. P., Endothelins, *FASEB J.*, 5, 2713-2720. 1991.

Saida, K., Mitsui, Y. and Ishida, N., A novel peptide, vasoactive intestinal contractor, of a new (endothelin) peptide family, *J. Biol. Chem.*, 264, 14613-14616. 1989.

Sakamoto, A., Yanagisawa, M., Nakao, K., Toyo-oka, T., Yano, M. and Masaki, T., The ligand-receptor interactions of the endothelin systems are mediated by distinct "message" and "address" domains, *J. Cardiovasc. Pharmacol.*, 22 (Suppl. 8), S113-S116. 1993.

Sakamoto, A., Yanagisawa, M., Sakurai, T., Takuwa, Y., Yanagisawa, H. and Masaki, T., Cloning and functional expression of human cDNA for the ET_B endothelin receptor, *Biochem. Biophys. Res. Commun.*, 178, 656-663. 1991.

Sakamoto, A., Yanagisawa, M., Sawamura, T., Enoki, T., Ohtani, T., Sakurai, T., Nakao, K., Toyo-oka, T. and Masaki, T., Distinct subdomains of human endothelin receptors determine their selectivity to $endothelin_A$-selective antagonist and $endothelin_B$-selective agonists, *J. Biol. Chem.*, 268, 8547-8553. 1993.

Saeki, T., Ihara, M., Fukuroda, T., Yamagiwa, M. and Yano, M., [$Ala^{1,3,11,15}$]endothelin-1 analogues with ET_B agonistic activity, *Biochem. Biophys. Res. Commun.*, 179, 286-292. 1991.

Saito, Y., Mizuno, T., Itakura, M., Suzuki, Y., Ito, T., Hagiwara, H. and Hirose, S., Primary structure of bovine endothelin ET(B) receptor and identification of signal peptidase and metal proteinase cleavage sites, *J. Biol. Chem.*, 266, 23433-23437. 1991.

Sakurai, T., Morimoto, H., Kasuya, Y., Takuwa, Y., Nakauchi, H., Masaki, T. and Goto, K., Level of ET_B receptor messenger RNA is downregulated by endothelins through decreasing the intracellular stability of messenger RNA molecules, *Biochem. Biophys. Res. Commun.*, 186, 342-347. 1992.

Sakurai, T., Yanagisawa, M., Takuwa, Y., Miyazaki, H., Kimura, S., Goto, K. and Masaki, T., Cloning of a cDNA encoding a non-isopeptide-selective subtype of the endothelin receptor, *Nature*, 348, 732-735. 1990.

Samson, W. K., Skala, K. D., Alexander, B. D. and Huang, F. S., Pituitary site of action of endothelin: selective inhibition of prolactin release *in vitro*, *Biochem. Biophys. Res. Commun.*, 169, 737-743. 1990.

Schvartz, I., Ittoop, O. and Hazum, E., Identification of endothelin receptors by chemical cross-linking, *Endocrinology*, 126, 1829-1833. 1990.

Schvartz, I., Ittoop, O. and Hazum, E., Identification of a single binding protein for endothelin-1 and endothelin-3 in bovine cerebellum membranes, *Endocrinology*, 128, 126-130. 1991a.

Schvartz, I., Ittoop, O. and Hazum, E., Direct evidence for multiple endothelin receptors, *Biochemistry*, 30, 5325-5327. 1991b.

Schwyzer, R., ACTH: a short introductory review, *Ann. N.Y. Acad. Sci.*, 297, 3-26. 1977.

Spindel, E. R., Giladi, E., Brehm, P., Goodman, R. H. and Segerson, T. P., Cloning and functional characterization of a complementary DNA encoding the murine fibroblast bombesin/gastrin-releasing peptide receptor, *Mol. Endocrinol.*, 4, 1956-1963. 1990.

Spinella, M. J., Kottke, R., Magazine, H. I., Healy, M. S., Catena, J. A., Wilken, P. and Andersen, T. T., Endothelin-receptor interactions. Role of a putative sulfhydryl on the endothelin receptor, *FEBS Lett.*, 328, 82-88. 1993.

Stavros, F. D., Hasel, K. W., Okun, I., Baldwin, J. and Freriks, K., COS-7 cells stably transfected to express the human ET_B receptor provide a useful screen for endothelin receptor antagonists, *J. Cardiovasc. Pharmacol.*, 22 (Suppl. 8), S34-S37. 1993.

Sugiura, M., Snajdar, R. M., Schwartzberg, M., Badr, K. F. and Inagami, T., Identification of two types of specific endothelin receptors in rat mesangial cell, *Biochem. Biophys. Res. Commun.*, 162, 1396-1401. 1989.

Takasuka, T., Horii, I., Furuichi, Y. and Watanabe, T., Detection of an endothelin-1-binding protein complex by low temperature SDS-PAGE, *Biochem. Biophys. Res. Commun.*, 176, 392-400. 1991.

Takasuka, T., Akiyama, N., Horii, I., Furuichi, Y. and Watanabe, T., Different stability of ligand-receptor complex formed with two endothelin receptor species, ET_A and ET_B, *Biochem. J.*, 111, 748-753. 1992.

Takasuka, T., Sakurai, T., Goto, K., Furuichi, Y. and Watanabe, T., Human endothelin receptor ET_B. Amino acid sequence requirements for super stable complex formation with its ligand, *J. Biol. Chem.*, 269, 7509-7513. 1994.

Terada, Y., Tomita, K., Nonoguchi, H. and Marumo, F., Different localization of two types of endothelin receptor messenger RNA in microdissected rat nephron segments using reverse transcription and polymerase chain reaction assay, *J. Clin. Invest.*, 90, 107-112. 1992.

Topouzis, S., Huggins, J. P., Pelton, J. T. and Miller, R. C., Modulation by endothelium of the responses induced by endothelin-1 and by some of its analogues in rat isolated aorta, *Br. J. Pharmacol.*, 102, 545-549. 1991.

Trumpp-Kallmeyer, S., Hoflack, J., Bruinvels, A. and Hibert, M. F., Modeling of G-protein-coupled receptors: applications to dopamine, adrenaline, serotonin, acetylcholine and mammalian opsin receptors, *J. Med. Chem.*, 35, 3448-3462. 1992.

Vilaro, M. T., Palacios, J. M. and Menogod, G., Localization of m5 muscarinic receptor mRNA in rat brain examined by *in situ* hybridization histochemistry, *Neurosci. Lett.*, 114, 154-159. 1990.

Watanabe, H., Miyazaki, H., Kondoh, M., Masuda, Y., Kimura, S., Yanagisawa, M., Masaki, T. and Murakami, K., Two distinct types of endothelin receptors are present on chick cardiac membranes, *Biochem. Biophys. Res. Commun.*, 161, 1252-1259. 1989.

Wu, L.-H., Zhu, G., Egloff, A. M., Mauzy, C. A., Rapundalo, S., Mirzadegan, T., Cody, W., Oxender, D. L., and Chung, F.-Z., Substitution of lysine-181 to aspartic acid in the third transmembrane region of the endothelin type B receptor (ET_B) abolishes its binding to an ET_B selective agonist-sarafotoxin S6C, *FASEB J.*, 6, 1563. 1992.

Yanagisawa, M., Kurihara, H., Kimura, S., Tomobe, Y., Kobayashi, M., Mitsui, Y., Yazaki, Y., Goto, K. and Masaki, T., A novel potent vasoconstrictor peptide produced by vascular endothelial cells, *Nature*, 332, 411-415. 1988.

Yang, H., Tabuchi, H., Furuichi, Y. and Miyamoto, C., Molecular characterization of the 5′-flanking region of human genomic ET_A gene, *Biochem. Biophys. Res. Commun.*, 190, 332-339. 1993.

Zaidi, S. H. E. and Malter, J. S., Amyloid precursor protein mRNA stability is controlled by a 29-base element in the 3′-untranslated region, *J. Biol. Chem.*, 269, 24007-24013. 1994.

Zhu, G., Wu, L.-H., Mauzy, C., Egloff, A. M., Mirzadegan, T. and Chung, F.-Z., Replacement of lysine-181 by aspartic acid in the third transmembrane region of endothelin type B receptor reduces its affinity to endothelin peptides and sarafotoxin S6c without affecting G protein coupling, *J. Cell. Biochem.*, 50, 159-164. 1992.

Chapter **3**

Distribution of Endothelin Receptors

Anthony P. Davenport

CONTENTS

0-8493-6975-4/97/$0.00+$.50

I. INTRODUCTION

ET-1 is one of the most potent constrictors of human blood vessels *in vitro* (Davenport et al., 1989) and *in vivo* (Vierhapper et al., 1990, Gasic et al., 1992). ET-like immunoreactivity has been detected in the endothelial cells of all vessels from a wide range of vascular beds (Howard et al., 1992), consistent with a proposed role as a ubiquitous endothelium-derived vasoactive peptide, contributing to systemic and peripheral vascular tone. ET receptors would be expected to be equally widely expressed in all tissues receiving a blood supply. However, in humans, receptors are also localized to nonvascular structures such as epithelial cells or glia in the CNS, indicating a possibly wider role than simply vasoconstriction.

A. ET_A- and ET_B-Receptors

Two ET receptor subtypes, ET_A and ET_B, have been isolated and cloned from human tissue (Cyr et al., 1991; Ogawa et al., 1991). These are currently classified according to the rank order of their affinity for the three ET isoforms (Table 1). ET-1 and ET-2 show similar affinity for the ET_A subtype while ET-3 has little or no affinity. All three isoforms have similar affinity for the ET_B-receptor. ET receptors are unusual in that subtypes were cloned before the development of subtype-selective ligands (see Chapters 1 and 2).

B. ET_C-Receptor

The existence of a third subtype, the ET_C-receptor, has been suggested from pharmacological studies in which ET-3 is more potent than ET-1. Recently, the cloning of an ET-3-specific receptor has been described from *Xenopus laevis* dermal melanophores (see, for example, Karne et al., 1993). The predicted mature receptor sequence is 424 amino acids with a heptahelical structure similar to the G-protein-coupled receptor superfamily. The amino acid sequence had 47% homology with the bovine ET_A subtype and 52% with the rat ET_B-receptor. However, the K_i value for unlabeled ET-1 displacing $[^{125}I]$ET-1 from HeLa cells transiently expressing the receptor was only 2.5 times lower than the K_i value for unlabeled ET-3. It is possible that

TABLE 1

Classification of ET Receptor Subtypes and Radioligands Used in Binding Assays

	ET_A	ET_B
Potency order	ET-1 = ET-2 ≫ ET-3	ET-1 = ET-2 = ET-3
Radioligands:		
Nonselective	[125I]ET-1	[125I]ET-1
	[125I]ET-2	[125I]ET-2
	[125I]VIC	[125I]VIC
	[125I]STX-b	[125I]STX-b
Selective	[125I]PD 151,242	[125I][Ala^{1,3,11,15}]ET-1
		[125I]BQ 3,020
		[125I]IRL 1,620
		[125I]ET-3 (weak selectivity)

Note: BQ 3,020: [Ala^{11,15}]Ac-ET-1(6-21); IRL 1,620: Suc[Glu^9,Ala^{11,15}]ET-1(8-21); PD 151,242: (*N*-[(hexahydro-1-azepinyl)carbonyl])L-Leu(1-Me)D-Trp-D-Tyr.

the receptor is an amphibian form of the ET_B subtype. Molecular biology techniques have not yet detected a similar sequence in human tissues (O'Reilly and Davenport, unpublished) and more information is needed concerning the possible existence of this receptor.

II. CHARACTERIZATION OF ET RECEPTORS

A. ET and Sarafotoxin Radioligands

The majority of studies characterizing and localizing ET receptors in human tissues have used [125I]ET-1, directly labeled via Tyr^{13} (Tables 1 and 2). This ligand appears very stable; under nonphysiological binding conditions no degradation of labeled ET-1 has been detected by HPLC when using human tissue (but see Chapter 7). ET-2 and VIC (which is thought to be the murine isoform of ET-2) have also been labeled in the same position. ET-3 can be labeled at Tyr^6, Tyr^{13}, and Tyr^{14}. The ligand labeled at Tyr^6 is generally used as it is easier to identify following HPLC separation. However, all three ET-3 ligands have similar affinities in human cardiac tissue (Peter and Davenport, unpublished). STX-b is the only peptide in the family of four isoforms found in snake venom with a Tyr that can be iodinated.

The distribution of binding sites for all five [125I]ligands has been compared in human kidney and heart. In these tissues, quantitative autoradiography showed the distribution and density of receptors labeled with a fixed concentration of [125I]ET-1, [125I]ET-2, [125I]VIC, and [125I]STX-b were similar, consistent with these ligands being nonselective for the two subtypes and labeling a similar population of receptors in the heart. The density of [125I]ET-3 binding was significantly lower, with the ligand mainly binding to ET_B-receptors (Davenport et al., 1991). However, since the selectivity of ET-3 for ET_B- vs. ET_A-receptors is often about 100-fold, the iodinated peptide is less useful than more selective ligands such as [125I]BQ 3,020.

The 38-amino-acid precursor, big ET-1, also has detectable vasoconstrictor activity, for example, in isolated human coronary arteries (although at higher concentrations compared to ET-1), but it is thought that the precursor must be metabolized to ET-1 for biological activity (Howard et al., 1992). [125I]Big ET-1 binding has been reported in animal tissues, suggesting the existence of specific receptors, but binding has not been detected in human tissue (Davenport et al., 1990).

TABLE 2

Receptor Binding Affinities (K_D) and Densities (B_{max}) for [^{125}I]ET-1 in Normal and Pathological Human Tissues

	K_D (pM)	B_{max} (fmol/mg protein)	Ref.
Cardiovascular System			
Arteries (media)			
Aorta	507	9	Davenport et al., 1993
Pulmonary	845	15	Davenport et al., 1993
Coronary	141	71	Davenport et al., 1993
Internal mammary	340	2	Kuc and Davenport, unpublished
Vein (media)			
Saphenous	280	53	Kuc and Davenport, unpublished
Heart muscle			
Atrium	595	153	Molenaar et al., 1993
	350	1166	Takyanaga et al., 1991
Ventricle	354	64	Molenaar et al., 1993
	30	7	Hemsen, 1991
Renal System			
Kidney			
Medulla	139	360	Nambi et al., 1992
	3460	4[a]	Grone et al., 1990
	170	58	Karet et al., 1993
Cortex	91	165	Nambi et al., 1992
	5590	3[a]	Grone et al., 1990
Glomeruli	42		Rebibou et al., 1992
Bladder			
Base	4	3	Kondo et al., 1992
Dome	7	47	Kondo et al., 1992
Respiratory System			
Lung	130	9610	Brink et al., 1991
	153	6	Hemsen, 1991
Central Nervous System			
Brain			
Brain stem	45	5052	Takahashi et al., 1991
Hypothalamus	34	963	Takahashi et al., 1991
Hippocampus	34	278	Williams et al., 1991
Cortex	1500	15	Kurihara et al., 1990
	25	115	Fernandez-Durango et al., 1994
Brain tumors			
Astroglioma	2100	30	Kurihara et al., 1990
Glioblastoma	2500	60	Kurihara et al., 1990
Spinal cord (thoracic)			
Laminae I-III	194	12	Niwa et al., 1992
Lamina X	256	14	Niwa et al., 1992
Intermediolateral nucleus	265	12	Niwa et al., 1992
Endocrine Glands			
Adrenal			
Cortex (normal)	65	60	Imai et al., 1992
Adenoma (tumor)	70	226	Imai et al., 1992
Pituitary[b]	59	418	Takahashi et al., 1991
	652	1717	

TABLE 2 (continued)

Receptor Binding Affinities (K_D) and Densities (B_{max}) for [^{125}I]ET-1 in Normal and Pathological Human Tissues

	K_D (pM)	B_{max} (fmol/mg protein)	Ref.
Thyroid	200[c]		Jackson et al., 1994
Parathyroid	62	77	Eguchi et al., 1992
Gastrointestinal Tract			
Colon			
Myenteric plexus	350	92[d]	Inagati et al., 1991
Mucosa	41	60[d]	Inagati et al., 1991
Liver	100	361	Takayanagi et al., 1991
Skeletal Tissue			
Synovium vessels	162	225[d]	Wharton et al., 1992
Reproductive Tissue			
Placenta	80	113	Hemsen, 1991
	36	185	Wilkes et al., 1990
	24	240	Fischi et al., 1989
	34	93	Kalina et al., 1992
Stem villi			
Vessels	26	681	Mondon et al., 1993
Arteries	45	602	Robaut et al., 1991
Veins	45	619	Robaut et al., 1991
Microvillus membranes	26	50	Mondon et al., 1993
Basal plasma membrane	12	150	Mondon et al., 1993
Uterus			
Myometrium	1190	37	Bacon et al., 1994
Endometrium	1390	181	Bacon et al., 1994

[a] fmol/mg tissue equivalents;
[b] two sites;
[c] 4°C;
[d] amol/mm^2.

B. Receptor Subtype Selective Radioligands

1. ET_A Selective

Although ET_A-selective antagonists such as BQ 123 and FR 139,317 (see Doherty, 1992; Huggins et al., 1993) have five to six orders of magnitude of selectivity for ET_A- over ET_B-receptors in human tissue (Table 3) they do not have a Tyr residue suitable for iodination. [^{125}I]PD 151,242 is a linear tetrapeptide (structurally related to FR 139,317) that binds with high nanomolar affinity to the ET_A-receptor and has about 10,000-fold selectivity for this subtype in human and animal tissues (Table 4; Davenport et al., 1994a,b; Peter and Davenport, 1994a). Hill slopes are close to unity. Binding of the ligand to human tissue is time dependent and reaches equilibrium at room temperature after 2 h with an association rate constant of 1.3×10^8 M^{-1} min^{-1} in human aortae. Although it is one of the shortest amino acid sequences of any current peptidergic radioligand, there is no evidence for metabolism in binding assays. The Trp and Tyr residues are nonnaturally occurring D-isomers, the N-terminus is blocked, and the C-terminus is iodinated. The unlabeled peptide is an antagonist in isolated human vessels (Davenport et al., 1994). ET_A-selective agonists have not yet been developed.

TABLE 3

Receptor Binding Affinities (K_D) for Unlabeled Receptor Subtype Selective Ligands Competing for [^{125}I]ET-1 Binding and Ratios of ET_A and ET_B Receptors in Human Tissue

ET_A-Selective	K_D ET_A (nM)	K_D ET_B (μM)	ET_A:ET_B (%)	Ref.
BQ 123				
Arteries				
Coronary	0.85	7.58	89:11	Davenport et al., 1993
Aorta	0.80	2.67	89:11	Davenport et al., 1993
Pulmonary	0.27	24.6	92:8	Davenport et al., 1993
Veins				
Saphenous	0.55	14.4	85:15	Davenport et al., 1994b
Heart				
Ventricle	0.7	24.3	57:43	Molenaar et al., 1993
Myocytes	2.2	179.0	92:8	Molenaar et al., 1993
Kidney				
Medulla	11.8	32.8	24:76	Karet et al., 1993
Cortex	26.0	6.5	26:74	Karet et al., 1993
Uterus				
Myometrium	1.4	39.9	79:21	Bacon et al., 1995
FR 139,317				
Heart				
Ventricle	1.2	287.0	62:38	Peter and Davenport, 1994
Uterus				
Myometrium	2.5	89.8	73:27	Bacon et al., 1995
ET_B-Selective	**K_D ET_A (μM)**	**K_D ET_B (nM)**	**ET_A:ET_B (%)**	**Ref.**
BQ 3,020				
Arteries				
Coronary	0.2	0.8	62:38	Davenport et al., 1993
Heart				
Ventricle	2.0	1.4	67:33	Molenaar et al., 1993
Kidney				
Medulla	5.0	3.0	37:63	Karet et al., 1993
Cortex	2.1	4.9	37:63	Karet et al., 1993
[$Ala^{1,3,11,15}$]ET-1				
Heart				
Ventricle	4.5	1.8	76:24	Molenaar et al., 1993
IRL 1,038				
Heart				
Ventricle	32.6	4860	63:37	Peter and Davenport, 1994

2. *ET_B Selective*

Human ET_B-receptors have been characterized using a linear analogue of ET-1 where the disulfide bridges have been removed by substitution of Ala for Cys residues, [^{125}I][$Ala^{1,3,11,15}$]ET-1, and a truncated Ala analogue [^{125}I]BQ 3,020 (Molenaar et al., 1993; Table 1). Both compounds have similar characteristics: binding with high nanomolar affinity to the ET_B-receptor and having at least 1500-fold selectivity for this subtype over the ET_A-receptor (Table 3). Hill slopes for both peptides are close to unity. Binding of the ligands to human ventricle was time dependent and reached equilibrium at room temperature after 2 h. A third truncated analogue, [^{125}I]IRL 1,620, has also been used to characterize ET_B-receptors in animal tissues (Watakabe et al., 1992). All three unlabeled peptides were originally developed as selective ET_B agonists.

TABLE 4

Examples of Receptor Binding Affinities (K_D), Densities (B_{max}) and Hill Slopes (nH) for ET_A- and ET_B-Receptor-Selective Radioligands in Human Tissues

	K_D (nM)	B_{max} (fmol/mg protein)	nH	Ref.
ET_A:[^{125}I]PD 151,242				
Arteries				
Coronary	0.51 ± 0.07	44.9 ± 1.67	0.84 ± 0.10	Davenport et al., 1993
Aorta	0.76 ± 0.17	6.0 ± 1.56	1.00 ± 0.01	Davenport et al., 1993
Pulmonary	1.75 ± 0.20	12.8 ± 1.39	1.00 ± 0.03	Davenport et al., 1993
Veins				
Saphenous	2.10 ± 0.41	32.8 ± 8.90	1.00 ± 0.01	Kuc and Davenport, *unpublished*
Heart				
Ventricle	1.07 ± 0.08	29.8 ± 4.2	0.98 ± 0.04	Peter and Davenport, 1995
Kidney	0.75 ± 0.07	48.4 ± 1.6	0.89 ± 0.07	Davenport et al., 1994
Uterus				
Myometrium	0.93 ± 0.08	138.7 ± 0.1	0.89 ± 0.02	Bacon et al., 1995
ET_B:[^{125}I]BQ 3,020				
Arteries				
Coronary	0.60 ± 0.31	4.5 ± 2.1	0.83 ± 0.06	Bacon et al., 1994
Heart				
Atrium	0.15 ± 0.04	25.7 ± 6.8	0.46 ± 0.14	Molenaar et al., 1992
Ventricle	0.11 ± 0.01	18.9 ± 4.8	0.43 ± 0.04	Molenaar et al., 1992
Kidney	0.36 ± 0.06	30.0 ± 5.0	0.77 ± 0.04	Karet et al., 1993
Uterus				
Myometrium	0.62 ± 0.07	44.5 ± 1.1	0.92 ± 0.01	Bacon et al., 1995
[^{125}I][$Ala^{1,3,11,15}$]ET-1				
Heart				
Atrium	0.24 ± 0.04	22.2 ± 3.4	0.77 ± 0.01	Molenaar et al., 1992
Ventricle	0.20 ± 0.03	12.4 ± 2.2	0.72 ± 0.03	Molenaar et al., 1992

At present, the putative ET_B selective antagonist in animals, IRL 1,038, containing the C-terminal sequence of ET-1(11 to 21) (Urade et al., 1992), has only micromolar affinity and about 10-fold selectivity for the human ET_B-receptor (Peter and Davenport, 1994b; Table 3).

C. Receptor Binding Assays

ET receptors have been studied in a range of human tissues (Table 2) using [^{125}I]ET-1 in saturation binding experiments. In these assays, two types of tissue preparation are used: partially purified plasma membrane fractions from tissue homogenates or cryostat sections of fresh-frozen tissue mounted on microscope slides. In each case, samples are incubated with increasing concentrations of [^{125}I] ET-1. An appropriate incubation period to allow the assay to reach equilibrium should be determined (although this information is not always given), but about 2 h at 25°C is usually sufficient. The equilibrium is rapidly broken by washing to separate free [^{125}I]ET-1 and the amount of radioactivity bound to the tissue is measured. Nonspecific binding is determined using a 1000-fold excess of unlabeled ET-1 over the K_D (usually 1 μM). Hill slopes are derived from Scatchard plots. The equilibrium dissociation constant, K_D, and density, B_{max}, are calculated using iterative curve-fitting programs such as LIGAND (see, for example, Davenport et al., 1993; Molenaar et al., 1993).

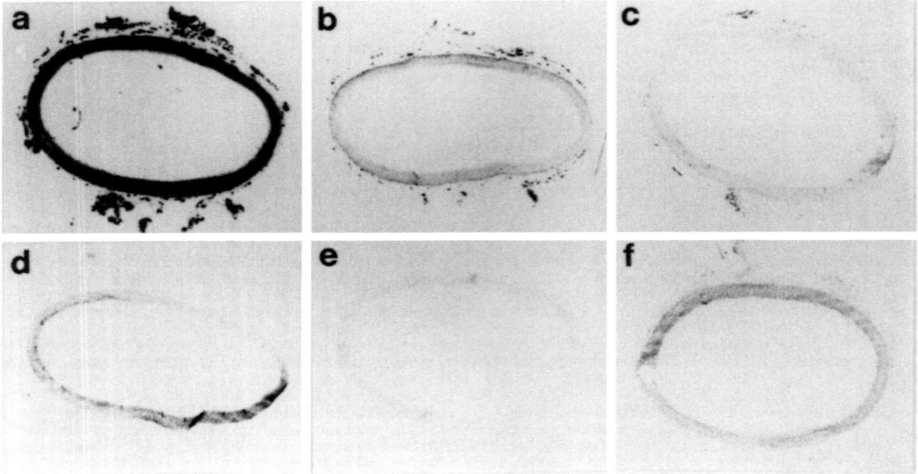

FIGURE 1
Autoradiographical localization of ET receptors in human aorta. (a) Total [^{125}I]ET-1 binding showing the distribution of all ET receptors. (b) ET_A-receptors localized using [^{125}I]PD 151,242. (c) ET_B-receptors localized with [^{125}I]BQ 3,020. Nonspecific binding was determined using the corresponding unlabeled peptide (d,e,f).

Binding parameters may be altered by many factors including the use of surgical samples vs. post-mortem material, and the method and time taken to prepare the tissue. No standard conditions have emerged for binding assays. The type and ionic strength of buffer, presence of divalent ions, and temperature may also affect binding.

Having determined the affinity of [^{125}I]ET-1 in a particular tissue, this information can be used to test the ability of unlabeled ligands over a much wider concentration range (typically 10 pM to 100 μM) to compete for binding of the labeled peptide. This allows the selectivity (if any) of a particular ligand to be calculated for each subtype (Table 3).

D. Anatomical Localization

ET-receptor subtypes can be localized by incubating tissue sections with [^{125}I]ET-1 in the presence of an unlabeled ET_B-selective ligand. A concentration that blocks [^{125}I]ET-1 binding to the ET_B site is calculated and used to show the distribution of ET_A-receptors. The reverse experiment is carried out in an adjacent section where [^{125}I]ET-1 is incubated in the presence of an ET_A ligand to show the distribution of the ET_B subtype (see Molenaar et al., 1993; Figure 3). Alternatively, adjacent sections can be incubated with [^{125}I]PD 151,242 or [^{125}I]BQ 3,020 to permit comparison of the ET_A and ET_B subtypes (see Figures 1, 6, and 7).

Macroautoradiography — Most human tissue will contain a heterogeneous population of cells and autoradiography is used to map receptors. After incubating sections with a fixed concentration of ligand, the equilibrium is broken by washing. The dry sections are exposed to radiation-sensitive film. The resulting autoradiograms can be used as a negative to produce photographs showing a qualitative distribution (see, for example, Figure 7). If standards are co-exposed with the labeled tissue sections, receptor densities can be measured in discrete areas using computer-assisted image analysis (see Davenport and Nunez, 1990).

Microautoradiography — The resolution of macroautoradiography is usually limited to groups of cells. An alternative is to use microautoradiography, where slide-mounted sections of radiolabelled tissue are dipped in a nuclear emulsion. After development, the distribution of the ligand, and therefore the ET receptors, is detected as the pattern of silver grains overlying individual cells when viewed under the microscope using darkfield illumination (e.g., Karet et al., 1993).

Electron Microscopy (EM) — The distribution of ET receptors has not been studied in detail by EM autoradiography in human tissues. A small number of studies have been carried out in rats where [^{125}I]ET-1 was infused into the animal prior to cutting sections for autoradiography (see, for example, Furuya et al., 1992). Barriers may exist to the penetration of [^{125}I]ET-1 when injected intravenously and this technique may not detect all receptors revealed by *in vitro* autoradiography. Thus, while [^{125}I]ET-1 binding was localized to endothelial cells in rat kidney, as expected, receptors were difficult to locate on smooth muscle of arterioles.

E. Nonradioactively Labeled Ligands

A number of ET peptides have been labeled using biotin. One of the most effective of these derivatives is ET-1 biotinylated at Lys^9 (see, for example, Natarajan et al., 1992). This has been shown to have almost the same ability to inhibit the binding of [^{125}I]ET-1 as the unlabeled peptide, suggesting that this may be of value for localizing receptors, although the peptide has not yet been used for this purpose.

F. Antibodies to Receptors

Antibodies have been generated to specific sequences of ET receptors. Antibodies to the C-terminus of the ET_B detected the subtype in isolated membranes from human heart, immobilized to nitrocellulose filters by dot immunoblotting analysis. The antibodies did not cross-react with the ET_A sequence (Persidis et al., 1993).

G. Localization of mRNA Encoding Human ET_A- and ET_B-Receptors

The reverse transcriptase polymerase chain reaction (RT-PCR) has been used to detect mRNA encoding ET_A- or ET_B-receptors following extraction of total RNA from tissue homogenates (O'Reilly et al., 1992, 1993a,b; Molenaar et al., 1993; Davenport et al., 1993, 1994b; Karet et al., 1994). An advantage of this method is that oligonucleotide primer pairs can be designed to give PCR products of different sizes when separated by agarose gel electrophoresis, allowing the detection of mRNA encoding both subtypes in the same reaction mix.

Using the same RT-PCR assay conditions with isolated cells, only ET_B-receptor mRNA was found in human vascular endothelial cells, whereas in rat aortic smooth muscle cells only ET_A-receptor mRNA and receptor protein could be detected (Molenaar et al., 1993; Davenport et al., 1994b). However, in isolated myocytes, mRNA encoding both subtypes was found. BQ 123 inhibited binding of iodinated ET-1 biphasically, confirming both receptor subtypes were present in these cells (Molenaar et al., 1993). In addition, the amount of mRNA encoding the two subtypes can now be quantified by using a synthetic cRNA as internal standard (Karet et al., 1994).

Molecular biology techniques can therefore provide further confirmation for the presence or absence of mRNA encoding ET receptors in a particular cell type. Cloning and sequencing of the PCR products has confirmed 100% homology with the expected sequences for ET_A- or ET_B-receptor mRNA (O'Reilly et al., 1992). At present, only single bands corresponding to the expected position of ET_A- or ET_B-receptor mRNA have been detected and there is no evidence for further subtypes using these particular oligonucleotide primers in human tissues.

The anatomical distribution of ET_A- and ET_B-receptor mRNA has also been compared by *in situ* hybridization using [^{35}S]labeled antisense and sense (control) RNA probes (Molenaar et al., 1993). The amount of probe specifically hybridized can be determined by digitally subtracting the autoradiographical image of the sense control probe from that of the antisense probe using computer-assisted densitometry (Davenport and Nunez, 1990; Molenaar et al., 1993).

III. ORGAN SYSTEMS

A. Circulatory System

1. Vasculature

In animals, ET-1 can elicit constriction through both ET_A- and ET_B-receptors, although the pattern is complex: the subtype ratio varies in different vascular beds in the same species and in the same vascular bed from different species (Davenport and Maguire, 1994). ET-1 and ET-2 are the most abundant isoforms expressed by the human cardiovascular system (Plumpton et al., 1993). mRNA encoding both peptides and their peptide precursors (but not those of ET-3) have been detected in human endothelial cells (O' Reilly et al., 1993a,b; Howard et al., 1992). A general expression of ET receptors in smooth muscle of the vessel wall would be expected to correspond to the widespread distribution of ET-1 (and possibly ET-2). ET-3 is the only endogenous isoform that is able to distinguish between the two receptor subtypes. The lack of expression of ET-3 in blood vessels suggests the ET_A subtype should be the predominant receptor present in the smooth muscle of the human vasculature.

To test this hypothesis, ET receptors have been characterized in detail in the medial layer (containing mainly smooth muscle) of a range of arteries (aorta, pulmonary, coronary, internal mammary) and saphenous veins. In each case, the endothelium (which may express ET_B-receptors) and adventitia were removed. [^{125}I]ET-1 bound with a similar subnanomolar affinity to the media of each vessel although the density of binding is comparatively low compared to other human tissue (Table 2). Over the concentration range tested, Hill slopes were close to unity and a one-site fit was preferred to a two-site model, in agreement with other types of human muscle. These results suggest the presence of either a single population of receptors or a heterogeneous population with the same affinity for the peptide. ET_A-selective ligands such as BQ 123 inhibited [^{125}I]ET-1 binding in a biphasic manner in both arteries and vein, with a high-affinity ET_A site accounting for more than 85% of the receptors (Table 3). A small population of ET_B-receptors could be confirmed by the low density of [^{125}I]BQ 3,020 binding to the vessel wall compared to [^{125}I]PD 151,242, as illustrated in Figure 1 with sections from aorta. Expression of receptor protein correlated with the presence of receptor mRNA. A consistent pattern emerged with the production of a single PCR product corresponding to the expected size for ET_A- and ET_B-receptor mRNAs from the media of all vessels. Localization of ET_B- as

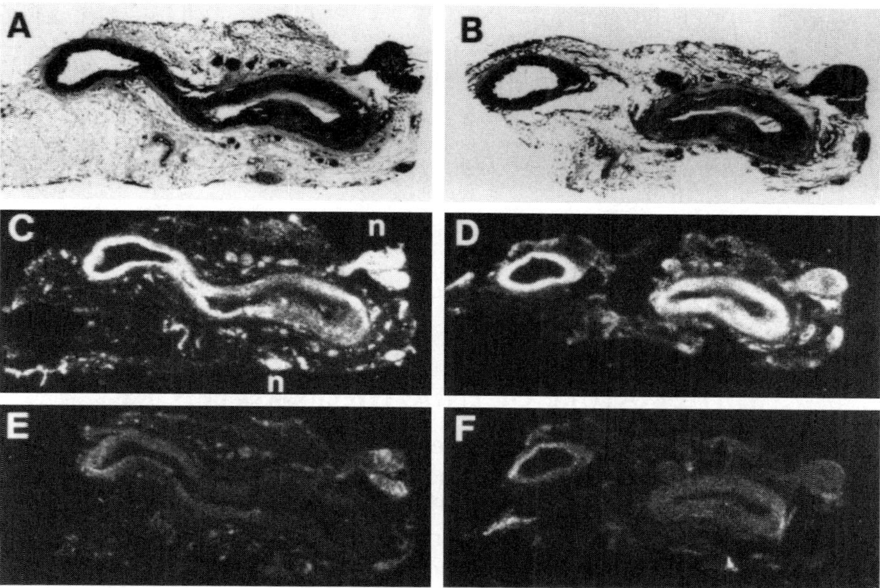

FIGURE 2
Autoradiographical images showing the distribution of antisense [^{35}S] labeled probes for ET_A (C) and ET_B (D) mRNA in coronal sections of human coronary artery. Adjacent sections incubated with the corresponding control; sense probes are shown in (E) and (F), respectively. (A) and (B) Brightfield micrographs of hematoxylin and eosin sections (n = nerves).

well as ET_A-receptor mRNA to vascular smooth muscle cells was confirmed by *in situ* hybridization in sections of coronary arteries (Davenport et al., 1993, Figure 2).

These studies show that human vessels express a small population of ET_B-receptors. Although the ET_B agonists BQ 3,020 and [Ala^{1,3,11,15}]ET-1 are potent constrictors of some rat vascular beds (Bigaud and Pelton, 1992; Gardiner et al., 1994), they had no detectable action on isolated vessels at concentrations up to 3 μM. The nonendogenous ET_B agonist in mammals, STX-c, was also inactive in aorta and pulmonary artery. This peptide does cause vasoconstriction in some human vessels such as saphenous vein and internal mammary artery but these responses were variable, occurring in less than 50% of individuals (Davenport and Maguire, 1994). Although the constrictor actions of this snake venom are potent, the magnitude of the response is much less than that of ET-1. Therefore, while some individuals may have functional constrictor ET_B-receptors, the poor response to ET-3 in these vessels (Davenport et al., 1993; Maguire and Davenport, 1994a; Davenport and Maguire, 1994) would suggest that these receptors may have limited physiological importance in the cardiovascular system.

Functional studies have shown that ET_A-selective antagonists in human vessels always cause a parallel, rightward shift of the ET-1 concentration response curve and Schild slopes were not significantly different from unity (Maguire and Davenport, 1994a,b; Maguire et al., 1994a,b,c). In the human vasculature, a consistent pattern is emerging: although both receptors can be detected by ligand binding, *in vitro* pharmacological experiments show that vasoconstriction in humans is mediated mainly via the ET_A subtype. These results suggest the ET_A-receptors must be blocked in humans to produce a beneficial vasodilatation in pathophysiological conditions where there are elevated concentrations of the peptide or increases in receptor density.

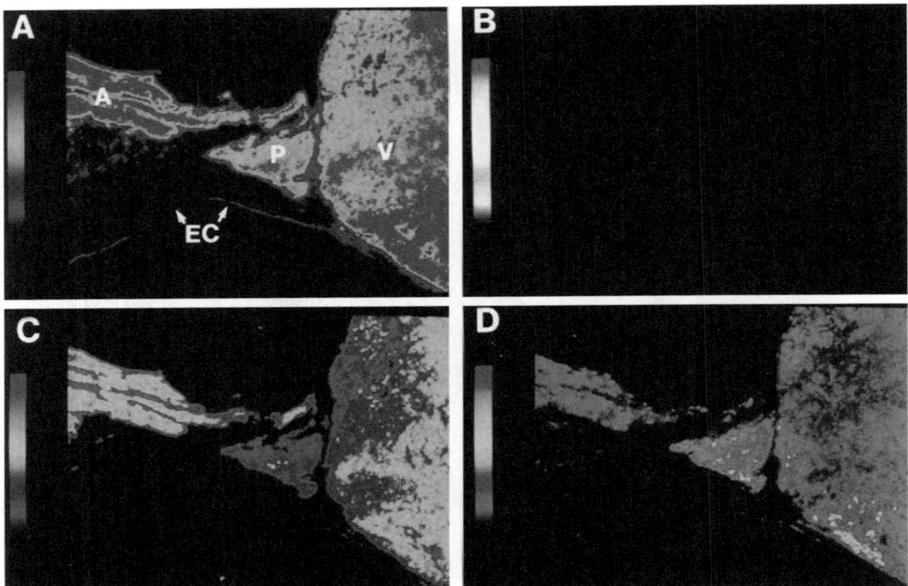

FIGURE 3
Color-coded autoradiographical images showing the distribution of ET receptors in serial sections of atria (A), ventricle (V), penetrating bundle of His (P), and endocardial cells (EC) from human heart. Highest densities are shown in red, intermediate values coded according to the scale bars and black is below the level for detection. (A) Total [^{125}I]ET-1 binding showing the distribution of all ET receptors. (B) Nonspecific binding. (C) ET_A-receptors visualized by inhibiting [^{125}I]ET-1 binding to the ET_B subtype with BQ 3,020. (D) ET_B-receptors visualized by inhibiting [^{125}I]ET-1 to the ET_A subtype with BQ 123. (For color version, see Plate 2 following page 70.)

2. *Human Heart*

a. Myocardium and the Atrioventricular Conducting System

ET-1 is a potent positive inotropic agent on isolated human heart muscle (Davenport et al., 1989b). [^{125}I]ET-1 binds with high subnanomolar affinity to both atria and ventricles, with a higher density in the former (Table 2). Binding is time dependent and reached equilibrium after 2 h. Hill coefficients were close to unity, indicating [^{125}I]ET-1 bound with a similar affinity to all binding sites (Molenaar et al., 1993).

Autoradiography using [^{125}I]ET-1 in the absence or presence of BQ 3,020 or BQ 123 demonstrated that both ET_A- and ET_B-receptors were located in the atrioventricular conducting system and surrounding myocardium. Both receptor subtypes were localized to the atrioventricular node, penetrating and branching bundles of His, the left bundle branch, atrial and ventricular myocardium, and endocardial cells (Figure 3). Receptors were not localized to the central fibrous body or adipose tissue. The distribution was not uniform. There was a higher proportion of ET_B-receptors in the atrioventricular conducting system compared to surrounding atrial and ventricular myocardium. A similar density of ET_B-receptors was found in all components of the atrioventricular conducting system studied and the interatrial septum, but a lower density in the interventricular septum. There was a lower density of ET_A-receptors in the atrioventricular conducting system compared with interatrial and ventricular septa.

In situ hybridization showed the distribution of ET_A- and ET_B-receptor mRNA matched that of receptor protein detected by ligand binding (Figure 4). RT-PCR confirmed the presence of both ET_A- and ET_B-receptor mRNA in the ventricle as well

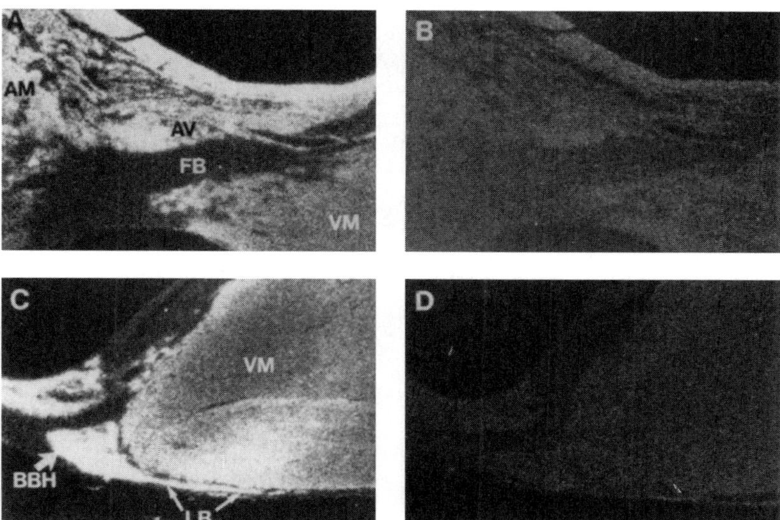

FIGURE 4
Autoradiographical images of antisense (A,C) and sense control probe for ET_A-receptor mRNA in adjacent sections (B,D) of human heart. (A) Atrioventricular node (AV) and (C) branching bundle of His (BBH), left bundle branch (LB), ventricular (VM), and atrial myocardium (AM). mRNA was localized to all areas of the heart excluding the central fibrous body (FB). (From Molenaar, P. et al., *Circ. Res.*, 72, 535, 1993. With permission.)

as isolated atrial myocytes, where ligand binding detected a high proportion of ET_A-receptors (90%, Table 3).

b. Fetal Heart

In fetal heart, like the adult, [^{125}I]ET-1 binding was detected in atria and ventricles. Interestingly, high levels of both [^{125}I]ET-1 and [^{125}I]ET-3 bound to valve cusps, which are one of the few avascular tissues. Microautoradiography localized [^{125}I]ET-1 binding to the wall of the aorta, pulmonary and coronary arteries, as well as associated nerve trunks, myocardium, ventricular conduction system, endocardium, and endothelial lining of valve cusps (Wharton et al., 1991).

B. Renal System

ET-1 has two major actions in the kidney, vasoconstriction and natriuresis, which may be mediated by different subtypes. In animals, ETs are thought to alter salt and water balance by interacting with receptors located on the tubule and collecting duct, which in humans are mainly of the ET_B subtype (see Karet and Davenport, 1994). The renal vasculature is particularly sensitive to the constrictor action of this peptide, producing falls in plasma flow, glomerular filtration rate, and urine production, which occurs mainly via ET_A or both subtypes depending on species. ET-1 is a potent vasoconstrictor in human kidney, decreasing renal plasma flow by about 30% and increasing vascular resistance by 50% when infused systemically (Gasic et al., 1992). High-affinity ET receptors have been detected in human kidney (Davenport et al., 1989a,b; Jones et al., 1989; Waeber et al., 1990; Karet et al., 1993; Table 2). High densities were present on the medulla, especially on the medullary rays, but lower densities were found in the cortex. Blood vessels in both regions were also labeled. In human kidney, binding to glomeruli was of a similar density to the surrounding

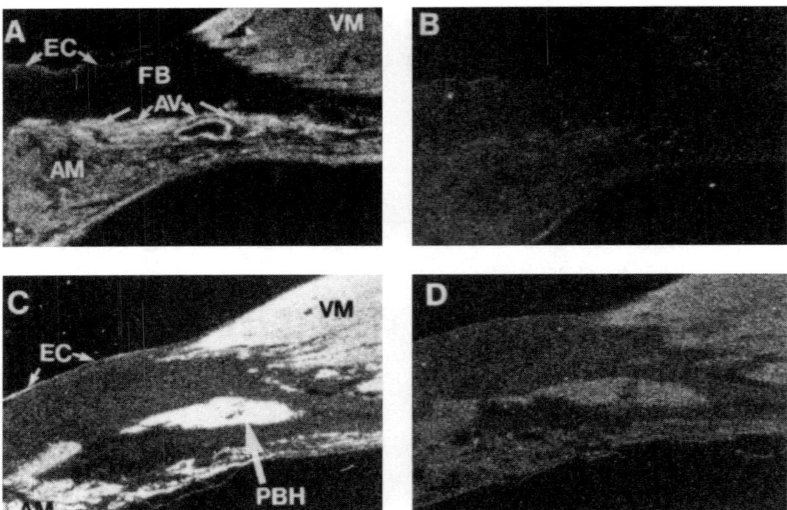

FIGURE 5
Autoradiographical images of antisense (A,C) and sense control (B,D) probe for ET_B-receptor mRNA in human heart. (A) Atrioventricular node and (C) penetrating bundle of His. Abbreviations are as Figure 4. mRNA was also localized to all areas of the heart excluding the central fibrous body (FB). (From Molenaar, P. et al., *Circ. Res.*, 72, 536, 1993. With permission.)

cortex, in contrast to rat kidney where high levels were detected. The distribution and density of [^{125}I]ET-1,[^{125}I]ET-2, and [^{125}I]STX-b binding was similar, but the density of [^{125}I]ET-3 binding was lower (Davenport et al., 1989a,b; 1991). Subtype-selective ligands competed for the binding of [^{125}I]ET-1 in a biphasic manner with a ratio of about 70% ET_B: 30% ET_A in both cortex and medulla (Karet et al., 1993).

Both [^{125}I]PD 151,242 and [^{125}I]BQ 3,020 bind with similar high affinity to human kidney (Table 4). Autoradiography using [^{125}I]PD 151,242 revealed that ET_A-receptors were mainly localized to blood vessels (Figure 6A) including those in the medulla as well as arcuate arteries and adjacent veins at the corticomedullary junction. Vessels with a diameter down to about 0.2 mm could also be delineated; this being characteristic of intrarenal resistance vessels. Microautoradiography clearly showed arterioles supplying the glomerulus expressed the ET_A subtype (Karet et al. 1993). In contrast, the pattern of binding visualized using the ET_B-selective [^{125}I]BQ 3,020 binding was more heterogeneous and concentrated in medulla, particularly to nonvascular structures such as the tubules and collecting ducts (Figure 6B). Binding of this ligand to renal vessels clearly delineated with [^{125}I]PD 151,242 was difficult to detect (Davenport et al., 1994b; Maguire et al. 1994b).

In agreement with other human vessels, ET-1-induced vasoconstriction in isolated renal arteries and veins is mediated mainly via the ET_A subtype (Maguire et al., 1994a). While the kidney is thought to be partly responsible for excreting ET in humans, ET-1 peptide has been detected in medulla and cortex (Karet and Davenport, 1994), suggesting a role as a locally acting renal peptide.

C. Respiratory System

The human lung contains one of the highest densities of ET receptors compared with other tissues (Table 2). These are present on the airways, vascular smooth muscle (including pulmonary artery), and nerve trunks. Little or no binding has been detected on connective tissues, submucosal layer, and cartilage (McKay et al., 1991;

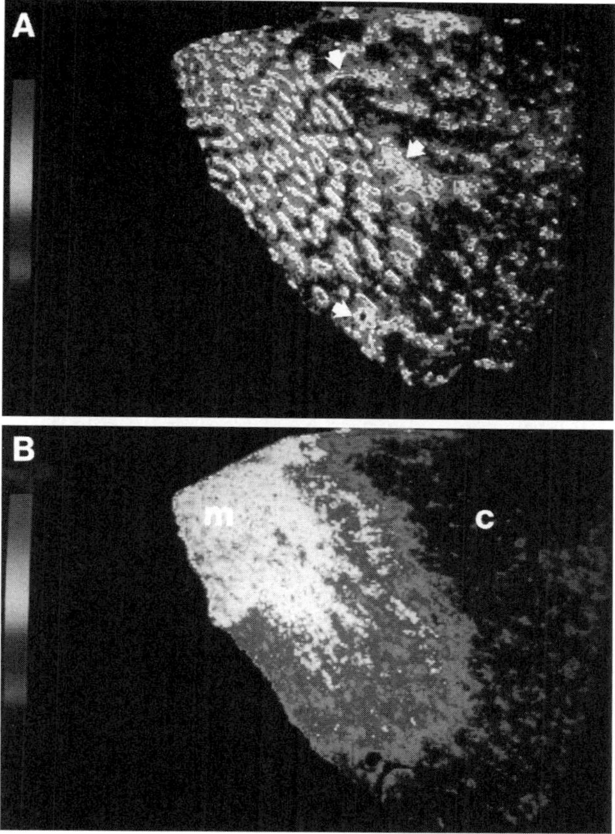

FIGURE 6
Color-coded autoradiographical images showing the distribution of ET_A (A) receptors (visualized using [^{125}I]PD 151,242) that are mainly present on intrarenal blood vessels including the arcuate arteries and adjacent veins (indicated by arrows) in human kidney. (B) High densities of ET_B-receptors (visualized using [^{125}I]BQ 3,020) are present in the medulla (m), especially on medullary rays, with lower densities in the cortex (c). (For color version, see Plate 3 following page 70.)

Henry et al., 1990; Marciniak et al., 1992). In one study, [^{125}I]ET-1 receptors were also localized to the lung epithelial cells (Marciniak et al., 1992) as has also been found in the endometrium of the uterus.

McKay et al. (1991) detected no difference in the binding density or location of [^{125}I]ET-1 or [^{125}I]ET-2 to bronchus, pulmonary artery, or alveolar walls, consistent with both peptides labeling a similar population of receptors. In addition, binding sites for both forms of the peptide were localized to parasympathetic ganglia within the airways, where muscarinic receptors were also detected. Microautoradiography confirmed that [^{125}I]ET-1 binding sites were present on the smooth muscle band, with very low levels of binding associated with cartilage and submucosal cells (Rigby et al., 1991). The rank order of the density of [^{125}I]ET-1, ET-2, and ET-3 binding was lung parenchyma > airway smooth muscle > airway epithelia (Marciniak et al., 1992).

Intense staining of mature ET-like immunoreactivity has been found in both the endothelial cells of the pulmonary vasculature and the epithelia of all airways (including bronchioles and small bronchi) and submucosal glands throughout the lung (Marciniak et al., 1992). In the airway epithelia immunoreactive big ET-1 and big ET-3 were detected, while immunoreactivity of all three isoforms was present in submucosal glands. Exogenously applied ET-1 causes long-lasting vaso- and bronchoconstriction.

Binding and functional studies suggest vasoconstriction of smooth muscle in pulmonary arteries is via the ET_A subtype whereas constrictor actions on airway smooth muscle is via ET_B (Hay et al., 1993). If ETs within the epithelia of the bronchi and endothelium of the vasculature are released *in vivo*, they may mediate these actions via different subtypes. ET_A antagonists may have a role in primary pulmonary hypertension, where current treatment is difficult, and in other forms of pulmonary hypertensive disease.

D. Central Nervous System

1. Brain

ET-1 was originally thought to be primarily involved in the cardiovascular system. However, both human and animal brains have some of the highest reported densities of $[^{125}I]$ET-1 receptors — up to 5000 fmol/mg protein (see, for example, Jones et al., 1989; Kurihara et al., 1990; Takahashi et al., 1991). $[^{125}I]$ET-1 does not cross the blood-brain barrier (at least in the rat), suggesting that these receptors may respond to ET-like immunoreactivity that is synthesized within cells of the CNS including neurons, rather than to peripherally released peptide. $[^{125}I]$ET-1 binding is localized to specific brain regions with a trend for the density to decrease from the hindbrain to the forebrain. High receptor densities have been detected in the brainstem, cerebellum, and hippocampus, with lower densities in the hypothalamus and cortex (Jones et al., 1989; Takahashi et al., 1991; Table 2). In the cortex, $[^{125}I]$ET-1 bound to a single high-affinity site. Unlabeled ET-1, ET-2, and ET-3 inhibited $[^{125}I]$ET-1 binding with similar potencies, suggesting the presence of mainly ET_B-receptors.

Fernandez-Durango et al. (1994) have suggested only ET_B-receptors are present in human cerebral cortex. This is a surprising conclusion since their binding assays were carried out using homogenates which would be expected to contain a significant amount of blood vessels expressing ET_A-receptors. In rat cerebellum, for example, BQ 3,020 competes for $[^{125}I]$ET-1 in a biphasic manner, giving a ratio of about 20% ET_A to 80% ET_B (Davenport et al., 1992b).

a. Glia

In animals, tissue culture studies suggest that ET receptors are located on glial cells, the nonneuronal elements of the CNS. In functional studies, ET-3 was equally effective as ET-1 in stimulating intracellular free calcium, suggesting that glial cells express the ET_B subtype. The precise localization of ET receptors in human brain has not been extensively studied by means of tissue culture. However, a homogeneous distribution of $[^{125}I]$ET-1 binding has been detected in astrocytomas (tumors mainly composed of astrocytes) and pleomorphic astrocytes within glioblastomas. These results suggest astrocytes express ET receptors, at least when proliferating. Since ET-3 was equipotent to ET-1 in inhibiting $[^{125}I]$ET-1 binding, astrocytes in human brain also appear to express this subtype (Kurihara et al., 1990).

b. Neurons

ET receptors can be detected on a proportion of neurons isolated from rat cerebellum growing in culture. ET-1 and ET-2 but not ET-3 increased levels of intracellular calcium, suggesting neurons express ET_A-receptors (Davenport and Morton 1991; Morton and Davenport, 1992). ET receptors have also been localized to parasympathetic ganglia within human airways (McKay et al., 1991).

2. *Spinal Cord*

[^{125}I]ET-1 was found to bind to a single high-affinity site in the sixth thoracic segment of human spinal cord. Binding was highly concentrated in the dorsal horn (Rexed's laminae I-III, innervated by nerve endings originating from the dorsal root ganglia that also contain immunoreactive ET), an area around the central canal (lamina X), and the principal part of the intermediolateral nucleus where cell bodies of the preganglionic sympathetic neurons are located and substance P receptors have been detected (Niwa et al., 1992). There were no differences between the potencies of unlabeled ET-1, ET-2, ET-3, and STX-b in inhibiting [^{125}I]ET-1 binding, suggesting that mainly ET_B-receptors are expressed in these areas. This was supported by chemical cross-linking studies: [^{125}I]ET-1 and [^{125}I]ET-3 mainly bound to a protein with a molecular mass of 43 kDa in the thoracic spinal cord. The presence of both ETs and their receptors within the spinal cord suggests possible roles for these peptides as neurotransmitters or modulators of substance P.

E. Endocrine Glands

1. *Adrenal*

In normal adrenal, similar densities of receptors labeled with [^{125}I]ET-1 have been detected in the three regions of the cortex (zona glomerulosa, fasiculata, and reticularis) that secrete steroid hormones as well as the catecholamine secreting medulla (Davenport et al., 1989). Subtype-selective ligands clearly demonstrate a differential distribution, with the ET_A subtype being localized to the zona glomerulosa and small arteries of the capsular plexus (supplying blood to the gland), whereas ET_B-receptors are present in the zona fasiculata, reticularis, and medulla (Figure 7; Davenport et al., 1992b). In agreement, Hinson et al. (1991) found ET-1 to be more potent (threshold 10 fM) than ET-3 (100 fM) in stimulating aldosterone (exclusively released by the zona glomerulosa) in isolated human adrenals. In contrast, both are equipotent in stimulating cortisol, a marker of zona fasiculata and reticularis activity, consistent with activation of ET_B-receptors. Immunoreactive ET has been localized to the endothelial cells of the vessels supplying the separate layers of the cortex (Davenport, unpublished observations). ET-1 may function as an intraglandular factor stimulating steroid secretion; increases in blood flow through the adrenal gland releases ET from the endothelial cells of the capsular plexus.

2. *Pituitary*

Unlike virtually all other tissue, in homogenates of human pituitary glands obtained at post-mortem [^{125}]ET-1 bound with apparently two affinities (Table 2). In this tissue, ET-2 and ET-3 were nearly equipotent in inhibiting ET-1 binding, suggesting a predominance of ET_B-receptors (Takahashi et al., 1992). It is not known whether the low- and high-affinity sites are functionally important or whether they are an artifact arising from degradation of the receptor protein (ET_B-receptors have a cleavable site). The distribution of ET receptors in the anterior vs. the posterior lobe in humans has not been studied. A role for these receptors is unclear, but in animals ET-1 is particularly effective in inhibiting prolactin release as well as stimulating the release of growth hormones.

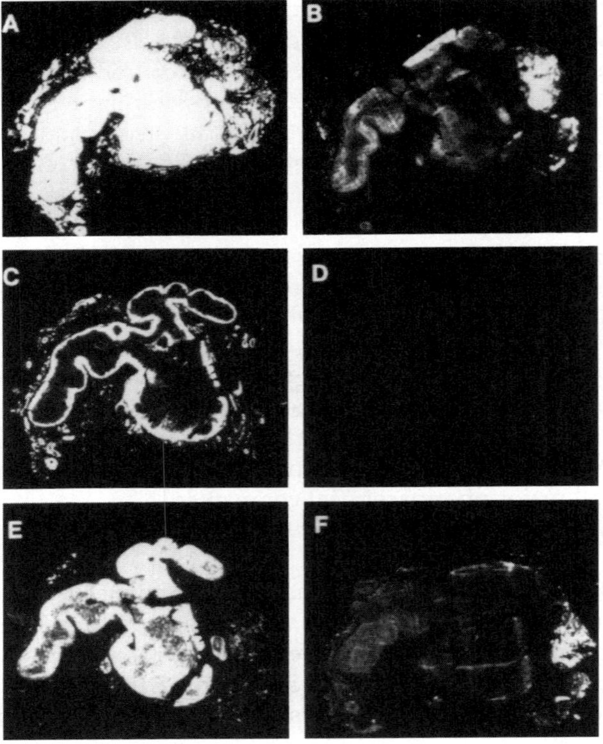

FIGURE 7
Autoradiographical localization of ET receptors in human adrenal. (A) Total [^{125}I]ET-1 binding showing the distribution of all ET receptors. (C) ET_A-receptors localized using [^{125}I]PD 151,242. (E) ET_B localized with [^{125}I]BQ 3202. Nonspecific binding was determined using the corresponding unlabeled peptide (B,D,F).

3. *Thyroid and Parathyroid*

[^{125}I]ET-1 bound with single high-affinity to thyrocytes cultured from goiter tissue but the unlabeled ligand only modestly inhibited thyroglobulin release (Jackson et al., 1992). A single class of [^{125}I]ET-1 receptors was also detected in parathyroid membranes prepared from patients with hyperparathyroidism; competition studies suggested both subtypes were present (Eguchi et al., 1992). There is no information about the anatomical distribution of ET receptor subtypes in normal thyroid, parathyroid, or pancreas.

F. Gastrointestinal Tract

Animal studies suggest ET is a potent stimulator of intestinal smooth muscle and intestinal epithelial secretion. Inagaki et al. (1991) localized ET receptors in human colon: [^{125}I]ET-1 bound to a single affinity site in mucosa and myenteric plexus (Table 2). ET-3 and STX-c (1 μM) did not completely inhibit [^{125}I]ET-1 binding, suggesting the presence of both ET_A- and ET_B-receptor subtypes in these two structures. High densities of [^{125}I]ET-1 binding were detected in the mucosa, especially the lamina propria which may have been binding to walls of small blood vessels or nerve fibers regulating mucosal function. Lower densities were also detected in the inner circular and outer longitudinal smooth muscle layers of the muscularis propria, which is responsible for peristaltic contraction. The inherent rhythmicity of the gut is modulated by parasympathetic ganglia organized into myenteric plexuses located between the two muscle layers, suggesting a possible role for ETs as neuropeptides, modulating motility and secretion in the human gut.

G. Skin

The skin is the largest organ in the body. ET receptors are mainly confined to the vasculature, which would be consistent with the peptide contributing to the regulation of blood flow and tone. Specific [^{125}I]ET-1 (but not [^{125}I]ET-3) binding was detected over the smooth muscle of arterioles and venules, suggesting these cells express mainly ET_A-receptors. Both labeled peptides bound to capillaries (which consist mainly of endothelial cells with occasional surrounding pericytes) of the subepidermal plexus and dermal papillae, as well as the capillary plexuses of hair follicles and sweat glands (Bull et al., 1991; Knock et al., 1993).

Competition binding studies suggested that the subepidermal plexus expressed both receptor subtypes, although it was not possible to identify which subtype was expressed by the endothelium or pericytes. However, ET_B-receptors were detected in human dermal microvascular endothelial cells cultured from neonate foreskin. Binding was detected in the basal cell layer of the epidermis (squamous epithelium) but not in the supporting dermis (fibrous-elastic tissue). The density of ET receptors was significantly higher in microvessels of skin from patients with systemic sclerosis (but not in patients with Raynaud's phenomenon). Lack of downregulation of ET receptors may contribute to the pathogenesis of vasospasm in this disease.

H. Skeletal Tissue

In synovial joints there is extensive movement of one joint upon another at articular surfaces, which is maintained in apposition by a fibrous capsule and ligaments. Wharton et al. (1992) showed that specific [^{125}I]ET-1 binding sites characteristic of the ET_A subtype were detected in the media of rheumatoid synovial blood vessels. Microautoradiography demonstrated ET receptors were localized to microvessels (20 to 40 μm external diameter), small arteries and arterioles (70 to 200 μm external diameter) in rheumatoid, osteoarthritic, and normal synovium. Silver grains were also detected overlying endothelial and synovial cells. No difference was found in vascular binding site density in rheumatoid and osteoarthritic synovium. ET-like immunoreactivity was localized to endothelial cells in blood vessels displaying [^{125}I]ET-1 binding sites, suggesting that ETs might act locally, modulating synovial perfusion and exacerbating hypoxia in chronic arthritis.

I. Reproductive Tissue

1. *Uterus*

The mucosal lining of the endometrium provides the environment for fetal development, and the thick smooth-muscle wall, the myometrium, protects then subsequently expels the fetus during birth (see Cameron et al., 1992). In agreement with vascular smooth muscle, the ET_A subtype predominates (80%) in this tissue (Table 3). ET-1 increases the contractility of isolated myometrium via this subtype, suggesting the peptide may have a role in parturition (Bacon et al., 1995).

In the nonpregnant state, the endometrium undergoes cyclical changes of proliferation then secretion to provide an environment suitable for implantation of a fertilized ovum. In the proliferative phase mainly ET_A-receptor mRNA was detected, but the amount of ET_B-receptor mRNA increased during the secretory phase (O'Reilly et al., 1992). In the absence of implantation, the endometrium is shed during menstruation. This is preceded by intense vasoconstriction of the spiral arterioles, caused by an as-yet uncharacterized compound. Importantly, in the endometrium, hemostasis does not occur (as in most other tissues) by deposition of platelet fibrinogen plugs but by vasoconstriction of the remaining arterioles, indicating the need for a powerful, locally produced mediator such as ET. The presence of intense staining for immunoreactive ET in uterine epithelial cells and vascular endothelium (Cameron et al., 1991) together with the detection of [^{125}I]ET-1 binding in the uterine vascular bed (Davenport et al., 1991a) suggests the peptide could act locally on spiral arteries before or during menstruation. High densities of [^{125}I]ET-1 binding were also localized to the uterine glandular epithelial cells (Davenport et al., 1991a) which have been characterized as being mainly of the ET_B-receptor subtype (Bacon et al., 1995) in agreement with the proposed expression of ET_B-receptors on human endothelium.

2. *Placenta*

Fetoplacental blood flow is regulated by locally released or circulating vasoactive compounds since vessels do not receive autonomic innervation. The powerful and long-lasting constrictor action of ET-1 makes this peptide an ideal candidate for closing blood supply in the umbilical vessels (Rath et al., 1993).

During pregnancy, the endometrium becomes highly vascular and is known as the decidua (forming the maternal placenta), and continues to express high densities of ET receptors that are mainly of the ET_B subtype. The presence of immunoreactive ET in decidual cells (van Papendorp et al., 1991) suggests that ETs may have a role in nonvascular tissue of placenta such as stimulating growth. ET_B-receptors were also concentrated on intermediate and terminal villi and fetal membranes (Rutherford et al., 1993). [^{125}I]ET-1 was detected in the media of arteries, arterioles, veins and venules throughout the chorionic plate and villous tree where fetoplacental exchange occurs, suggesting a role in regulating blood flow (Table 2; Robaut et al., 1993; Rutherford et al., 1993). The ratio of subtypes appears to vary. ET_A-receptors predominate (80%) on fetal veins and arteries in the chorionic plate, whereas veins in stem villi and blood vessels in distal regions of the villous tree express a higher proportion of ET_B-receptors (Rutherford et al., 1993).

IV. CONCLUSION

ET receptors are widely distributed in human tissues. [^{125}I]ET-1 binds with reported affinities ranging from about 10 pM to 10 nM with densities of 10 to 10,000 fmol/mg protein. The lungs and brain have the highest density of receptors although labeled ET-1 binds with lower affinity. The cardiovascular system contains predominantly ET_A-receptors, whereas in placenta, kidney, lung, adrenal cortex, and brain, ET_B-receptors are more abundant. A pattern is emerging of a differential distribution of receptor subtypes with ET_A-receptors predominating in human blood vessels whereas ET_B-receptors are expressed in high densities on tissues such as epithelium. The majority of receptors present in vascular smooth muscle are of the ET_A subtype. ET-1-induced vasoconstriction in isolated blood vessels is predominantly via this receptor. This difference may be exploited by the use of ET_A-receptor antagonists to produce a beneficial vasodilatation in pathophysiological conditions where there are elevated concentrations of ET peptide or increases in receptor density, avoiding blocking ET_B-receptors, whose function is less well understood.

ACKNOWLEDGMENTS

I thank Fiona Karet and Fraser Russell for critically reading the manuscript. Supported by grants from the British Heart Foundation, Royal Society, and Isaac Newton Trust.

REFERENCES

Bacon, C. R. and Davenport, A. P., Endothelin receptors in human arteries characterized by subtype selective ligands, *Br. J. Pharmacol.*, 113, P457. 1994.

Bacon, C. R., Morrison, J. J., Cameron, I. T. and Davenport, A. P., ET_A and ET_B endothelin receptors in human myometrium characterized by subtype selective ligands BQ123, BQ3020, FR139317 and PD151242, *J. Endocrinol.*, 144, 127-134. 1995

Brink, C., Gillard, V., Roubert, P., Mencia-Huerta, J. M., Chabrier, P. E., Braquet, P. and Verley, J., Effects and specific binding sites of endothelin in human lung preparations, *Pulm. Pharmacol.*, 4, 54-59. 1991.

Bigaud, M. and Pelton, J. T., Discrimination between ET_A- and ET_B-receptor mediated effects of endothelin-1 and [Ala^{1,3,11,15}]endothelin-1 by BQ123 in the anesthetized rat, *Br. J. Pharmacol.*, 107, 912-917. 1991.

Bull, H. A., Bunker, C. B., Terenghi, G., Springall, D. R., Zhao, Y., Polak, J. M. and Dowd, P. M., Endothelin-1 in human skin: immunolocalization, receptor binding, mRNA expression, and effects on cutaneous microvascular endothelial cells, *J. Invest. Dermatol.* 97, 618-623. 1991.

Cameron, I. T. and Davenport, A. P., Endothelins in reproduction, *Reprod. Med. Rev.*, 1, 99-113. 1992.

Cameron, I. T., Davenport, A. P., van Papendorp, C., Barker, P. J., Huskisson, N. S., Gilmour, R. S., Brown, M. J. and Smith, S. K., Endothelin-like immunoreactivity in human endometrium. *J. Reprod. Fertil.*, 95, 623-628. 1992.

Cyr, C., Huebener, K., Druck, T. and Kris, R., Cloning and chromosomal localization of the human endothelin ET_A receptor, *Biochem. Biophys. Res. Commun.*, 181, 184-190. 1991.

Davenport, A. P., Cameron, I. T., Brown, M. J. and Smith, S. K., Binding sites for iodinated endothelins-1, endothelin-2 and endothelin-3 demonstrated on human uterine glandular epithelial cells by quantitative high-resolution autoradiography, *J. Endocrinol.*, 129, 149-154. 1991a.

Davenport A. P., Kuc, R. E., Fitzgerald, F., Maguire, J. J., Berryman K. and Doherty, A. M., [^{125}I]-PD151242: a selective radioligand for human ET_A receptors, *Br. J. Pharmacol.*, 111, 4-6. 1994a.

Davenport, A. P., Kuc, R. E., Hoskins, S. L., Fiona E., Karet, F. E. and Fitzgerald, F., [^{125}I]-PD151242 is selective for endothelin ET_A receptors in human kidney and localizes to renal vasculature, *Br. J. Pharmacol.*, 114, 1110-1116. 1994b.

Davenport, A. P., Kuc, R. E., Howard, P., Plumpton, C., Molenaar, P., Hiley, C. R. and Brown, M. J., Distribution of binding sites for the proposed ET_B selective agonist ^{125}I [Ala^{1,3,11,15}] endothelin-1 in cardiovascular, renal and central nervous system: comparison with ^{125}I endothelin-1 binding, *J. Vasc. Res.* 29, 100. 1992a.

Davenport, A. P. and Maguire, J. J., Is endothelin-induced vasoconstriction mediated only by ET_A receptors in man?, *Trends Pharmacol. Sci.*, 15,9-11. 1994.

Davenport, A. P., Molenaar, P. and Kuc, R. E., BQ123 and BQ3020 reveal endothelin ET_A and ET_B receptor subtypes in human cardiac ventricle and rat cerebellum, *Br. J. Pharmacol.*, 107S, 304. 1992b.

Davenport, A. P. and Morton, A. J., Binding sites for ^{125}I ET-1, ET-2, ET-3 and vasoactive intestinal contractor are present in adult rat brain and neurone-enriched primary cultures of embryonic brain cells, *Brain Res.*, 554, 278-285. 1991.

Davenport, A. P., Morton, A. J. and Brown, M. J., Localization of endothelin(1-3), mouse VIC and sarafotoxin S6b binding sites in mammalian heart and kidney, *J. Cardiovasc. Pharmacol.*, 17,7, S152-S155. 1991b.

Davenport, A. P. and Nunez, D. J., Quantitative image analysis: methodology and application to techniques in *in situ* hybridization, in *Quantitative Methods in Neuroanatomy*, Stewart, M. G. Ed., John Wiley and Sons, Chichester, 65-80. 1992.

Davenport, A. P., Nunez, D. J., and Brown, M. J., Quantitative autoradiography reveals binding sites for [^{125}I] endothelin-1 in kidneys with a differential distribution among rat, pig and man, *Clin. Sci.*, 77, 129-131. 1989a.

Davenport, A. P., Nunez, D. J. and Brown, M. J., Localization of binding sites for iodinated endothelin and sarafotoxin peptides using quantitative receptor autoradiography, *Eur. J. Pharmacol.*, 183, 2153. 1990.

Davenport, A. P., Nunez, D. J., Hall, J. A., Kaumann, A. J. and Brown, M. J., Autoradiographical localization of binding sites for [^{125}I] endothelin-1 in humans, pigs and rats: functional relevance in man, *J. Cardiovasc. Pharmacol.*, 13 (5), S166-170. 1989b.

Davenport, A. P., O'Reilly, G., Molenaar, P., Maguire, J. J., Kuc, R. E., Sharkey, A., Bacon, C. R. and Ferro, A., Human endothelin receptors characterized using reverse transcriptase-polymerase chain reaction, *in situ* hybridization and subtype selective ligands BQ123 and BQ3020: evidence for expression of ET_B receptors in human vascular smooth muscle, *J. Cardiovasc. Pharmacol.*, 22 (S8), 22-25. 1993.

Doherty, A. M., Endothelin: a new challenge, *J. Med. Chem.*, 35, 1493-1508. 1992.

Eguchi, S., Hirata, Y., Imai, T., Kanno, K., Akiba, T., Sakamoto, A., Yanagisawa, M., Masaki, M. and Marumo, F., Endothelin receptors in human parathyroid gland, *Biochem. Biophys. Res. Commun.*, 184,1448-1455. 1992.

Fernandez-Durango, R., de Juan, J. A., Zimman, H., Moya, F. J., Garcia de la Coba, M. and Fernandez-Cruz, A., Identification of endothelin receptor subtype (ET_B) in human cerebral cortex using subtype selective ligands, *J. Neurochem.*, 62, 1482-1488. 1994.

Fischli, W., Clozel, M. and Guilly, C., Specific receptors for endothelin on membranes from human placenta. Characterization and use in a binding assay, *Life Sci.*, 44, 1429-1436. 1989.

Gasic, S., Wagner, O. F., Vierhapper, H., Nowotny, P. and Waldhausl, W., Regional hemodynamic effects and clearance of endothelin-1 in humans. Renal and peripheral tissues may contribute to the overall disposal of the peptide, *J. Cardiovasc. Pharmacol.*, 19, 176-180. 1992.

Gardiner, S. M., Kemp, P. A., March, J. E., Bennett, T., Davenport, A. P. and Edvinsson, L., Effects of an ET1-receptor antagonist, FR139317 on regional haemodynamic responses to endothelin-1 and [Ala^{11,15}] Ac-endothelin-1(621) in conscious rats, *Br. J. Pharmacol.*, 112, 477-486. 1994.

Grone, H. J., Laue, A. and Fuchs, E., Localization and quantification of [^{125}I]-endothelin binding sites in human fetal and adult kidneys-relevance to renal ontogeny and pathophysiology, *Klin. Wochenschr.*, 68, 758-767. 1990.

Hay, D. W. P., Luttmann, M. A., Hubbard, W. C. and Undem, B. J., Endothelin receptor subtypes in human and guinea-pig pulmonary tissues, *Br. J. Pharmacol.*, 110, 1175-1183. 1993.

Hemsen, A., Biochemical and functional characterization of endothelin peptides with special reference to vascular effects, *Acta Physiol. Scand. (Suppl.)*, 602, 1-61. 1991.

Henry, P. J., Rigby, P. J., Self, G. J., Preuss, J. M. and Goldie, R. G., Relationship between endothelin-1 binding site densities and constrictor activities in human and animal airway smooth muscle, *Br. J. Pharmacol.* 100, 786-792. 1990.

Howard, P. G., Plumpton, C. and Davenport, A. P., Anatomical localization and pharmacological activity of mature endothelins and their precursors in human vascular tissue, *Hypertension*, 10, 1379-1386. 1992.

Huggins, J. P., Pelton, J. T., and Miller, R. C., The structure and specificity of endothelin receptors. Their importance in physiology and medicine, *Pharmacol. Ther.*, 59, 55-123. 1993.

Ihara, M., Noguchi, K., Saeki, T., Fukuroda, T., Tsuchida, S., Kimura, S., Fukami, T., Ishikawa, K., Nishikibe, M. and Yano, M., Biological profiles of highly potent novel endothelin antagonists selective for the ET(A) receptor, *Life Sci.*, 50, 247-55. 1992.

Imai, T., Hirata, Y., Eguchi, S., Kanno, K., Ohta, K., Emori, T., Sakamoto, A., Yanagisawa, M., Masaki, M. and Marumo, F., Concomitant expression of receptor subtype and isopeptides of endothelin by human adrenal gland, *Biochem. Biophys. Res. Commun.*, 182, 1115-1121. 1992.

Inagaki, H., Bishop, A. E., Yura, J. and Polak, J. M., Localization of endothelin-1 and its binding sites to the nervous system of the human colon, *J. Cardiovasc. Pharmacol.* 17(7), S455-457, 1991.

Jackson, S., Tseng, Y. C., Lahiri, S., Burman, K. D. and Wartofsky, L., Receptors for endothelin in cultured human thyroid cells and inhibition by endothelin of thyroglobulin secretion, *J. Clin. Endocrinol. Metab.* 75, 388-392. 1992.

Jones, C. R., Hiley, C. R., Pelton, J. T. and Miller, R. C. Autoradiographic localization of endothelin binding sites in kidney, *Eur. J. Pharmacol.*, 163, 379-382. 1989.

Jones, C. R., Hiley, C. R., Pelton, J. T. and Miller, R. C., Endothelin receptor heterogeneity; structure activity, autoradiographic and functional studies, *J. Receptor Res.*, 11, 299-310. 1991.

Kalina, B. and Loffler, B. M., Cross linking analysis of an endothelin receptor protein from human placenta, *Biochem. Int.*, 27, 735-744. 1992.

Karet, F. E., and Davenport, A. P., Human kidney. Endothelin isoforms revealed by HPLC with radioimmunoassay, and receptor subtypes detected using ligands BQ123 and BQ3020, *J. Cardiovasc. Pharmacol.*, 22(S8), 29-33. 1993.

Karet, F. E. and Davenport, A. P., Endothelin and the human kidney. A potential target for new drugs, *Nephrol. Dialysis Transplant.*, 9, 465-468. 1994.

Karet, F. E., Kuc, R. E. and Davenport, A. P., Novel ligands BQ123 and BQ3020 characterise endothelin receptor subtypes ET_A and ET_B in human kidney, *Kidney Int.*, 44, 36-42. 1994.

Karet, F. E., Jones, D. S. C., Harrison-Woolrych, M., O'Reilly, G. R., Davenport, A. P. and Smith, S. K., Validation of mRNA quantification using a novel fluorescent nested reverse transcriptase polymerase chain reaction, *Anal. Biochem.*, 220, 384-390. 1994.

Karne, S., Jayawickreme, C. K., and Lerner, M. R., Cloning and characterization of an endothelin-3 specific receptor (ET_C receptor) from *Xenopus laevis* dermal melanophores, *J. Biol. Chem.*, 25, 19126-19133. 1993.

Knock, G. A., Terenghi, G., Bunker, C. B., Bull, H. A., Dowd, P. M. and Polak, J. M. Characterization of endothelin-binding sites in human skin and their regulation in primary Raynaud's phenomenon and systemic sclerosis, *J. Invest. Dermatol.* 101, 73-78. 1993.

Kondo, S., Fushimi, E., Morita, T. and Tashima, Y., Direct measurement of endothelin receptor in human bladder base and dome using ^{125}I-endothelin, *Tohoku J. Exp. Med.*, 167, 159-161. 1992.

Kurihara, M., Ochi, A., Kawaguchi, T., Niwa, M., Kataoka, Y. and Mori, K., Localization and characterization of endothelin receptors in human gliomas: a growth factor?, *Neurosurgery*, 27, 275-281. 1990.

Maguire, J. J. and Davenport, A. P., Endothelin-induced vasoconstriction in human isolated vasculature is mediated predominantly via activation of ET_A receptors, *Br. J. Pharmacol.*, 110S, P47. 1993a.

Maguire, J. J. and Davenport, A. P., Pre- or post-administration of BQ123 and FR139317 antagonises endothelin (ET-1)-induced contraction of human blood vessels in vitro, *Br. J. Pharmacol.*, 111S, P149. 1993b.

Maguire, J. J., Bacon, C. R., Fuijimoto, M. and Davenport, A. P., Myricerone caffeoyl ester 50-235, is a nonpeptide antagonist selective for human ET_A receptors, *J. Hypertension*, 12, 675-680. 1994a.

Maguire, J. J., Kuc, R. E., O'Reilly, G. and Davenport, A. P., Vasoconstrictor endothelin receptors characterised in human renal artery and vein in vitro, *Br. J. Pharmacol.*, 113, 49-54. 1994b.

Maguire, J. J., Kuc, R. E., O'Reilly, G. and Davenport, A. P., Potency of the novel orally active antagonist Ro46-2005 for endothelin receptors in human vascular smooth muscle, *Br. J. Pharmacol.*, 112, 552P. 1994c.

Marciniak, S. J., Plumpton, C., Barker, P. J., Huskisson, N. S. and Davenport, A. P., Localization of immunoreactive endothelin and proendothelin in the human lung, *Pulm. Pharmacol.*, 5, 175-182. 1993.

McKay, K. O. Black, J. L., Diment, L. M. and Armour, C. L., Functional and autoradiographic studies of endothelin-1 and endothelin-2 in human bronchi, pulmonary arteries, and airway parasympathetic ganglia, *J. Cardiovasc. Pharmacol.*, 17 (S7), S206-S209. 1991.

Molenaar, P., Kuc, R. E. and Davenport, A. P., Characterization of two new ET_B selective radioligands, [^{125}I]-BQ3020 and [^{125}I]-[Ala^{1,3,11,15}]ET-1 in human heart, *Br. J. Pharmacol.*, 107, 637-639. 1992.

Molenaar, P., O'Reilly, G., Sharkey, A., Kuc, R. E., Harding, D. P., Plumpton, C., Gresham, G. A. and Davenport, A. P., Characterization and localization of endothelin receptor subtypes in the human atrioventricular conducting system and myocardium, *Circ. Res.*, 72, 526-538. 1993.

Mondon, F., Malassine, A., Robaut, C., Vial, M., Bandet, J., Tanguy, G., Rostene, W., Cavero, I. and Ferre, F., Biochemical characterization and autoradiographic localization of [^{125}I]endothelin-1 binding sites on trophoblast and blood vessels of human placenta, *J. Clin. Endocrinol. Metab.*, 76, 237-244. 1993.

Morton, A. J. and Davenport, A. P., Neurons respond differentially to endothelins and sarafotoxin S6b in primary cultured cerebellum, *Brain Res.*, 581, 299-306. 1992.

Nambi, P., Pullen, M., Wu, H. L., Aiyar, N., Ohlstein, E. H. and Edwards, R. M., Identification of endothelin receptor subtypes in human renal cortex and medulla using subtype-selective ligands, *Endocrinology*, 131, 1081-1086. 1992.

Natarajan, S., Festin, S. M., Hedberg, A., Liu, E. C. -K., Floyd, D. and Hunt, J. T., Site specific biotinylation, *Int. J. Pept. Protein Res.*, 40, 567-574. 1992.

Niwa, M., Kawaguchi, T., Himeno, A., Fujimoto, M., Kurihara, M., Yamashita, K., Kataoka, Y., Shigematsu, K. and Taniyama, K., Specific binding sites for ^{125}I-endothelin-1 in the porcine and human spinal cord, *Eur. J. Pharmacol.*, 225, 281-289. 1992.

Ogawa, Y., Nakao, K., Arai, H., Nakagawa, O., Hosoda, K., Suga, S., Nakanishi, S. and Imura, H., Molecular cloning of a non-isopeptide-selective human endothelin receptor, *Biochem. Biophys. Res. Commun.*, 178, 248-255. 1991.

O'Reilly, G., Charnock-Jones, D. S., Cameron, I. T., Smith, S. K. and Davenport, A. P., Endothelin-2 mRNA splice variants detected by RT-PCR in cultured human vascular smooth muscle and endothelial cells, *J. Cardiovasc. Pharmacol.*, 22(S8), 18-21. 1993a.

O'Reilly, G., Charnock-Jones, D. S., Davenport, A. P., Cameron, I. T. and Smith, S. K., Presence of mRNA for endothelin-1, endothelin-2, and endothelin-3 in human endometrium, and a change in the ratio of ET_A and ET_B receptor subtype across the menstrual cycle, *J. Clin. Endocrinol. Metab.*, 75,1545-1549. 1992.

O'Reilly, G., Charnock-Jones, D. S., Morrison, J. J., Cameron, I. T., Davenport, A. P. and Smith, S. K., Alternatively spliced mRNAs for human endothelin-2 and their tissue distribution, *Biochem. Biophys. Res. Commun.*, 193, 834-840. 1993b.

Peter, M. G. and Davenport A. P., [125I]-PD151242 as an endothelin ET_A selective radioligand in the human, rat and porcine heart, *Br. J. Pharmacol.* 114, 297-302. 1994a.

Peter, M. G. and Davenport, A. P., Characterisation of the endothelin ET_A antagonists BQ123 and FR139317 in human, rat and porcine hearts, and the ET_B selective antagonist IRL1038 in human tissue, *Br. J. Pharmacol.*, 112, 122P. 1994b.

Persidis, A., Harcombe, A. A., Davenport, A. P., Kuc, R. E., Plumpton, C. and Weissberg, P. L., Isolation of human cardiac endothelin receptors by peptide/mobility shift assay, *Clin. Sci.*, 85, 169-173. 1993.

Plumpton, C., Champeney, R., Ashby, M. J., Kuc, R. E., and Davenport, A. P., Characterisation of endothelin isoforms in human heart. Endothelin 2 demonstrated, *J. Cardiovasc. Pharmacol.*, 22(S8), 26-28. 1993.

Rath, W., Osterhage, G., Kuhn, W., Grone, H. J. and Fuchs, E., Visualization of ^{125}I-endothelin-1 binding sites in human placenta and umbilical vessels, *Gynecol. Obstet. Invest.* 35, 209-213. 1993.

Rebibou, J. M., He, C. J., Delarue, F., Peraldi, M. N., Adida, C., Rondeau, E. and Sraer, J. D., Functional endothelin 1 receptors on human glomerular podocytes and mesangial cells, *Nephrol. Dialysis Transplant.* 7, 288-292. 1992.

Robaut, C., Mondon, F., Bandet, J., Ferre, F. and Cavero, I., Regional distribution and pharmacological characterization of [^{125}I]endothelin,1 binding sites in human fetal placental vessels, *Placenta*, 12, 55-67. 1991.

Rutherford, R. A. D, Wharton, J., McCarthy, A., Gordon, L., Sullivan, M. H. F, Elder, M. G. and Polak, J. M., Differential localization of endothelin ET(A) and ET(B) binding site in human placenta, *Br. J. Pharmacol.*, 109, 544-552. 1993.

Takahashi, K., Ghatei, M. A., Jones, P. M., Murphy, J. K., Lam, H. C., O'Halloran, D. J. and Bloom, S. R., Endothelin in human brain and pituitary gland: presence of immunoreactive endothelin, endothelin messenger ribonucleic acid, and endothelin receptors, *J. Clin. Endocrinol. Metabol.*, 72, 693-699. 1991.

Takayanagi, R., Ohnaka, K., Takasaki, C., Ohashi, M. and Nawata, H., Multiple subtypes of endothelin receptors in human and porcine tissues: characterization by ligand binding, affinity labeling, and regional distribution, *J. Cardiovasc. Pharmacol.*, 17(7), S127-130. 1991.

Urade, Y., Fuitani, Y., Oda, K., Watakabe, T., Umemura, I., Takai, M., Okada, T., Sakata, K. and Karaki, H., An endothelin B receptor-selective antagonist: IRL1038 [Cys11-Cys15]-endothelin(11-21), *FEBS Lett.* 311, 12-16. 1992.

Vierhapper, H., Wagner, O., Nowotny, P. and Waldhausl, W., Effect of endothelin in man, *Circulation*, 81, 1415-1418. 1990.

Waeber, C., Hoyer, D. and Palacios, J. M., Similar distribution of (^{125}I)sarafotoxin-6b and (^{125}I)endothelin-1,-2,-3 binding sites in the human kidney, *Eur. J. Pharmacol.*, 176, 233-236. 1990.

Watakabe, T., Urade, Y., Takai, M., Umemura, I. and Okada, T. A., Reversible radioligand specific for the ET(B) receptor: (^{125}I) Tyr^{13}-Suc(Glu^{9},Ala^{11,15})-endothelin-1(8-21), (^{125}I)IRL 1620, *Biochem. Biophys. Res. Commun.*, 185, 867-873. 1992.

Wilkes, B. M., Mento, P. F., Hollander, A. M., Maita, M. E., Sung, S. and Girardi, E. P., Endothelin receptors in human placenta: relationship to vascular resistance and thromboxane release, *Am. J. Physiol.*, 258, E864-870. 1990.

Williams, D. L., Jones, K. L., Colton, C. D. and Nutt, R. F., Identification of high affinity ET-1 receptor subtypes in human tissues, *Biochem. Biophys. Res. Commun.* 180, 475-480. 1991.

Wharton, J., Rutherford, R. A. D., Gordon, L., Moscoso, G., Schiemberg, I., Gaer, J. A. R., Taylor, K. M. and Polak, J. M., Localization of endothelin binding sites and endothelin-like immunoreactivity in human fetal heart, *J. Cardiovasc. Pharmacol.* 17(S7), S378-S384. 1991.

Wharton, J., Rutherford, R. A. D., Walsh, D. A., Mapp, P. I., Knock, G. A., Blake D. R. and Polak, J. M., Autoradiographic localization and analysis of endothelin-1 binding sites in human synovial tissue, *Arthritis Rheum.*, 35, 894-899. 1992.

van Papendorp, C., Cameron, I. T., Davenport, A. P., King, A. J., Brown, M. J. and Smith, S. K., Localization and endogenous concentration of endothelin-like immunoreactivity in human placenta, *J. Endocrinol.*, 131, 507-511. 1991.

Chapter **4**

Peptide Receptor Interactions

Denis Guedin, Christian Frelin, and Paul Vigne

CONTENTS

I. INTRODUCTION

Early studies on the contractile action of ET-1 in aortic strips outlined two important features: (1) contractions develop very slowly compared with other vasoconstrictors, and (2) contractions are almost irreversible and washing unbound ET off the preparations hardly reverses contractions (Yanagisawa et al., 1988). These

0-8493-6975-4/97/$0.00+$.50

features are believed to reflect in some way the properties of interactions of ET with its receptors, but the relationship between a molecular event — the binding reaction — and the physiological response — the development of a contraction — is not simple and needs to be evaluated properly in order to define the physiopathological role of ETs.

Any bimolecular reaction of the type

$$R + L \underset{k_{-1}}{\overset{k_1}{\longleftrightarrow}} RL$$

is governed by two rate constants: k_1, the second order rate constant of association, and k_{-1}, the first order rate constant of dissociation. The half time of dissociation of receptor-ligand complexes ($t_{1/2}$) is $\ln(2)/k_{-1}$. At equilibrium, the concentrations of the different reactants are linked by the well-known law of mass action

$$K_d = [RL]/[L][R] \tag{1}$$

where K_d is the equilibrium dissociation constant. K_d can also be derived from kinetic parameters using the relationship $K_d = k_{-1}/k_1$. All three parameters defined above (k_1, k_{-1}, and K_d) describe important properties of the interaction of the ligand with its receptor and need to be known with precision in order to assess the functional role(s) of the ligand.

In this chapter we review the main properties of interactions of ET isopeptides with their receptors and analyze some of the problems encountered when defining these properties. Finally, we analyze the functional consequences of these properties and discuss some practical implications.

II. THE ASSOCIATION REACTION

A. Second Order Rate Constant of Association (k_1)

The k_1 values for the association of ET-1 with its receptors have been determined in some membrane preparations. Table 1 presents a non-exhaustive list of values reported in the literature ranging from $7 \cdot 10^5$ to $7 \cdot 10^7$ M^{-1} s^{-1}. Variability can largely be accounted for by experimental errors. There is no obvious difference between k_1 values for ET-1 association with ET_A and ET_B receptor subtypes.

The k_1 value for the association of ET-1 to its receptors is much lower than the value expected for a diffusion-controlled process (10^9 M^{-1} s^{-1}). This suggests that the binding of ET-1 is a two-step process. First, ET-1 reaches receptors by simple diffusion; it is then locked onto the receptor by a slower process that probably involves changes in the conformations of either (or both) ET-1 or its receptor. A putative locking process is suggested by the results of three dimensional modeling (Huggins et al., 1993).

B. Functional Implications

Aortic strips respond to ET-1 by contractions that develop much more slowly than contractions induced by other vasoconstrictors. A common belief is that a slower rate of association is responsible for the slower contractile action. In fact this is not true. The k_1 values for the association of angiotensin II to type 1 angiotensin II receptors ($4 \cdot 10^5$ M^{-1} s^{-1}, Gunther et al., 1982) and of vasopressin to V_1 receptors ($2 \cdot 10^5$

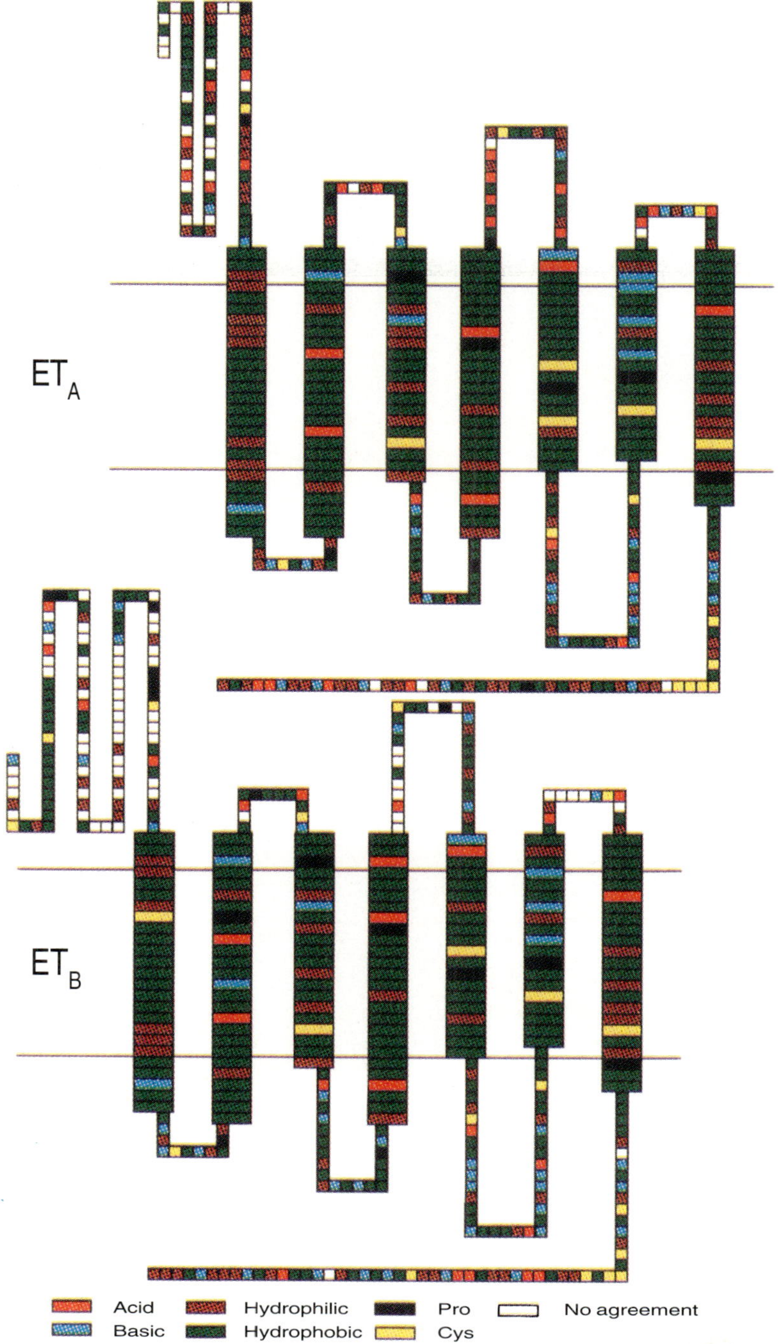

PLATE 1. A schematic representation of ET_A- and ET_B-receptors in which residues that are consistently of a particular type in each subtype are color-coded according to their properties as shown in the key. The N-terminal, extracellular amino acid (top left in each figure) is that directly after the signal cleavage site.

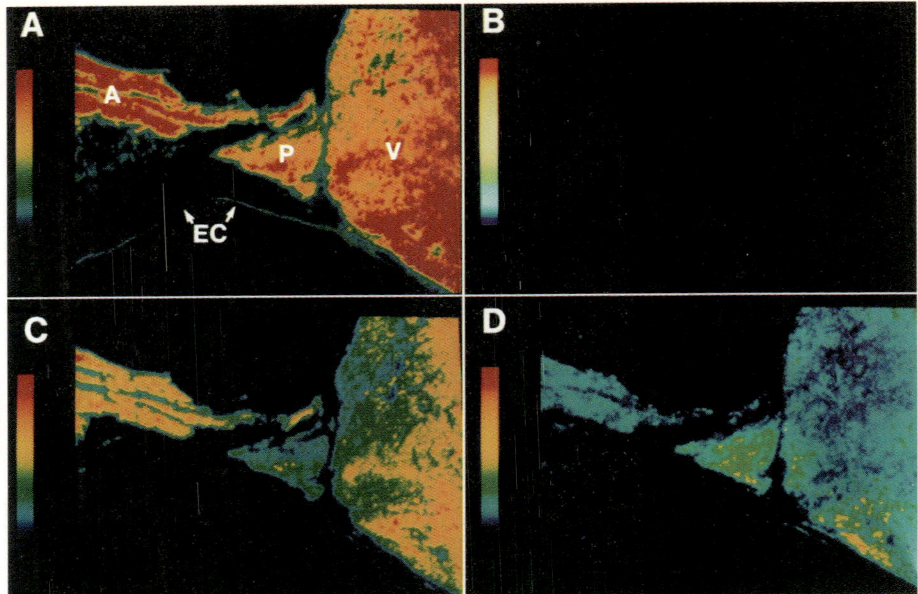

PLATE 2. Color-coded autoradiographical images showing the distribution of ET receptors in serial sections of atria (A), ventricle (V), penetrating bundle of His (P) and endocardial cells (EC) from human heart. Highest densities are shown in red, intermediate values coded according to the scale bars and black is below the level for detection. (A) Total [^{125}I]-ET-1 binding showing the distribution of all ET receptors. (B) Nonspecific binding. (C) ET_A receptors visualized by inhibiting [^{125}I]-ET-1 binding to the ET_B sub-type with BQ3020. (D) ET_B receptors visualized by inhibiting [^{125}I]-ET-1 to the ET_A sub-type with BQ123.

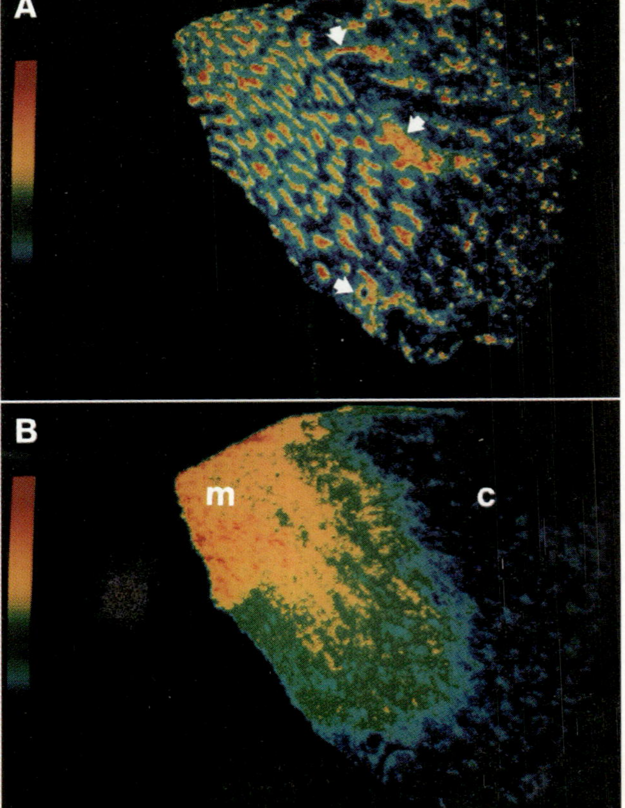

PLATE 3.
Color-coded autoradiographical images showing the distribution of ET_A (A) receptors (visualized using [^{125}I]-PD151242) that are mainly present on intra-renal blood vessels including the arcuate arteries and adjacent veins (indicated by arrows) in human kidney. (B) High densities of ET_B receptors (visualized using [^{125}I]-BQ30920) are present in the medulla (m) especially on medullary rays, with lower densities in the cortex (c).

TABLE 1

Rate Constants of Association

Cell Preparations	Receptor Subtype	k_1 (M^{-1} s^{-1})	Ref.
Hamster ventricle	NS	$7{\cdot}10^5$	1
Rat cerebellum	ET_B	$8{\cdot}10^5$	2
Human placenta	$ET_A = ET_B$	$1{\cdot}10^6$	3
Rat aortic SMC	ET_A	$1{\cdot}10^6$	4
Renal cells	ET_A	$1{\cdot}10^6$	5
BCEC	ET_A	$2{\cdot}10^6$	6
Human placenta	ET_B	$3{\cdot}10^6$	7
Hamster brain	NS	$3{\cdot}10^6$	1
Hamster lung	NS	$4{\cdot}10^6$	1
Rat atria	$ET_A > ET_B$	$4{\cdot}10^6$	6
Hamster kidney	NS	$5{\cdot}10^6$	1
Porcine aortic SMC	ET_A	$5{\cdot}10^6$	8
Rat lung	$ET_B > ET_A$	$5{\cdot}10^6$	8
Rat heart	$ET_A > ET_B$	$8{\cdot}10^6$	8
Rat cerebellum	ET_B	$1{\cdot}10^7$	8

Note: When more than one receptor subtype is expressed in the preparation, their relative abundance is indicated. NS: not specified, SMC: smooth muscle cells, BCEC: brain capillary endothelial cells. 1: Bolger et al., 1992, 2: Hiley et al., 1990, 3: Kilpatrick et al., 1993, 4: Marsault et al., 1993, 5: Wilkes et al., 1991, 6: Marsault et al., 1991, 7: Wilkes et al., 1990, 8: Waggoner et al., 1992.

to $7{\cdot}10^5$ M^{-1} s^{-1}, Roy and Ausiello, 1981) are similar to (or even smaller than) the values observed for ET-1 (Table 1). This means that at identical concentrations, angiotensin II, vasopressin, and ET-1 bind to their receptors at comparable rates. The obvious conclusion is that differences in the time courses of the contractile actions of these peptides are not due to differences in the rate of association. They may be accounted for by different intracellular signaling pathways. At this time, however, no clear difference has been observed in the intracellular signaling pathways that are controlled by these three peptides. They all activate phospholipase C, phospholipase A_2, depolarize the plasma membrane, and indirectly open L-type Ca^{2+} channels (Van Renterghem et al., 1988, Wallnofer et al., 1989).

It is often believed that a slow rate of association of ET-1 with its receptor determines slow contractions in isolated vessel preparations. It is true that in binding experiments the association of ET-1 with its receptors proceeds slowly. However, binding experiments are performed at picomolar concentration of ET-1 that are at least three orders of magnitude lower than the concentrations necessary to induce contractions in isolated vessel preparations. At high concentrations, the association process is accelerated and binding sites are occupied more rapidly. It can be calculated, for instance, that at a concentration of 50 nM, which produces maximum vasoconstriction in isolated rat aortic strips, saturation of ET_A receptors is achieved within less than 30 s (Marsault et al., 1991). In comparison, the half time for tension development in isolated aortic rings is 5.3 min (Marsault et al., 1991). The obvious conclusion, therefore, is that tension development in isolated aortic ring preparations is not limited by the binding reaction but by an intracellular signaling mechanism that activates slowly (Marsault et al., 1991). This mechanism has not yet been definitely identified. It was proposed that receptor externalization following its internalization could be rate limiting for tension development in rat aortic strips (Marsault et al.,

1993). It should also be noted that slow contractions in response to ETs are a unique property of vascular cells. Smooth muscle cells from the uterus and the intestine respond to ET-1 by much faster contractions (Maggi et al., 1991, Yoshinaga et al., 1992). This stresses the point that excitation-contraction coupling mechanisms in smooth muscle cells are still poorly understood and that a detailed comparison of the biochemical actions of ETs in different smooth muscle preparations may be worthwhile.

III. THE DISSOCIATION REACTION

A. First Order Rate Constant of Dissociation (k_{-1})

It is well known that ETs and their receptors form tight complexes that hardly dissociate and that may survive in the presence of 1% SDS at reduced temperature (Takasuka et al., 1991). It has even been proposed that ET-1 and its receptors form covalent complexes by thiol-disulfide exchange (Spinella et al., 1993). Although the data presented suggest that a critical sulfhydryl group on receptors is essential for ET-1 binding, covalent complexes are unlikely to be necessary for ET-1 to bind. This is simply because ETs can be induced to dissociate from their receptors when the pH of incubation solutions is reduced, or under strong denaturing (but nonreducing) conditions. A tight association of ETs with their receptors should not be surprising as noncovalent protein-protein interactions may be extremely stable. The best example is the complex formed by trypsin and trypsin inhibitors which dissociates with an estimated half life of 4.5 months (Vincent and Lazdunski, 1972)!

Dissociation of ET receptor complexes is best studied using membrane preparations. Dissociation experiments using cultured cells are less reliable, for the internalization of receptor ligand complexes (Resink et al., 1990; Marsault et al., 1993) may lead to artificially low rates of dissociation. Conversely, artificially fast rates of dissociation are obtained if internalized ET is degraded and if degradation products are released into the culture medium.

In many cases, the dissociation of ET-1 from its receptors does not follow a single exponential time course (Sokolovsky, 1993). For practical reasons, only the fast component of the dissociation is measured. Table 2 presents a nonexhaustive list of $t_{1/2}$ values for ET-1 receptor complexes. It shows a large scatter of values from 18 min to 206 h. This variability cannot be accounted for by experimental errors. In some cases, a heterogeneity of receptors with respect to the dissociation of their ligand was suggested. ET-1 ET_B receptor complexes are much less stable in the guinea-pig ileum than in the rat cerebellum (Galron et al., 1991). Similarly, IRL 1620, a high-affinity ET_B receptor-selective agonist, binds almost irreversibly to rat tissues but reversibly in dog and human tissues (Watakabe et al., 1992; Nambi et al., 1994). The molecular basis of these differences is not yet known.

ET-1 forms more stable complexes with ET_B receptors than with ET_A receptors, as evidenced by their stability in mild denaturing conditions (Takasuka et al., 1994). Further experiments using site-directed mutagenesis experiments have shown that Asp75 and Pro93 in the human ET_B receptor sequence are probably responsible for the tight association with ET-1 (Takasuka et al., 1994).

The binding properties of other agonists and antagonists of ET receptors have been less thoroughly analyzed but some trends may be outlined. In rat cerebellar membranes, ET-3 and ET_B receptors form tighter complexes than do ET-1 and ET_B receptors. Conversely, STX-b forms less stable complexes than does ET-1 with ET_B receptors (Galron et al., 1991). Circumstantial evidence also suggests that the binding of BQ 123 to ET_A receptors is more reversible than that of ET-1 (Vigne et al., 1993).

TABLE 2

Rate Constants of Dissociation

Cell Preparations	Receptor Subtype	$t_{1/2}$	Ref.
Rat cerebellum	ET_B	206 h	1
Rat lung	$ET_B > ET_A$	165 h	1
Rat heart	$ET_B > ET_A$	77 h	1
Porcine aortic SMC	ET_A	38 h	1
Renal cells	ET_A	5.4 h	2
Rat cerebellum	ET_B	2-3 h	3
Human placenta	ET_B	2.6 h	4
Rat atria	$ET_A > ET_B$	90 min	5
Human placenta	$ET_A = ET_B$	53 min	6
Rat atria	$ET_A > ET_B$	45-60 min	7
Guinea pig ileum	ET_B	18 min	8

Note: When more than one receptor subtype is expressed in the preparation, their relative abundance is indicated. SMC: smooth muscle cells. 1: Waggoner et al., 1992, 2: Wilkes et al., 1991, 3: Galron et al., 1991, 4: Wilkes et al., 1990, 5: Vigne and Frelin, unpublished results, 6: Kilpatrick et al., 1993, 7: Sokolovsky, 1993, 8: Galron et al., 1991.

B. Functional Implications

The irreversible binding nature of ET has been suggested as one of the reasons for the sustained effects observed with this peptide in many systems (Simonson and Dunn, 1990). Different pieces of evidence, however, indicate that irreversibility, although it may be true at the receptor level, is not true at a cellular level: (1) short exposures of rat aortic strips to ET-1 induce fast and transient contractile responses (Marsault et al., 1991), and (2) receptor antagonists such as BQ 123, which are unable to promote the dissociation of preformed ET-1-ET_A receptor complexes, do relax precontracted arteries (Marsault et al., 1993). The reversibility of the actions of ET-1 at a cellular level can be accounted for by a rapid internalization of ET receptor complex (Resink et al., 1990) and by the recycling of free receptors to the plasma membrane (Marsault et al., 1993). A consequence of these observations is that the irreversible contractions observed in aortic strips that are exposed for prolonged periods to ET-1 are the consequence of an irreversible (or slowly desensitizing) intracellular signal that controls tension and not the consequence of microscopic irreversibility (Marsault et al., 1991). The sustained property of the contractile actions of ET-1 in vascular smooth muscle, but not in other smooth muscle preparations such as uterus (Maggi et al., 1991) or intestine (Yoshinaga et al., 1992), again suggests that the specificity of the action of ET-1 is more likely to reside in its intracellular action rather than in its binding properties.

IV. THE EQUILIBRIUM DISSOCIATION CONSTANT

A. Experimentally Estimated Affinity Values

Three methods may be used to define the equilibrium dissociation constant of ET receptor complexes. For most ligands, affinity values obtained in different experiments are very close and good estimates of the true affinity value are available. This is not the case for ET-1. Reported affinity values are both highly variable and dependent on the method used to calculate them.

TABLE 3

K_d Values Determined from Kinetic Parameters

Tissue	Receptor Subtype	K_d (pM)	Ref.
Human placenta	$ET_A = ET_B$	60	1
Rat atria	$ET_A > ET_B$	30	2
Renal cells	ET_A	10	3
Human placenta	ET_B	10	4
Porcine aortic SMC	ET_A	1	5
Rat lung	$ET_B > ET_A$	0.2	5
Rat ventricule	$ET_A > ET_B$	0.3	5
Rat cerebellum	ET_B	0.09	5

Note: When more than one receptor subtype is expressed in the preparation, their relative abundance is indicated. SMC: smooth muscle cells. 1: Kilpatrick et al., 1993, 2: Vigne and Frelin, unpublished results, 3: Wilkes et al., 1991, 4: Wilkes et al., 1990, 5: Waggoner et al., 1992.

K_d values can first be obtained from a kinetic analysis and determination of k_{-1} and k_1 (Section I). Table 3 lists some of the values that have been reported. It shows a wide range of values of between 0.1 and 60 pM.

K_d values can also be derived from Scatchard analyses of saturation experiments and K_i values from competition experiments. Numerous values have been reported and it would take too long to list them all. However, it is generally recognized that these values are highly variable from one preparation to another. Some preparations are very sensitive to ET-1, with K_d values in the low picomolar range. Other preparations are much less sensitive, with apparent affinity values in the low nanomolar range. This holds both for ET_A and ET_B receptor subtypes and for endogenous and cloned receptors.

B. Technical Difficulties

The large variability of affinity values is unusual and is a major source of complications when trying to define what the physiological role(s) of the peptide could be. Three major technical reasons may contribute to the high variability of estimated K_d values.

1. Nonequilibrium Conditions

It is not often realized that meaningful data from Scatchard analyses or competition studies can only be obtained if the duration of the experiment is severalfold greater than the half time of dissociation of ligand-receptor complexes (so as to reach equilibrium). The dissociation of ET-1 from its receptors being slow (Table 2), this condition is not always fulfilled and may lead to overestimated K_d values (Waggoner et al., 1992).

2. Stoichiometric Binding Conditions

It is not often realized that the concentration of ET receptors ($[R_0]$) in preparations used for binding experiments is often high, relative to the K_d value of ET receptor complexes. Binding experiments are best performed under conditions of pseudo first-order kinetics, i.e., under conditions in which the free ligand concentration decreases by less than 10% during the association process. If the receptor concentration used in the experiments is comparable to, or higher than, the K_d value of ET

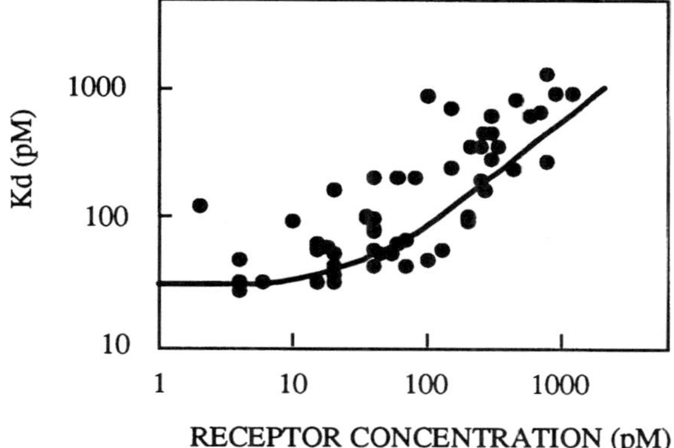

FIGURE 1
Relationship between K_d values and $[R_0]$, the receptor concentration used in binding experiments. Data used are taken from the literature. The line indicates the relationship predicted by Equation 2 for a K_d value of 30 pM.

receptor complexes (i.e., under conditions known as stoichiometric binding conditions), then an important fraction of the initial ligand is bound to receptors and this simple fact may be a source of complications when assessing affinity.

The important point to consider is that it is not possible to determine meaningful affinity values when the receptor concentration used in the experiments is larger than K_d. The reason for this is the following. In saturation experiments, the total concentration of ligand at which half of the sites are occupied (half saturation point, HSP) is always equal to

$$HSP = K_d + [R_0]/2 \tag{2}$$

If $[R_0]$ is negligible as compared to K_d, HSP is a good measure of K_d. If $[R_0] >> K_d$, this is no longer true; HSP are higher than K_d and in extreme cases HSP measures the receptor concentration rather than K_d! Scatchard analyses also become very difficult to perform in stoichiometric binding conditions: because the concentration of bound ligand is close to the total concentration of ligand, the free ligand concentration cannot be meaningfully derived by subtracting the bound ligand from the total ligand concentration. Finally, affinities for other ET isopeptides and affinity ratios determined in competition experiments may also be inaccurate. The Cheng-Prüsoff approximation, which is widely used to derive affinity values from competition experiments, is only valid under pseudo first-order kinetics and may lead to inaccurate K_d values if this condition is not fulfilled.

Using 48 articles that describe in detail the binding conditions used, it is possible to calculate $[R_0]$ and to compare this value to reported apparent K_d values. Figure 1 shows that K_d values are indeed related to $[R_0]$ with a relationship which is predicted by Equation 2. The asymptotic minimum observed at low $[R_0]$ values is 30 pM and is probably close to the true K_d value. This value corresponds to the super-high-affinity ET_B-receptor sites detected in brain membranes (Sokolovsky et al., 1992) and is of the same order of magnitude as K_d values determined from kinetic parameters of ET-1 binding (Table 3).

This stresses the point that because of the high affinity of ET-1 for its receptors, $[R_0]$ is an important parameter to consider in all experiments. If $[R_0]$ is in the low picomolar range, then meaningful K_d values are likely to be obtained. If it is higher, then higher apparent K_d values are likely to be obtained. The only way to avoid these problems is to dilute the membrane preparations until $[R_0]$ falls into the low picomolar range. A typical example, in an unrelated field, may be found in Barhanin et al. (1982). Unfortunately, diluting the receptor concentration decreases the signal and makes experiments more difficult.

3. *Degradation of the Ligand*

A necessary condition for performing binding experiments is that the ligand is not degraded during the course of the experiments. When membrane preparations are incubated at temperatures of 20 to 37°C, tissue peptidases may partially degrade the ligand, hence resulting in higher K_d or K_i values than expected. It is seldom verified that the free ligand remaining in the incubation mixtures is indeed ET at the end of binding studies. This point is illustrated, albeit in experiments not involving binding, by studies comparing the neurotoxicities of ET-1 and STX-b. STX-b is less potent than ET-1 at ET_A receptor subtypes and as potent as ET-1 at ET_B receptors (when measured in the presence of peptidase inhibitors), but is more toxic to animals when injected into the brain (Sokolovsky et al., 1990). This is due to the fact that ET-1 is more prone to degradation by brain neutral endopeptidase. As a consequence, for similar doses of the two peptides effective STX-b concentrations are higher than effective ET-1 concentrations. Analogous situations may be met in binding experiments.

C. Functional Implications

Effective concentrations of ET that produce half-maximum cellular responses (EC_{50} values) are often compared to K_d values estimated from binding experiments. Such comparisons may be misleading for the following reasons. Firstly, as discussed above, K_d values may be overestimated. Secondly, K_d values describe the properties of interactions of the ligand and the receptor at equilibrium, i.e., after a time of equilibration that exceeds the half-life of ligand-receptor complexes, which may be quite long (Table 2). Physiological responses are usually analyzed over a much shorter time scale. Under these conditions dissociation of ET-1 from its receptor is negligible, the binding can be considered as irreversible, and the association of ET-1 (and its action) is only governed by k_1, the rate of association. This may have important practical implications. Consider, for instance, a ligand that binds to ET receptors at a faster rate than ET-1 (i.e., with a higher k_1 value) but that dissociates faster (i.e., with a higher k_{-1} value). Its K_d value may be similar to that of ET-1, but because biological actions are mainly determined by k_1, they are expected to occur at lower concentrations than those of ET-1. Conversely, two ligands that bind at the same rate (i.e., with identical k_1 values) but which dissociate at different rates (and hence have different K_d values) would be equally potent for triggering biological actions. They would not be equally potent in binding experiments. A practical consequence is that screening new molecules on the basis of equilibrium binding experiments may overlook potentially useful substances.

Another consequence of the irreversibility of the binding of ET receptor ligands is that their action is strongly dependent on their mode of application. This is illustrated by studies analyzing the antagonistic action of BQ-123 on ET-1-induced intracellular Ca^{2+} mobilization in brain endothelial cells. When cells are exposed to BQ-123 and ET-1 at the same time (to favor kinetic competition), BQ-123 appears to antagonize the action of ET-1 in a competitive manner. When BQ-123 is preequilibrated with cells prior to the stimulation by ET-1, a noncompetitive type of inhibition is observed (Vigne et al., 1993).

V. INTERACTION WITH G PROTEINS

The intracellular actions of endothelins are mediated by G proteins, and guanine nucleotides are well known to affect the binding properties of agonists to seven-transmembrane-domain receptors. Their action reflects the process involved in the conformational switch of the G protein between its inactive and active forms. Receptor agonist complexes bind tightly to the transition-state unloaded G protein, but not with forms of G proteins that are bound to GTP or GDP. The unloaded G protein in turn stabilizes the active state of the receptor, for which agonists have the highest affinity. GTP analogues, by provoking the departure of the G protein, disrupt the high-affinity binding of agonists and, in binding experiments, transform biphasic competition curves for agonists into monophasic low-affinity curves. They have no action on competition curves produced by antagonists and this property is widely used to identify the agonist or antagonist properties of compounds. For many receptors, GTP analogues substantially increase the rate of dissociation of the ligand. The same mechanism may hold for ET receptors but it is probably masked by the slow rate of dissociation of ET-1 receptor complexes. An increase in the dissociation rate could be difficult to see during the usual short-lived experiments. Indeed, all short-term experiments have indicated that GTPγS does not decrease ET binding. When GTPγS is incubated for longer periods (24 h to favor a substantial dissociation of ET-1 receptor complexes), it decreases ET-1 binding (Takuwa et al., 1990). Because of these kinetic reasons, only high-affinity states of ET receptors are analyzed in binding experiments, even in the presence of GTP analogues, and competition curves for agonists and antagonists have similar profiles characterized by Hill coefficients close to unity.

VI. CONCLUSIONS

The main conclusions of this review are

1. ET-1 associates with its receptors at a fast rate, but this rate is not a diffusion-controlled process. In turn, this rate is not responsible for the slow contractile action of ET-1 in vessel preparations and for differences in the contractile actions of different vasoconstricting peptides.
2. Endothelins form stable, noncovalent complexes with their receptors but the stability of the complexes shows large variations from one preparation to another. For most physiological and biochemical experiments, the association can be considered as irreversible. At a cellular level, however, and because of the internalization of endothelin receptor complexes and the ensuing dissociation, the binding can be considered as dissociable.
3. Equilibrium dissociation constants are highly variable, ranging from subpicomolar to nanomolar concentrations. One reason may be that high receptor concentrations used in the experiments prevent an accurate determination of K_d. The true K_d value for ET-1 is not yet known with precision. It is probably in the low picomolar range.

REFERENCES

Barhanin, J., Giglio, J.R., Léopold, P., Schmid, A., Sampaio, S.V. and Lazdunski, M., *Tityus serrulatus* venom contains two classes of toxins, *J. Biol. Chem.*, 257, 12553-12558. 1982.

Bolger, G.T., Berry, R., Liard, F., Garneau, M. and Jaramillo, J., Cardiac responses and binding sites for endothelin in normal and cardiomyopathic hamsters, *J. Pharmacol. Exp. Ther.*, 260, 1314-1322. 1992.

Galron, R., Bdolah, A., Kochva, E., Wollberg, Z., Kloog, Y. and Sokolovsky, M., Kinetic and cross linking studies indicate different receptors for endothelins and sarafotoxins in the ileum and cerebellum, *FEBS Lett.*, 283, 11-14. 1991.

Gunther, S., Alexander, R.W., Atkinson, W.J. and Gimbrone, M.A., Functional angiotensin II receptor in cultured vascular smooth muscle cells, *J. Cell Biol.*, 92, 289-298. 1982.

Hiley, C.R., Jones, C.R., Pelton, J.T. and Miller, R.C., Binding of [^{125}I]endothelin-1 to rat cerebellar homogenates and its interaction with some analogues, *Br. J. Pharmacol.*, 101, 319-325. 1990.

Huggins, J.P., Trumpp-Kallmeyer, S., Hibert, M.F., Hoflack, J.M., Fanger, B.O. and Jones, C.R., Modeling and modification of the binding site of endothelin and other receptors, *Eur. J. Pharmacol.*, 245, 203-214. 1993.

Kilpatrick, S.J., Roberts, J.M., Lykins, D.L. and Taylor, R.N., Characterization and ontogeny of endothelin receptors in human placenta, *Am. J. Physiol.*, 264, E367-E372. 1993.

Maggi, M., Vanelli, G.B., Peri, A., Brandi, M.L., Fantoni, G., Giannini, S., Torrisi, C., Guardabasso, V., Barni, T., Toscano, V., Massi, G. and Serio, M., Immunolocalization, binding, and biological activity of endothelin in rabbit uterus: effect of ovarian steroids, *Am. J. Physiol.*, 260, E292-E305. 1991.

Marsault, R., Feolde, E. and Frelin, C., Receptor externalization determines sustained contractile responses to endothelin-1 in the rat aorta, *Am. J. Physiol.*, 264, C687-C693. 1993.

Marsault, R., Vigne, P. and Frelin, C., The irreversibility of endothelin action is a property of a late intracellular signaling event, *Biochem. Biophys. Res. Commun.*, 179, 1408-1413. 1991.

Marsault, R., Vigne, P., Breittmayer, J.P. and Frelin, C., Kinetics of the vasoconstrictor action of endothelins, *Am. J. Physiol.*, 261, C986-C993. 1991.

Nambi, P., Pullen, M. and Spielman, W., Species differences in the binding characteristics of [^{125}I]IRL-1620, a potent agonist specific of endothelin B receptors, *J. Pharmacol. Exp. Ther.*, 268, 202-207. 1994.

Resink, T.J., Scott-Burden, T., Boulanger, C., Weber, E. and Bühler, F., Internalization of endothelin by cultured human vascular smooth muscle cells. Characterization and physiological significance, *Mol. Pharmacol.*, 38, 244-252. 1990.

Roy, C. and Ausiello, D.A., Characterization of (8-lysine) vasopressin binding sites on a pig kidney cell line (LLC-PK1), *J. Biol. Chem.*, 256, 3415-3422. 1981.

Simonson, M.S. and Dunn, M.J., Cellular signaling by peptides of the endothelin family, *FASEB J.*, 4, 2989-3000. 1990.

Sokolovsky, M., Ambar, I. and Galron, R., A novel subtype of endothelin receptor, *J. Biol. Chem.*, 267, 20551-20554. 1992.

Sokolovsky, M., BQ-123 identifies heterogeneity and allosteric interactions at the rat heart endothelin receptor, *Biochem. Biophys. Res. Commun.*, 196, 32-38. 1993.

Sokolovsky, M., Galron, R., Kloog, Y., Bdolah, A., Indig, F.E., Blumberg, S. and Fleminger, G., Endothelins are more sensitive than sarafotoxins to neutral endopeptidase: possible physiological significance, *Proc. Natl. Acad. Sci. U.S.A.*, 87, 4702-4706. 1990.

Spinella, M.J., Kottke, R., Magazine, H.I., Healy, M.S., Catena, J.A., Wilken, P. and Andersen, T.T., Endothelin-receptor interactions. Role of a putative sulfhydryl on the endothelin receptor, *FEBS Lett.*, 328, 82-88. 1993.

Takasuka, T., Horii, I., Furuichi, Y. and Watanabe, T., Detection of an endothelin-1 binding protein complex by low temperature SDS-PAGE, *Biochem. Biophys. Res. Commun.*, 176, 392-400. 1991.

Takasuka, T., Sakurai, T., Goto, K., Furuichi, Y. and Watanabe, T., Human endothelin receptor ET_B. Amino acid sequence requirements for superstable complex formation with its ligand, *J. Biol. Chem.*, 269, 7509-7513. 1994.

Takuwa, Y., Kasuya, Y., Takuwa, N., Kudo, M., Yanagisawa, M., Goto, K., Masaki, T. and Yamashita, K., Endothelin receptor is coupled to phospholipase C via a pertussis toxin insensitive guanine nucleotide binding regulatory protein in vascular smooth muscle cells, *J. Clin. Invest.*, 85, 653-658. 1990.

Van Renterghem, C., Vigne, P., Barhanin, J., Schmid-Alliana, A., Frelin, C. and Lazdunski, M., Molecular mechanism of action of the vasoconstrictor peptide endothelin, *Biochem. Biophys. Res. Commun.*, 157, 977-985. 1988.

Vigne, P., Breittmayer, J.P. and Frelin, C., Competitive and non competitive interactions of BQ-123 with endothelin ET_A receptors. *Eur. J. Pharmacol.*, 245, 229-232. 1993.

Vincent, J.P. and Lazdunski, M., Trypsin-pancreatic trypsin inhibitor association. Dynamics of the interaction and role of disulfide bridges, *Biochemistry*, 11, 2967-2977. 1972.

Waggoner, W.G., Genova, S.L. and Rash, V.A., Kinetic analyses demonstrate that the equilibrium assumption does not apply to [^{125}I]endothelin-1 binding data, *Life Sci.*, 51, 1869-1876. 1992.

Wallnofer, A., Weir, S., Ruegg, U. and Cauvin, C., The mechanism of action of endothelin-1 as compared with other agonists in vascular smooth muscle cells, *J. Cardiovasc. Pharmacol.*, 13(Suppl. 5), S23-S31. 1989.

Watakabe, T., Urade, Y., Takai, M., Umemura, I. and Okada, T., A reversible radioligand for the ET_B receptor: [^{125}I]Tyr^{13}-Suc-[Glu^{9},Ala^{11,15}]-endothelin-1 (8-21), [^{125}I]IRL 1620, *Biochem. Biophys. Res. Commun.*, 185, 867-873. 1992.

Wilkes, B.M., Mento, P.F., Hollander, A.M., Maita, M.E., Sung, S. and Girardi, E.P., Endothelin receptors in human placenta: relationship to vascular resistance and thromboxane release, *Am. J. Physiol.*, 258, E864-E870. 1990.

Wilkes, B.M., Ruston, A.S., Mento, P., Girardi, E., Hart, D., Vander Molen, M., Barnett, R. and Nord, E.P., Characterization of endothelin-1 receptor and signal transduction mechanisms in rat medullary interstitial cells, *Am. J. Physiol.*, 260, F579-F589. 1991.

Yanagisawa, M., Kurihara, H., Kimura, S., Mitsui, Y., Kobayashi, M., Watanabe, T.X. and Masaki, T., A novel potent vasoconstrictor peptide produced by vascular endothelial cells, *Nature*, 332, 411-415. 1988.

Yoshinaga, M., Chijiiwa, Y., Misawa, T., Harada, N. and Nawata, H., Endothelin B receptor on guinea-pig small intestinal smooth muscle cells, *Am. J. Physiol.*, 262, G308-G311. 1992.

Chapter **5**

Endothelin-Induced Intracellular Signaling Pathways

Herbert Schramek and Michael J. Dunn

CONTENTS

0-8493-6975-4/97/$0.00+$.50

I. INTRODUCTION

Since their discovery in 1988 (Yanagisawa et al., 1988), peptides of the endothelin (ET) family have been implicated in various physiological and pathophysiological events. Besides others, physiological actions of ET include its influence on transepithelial solute and water transport in the kidney and gastrointestinal tract, its regulatory function on vascular tone and the state of contraction of numerous nonvascular smooth muscle cells (airway, gastrointestinal, uterine), as well as its stimulatory effects on mitogenesis. The fact that a single peptide is able to accomplish a variety of different functions can, in principle, be due to the presence and specificity of different isopeptides, to the cell-specific expression of distinct receptor subtypes, and/or to the availability of specific intracellular signaling pathways in certain cells. The purpose of this review is to summarize the present knowledge about ET-induced intracellular signaling pathways. We will therefore discuss ET-derived signals, which are transduced by channel and transport proteins, summarize the current knowledge concerning phospholipases activated by these peptides, and analyze recent insights into ET-stimulated protein kinase cascades as well as their putative regulation by protein phosphatases. Finally, we will focus on nuclear targets of these intracellular signals resulting from ET binding to its receptors on the cell membrane.

II. ENDOTHELIN-DERIVED SIGNALS TRANSDUCED BY CHANNEL- AND TRANSPORT-PROTEINS

A. Ion Channels

1. Ca^{2+} Channels

ET usually causes a biphasic rise in $[Ca^{2+}]_i$, consisting of a rapid, transient increase followed by a lesser but sustained increment lasting up to 20 min (Badr et al., 1989; for review see Simonson, 1993). This ET receptor-activated increase in $[Ca^{2+}]_i$ most likely results from both a release of Ca^{2+} from intracellular stores (see Section 3.1.2.) as well as an increased net influx of Ca^{2+} across the plasma membrane, which is primarily responsible for the sustained phase (Simonson and Dunn, 1990a; Simonson, 1993).

The ET-induced Ca^{2+} influx from the extracellular space is due to the activation of multiple types of Ca^{2+} channels in the plasma membrane (Figure 1). Current evidence demonstrates that ET peptides can activate both voltage-sensitive calcium channels (VSCC) and receptor-operated calcium channels (ROCC). Early evidence that ET activates VSCC came from studies showing that dihydropyridine channel

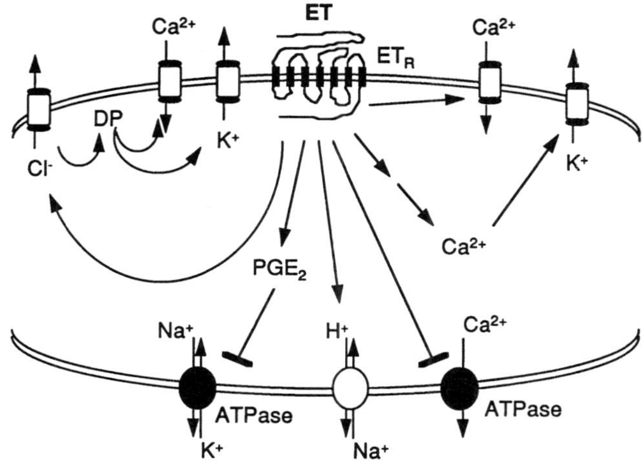

FIGURE 1
Ion channels and transport proteins activated by endothelins. ET peptides activate both voltage-sensitive calcium entry and receptor-operated calcium entry from the extracellular space via specific channel proteins. Furthermore, ETs, via binding to their receptors, can activate chloride channels, calcium-sensitive and voltage-sensitive potassium channels, and are able to stimulate Na^+/H^+ exchange. Finally, ETs seem to inhibit Na^+/K^+-ATPase by a PGE_2-dependent pathway as well as Ca^{2+}-ATPase. DP, membrane depolarization; ET, endothelin; ET_R, endothelin receptor; PGE_2, prostaglandin E_2.

blockers inhibit ET-induced contraction (VanRenterghem et al., 1988; Goto et al., 1989; Highsmith et al., 1989) as well as increments in $[Ca^{2+}]_i$ (Goto et al., 1989) and $^{45}Ca^{2+}$ uptake (Highsmith et al., 1989) in vascular smooth muscle cells. These findings were supported by direct measurements of Ca^{2+} conductance utilizing patch-clamp techniques which revealed that ET-1 increases VSCC activity by an indirect mechanism (Goto et al., 1989; Silberberg et al., 1989; Inoue et al., 1990). Although the ability of bath-applied ET-1 to activate VSCC in the cell-attached mode of patch-clamp experiments strongly argues that ET acts via a diffusible second messenger system (Silberberg et al., 1989; Inoue et al., 1990), it is still unclear by which molecular mechanism ET activates VSCC. Beyond that, there exists evidence that ET affects Ca^{2+} signaling through mechanisms regulated by membrane depolarization but distinct from dihydropyridine-sensitive L-type VSCC (Blackburn and Highsmith, 1990).

In mesangial cells and many other cell types, ET also acts via ROCC, which are independent of membrane depolarization and insensitive to dihydropyridine Ca^{2+} channel blockers such as nifedipine (Auguet et al., 1988; D'Orleans-Juste et al., 1989; Mitsuhashi et al., 1989; Simonson et al., 1989; Edwards et al., 1990; Simonson et al., 1990; Simonson and Dunn, 1991). Recent experiments in A7r5 cells furthermore revealed that three different dihydropyridine Ca^{2+} channel blockers (manidipine, nifedipine, and verapamil), when used at their maximal inhibitory dose, only blocked about 50% of ET-1-induced Ca^{2+} influx (Huang et al., 1993). These data add further support to the view that dihydropyridine-insensitive pathways are also involved in ET-1-induced Ca^{2+} influx. In the same study, preincubation of A7r5 cells with manidipine (10 μM) strongly potentiated ET-1-induced *c-fos* and *c-jun* expression (Huang et al., 1993). Similar effects were obtained with nifedipine, suggesting that the potentiation of ET-1-induced *c-fos* and *c-jun* expression by manidipine is related to its Ca^{2+} channel-blocking effect. In contrast to these results, nifedipine, when used at a concentration of 1 μM, has been reported to completely inhibit both the elevation of intracellular calcium as well as the stimulatory effect of ET-1 on DNA synthesis in vascular smooth muscle cells (Hirata et al., 1989; Nakaki et al., 1989). Beyond that, it was recently shown that carvedilol, another antagonist of VSCC, inhibited ET-1-induced [^{3}H]thymidine incorporation in rat vascular smooth muscle cells by 95% (Sung et al., 1993). Together, this evidence not only supports the involvement of extracellular calcium influx in mediating some short-term effects of ET such as vasoconstriction and secretion (Masaki et al., 1990; Highsmith et al., 1992; Stojikovic and Catt, 1992; Simonson and Dunn, 1993) but also in ET's long-term effects such as DNA synthesis and cell proliferation.

2. K^+ Channels

Besides its effects on Ca^{2+} channels, ET-1 has also been reported to act on K^+ channels (Figure 1). Ca^{2+}-dependent K^+ channels, for example, are activated after ET-1 application in both A7r5 cells and smooth muscle cells from pig coronary artery (VanRenterghem et al., 1988; Hu et al., 1991). VanRenterghem et al. furthermore reported that in A7r5 cells, a nonselective cation channel was opened during a second phase of the ET-1 action. In the pulmonary vasculature, a significant reduction in the dilator response to ET-1 was demonstrated when the preparation was treated with the K^+ channel inhibitor glibenclamide, indicating that the mechanism responsible for the dilator response to ET-1 may involve stimulation of potassium conductance and membrane hyperpolarization (Hasunuma et al., 1990; Lippton et al., 1991). Injection of cardiac polyA$^+$ RNA from neonatal mouse heart into *Xenopus* oocytes leads to the expression of a slowly activating, voltage-dependent K^+ channel as well as to the expression of purinergic P_2 and ET receptors (Honore et al., 1992). This slowly activating, voltage-dependent K^+ channel was stimulated by administration of either ATP or ET-1. The regulation of this channel could be important in cardiac cells where it would lead to alterations of both the duration of the action potential as well as the heart rate (Honore et al., 1992). In contrast, when *Xenopus* oocytes were co-transfected with cDNA encoding the RH1 delayed rectifier K^+ channel from rat heart together with cDNA encoding either the ET_A or the ET_B receptor, exposure to 100 nM ET-1 inhibited channel openings regardless of the receptor subtype present (Ishii et al., 1992). This study is supported by the finding that ET-1 inhibits a delayed rectifier K^+ current in a protein kinase C (PKC)-dependent manner in rat cardiac myocytes (Damron et al., 1993).

3. Na^+ Channels

As the 21-amino-acid sequence of ET shows some homology to scorpion α-toxins, which are thought to prevent inactivation of voltage-dependent Na^+ channels, the possibility existed that ET could act at voltage-dependent Na^+ channels either in presynaptic nerve terminals or in smooth muscle cells. ET-induced contractions of isolated rat aorta and portal vein, however, are unaffected by tetrodotoxin or by removal of sodium chloride from the solution bathing these isolated vessel preparations (Borges et al., 1989). Moreover, the ET responses in isolated rat aorta and portal vein are entirely dependent on the presence of extracellular calcium and can be blocked by nitrendipine (Borges et al., 1989). These findings were confirmed in studies of glomerular mesangial and vascular smooth muscle cells (Iijima et al., 1991). In these cells ET-1 caused an immediate and sustained depolarization, which could not be attributed to Na^+ influx since Na^+-free medium did not abolish the ET-1-induced membrane depolarization (Iijima et al., 1991). It is thus very unlikely that voltage-activated Na^+ channels are a primary site of ET action.

4. Cl^- Channels

Effects of ET-1 on $[Ca^{2+}]_i$, $[Cl^-]_i$, as well as on membrane potential have been studied in both glomerular mesangial and vascular smooth muscle cells (Iijima et al., 1991). ET-1 releases IP_3 which results in both mobilization of intracellular Ca^{2+} stores and activation of Cl^- channels. The ensuing Cl^- efflux causes membrane depolarization and, in turn, activation of VSCC, resulting in a sustained elevation of $[Ca^{2+}]_i$ which is indispensable for the full-scale contraction produced by ET-1 (Iijima et al., 1991; Takenaka et al., 1992). These conclusions are based on the following results: (1) the Cl^- channel inhibitor indanyloxyacetic acid (IAA-94), abolishes changes in 6-methoxy-*N*-(sulfopropyl)quinolinium (SPQ) fluorescence, curtails the sustained phase of membrane depolarization and Ca^{2+} elevation and attenuates ET-1-induced contraction of aortic rings (Iijima et al., 1991), and (2) IAA-94 markedly diminishes ET-1-induced constriction of afferent arterioles in isolated perfused hydronephrotic kidneys (Takenaka et al., 1992). Therefore, the contractile effect of ET-1 can be regulated through plasma membrane ion fluxes, namely, Cl^- efflux, a change in membrane potential and Ca^{2+} entry via VOCC or ROCC (Figure 1).

B. Transporters

1. Na^+/K^+-ATPase

In freshly prepared suspensions of inner medullary collecting duct (IMCD) cells, but not in proximal tubule (PT) cells, ET-1 at concentrations above 1 pM was reported to inhibit transport-dependent oxygen consumption (QO_2), indicating that ET-1 reduces IMCD Na^+ transport (Zeidel et al., 1989). In this study, several lines of evidence revealed that ET-1 inhibits an ouabain-sensitive ATPase:

1. The inhibitory effect of ET-1 on QO_2 was present when Na^+ entry was maximally stimulated with amphotericin B.
2. ET-1 and ouabain stimulated net K^+ efflux from IMCD cells, indicating that each had inhibited the main pathway for K^+ entry, namely, Na^+/K^+-ATPase.
3. ET-1 inhibited the maximal rate of ouabain-sensitive $^{86}Rb^+$ uptake in IMCD cells. Inhibition of QO_2 by ET-1 was unaffected by removal of Ca^{2+}, but was abolished in the presence of ibuprofen, an inhibitor of PGE_2 formation (Zeidel et al., 1989).

Ibuprofen also blocked the inhibitory effect of ET-1 on ouabain-sensitive $^{86}Rb^+$ uptake. The effects of ET-1 and PGE_2 on $^{86}Rb^+$ uptake were equivalent and nonadditive. Together these results demonstrate that ET-1 inhibits IMCD Na^+ transport by stimulation of PGE_2 accumulation and inhibition of Na^+/K^+-ATPase (Figure 1) (Zeidel et al., 1989). Compared to this study, Garvin and Sanders found a small but significant inhibition of Na^+/K^+-ATPase activity in isolated straight proximal tubules of the rat by measuring enzymatic activity in permeabilized tubules (Garvin and Sanders, 1989). In the A10 cell line of vascular smooth muscle, however, a stimulation of both the Na^+/K^+-pump and the $Na^+/K^+/Cl^-$ cotransporter was found when cells were stimulated with ET-1 concentrations higher than 100 pM (Garay et al., 1989; Rosati et al., 1990).

2. Na^+/H^+ Antiporter

Different laboratories, including ours, have shown that ET-1 activates an amiloride-inhibitable Na^+/H^+ antiporter, causing marked alterations of intracellular pH (Badr et al., 1989; Gardner et al., 1989; Meyer-Lehnert et al., 1989; Simonson et al.,

1989; Kramer et al., 1991; Vigne et al., 1991). In mesangial cells loaded with the pH-sensitive probe BCECF, concentrations of ET-1 that increase phospholipase C activity also increase Na^+/H^+ antiport (Simonson et al., 1989). ET-1 causes a concentration-dependent net alkalinization of 0.1 to 0.3 pH units. Peak alkalinization occurs 3 to 4 min after addition of ET-1 (Badr et al., 1989; Simonson et al., 1989) and at high concentrations of ET-1 (10 and 100 nM) intracellular alkalinization is preceded by a transient acidification (Simonson et al., 1989). Furthermore, ET-1 has been reported to exert a concentration-dependent stimulation of the apical Na^+/H^+ antiporter in brush border membrane vesicles as well as the basolateral Na^+/HCO_3^- cotransporter in basolateral membrane vesicles prepared from rabbit renal cortex (Eiam-Ong et al., 1992). The stimulation of the apical Na^+/H^+ antiporter and the basolateral Na^+/HCO_3^- cotransporter by ET-1 displayed specificity as indicated by the lack of effects on the activity of the apical Na^+-glucose transporter and the basolateral Na^+-succinate transporter (Eiam-Ong et al., 1992). Although the precise role for Na^+/H^+ exchange in signal transduction remains to be clarified, it seems possible that under certain conditions ET-1-induced Na^+/H^+ exchange might facilitate the mitogenic action of ET-1 through regulation of cytosolic pH.

III. INTRACELLULAR SIGNALING PATHWAYS

A. Phospholipases

In addition to activating a number of ion channels and transport proteins, ET also stimulates different phospholipases, such as phospholipase A_2 (PLA_2), phospholipase C (PLC), and phospholipase D (PLD) which, depending on the enzymatic capabilities of a particular cell type, lead to the generation of distinct intracellular signals (Figure 2).

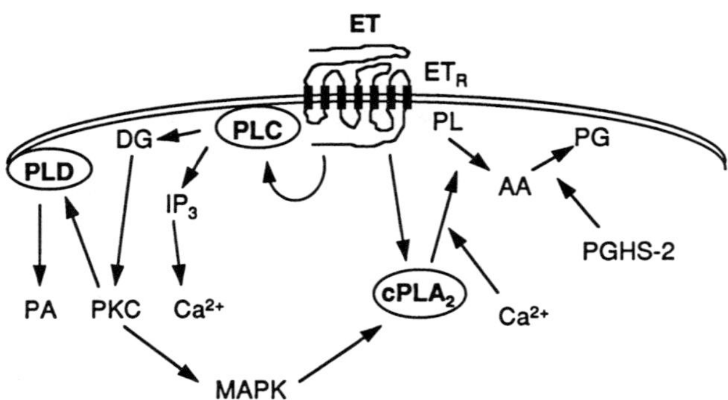

FIGURE 2
Endothelin-induced intracellular phospholipases. ETs have been reported to stimulate cytosolic phospholipase A_2 ($cPLA_2$), phospholipase C (PLC), and phospholipase D (PLD), leading to the generation of various intracellular signaling molecules such as arachidonic acid (AA), 1,2-diacylglycerol (DG), inositol 1,4,5-trisphosphate (IP_3), and phosphatidic acid (PA). ET-induced activation of prostaglandin endoperoxide synthase-2 (PGHS-2) leads to the conversion of AA to prostaglandins (PG). Ca^{2+}, intracellular Ca^{2+} concentration; ET_R, endothelin receptor; MAPK, mitogen-activated protein kinase; PKC, protein kinase C; PL, phospholipids.

1. *Phospholipase A_2*

Assuming that arachidonate is mainly derived from the *sn*-2 position of plasma membrane phospholipids, PLA_2 is an important pathway that can provide arachidonate (Mené et al., 1989; Smith 1992). PLA_2 cleaves an acyl group from the *sn*-2 position of either phosphatidylinositol, phosphatidylethanolamine, or phosphatidylcholine and thereby releases fatty acids (e.g., arachidonic acid) and lysophospholipids. These products may themselves serve as intracellular second messengers or can be further metabolized to lipid mediators like prostaglandins, leukotrienes, hydroxyeicosatetraenoic acids, or platelet-activating factor (PAF) depending on the enzymatic capabilities of target cells. From studies in different cell types indirect evidence exists for an ET-induced phospholipase A_2 activation. In mesangial and endothelial cells, for example, ET-1 leads to the release of both prostaglandin E_2 (PGE_2) and stable metabolites of prostaglandin I_2 (PGI_2) or thromboxane A_2 (TXA_2) (Simonson and Dunn, 1990b; Zoja et al., 1990; Filep et al., 1991; Fukunaga et al., 1991).

An ET-1-stimulated [^{3}H]arachidonic acid release has been shown in vascular smooth muscle cells (Resink et al., 1989; Reynolds et al., 1989), glomerular mesangial cells (Simonson and Dunn, 1990b) and, more recently, in Chinese hamster ovary (CHO) cells stably expressing the bovine ET_A or the rat ET_B receptor (Aramori and Nakanishi, 1992). Furthermore, in rat renal medullary interstitial cells, ET-1-mediated PGE_2 production has been found to be a Ca^{2+}-dependent but PKC-independent process (Barnett et al., 1994). PLA_2s, however, comprise a family of enzymes that can be subdivided into secretory and intracellular forms (Bonventre, 1992; Mayer and Marshall, 1993). One group contains multiple secretory phospholipases A_2 ($sPLA_2$) which have molecular masses in the range of about 14 kDa with 7 disulfide bonds sensitive to dithiothreitol, depend upon millimolar concentrations of calcium for maximal activity, and do not demonstrate preference for the fatty acid present in the *sn*-2 position of phospholipids. Another form, an intracellular, high molecular weight cytosolic phospholipase A_2 ($cPLA_2$), is dithiothreitol insensitive and depends upon micromolar concentrations of calcium for maximal activity (Bonventre, 1992; Mayer and Marshall, 1993). Using a direct measurement of enzymatic activity and CHO cells stably expressing the human ET_A or the human ET_B receptor, our laboratory was able to demonstrate that ET-1 rapidly stimulates cytosolic phospholipase A_2 ($cPLA_2$) via both human ET receptor subtypes (Schramek et al., 1994a).

In rat glomerular mesangial cells, ET-1 rapidly increases $cPLA_2$ activity without any change in $cPLA_2$ mRNA level or $cPLA_2$ protein mass (Schramek et al., 1994b). This rapid activation of $cPLA_2$ is inhibitable by potato acid phosphatase, suggesting that its regulation occurs through phosphorylation at a posttranslational level (Schramek and Dunn, 1994). Long-term incubations in the presence of ET-1 induce an increase in $cPLA_2$ mRNA expression, which is accompanied by both an increase in $cPLA_2$ protein mass as well as enzymatic activity (Schramek et al., 1994b). As herbimycin A, an inhibitor of cellular protein tyrosine kinases, completely blocks the ET-1-induced $cPLA_2$ mRNA expression, it is most likely that this ET-1-stimulated $cPLA_2$ mRNA increase occurs predominantly via tyrosine phosphorylation-signaling cascades (Schramek et al., 1994b). Thus, ET-1 can stimulate $cPLA_2$ in a biphasic manner: acute stimulation by ET-1 is most likely due to posttranslational modification of the enzyme, while chronic activation also involves regulation at a transcriptional level. These results are in line with our recent experiments showing that ET-1 also induces prostaglandin endoperoxide synthase-2 (PGHS-2) mRNA expression, protein formation, and bioactivity (Kester et al., 1994), thereby enhancing cellular capacity not only to release arachidonic acid but also to convert arachidonate to biologically active PGs such as PGE_2. In support of earlier reports (DeNucci et al., 1988; Schramek et al., 1993; Simonson, 1993), we therefore suggest that ET-1-induced short-term and long-term activation of $cPLA_2$ plays an important role in the ET-1-mediated release of PGs and thereby serves as an autocrine and/or paracrine negative regulator for vasoconstrictive and mitogenic stimuli. Moreover, prostanoids most likely also exert inhibitory effects on the synthesis of peptides of the ET family (Yokokawa et al., 1991; Sakamoto et al., 1992).

2. *Phospholipase C*

The best characterized intracellular effector system linked to ET receptors is phospholipase C (PLC) (for review see Simonson and Dunn, 1990a; Simonson, 1993). ET-induced PLC leads to the production of at least two second messengers: it catalyzes the hydrolysis of an inositol lipid precursor stored in the plasma membrane (phosphatidylinositol 4,5-bisphosphate) and releases (1) inositol 1,4,5-trisphosphate [Ins(1,4,5)P_3], which diffuses to specific receptors on specialized compartments of the endoplasmatic reticulum to release intracellular Ca^{2+} $[Ca^{2+}]_i$, and (2) neutral 1,2-diacylglycerol (1,2-DG), which remains in the plasma membrane and activates protein kinase C (PKC). 1,2-DG can also be formed by the PLC-mediated hydrolysis of other phospholipids such as phosphatidylcholine and phosphatidylethanolamine (MacNulty et al., 1990; Resink et al., 1990; Kester et al., 1994), but the exact role of this putative signaling pathway remains to be elucidated.

The increase in $[Ca^{2+}]_i$ itself activates a family of Ca^{2+}-dependent calmodulin protein kinases that, similar to the action of PKC, regulate the activity and function of target proteins by reversible phosphorylation of serine or threonine residues. ET peptides rapidly induce a concentration-dependent increase in phosphatidylinositol turnover in vascular smooth muscle cells, fibroblasts, atrial cells, myocytes, glomerular mesangial cells, renal medullary interstitial cells, glial cells, and some but not all endothelial cells as well as in intact tissues such as aorta, coronary artery strips, and rat atria and brain (for review see Simonson, 1993). Experiments with CHO cells stably transfected with either the bovine ET_A or the rat ET_B receptor revealed that both ET receptor subtypes activate PLC (Aramori and Nakanishi, 1992). In rat hepatocytes the Ca^{2+} signal provoked by ETs is not only secondary to activation of PLC but also to inhibition of the plasma membrane Ca^{2+} pump (Figure 1), each effector being coupled to ET_B receptors by different G proteins, G_q and G_s (Jouneaux et al., 1994).

At least two G proteins, differentiated by pertussis toxin sensitivity, appear to mediate ET-induced PLC activity. In rat mesangial cells (Thomas et al., 1991) and vascular smooth muscle cells from rabbit renal artery (Reynolds et al., 1989), ET activates PLC via a pertussis toxin-sensitive G protein. In contrast, in other vascular smooth muscle cells (Mitsuhashi et al., 1989; Takuwa et al., 1990), astrocytes (Marsault et al., 1990), and Rat-1 fibroblasts (Muldoon et al., 1989), ET-1 activates PLC via a pertussis toxin-insensitive G protein. However, even in cells demonstrating pertussis toxin-sensitive PLC activation, the toxin inhibits activity by only 50% or less, suggesting that other toxin-insensitive G proteins (i.e., G_q or G_{11}) contribute to enzyme activation (Serradeil-Le Gal et al., 1991; Thomas et al., 1991). Further evidence for G protein involvement comes from experiments demonstrating that the nonhydrolyzable analogue of GTP, guanosine 5′-O-(3-thiotriphosphate), potentiates ET's stimulatory effect on PLC (Takuwa et al., 1990; Thomas et al., 1991).

3. *Phospholipase D*

Several investigators have demonstrated that phospholipase D (PLD)-mediated phosphatidylcholine hydrolysis is stimulated after activation of ET receptors (MacNulty et al., 1990; Resink et al., 1990; Pai et al., 1991; Kester et al., 1992). In C6-glioma cells and in Rat-1 and Swiss 3T3 fibroblasts ET-1 binds to ET_A receptors which results in activation of PLC and PLD (Ambar and Sokolovsky, 1993). In all three cell lines activation of PLD by ET-1 was reported to involve PKC-dependent and PKC-independent pathways (Ambar and Sokolovsky, 1993). In rat glomerular mesangial cells, ET-1 stimulates a PKC-regulated $[Ca^{2+}]_i$-insensitive PLD activity that forms phosphatidic acid (PA) (Kester et al., 1992). ET-1 also induces 1,2-DG formation through a $[Ca^{2+}]_i$-sensitive PLC hydrolysis of phosphatidylcholine that is distinct from the PKC-regulated $[Ca^{2+}]_i$-insensitive PLD activity which generates PA (Baldi et al., 1994). Both phosphatidylcholine-derived 1,2-DG formation via PKC-dependent and -independent mechanisms as well as PA- or lysoPA-formation have been implicated in the cellular proliferative response. PA and lysoPA, for example, have been shown to stimulate glomerular mesangial cell proliferation to an extent similar to that of ET (Kester et al., 1989; Simonson et al., 1989). Furthermore, as shown in A10 vascular smooth muscle cells, stimulation of PLD by ET-1 is dependent on both PKC and protein tyrosine kinase (PTK), and the incorporation of $[^3H]$thymidine is dependent on PLD-derived PA (Wilkes et al., 1993). ET-stimulated phosphatidylcholine hydrolysis thus generates diverse species of lipid second messengers that may augment an activated cell phenotype.

B. Protein Kinases

The intracellular signals generated by the ET stimulation of phospholipases can be amplified by diverse protein kinase cascades (Figure 3), which finally lead to altered programs of gene expression by activating, among others, AP-1 and SRF transcription factors (see Section IV).

1. *Protein Kinases A and G*

Several reports exist in the literature which suggest that ETs cannot activate adenylate cyclase (Ohlstein et al., 1989; Sugiura et al., 1989; Takuwa et al., 1989; Vigne et al., 1989; Simonson et al., 1990b). Although ET-1 does not alter basal levels of cAMP in mesangial cells, it amplifies β-adrenergic-stimulated cAMP accumulation by a PGE_2-dependent mechanism (Simonson and Dunn, 1990b). cAMP levels were reported to be similarly unaffected by application of ET to intact preparations of rabbit aorta (Ohlstein et al., 1989) and rat atria (Vigne et al., 1989). In contrast to the inability of native ET receptors to stimulate adenylate cyclase, this enzyme can be activated by ET_A receptors transfected into CHO cells (Aramori and Nakanishi, 1992). Binding of ET to ET_B receptors stably transfected into CHO cells, on the other hand, was linked to downregulation of forskolin-stimulated adenylate cyclase (Aramori and Nakanishi, 1992). Beyond that, inhibition of adenylate cyclase by ET-1 has also been demonstrated in cardiac myocytes (Hilal-Dandan et al., 1992).

The actions of ET on membrane-bound or soluble guanylate cyclase are controversial (Ohlstein et al., 1989; Emori et al., 1990; Reiser, 1990). Some studies, however, support the concept of an important role for nitric oxide (NO)-derived cGMP as a negative feedback signal regulating the hemodynamic effects of ET. The ET-3-induced stimulation of intracellular cGMP production in cultured bovine endothelial cells,

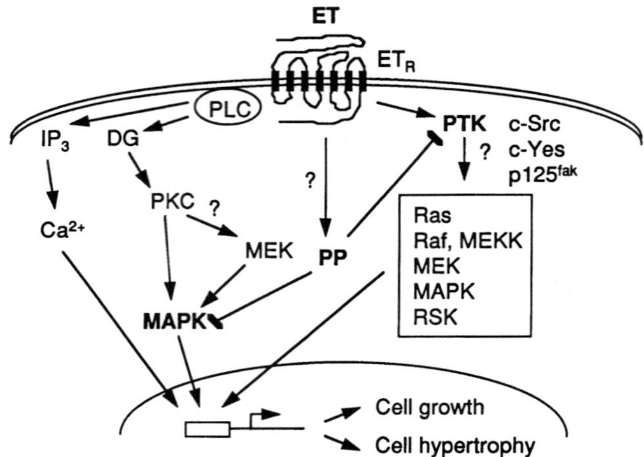

FIGURE 3
Protein kinase cascades activated by endothelins. ETs activate various protein kinases such as protein kinase C (PKC), mitogen-activated protein kinase (MAPK) and nonreceptor protein-tyrosine kinases (PTK). Furthermore, ET-induced protein phosphatases (PP) might play an equally important role in ET-induced protein kinase-mediated cellular responses such as modulation of cell growth and initiation of cell hypertrophy. Ca^{2+}, intracellular Ca^{2+} concentration; DG, 1,2-diacylglycerol; ET_R, endothelin receptor; IP_3, inositol 1,4,5-trisphosphate; MEK, MAPK/ERK kinase; MEKK, MAPK/ERK kinase kinase; PLC, phospholipase C; RSK, ribosomal S6 kinase.

for example, is abolished by N^G-monomethyl-L-arginine and methylene blue, suggesting that in these cells ET-3 stimulates NO synthesis from L-arginine, which then activates soluble guanylate cyclase to increase intracellular cGMP (Emori et al., 1991). Recently, an ET-induced increase of cGMP was also described in intact rat glomeruli (Edwards et al., 1992). ET-1, ET-2, ET-3, and sarafotoxin 6c (STXc) increased glomerular cGMP levels in a concentration-dependent manner and with similar potencies. The stimulation of cGMP accumulation by ET-3 was dependent on extracellular Ca^{2+} and inhibited by both N^G-nitro-L-arginine and methylene blue.

2. *Protein Kinase C*

ETs have been shown to elicit the generation of 1,2-DG in a variety of cell types, e.g., smooth muscle cells (Griendling et al., 1989; Lee et al., 1989; Muldoon et al., 1989; Sunako et al., 1990) and fibroblasts (Muldoon et al., 1989; Takuwa et al., 1989), leading to activation of PKC and finally to PKC-dependent protein phosphorylation (Figure 3). Translocation of PKC from the cytosol to membranes and activation of the enzyme in the membrane compartment has been demonstrated in bovine vascular smooth muscle cells (Lee et al., 1989) and in intact porcine carotid artery strips (Haller et al., 1990), where an excellent correlation between the time course of tension induced and kinase translocated has been obtained. Two substrates of PKC, an acidic 80-kDa protein in Swiss 3T3 cells and myosin light chains in pig carotid arteries, were found to be phosphorylated after exposure to ET-1 (Takuwa et al., 1989; Adam et al., 1990). Furthermore, PKC inhibitors and downregulation of PKC by long-term incubation with phorbol esters inhibit the growth factor action of ET-1 (Muldoon et al., 1990; Pai et al., 1991; Simonson and Herman, 1993) but not of EGF (Simonson and Herman, 1993). Phorbol ester-mediated activation of PKC alone, however, does not mimic the growth response seen with ET-1, suggesting that PKC is necessary but not sufficient to induce growth (Simonson and Herman, 1993). Beyond that, PKC activation contributes to ET-1-induced activation of mitogen-activated protein (MAP)

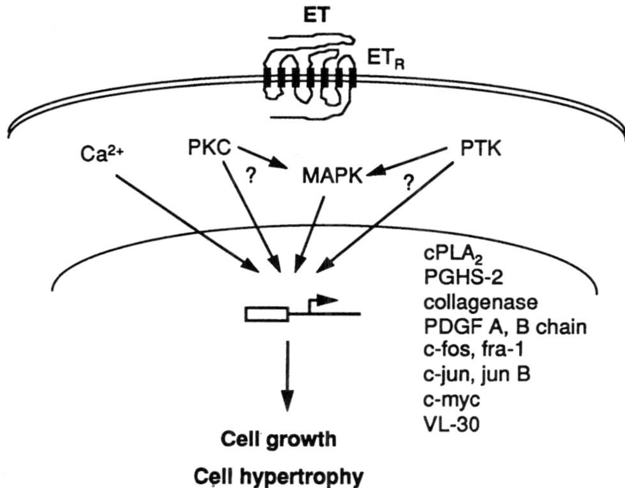

FIGURE 4
Intracellular signals leading to induction of certain genes by endothelins. ET-mediated induction of various genes has been shown to be dependent on intracellular Ca^{2+} concentration (Ca^{2+}), protein kinase C (PKC), nonreceptor protein tyrosine kinases (PTK), and is assumed to depend on the activation of mitogen-activated protein kinase (MAPK). PGHS-2, prostaglandin endoperoxide synthase-2; $cPLA_2$, cytosolic phospholipase A_2; PDGF, platelet-derived growth factor.

kinase (Wang et al., 1992) and induction of *c-fos* (Pribnow et al., 1992; Simonson and Herman, 1993), which is consistent with its role in mitogenic signaling by ET-1 (Figures 3 and 4). Recently, it has been reported that ET-1 activates PKC_ε in cardiac myocytes, which has been implicated in compensatory growth of the myocardium (Bogoyevitch et al., 1993b).

3. *Mitogen-Activated Protein Kinases*

Mitogen-activated protein (MAP) kinases (also known as extracellular signal-related kinases or ERKs) comprise a family of cytosolic effectors which are stimulated by a broad range of growth factors or differentiating agents that bind to receptor protein tyrosine kinases (PTK) or G-protein-coupled receptors (Ruderman, 1993; Johnson and Vaillancourt, 1994). MAP kinases are serine/threonine kinases, activated through phosphorylation on both tyrosine and threonine residues by diverse stimuli, and are thought to integrate growth factor-induced signals from serine/threonine and tyrosine kinases (Blenis, 1993; Ruderman, 1993). Genetic studies reveal that MAP kinase homologues in yeast are required for growth control by G-protein-coupled receptors (Ruderman, 1993). MAP kinase is activated by a dual specifity MAP kinase kinase (MEK, MAPK/ERK kinase) that phosphorylates MAP kinase on Tyr^{183} and Thr^{185} residues. Activation of MEK has been shown to occur by upstream kinases including c-mos, raf-1, and MEKK (MAPK/ERK kinase kinase) (Lange-Carter et al., 1993).

ET-1 has been reported to rapidly activate MAP kinase in glomerular mesangial cells (Wang et al., 1992; Sugimoto et al., 1993), cardiac myocytes (Bogoyevitch et al., 1993a), and aortic smooth muscle cells (Yoshimasa et al., 1992). Atrial natriuretic peptide (ANP) can inhibit the ET-1-induced activation of both p44 and p42 MAP kinases (Sugimoto et al., 1993). In mesangial cells stimulation of MAP kinase activity involves both PKC, as shown by the partial inhibition of ET-1-mediated MAP kinase activation by PKC downregulation, and protein tyrosine kinase (PTK), as indicated by the partial inhibition of ET-1-induced MAP kinase activation by genistein, an inhibitor of protein tyrosine kinases (Wang et al., 1992) (Figure 3). Exposure to 1 µM PMA for 24 h has been shown to downregulate PKC_α, PKC_δ, and PKC_ε, but not PKC_ζ in cardiac myocytes (Bogoyevitch et al., 1994). This maneuver abolished the activation of MAP kinase on reexposure to PMA, partially inhibited the effect of ET-1, but did not affect the response to fibroblast growth factor (FGF) (Bogoyevitch et al., 1994).

Treatment of quiescent mesangial cells with ET-1 revealed a biphasic stimulation of p42 MAP kinase (Wang et al., 1993). The kinetics of the immunoprecipitated p42 MAP kinase activity induced by ET-1 showed a 3.5 to 4.5-fold stimulation 5 min after addition of the agonist, and a smaller 2.5-fold increase of activity between 2 and 6 h (Wang et al., 1993). Neither peak of p42 MAP kinase activity induced by ET-1 was inhibited by pretreatment of the cells with either cycloheximide, a protein synthesis inhibitor, or actinomycin D, a transcriptional inhibitor (Wang et al., 1993). In addition, the kinetics of phosphorylation of p42 MAP kinase showed a significant ^{32}P incorporation into p42 at 5, 30, and 240 min after ET-1 stimulation, suggesting that the phosphorylation of p42 MAP kinase, rather than transcriptional or translational induction, is reponsible for the activation of p42 MAP kinase in mesangial cells (Wang et al., 1993). Experiments in Chinese Hamster Lung fibroblasts transfected with either the rat ET_A or the rat ET_B receptor revealed that MAP kinase activation by ET-1 can occur via both ET_A and ET_B receptor subtypes (Wang et al., 1994a). Furthermore, ET-1 and ET-3 can induce MAP kinase kinase (MAPKK or MEK) gel retardation, p42 MAP kinase activation, and [^{3}H]thymidine uptake in CHO cells stably transfected with either the human ET_A or the human ET_B receptor (Wang et al., 1994b). Therefore, ET isopeptides are able to induce the MAP kinase cascade as well as cell proliferation through either ET_A or ET_B receptors.

4. Nonreceptor Protein Tyrosine Kinases

Although protein tyrosine kinases (PTK) were not generally appreciated to be important components of signaling pathways stimulated by G protein-coupled receptors, there exists clear evidence that the activation of cellular PTKs is required for mitogenic signaling by ET-1. In mesangial cells, fibroblasts, and osteoblasts, ET-1 rapidly increases tyrosine phosphorylation (P-Tyr) of distinct cellular proteins (Force et al., 1991; Zachary et al., 1991a; Schvartz et al., 1992; Simonson and Herman, 1993), and PTK activity is markedly increased in P-Tyr immunoprecipitates from ET-1-treated cells (Zachary et al., 1991b) (Figure 3). Two chemically and mechanistically distinct PTK inhibitors, herbimycin A (blocks cellular PTKs) and genistein (blocks PTKs by competitive binding at their ATP binding site), block cell growth stimulated by ET-1, demonstrating that PTK activation is another signal required in concert with PKC activation to stimulate growth (Simonson and Herman, 1993). In *v-src* transformed Rat-1 fibroblasts, Ins(1,4,5)P_3 and Ca^{2+} signaling is amplified by ET-1 but not when thrombin is used as an agonist, suggesting that the selective interactions within the G protein system that couples the ET receptor to PLC are potential sites at which the *v-src* transformation process may act to amplify ET-dependent Ins(1,4,5)P_3 production (Mattingly et al., 1992). Moreover, PTK inhibitors attenuate ET-1-induced *c-fos* induction, DNA binding by AP-1 transcription factors, and transactivation directed by AP-1 complexes (Simonson and Herman, 1993; see also Section IV.). Although PTK activity is necessary for mitogenic signaling by ET-1, it is not yet known which of the many possible PTK activities are relevant. ET-1 activates at least two types of cytosolic PTKs: (1) nonreceptor PTKs of the *src* gene family such as c-Src and c-Yes (Force and Bonventre, 1992; Simonson and Herman, 1993), and (2) the recently described focal adhesion kinase, p125^{fak} (Zachary et al., 1992). As the concentration-response relationships for c-Src activation and DNA synthesis by ET-1 are similar, it is possible that the two might be functionally linked (Simonson and Herman, 1993).

C. Protein Phosphatases

In general, the net level of protein phosphorylation is determined by the balance between the activities of protein kinases and those of protein phosphatases. It is thus likely that the different families of inducible phosphatases play as important a role in signal transduction pathways, for example, leading to cell growth and/or differentiation as their phosphorylating counterparts, the respective protein kinases. But besides all the recent excitement about the characterization of dual specifity protein kinases and nonreceptor protein tyrosine kinases, little attention has been paid to the fact that the activation of most of these kinases is transient under many conditions. Very recently, however, several groups have identified and characterized phosphatases with dual specificity (protein tyrosine/threonine phosphatases or P[Tyr/Thr]Pases), which specifically and efficiently dephosphorylate and inactivate MAP kinase, or have focused on protein tyrosine phosphatases (P[Tyr]Pases) as well as protein serine/threonine phosphatases (P[Ser/Thr]Pases) (for recent reviews see Mauro and Dixon, 1994; Nebreda, 1994; Zolnierowicz and Hemmings, 1994).

The P[Tyr/Thr]Pase 3CH134, for example, has recently been shown to dephosphorylate tyrosine-phosphorylated MAP kinase 15- to 200-fold more rapidly than other tyrosine phosphorylated substrates, suggesting an important role for 3CH134 in MAP kinase regulation (Sun et al., 1993). Furthermore, angiotensin II, a potent vasoconstrictor and growth factor for vascular smooth muscle cells, has been reported to rapidly induce 3CH134 mRNA levels (Duff et al., 1993). This induction of 3CH134 by angiotensin II is partially inhibited after downregulation of PKC, but fully inhibited after chelating intracellular Ca^{2+} (Duff et al., 1993), suggesting that 3CH134 is an immediate early gene in vascular smooth muscle cells which might play an important role in regulating angiotensin II-stimulated events mediated by MAP kinases and tyrosine kinases. Although no comparable published data exist for ET so far, preliminary experiments in our laboratory suggest a role for 3CH134 in ET-1-induced intracellular signaling in glomerular mesangial cells (Figure 3) (Simonson and Dunn, unpublished data, 1994).

IV. INTRANUCLEAR SIGNALING

There exists increasing evidence that the three isoforms of ETs (ET-1, ET-2, and ET-3) regulate growth and/or differentiation in several types of cells (for review see Battistini et al., 1993; Simonson, 1994). To fullfill its long-term physiological functions in cell growth and/or differentiation, the intracellular signals generated by binding of ET peptides to their receptors in the cell membrane must finally lead to the regulation of target genes in the nucleus. Indeed, ETs have been shown to increase the expression of several genes in different cells. These ET-mediated changes in gene expression can, in principle, be regulated at multiple levels including transcriptional control, RNA processing, mRNA stability, and posttranslational modification of preexisting proteins.

A. Genes Induced by Endothelin

As depicted in Figure 4, experiments utilizing various cell models revealed that peptides of the ET family can elevate the expression of different genes including $cPLA_2$ (Schramek et al., 1994), prostaglandin endoperoxide synthase-2 (PGHS-2) (Kester et al., 1994), platelet-derived growth factor A and B chain (Jaffer et al., 1990), collagenase (Simonson et al., 1992a) as well as transcription factors or nuclear proteins such as *c-fos, fra-1, c-jun, jun B, c-myc,* and VL-30 (Komuro et al., 1988; Dubin et al., 1989; Muldoon et al., 1989; Simonson et al., 1989; Takuwa et al., 1989; Lenormand et al., 1990; Muldoon et al., 1990; Rodland et al., 1990; Mills et al., 1992; Simonson et al., 1992a). The finding that ET-1 increases *c-fos* and *c-jun* expression suggests that AP-1 transcription factors might contribute to nuclear signaling by ET peptides (for review see Simonson et al., 1992b; Simonson, 1994).

Using glomerular mesangial cells as a cell model, we were able to show that ETs differentially regulate expression of distinct *fos* and *jun* family genes whose protein products dimerize to form functional AP-1 complexes (Simonson et al., 1992a). Interestingly, ET-1, but not ET-3, induces *c-fos* and expression of c-Fos protein. These isopeptide-specific differences were also reflected in the ability of ET-1, but not ET-3, to increase AP-1 *cis*-element activity and to stimulate cell growth (Simonson et al., 1992a). Furthermore, different ET receptor subtypes appear to direct distinct patterns of *fos* and *jun* expression (Simonson et al., 1992a). Ca^{2+} signaling and PKC- as well as PTK-activation contribute to *c-fos* induction by ET-1 (Pribnow et al., 1992; Simonson and Herman, 1993). PKC- and PTK-dependent signaling mechanisms are also crucial for ET-1-stimulated transactivation by AP-1 as well as for ET-1-mediated $cPLA_2$ (Schramek et al., 1994) and PGHS-2 mRNA (Kester et al., 1994) induction. The exact signaling mechanisms by which ET receptors activate AP-1 are not known, but genetic analyses suggest that in other systems Src-kinases, Ras, and Raf-1 lie upstream of AP-1 in mitogenic signaling cascades (Smeal et al., 1992; Catling et al., 1993). Moreover, p42 MAP kinase, and MAP kinase-related proteins can associate with c-Jun and AP-1 dimers (Bernstein et al., 1994) and might thus directly phosphorylate c-Jun in the transactivation domain, thereby increasing transactivation by AP-1 (Pulverer et al., 1991).

Beyond that, other transcription factors are regulated by ET-1 and have been proposed to participate in ET-mediated long-term events. For example, ET-1 stimulates transcriptional regulation through the serum response element (SRE), which binds serum response factor and other auxiliary proteins (Simonson, 1994). Preliminary experiments reveal that activation of the SRE by ET receptors in mesangial cells requires PTK activity but is insensitive to treatments that deplete or inhibit PKC (Simonson, 1994).

V. SUMMARY AND PERSPECTIVES

We have summarized the pathways of intracellular signal transduction after binding of ET to its receptors on the cell membrane. Depending on the target cell, its function and its cell-specific expression of signaling molecules, ET can lead to vasodilatation (e.g., ET-mediated activation of $cPLA_2$ and generation of nitric oxide in endothelial cells), vasoconstriction (e.g., ET-induced PLC activation and Ca^{2+} entry in vascular smooth muscle cells), as well as to DNA synthesis, cell division, and hypertrophy (e.g., ET-stimulated activation of protein kinase cascades in glomerular mesangial cells).

The studies of many different laboratories summarized in this chapter already provide some answers on the intracellular mechanisms by which peptides of the ET family accomplish some of their functions. However, they also raise more questions about the precise signaling mechanisms, their integration, and interaction resulting in distinct, tightly regulated biological actions. Given the fact that ET also has potent long-term effects that might be of relevance to disease, it will be of major importance in the future to elucidate the pathways leading to certain nuclear signals by which ETs regulate the genetic program of cells in various tissues.

ACKNOWLEDGMENTS

This work was supported by a fellowship of the Max Kade Foundation Inc. (H.S.) and the NIH HL 22563 and DK 41684 (M.J. D).

REFERENCES

Adam, L. P., Milio, L., Brengle, B. and Hathaway, D. R., Myosin light chain and caldesmon phosphorylation in arterial muscle stimulated with endothelin-1, *J. Mol. Cell. Cardiol.*, 22, 1017-1023. 1990.

Ambar, I. and Sokolovsky, M., Endothelin receptors stimulate both phospholipase C and phospholipase D activities in different cell lines, *Eur. J. Pharmacol.*, 245, 31-41. 1993.

Aramori, I. and Nakanishi, S., Coupling of two endothelin receptor subtypes to different signal transduction in transfected Chinese hamster ovary cells, *J. Biol. Chem.*, 267, 12468-12474. 1992.

Auguet, M., Delaflotte, S., Chabrier, P.-E., Pirotzky, E., Clostre, F. and Braquet, P., Endothelin and Ca^{++} agonist BAY K 8644. Different vasoconstrictive properties, *Biochem. Biophys. Res. Commun.*, 156, 186-92. 1988.

Badr, K. F., Murray, J. J., Breyer, M. D., Takahashi, K., Inagami, T. and Harris, R. C., Mesangial cell, glomerular and renal vascular responses to endothelin in the rat kidney, *J. Clin. Invest.*, 83, 336-342. 1989.

Baldi, E., Musial, A. and Kester, M., Endothelin stimulates phosphatidylcholine hydrolysis through both phospholipase C and D pathways, *Am. J. Physiol.*, 266, F957-F965. 1994.

Barnett, R. L., Ruffini, L., Hart, D., Mancuso, P. and Nord, E. P., Mechanism of endothelin activation of phospholipase A_2 in rat renal medullary interstitial cells, *Am. J. Physiol.*, 266, F46-F56. 1994.

Battistini, B., Chailler, P., D'Orleans-Juste, P., Brière, N. and Sirois, P., Growth regulatory properties of endothelins, *Peptides*, 14, 385-399. 1993.

Bernstein, L. R., Ferris, D. K., Colburn, N. H. and Sobel, M. E., A family of mitogen-activated protein kinase-related proteins interacts *in vivo* with activator protein-1 transcription factor, *J. Biol. Chem.*, 269, 9401-9404. 1994.

Blackburn, K. and Highsmith, R. F., Nickel inhibits endothelin-induced contractions of vascular smooth muscle, *Am. J. Physiol.*, 258, C1025-C1030. 1990.

Blenis, J., Signal transduction via MAP kinases: proceed at your own RSK, *Proc. Natl. Acad. Sci. U.S.A.*, 90, 5889-5892. 1993.

Bogoyevitch, M. A., Glennon, P. E., Andersson, M. B., Clerk, A., Lazou, A., Marshall, C. J., Parker, P. J. and Sugden, P. H., Endothelin-1 and fibroblast growth factor stimulate the mitogen-activated protein kinase signaling cascade in cardiac myocytes, *J. Biol. Chem.*, 269, 1110-1119. 1994.

Bogoyevitch, M. A., Glennon, P. E. and Sugden, P. H., Endothelin-1, phorbol esters and phenylephrine stimulate MAP kinase activities in ventricular cardiomyocytes, *FEBS Lett.*, 317, 271-275. 1993a.

Bogoyevitch, M. A., Parker, P. J. and Sugden, P. H., Characterization of protein kinase C isotype expression in adult rat heart, *Circ. Res.*, 72, 757-767. 1993b.

Bonventre, J. V., Phospholipase A_2 and signal transduction, *J. Am. Soc. Nephrol.*, 3, 128-150. 1992.

Borges, R., Carter, D. V., von Grafenstein, H., Halliday, J. and Knight, D. E., Activation of sodium channels is not essential for endothelin induced vasoconstriction, *Pflügers Arch.*, 413, 313-315. 1989.

Damron, D. S., Van Wagoner, D. R., Moravec, C. S. and Bond, M., Arachidonic acid and endothelin potentiate Ca^{2+} transients in rat cardiac myocytes via inhibition of distinct K^+ channels, *J. Biol. Chem.*, 268, 27335-27344. 1993.

De Nucci, G., Thomas, R., D'Orleans-Juste, P., Antunes, E., Walder, C., Warner, T. D. and Vane, J. R., Pressor effects of circulating endothelin are limited by its removal in the pulmonary circulation and by the release of prostacyclin and endothelium-derived relaxing factor, *Proc. Natl. Acad. Sci. U.S.A.*, 85, 9797-9800. 1988.

D'Orleans-Juste, P., de Nucci, G. and Vane, J. R., Endothelin-1 contracts isolated vessels independently of dihydropyridine-sensitive Ca^{2+} channel activation, *Eur. J. Pharmacol.*, 165, 289-295. 1989.

Dubin, D., Pratt, R. E., Cooke, J. P. and Dzau, V. J., Endothelin, a potent vasoconstrictor, is a vascular smooth muscle mitogen, *J. Vasc. Med. Biol.*, 1, 150-154. 1989.

Edwards, R. M., Pullen, M. and Nambi, P., Activation of endothelin ET_B receptors increases glomerular cGMP via an L-arginine-dependent pathway, *Am. J. Physiol.*, 263, F1020-F1025. 1992.

Edwards, R. M., Trizna, W. T. and Ohlstein, E. H., Renal microvascular effects of endothelin, *Am. J. Physiol.*, 259, F217-F221. 1990.

Eiam-Ong, S., Hilden, S. A., King, A. J., Johns, C. A. and Madias, N. E., Endothelin-1 stimulates the Na^+/H^+ and Na^+/HCO_3^- transporters in rabbit renal cortex, *Kidney Int.*, 42, 18-24. 1992.

Emori, T., Hirata, Y., Kanno, K., Ohta, K., Eguchi, S., Imai, T., Shichiri, M. and Marumo, F., Endothelin-3 stimulates production of endothelium-derived nitric oxide via phosphoinositide breakdown, *Biochem. Biophys. Res. Commun.*, 174, 228-235. 1991.

Filep, J. G., Battistini, B., Côté, Y. P., Beaudoin, A. R. and Sirois, P., Endothelin-1 induces prostacyclin release from bovine aortic endothelial cells, *Biochem. Biophys. Res. Commun.*, 177, 171-176. 1991.

Force, T. and Bonventre, J. V., Endothelin activates Src tyrosine kinase in glomerular mesangial cells, *J. Am. Soc. Nephrol.*, 3, 491A. 1992.

Force, T., Kyriakis, J. M., Avruch, J. and Bonventre, J. V., Endothelin, vasopressin, and angiotensin II enhance tyrosine phosphorylation by protein kinase C-dependent and -independent pathways in glomerular mesangial cells, *J. Biol. Chem.*, 266, 6650-6656. 1991.

Fukunaga, M., Ochi, S., Takama, T., Yokoyama, K., Fujiwara, Y., Orita, Y. and Kamada, T., Endothelin-1 stimulates prostaglandin E_2 production in an extracellular calcium-independent manner in cultured rat mesangial cells, *Am. J. Hypertension*, 4, 137-143. 1991.

Gardner, J. P., Maher, E. and Aviv, A., Calcium mobilization and Na^+/H^+ antiport activation by endothelin in human skin fibroblasts, *FEBS Lett.*, 256, 38-42. 1989.

Garvin, J. and Sanders, K., Endothelin inhibits fluid and bicarbonate transport in part by reducing Na^+/K^+ ATPase activity in the rat proximal straight tubule, *J. Am. Soc. Nephrol.*, 2, 976-982. 1989.

Goto, K., Kasuya, Y., Matsuki, N., Takuwa, Y., Kurihara, T., Ishikawa, T., Kiumura, S., Yanagisawa, M. and Masaki, T., Endothelin activates the dihydropyridine-sensitive, voltage-dependent Ca^{2+} channel in vascular smooth muscle, *Proc. Natl. Acad. Sci. U.S.A.*, 86, 3915-3918. 1989.

Griendling, K. K., Tsuda, T. and Alexander, R. W., Endothelin stimulates diacylglycerol accumulation and activates protein kinase C in cultured vascular smooth muscle cells, *J. Biol. Chem.*, 264, 8237-8240. 1989.

Garay, R., Rosati, C., Braquet, M., Chabrier, P.-E. and Braquet, P., Stimulation of the Na^+, K^+ pump and the (Na^+, K^+, Cl^-) cotransport system by endothelin-1 in cultured vascular smooth muscle cells. Involvement of cyclo-oxygenase products, *J. Cardiovasc. Pharmacol.*, 13 (Suppl. 5), S213-S215. 1989.

Hasunuma, K., Rodman, D. M., O'Brien, R. F. and McMurtry, I. F., Endothelin-1 causes pulmonary vasodilation in rats, *Am. J. Physiol.*, 259, H48-H54. 1990.

Highsmith, R. F., Blackburn, K. and Schmidt, D. J., Endothelin and calcium dynamics in vascular smooth muscle, *Annu. Rev. Physiol.*, 54, 257-277. 1992.

Highsmith, R. F., Pang, D. C. and Rappoport, R. M., Endothelial cell-derived vasoconstrictors. Mechanisms of action in vascular smooth muscle, *J. Cardiovasc. Pharmacol.*, 13 (Suppl. 5), S36-44. 1989.

Hilal-Dandan, R., Urasawa, K. and Brunton, L. L., Endothelin inhibits adenylate cyclase and stimulates phosphoinositide hydrolysis in adult cardiac monocytes, *J. Biol. Chem.*, 267, 10620-10624. 1992.

Hirata, Y., Takagi, Y., Fukuda, Y. and Marumo, F., ET is a potent mitogen for rat vascular smooth muscle cells, *Atherosclerosis*, 78, 225-228. 1989.

Honore, E., Attali, B., Lesage, F., Barhanin, J. and Lazdunski, M., Receptor-mediated regulation of IsK, a very slowly activating, voltage-dependent K^+ channel in *Xenopus* oocytes, *Biochem. Biophys. Res. Commun.*, 184, 1135-1141. 1992

Hu, S., Kim, H. S. and Jeng, A. Y., Dual action of endothelin-1 on the Ca^{2+}-activated K^+ channel in smooth muscle cells of porcine coronary artery, *Eur. J. Pharmacol.*, 194, 31-36. 1991.

Huang, S., Simonson, M. S. and Dunn, M. J., Manidipine inhibits endothelin-1-induced $[Ca^{2+}]_i$ signaling but potentiates endothelin's effect on *c-fos* and *c-jun* induction in vascular smooth muscle and glomerular mesangial cells, *Am. Heart J.*, 125, 589-597. 1993.

Iijima, K., Lin, L., Nasjletti, A. and Goligorsky, M. S., Intracellular ramification of endothelin signal, *Am. J. Physiol.*, 260, C982-C992. 1991.

Inoue, Y., Oike, M., Kitamura, K. and Kuriyama, H., Endothelin augments unitary calcium channel currents on the smooth muscle cell membrane of guinea-pig portal vein, *J. Physiol. (London)*, 423, 171-191. 1990.

Ishii, K., Nunoki, K., Murakoshi, H. and Taira, N., Cloning and modulation by endothelin-1 of rat cardiac K channel, *Biochem. Biophys. Res. Commun.*, 184, 1484-1489. 1992.

Jaffer, F. E., Knauss, T. C., Poptic, E. and Abboud, H. E., Endothelin stimulates PDGF secretion in cultured human mesangial cells, *Kidney Int.*, 38, 1193-1198. 1990.

Johnson, G. L. and Vaillancourt, R. R., Sequential protein kinase reactions controlling cell growth and differentiation, *Curr. Opin. Cell Biol.*, 6, 230-238. 1994.

Jouneaux, C., Mallat, A., Serradeil-Le Gal, C., Goldsmith, P., Hanoune, J. and Lotersztajn, S., Coupling of endothelin B receptors to the calcium pump and phospholipase C via Gs and G_q in rat liver, *J. Biol. Chem.*, 269, 1845-1851. 1994.

Kester, M., Coroneos, E. J., Thomas, P. J. and Dunn, M. J., Endothelin stimulates prostaglandin endoperoxide synthase-2 mRNA expression and protein synthesis through a tyrosine kinase signaling pathway in rat mesangial cells. *J Biol Chem.*, 1994, 269, 22574-22580, 1994.

Kester, M., Simonson, M. S., McDermott, R. G., Baldi, E. and Dunn, M. J., Endothelin stimulates phosphatidic acid formation in cultured rat mesangial cells. Role of a protein kinase C-dependent phospholipase D, *J. Cell. Physiol.*, 150, 578-585. 1992.

Kester, M., Simonson, M. S., Mene, P. and Sedor, J. R., Interleukin-1 generates transmembrane signals from phospholipids through novel pathways in cultured rat mesangial cells, *J. Clin. Invest.*, 83, 718-723. 1989.

Komuro, I., Kurihara, H., Sugiyama, T., Takaku, F. and Yazaki, Y., Endothelin stimulates *c-fos* and *c-myc* expression and proliferation of vascular smooth muscle cells, *FEBS Lett.*, 238, 249-252. 1988.

Kramer, B. K., Smith, T. W. and Kelly, R. A., Endothelin and increased contractility in adult rat ventricular myocytes. Role of intracellular alkalosis induced by activation of the protein kinase C-dependent Na^+-H^+ exchanger, *Circ. Res.*, 68, 269-279. 1991.

Lange-Carter, C. A., Pleiman, C. M., Gardner, A. M., Blumer, K. J. and Johnson, G. L., A divergence in the MAP kinase regulatory network defined by MEK kinase and Raf, *Science*, 260, 315-319. 1993.

Lee, T.-E., Chao, T., Hu, K.-Q. and King, G. L., Endothelin stimulates a sustained 1,2 diacylglycerol increase and protein kinase C activation in bovine aortic smooth muscle cells, *Biochem. Biophys. Res. Commun.*, 162, 381-386. 1989.

Lippton, H. L., Cohen, G. A., McMurtry, I. F. and Hyman, A. L., Pulmonary vasodilation to endothelin isopeptides *in vivo* is mediated by potassium channel activation, *J. Appl. Physiol.*, 70, 947-952. 1991.

MacNulty, E. E., Plevin, R. and Wakelam, M. J. O., Stimulation of the hydrolysis of phosphatidylinositol 4,5-bisphosphate and phosphatidylcholine by endothelin, a complete mitogen for Rat-1 fibroblasts, *Biochem. J.*, 272, 761-766. 1990.

Marsault, R., Vigne, P., Breittmayer, J.-P. and Frelin, C., Astrocytes are target cells for endothelins and sarafotoxin, *J. Neurochem.*, 54, 2142-2144. 1990.

Masaki, T., Yanagisawa, M., Takuwa, Y., Kasuya, Y., Kimura, S. and Goto, K., Cellular mechanism of vasoconstriction induced by endothelin, *Adv. Second Messenger Phosphoprotein Res.*, 24, 425-428. 1990.

Mauro, L. J. and Dixon, J. E., "Zip codes" direct intracellular protein tyrosine phosphatases to correct cellular "address", *Trends Biochem. Sci.*, 19, 151-155. 1994.

Mayer, R. J. and Marshall, L. A., New insights on mammalian phospholipase A_2(s); comparison of arachidonoyl-selective and -nonselective enzymes, *FASEB J.*, 7, 339-348. 1993.

Mené, P., Simonson, M. S. and Dunn, M. J., Phospholipids in signal transduction of mesangial cells, *Am. J. Physiol.*, 256, F375-F386. 1989.

Meyer-Lehnert, H., Wanning, C., Predel, H.-G., Backer, A., Stelkens, H. and Kramer, H. J., Effects of endothelin on sodium transport mechanisms. Potential role in cellular Ca^{2+} mobilization, *Biochem. Biophys. Res. Commun.*, 163, 458-465. 1989.

Mitsuhashi, T., Morris, R. C. and Ives, H. E., Endothelin-induced increases in vascular smooth muscle Ca^{2+} do not depend on dihydropyridine-sensitive Ca^{2+} channels, *J. Clin. Invest.*, 84, 635-639. 1989.

Muldoon, L., Rodland, K. D., Forsythe, M. L. and Magun, B. E., Stimulation of phosphatidylinositol hydrolysis, diacylglycerol release, and gene expression in response to endothelin, a potent new agonist for fibroblasts and smooth muscle cells, *J. Biol. Chem.*, 264, 8529-8536. 1989.

Muldoon, L. L., Pribnow, D., Rodland, K. D. and Magun, B. E., Endothelin-1 stimulates DNA synthesis and anchorage-independent growth of rat-1 fibroblasts through a protein kinase C-dependent mechanism, *Cell Regul.*, 1, 379-390. 1990.

Nakaki, T., Nakayama, M., Yamamoto, S. and Kato, R., ET-mediated stimulation of DNA synthesis in VSMC, *Biochem. Biophys. Res. Commun.*, 158, 880-883. 1989.

Nebreda, A. R., Inactivation of MAP kinases, *Trends Biochem.* Sci., 19, 1-2. 1994.

Ohlstein, E. H., Horohonich, S. and Hay, D. W., Cellular mechanisms of endothelin in rabbit aorta, *J. Pharmacol. Exp. Ther.*, 250, 548-555. 1989.

Pai, J.-K., Dobek, E. A. and Bishop, W. R., Endothelin-1 activates phospholipase D and thymidine incorporation in fibroblasts overexpressing protein kinase C beta 1, *Cell Regul.*, 2, 897-903. 1991.

Pribnow, D., Muldoon, L. L., Fajardo, M., Theodor, L., Chen, L. S. and Magun, B. E., Endothelin induces transcription of fos/jun family genes. A prominent role for calcium ion, *Mol. Endocrinol.*, 6, 1003-1012. 1992.

Pulverer, B. J., Kyriakis, J. M., Avruch, J., Nikolakaki, E. and Woodgett, J. R., Phosphorylation of *c-jun* mediated by MAP kinases, *Nature*, 353, 670-674. 1991.

Reiser, G., Endothelin and a Ca^{2+} ionophore raise cyclic GMP levels in a neuronal cell line via formation of nitric oxide, *Br. J. Pharmacol.*, 101, 722-726. 1990.

Resink, T. J., Scott-Burden, T. and Bühler, F. R., Activation of phospholipase A_2 by endothelin in cultured vascular smooth muscle cells, *Biochem. Biophys. Res. Commun.*, 158, 279-286. 1989.

Resink, T. J., Scott-Burden, T. and Bühler, F. R., Activation of multiple signal transduction pathways by endothelin in cultured human vascular smooth muscle cells, *Eur. J. Biochem.*, 189, 415-421. 1990.

Reynolds, E. E., Mok, L. L. S. and Kurokawa, S., Phorbol ester dissociates endothelin-stimulated phosphoinositide hydrolysis and arachidonic acid release in vascular smooth muscle cells, *Biochem. Biophys. Res. Commun.*, 160, 868-873. 1989.

Rodland, K. D., Muldoon, L. L., Lenormand, P. and Magun, B. E., Modulation of RNA expression by intracellular calcium, *J. Biol. Chem.*, 265, 11000-11007. 1990.

Rosati, C., Jeanclos, E., Cavalier, S., Chabrier, P.-E., Hannaert, P., Braquet, P. and Garay, R., Stimulatory action of endothelin-1 on membrane Na^+ transport in vascular smooth muscle cells in culture, *Am. J. Hypertens.*, 3, 711-713. 1990.

Ruderman, J. V., MAP kinase and the activation of quiescent cells, *Curr. Opin. Cell Biol.*, 5, 207-213. 1993.

Sakamoto, H., Sasaki, S., Nakamura, Y., Fushimi, K. and Marumo, F., Regulation of endothelin-1 production in cultured rat mesangial cells, *Kidney Int.*, 41, 350-355. 1992.

Schramek, H. and Dunn, M. J., Unpublished data, 1994.

Schramek, H., Marsen, T. A. and Dunn, M. J., Endothelins: vasoactive peptides interacting with prostaglandin- and nitric oxide-induced signaling pathways, in *Thromboxane A_2 and Other Vasoconstrictors in Clinical Conditions*, Neri Serneri, G. G., Bonomini, V., Gensini, G. F., Pozzi, E. and Prisco, D., Eds., Scientific Press, Florence, 1993, 227-249.

Schramek, H., Wang, J., Konieczkowski, M., Rose, P. M., Sedor, J. R. and Dunn, M. J., Endothelin-1 stimulates cytosolic phospholipase A_2 in Chinese Hamster Ovary cells stably expressing the human ET_A or ET_B receptor subtype, *Biochem. Biophys. Res. Commun.*, 199, 992-997. 1994a.

Schramek, H., Wang, J., Konieczkowski, M., Simonson, M. S. and Dunn, M. J., Endothelin-1 stimulates cytosolic phospholipase A_2 activity and gene expression in rat glomerular mesangial cells, *Kidney Int.*, 46, 1644-1652. 1994b.

Schvartz, I., Ittoop, O., Davidai, G. and Hazum, E., Endothelin rapidly stimulates tyrosine phosphorylation in osteoblast-like cells, *Peptides*, 13, 159-163. 1992.

Serradeil-Le Gal, C., Jouneaux, C., Sanchez-Bueno, A., Raufaste, D., Roche, B., Preaux, A. M., Maffrand, J. P., Cobbald, P. H., Hanoune, J. and Lotersztajn, S., Endothelin action in rat liver. Receptors, Ca^{2+} oscillations, and activation in glycogenolysis, *J. Clin. Invest.*, 87, 133-138. 1991.

Silberberg, S. D., Poder, T. C. and Lacerda, A. E., Endothelin increases single-channel calcium currents in coronary arterial smooth muscle cells, *FEBS Lett.*, 247, 68-72. 1989.

Simonson, M. S., Endothelins: multifunctional renal peptides, *Physiol. Rev.*, 73, 375-411. 1993.

Simonson, M. S., Endothelin peptides and compensatory growth of renal cells, *Curr. Opin. Nephrol. Hypertension*, 3, 73-85. 1994.

Simonson, M. S. and Herman, W. H., Protein kinase C and protein tyrosine kinase activity contribute to mitogenic signaling by endothelin-1. Cross-talk between G protein-coupled receptors and $pp60^{c-src}$, *J. Biol. Chem.*, 268, 9347-9357. 1993.

Simonson, M. S. and Dunn, M. J., Cellular signaling by peptides of the endothelin gene family, *FASEB J.*, 4, 2989-3000. 1990a.

Simonson, M. S. and Dunn, M. J., Endothelin-1 stimulates contraction of rat glomerular mesangial cells and potentiates β-adrenergic-mediated cyclic adenosine monophosphate accumulation, *J. Clin. Invest.*, 85, 790-797. 1990b.

Simonson, M. S. and Dunn, M. J., Ca^{2+} signaling by distinct endothelin peptides in glomerular mesangial cells, *Exp. Cell Res.*, 192, 148-156. 1991.

Simonson, M. S. and Dunn, M. J., Unpublished data, 1994.

Simonson, M. S., Jones, J. M. and Dunn, M. J., Differential regulation of *fos* and *jun* gene expression and AP-1-*cis*-element activity by endothelin isopeptides, *J. Biol. Chem.*, 267, 8643-8649. 1992a.

Simonson, M. S., Osanai, T. and Dunn, M. J., Endothelin isopeptides evoke Ca^{2+} signaling and oscillations of cytosolic free $[Ca^{2+}]$ in human mesangial cells, *Biochim. Biophys. Acta*, 1055, 63-68. 1990.

Simonson, M. S., Wang, Y. and Dunn, M. J., Cellular signaling by endothelin peptides. Pathways to the nucleus, *J. Am. Soc. Nephrol.*, 2, S116-S125. 1992b.

Simonson, M. S., Wann, S., Mené, P., Dubyak, G., Kester, M., Nakazato, Y., Sedor, J. R. and Dunn, M. J., Endothelin stimulates phospholipase C, Na^+/H^+ exchange, *c-fos* expression, and mitogenesis in rat mesangial cells, *J. Clin. Invest.*, 83, 708-712. 1989.

Smith, W., Prostanoid biosynthesis and mechanisms of action, *Am. J. Physiol.*, 263, F181-F191. 1992.

Stojilkovic, S. S. and Catt, K. J., Neuroendocrine actions of endothelins, *Trends Pharmacol. Sci.*, 13, 385-391. 1992.

Sugimoto, T., Kikkawa, R., Haneda, M. and Shigeta, Y., Atrial natriuretic peptide inhibits endothelin-1-induced activation of mitogen-activated protein kinase in cultured rat mesangial cells, *Biochem. Biophys. Res. Commun.*, 195, 72-78. 1993.

Sugiura, M., Inagami, T., Harem, G. M. and Johns, J. A., Endothelin action: inhibition by protein kinase C inhibitor and involvement of phosphoinositols, *Biochem. Biophys. Res. Commun.*, 158, 170-176. 1989.

Sun, H., Charles, C. H., Lau, L. F. and Tonks, N. K., MKP-1 (3CH134), an immediate early gene product, is a dual specificity phosphatase that dephosphorylates MAP kinase *in vivo*, *Cell*, 75, 487-493. 1993.

Sunako, M., Kawahara, Y., Hirata, K., Tsuda, T., Yokoyama, M., Fukuzaki, H. and Takai, Y., Mass analysis of 1,2-diacylglycerol in cultured rabbit vascular smooth muscle cells-comparison of stimulation by angiotensin-II and endothelin, *Hypertension*, 15, 84-88. 1990.

Sung, C.-P., Arleth, A. J. and Ohlstein, E. H., Carvedilol inhibits vascular smooth muscle cell proliferation, *J. Cardiovasc. Pharmacol.*, 21, 221-227. 1993.

Takenaka, T., Epstein, M., Forster, H., Landry, D. W., Iijima, K. and Goligorsky, M. S., Attenuation of endothelin effects by a chloride channel inhibitor, indanyloxyacetic acid, *Am. J. Physiol.*, 262, F799-F806. 1992.

Takuwa, N., Takuwa, Y., Yanagisawa, M., Yamashita, K. and Masaki, T., A novel vasoactive peptide endothelin stimulates mitogenesis through inositol lipid turnover in Swiss 3T3 fibroblasts, *J. Biol. Chem.*, 264, 7856-7861. 1989.

Takuwa, Y., Kasuya, Y., Takuwa, N., Kudo, M., Yanagisawa, M. and Masaki, T., A novel vasoactive peptide endothelin stimulates mitogenesis through inositol lipid turnover in Swiss 3T3 fibroblasts, *J. Biol. Chem.*, 264, 7856-7861. 1989.

Takuwa, Y., Kasuya, Y., Takuwa, N., Kudo, M., Yanagisawa, M., Goto, K., Masaki, T. and Yamashita, K., Endothelin receptor is coupled to phospholipase C via a pertussis toxin-insensitive guanine nucleotide-binding regulatory protein in vascular smooth muscle cells, *J. Clin. Invest.*, 85, 653-658. 1990.

Thomas, C. P., Kester, M. and Dunn, M. J., A pertussis toxin sensitive GTP-binding protein couples endothelin to phospholipase C in rat mesangial cells, *Am. J. Physiol.*, 260, F347-F352. 1991.

VanRenterghem, C., Vigne, P., Barhanin, J., Schmid-Alliana, A., Frelin, C. and Lazdunski, M., Molecular mechanism of action of the vasoconstrictor peptide endothelin, *Biochem. Biophys. Res. Commun.*, 157, 977-985. 1988.

Vigne, P., Ladoux, A. and Frelin, C., Endothelins activate Na^+/H^+ exchange in brain capillary endothelial cells via a high affinity endothelin-3 receptor that is not coupled to phospholipase C, *J. Biol. Chem.*, 266, 5925-5928. 1991.

Vigne, P., Lazdunski, M. and Frelin, C., The inotropic effect of endothelin-1 on rat atria involves hydrolysis of phosphatidylinositol, *FEBS Lett.*, 249, 143-146. 1989.

Wang, Y., Pouysségur, J. and Dunn, M. J., Endothelin stimulates mitogen-activated protein kinase p42 activity through the phosphorylation of the kinase in rat mesangial cells, *J. Cardiovasc. Pharmacol.*, 22, S164-S167. 1993.

Wang, Y., Pouysségur, J. and Dunn, M. J., Endothelin stimulates mitogen-activated protein kinase activity in mesangial cells through ET_A, *J. Am. Soc. Nephrol.*, 5, 1074-1080. 1994a.

Wang, Y., Rose, P. M., Webb, M. L. and Dunn, M. J., Endothelins stimulate the mitogen-activated protein kinase cascade and Chinese Hamster Ovary cell proliferation through either ET_A or ET_B, *Am. J. Physiol.*, 267, C1130-C1135. 1994b.

Wang, Y., Simonson, M. S., Pouysségur, J. and Dunn, M., Endothelin rapidly stimulates mitogen-activated protein kinase activity in rat mesangial cells, *Biochem. J.*, 287, 589-594. 1992.

Wilkes, L. C., Patel, V., Purkiss, J. R. and Boarder, M. R., Endothelin-1 stimulated phospholipase D in A10 vascular smooth muscle derived cells is dependent on tyrosine kinase, *FEBS Lett.*, 322, 147-150. 1993.

Yanagisawa, M., Kurihara, H., Kimura, S., Tomobe, Y., Kobayashi, M., Mitsui, Y., Yazaki, Y., Goto, K. and Masaki, T., A novel potent vasoconstrictor peptide produced by vascular endothelial cells, *Nature*, 332, 411-415. 1988.

Yoshimasa, T., Nakao, K., Suga, S.-I., Kishimoto, I., Kiso, Y. and Imura, H., Identification and endothelin-induced activation of multiple extracellular signal-regulated kinases in aortic smooth muscle cells, *FEBS Lett.*, 311, 195-198. 1992.

Zachary, I., Gil, J., Lehmann, W., Sinnett-Smith, J. and Rozengurt, E., Bombesin, vasopressin, and endothelin rapidly stimulate tyrosine phosphorylation in intact Swiss 3T3 cells, *Proc. Natl. Acad. Sci. U.S.A.*, 88, 4577-4581. 1991a.

Zachary, I., Sinnett-Smith, J. and Rozengurt, E., Stimulation of tyrosine kinase activity in anti-phosphotyrosine immune complexes of Swiss 3T3 cell lysates occurs rapidly after addition of bombesin, vasopressin, and endothelin to intact cells. *J. Biol. Chem.*, 266, 24126-24133. 1991b.

Zachary, I., Sinnett-Smith, J. and Rozengurt, E., Bombesin, vasopressin, and endothelin stimulation of tyrosine phosphorylation in Swiss 3T3 cells. Identification of a novel tyrosine kinase as a major substrate, *J. Biol. Chem.*, 267, 19031-19034. 1992.

Zeidel, M. L., Brady, H. R., Kone, B. C., Gullans, S. R. and Brenner, B. M., Endothelin, a peptide inhibitor of Na^+-K^+-ATPase in intact renal tubular epithelial cells, *Am. J. Physiol.*, 257, C1101-C1107. 1989.

Zoja, C., Benigni, A., Renzi, D., Piccinelli, A., Perico, N. and Remuzzi, G., Endothelin and eicosanoid synthesis in cultured mesangial cells, *Kidney Int.*, 37, 927-933. 1990.

Zolnierowicz, S. and Hemmings, B. A., Tethering, targeting and triggering of protein phosphatases, *Trends Cell Biol.*, 4, 61-64. 1994.

Part II

Endothelin Synthesis and Degradation

Chapter **6**

ENDOTHELIN CONVERTING ENZYMES AND RELATED INHIBITORS

Marc Bigaud and Hugues D'Orchymont

CONTENTS

0-8493-6975-4/97/$0.00+$.50

I. INTRODUCTION

In their historical publication concerning the discovery of endothelin, Yanagisawa et al. (1988) predicted that the endothelin gene product would be processed in a manner similar to that of many peptide hormones and neuropeptides (for review see Rang and Dall, 1991). Based on cDNA sequencing of a cloned porcine endothelin gene, they proposed that the posttranslational synthetic pathway of endothelin would result from the intracellular proteolytic processing of a 203-amino-acids precursor peptide, preproendothelin. Similarly to precursors of other peptide hormones, it presented a hydrophobic N-terminal signal, important for inserting proteins into the endoplasmic reticulum, and pairs of basic amino acids demarcating 39 amino acids containing the endothelin sequence. It was proposed that the peptide resulting from the removal of the signal sequence would be called proendothelin and that the 39-amino-acid fragment resulting from the processing by conventional dibasic endopeptidase be called big endothelin. At that stage, an uncommon cleavage by a putative specific endothelin-converting enzyme was suspected to generate the mature endothelin.

It rapidly became evident that specific inhibitors of endothelin synthesis might provide tools to elucidate the biological roles of endothelins and eventually lead to new therapeutic approaches to the treatment of some diseases possibly related to increased synthesis of endothelins, such as hypertension, asthma, congestive heart failure, renal failure, and cerebral vasospasms (for reviews see Doherty, 1992; Opgenorth et al., 1992). Therefore, immense efforts in basic research have been devoted to assess the hypothesis of Yanagisawa et al. (1988) concerning the synthetic pathway of endothelin peptides.

II. SYNTHETIC PATHWAY OF ENDOTHELIN PEPTIDES

A. Structure and Processing of Preproendothelins

The cloning and sequencing of a human endothelin-related gene revealed that the deduced amino acid sequences of human and porcine preproendothelins were very much conserved (Itoh et al., 1988) and it was then suggested that the porcine and human preproendothelins would probably be matured in a similar manner. Three different human endothelin gene products have so far been identified: the vasoactive peptide described initially by Yanagisawa et al. (1988), now being referred to as endothelin-1 (ET-1), and two other closely related peptides, designated endothelin-2 (ET-2) and endothelin-3 (ET-3) (Inoue et al., 1989). Although these three peptides present high degrees of similarities, marked differences were observed between the predicted amino acid sequences of human preproET-1, preproET-2 and preproET-3 (Inoue et al., 1989; Bloch et al., 1989; Ohkubo et al., 1990). The predicted amino acid sequences of these three prepropeptides are presented in Figure 1.

Preproendothelin-1 :

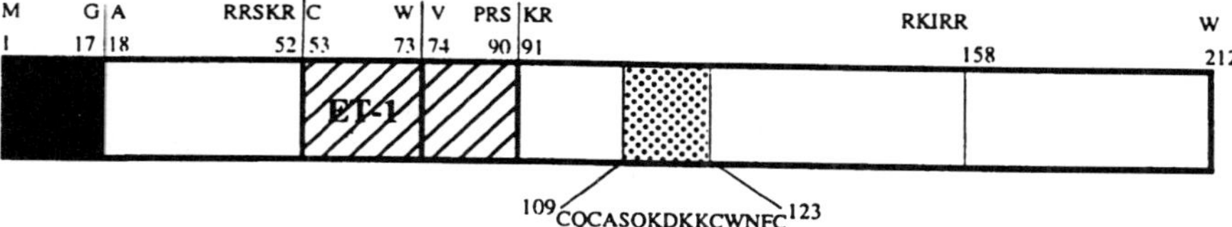

Preproendothelin-2 :

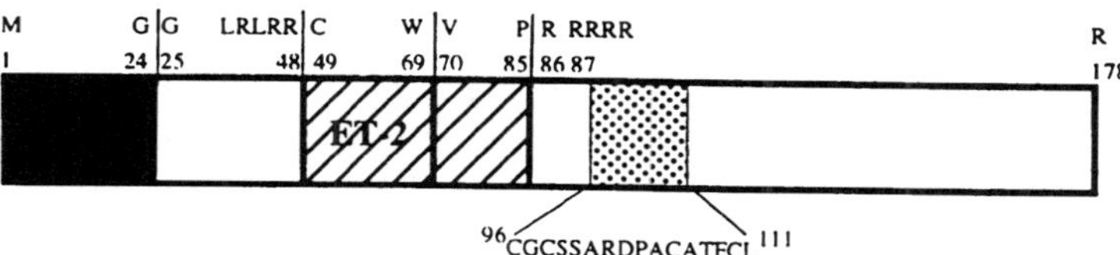

Preproendothelin-3 :

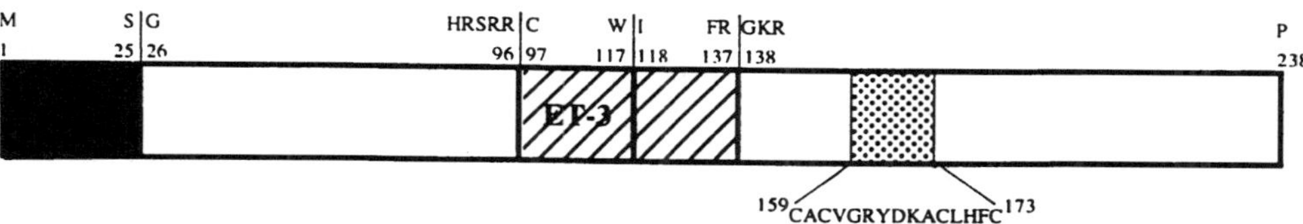

FIGURE 1
Schematic representation of the human family of preproendothelins according to the amino acid sequences from deduced human preproendothelin cDNA sequencing (Itoh et al., 1988; Ohkubo et al., 1990; Bloch et al., 1989). Signal fragments are represented as solid black boxes, big endothelin fragments as hatched regions and endothelin-like fragments as dotted regions. For each peptide, the total number of amino acids is given and putative cleavage sites for proprotein convertases and ECEs are indicated.

The first 17, 24, and 25 amino acids, respectively, of preproET-1, preproET-2, and preproET-3 contain sequences characteristic of a secretory signal, i.e., a hydrophobic core followed by a sequence carrying small polar side chains (Walter and Blobel, 1981). Fabbrini et al. (1991a) demonstrated that the peptide generated by a synthetic RNA coding for human preproET-1 lacking the first 17 amino acids could no longer be inserted into membranes of the endoplasmic reticulum and translocate into microsomes. They also observed that the first 17 amino acids of preproET-1 were removed upon insertion into microsomes. Taken together, these results suggested that the first 17 amino acids of human preproET-1 indeed constitute a functional signal sequence and, as for many peptide hormones (Walter and Blobel, 1981), the first step in the *in vivo* synthesis of endothelins consists of insertion of the nascent protein chains into the endoplasmic reticulum and the concomitant cleavage of the signal sequence. Proendothelin peptides may then enter the secretory pathway.

B. Structure and Processing of Proendothelins

Upon examination of the amino acid sequence of porcine proET-1, Yanagisawa et al. (1988) identified five paired basic amino acids, Lys-Arg or Arg-Arg, as putative recognition sites for cleavage by conventional dibasic proprotein endopeptidase. It might then be expected that proET-1 would be processed into 6 possible fragments, but only 2 fragments merited special attention: those defining big ET-1, a 39-amino-acid fragment flanked by Lys51-Arg52 and Arg92-Arg93 and containing the 21-amino-acid sequence of ET-1, and another fragment inserted between Arg106-Arg107 and Lys139-Arg140 containing a second cysteine-rich "endothelin-like structure" of 15 amino acids. The processing of porcine proET-1 by dibasic endopeptidases was supported by the detection, in the supernatant of cultured porcine vascular endothelial cells, of peptides presenting an identical sequence to the N-terminal fragment of big ET-1 and containing the 15-amino-acid cysteine-rich ET-1-like sequence (Nakagawa et al., 1991).

Human proET-1, proET-2, and proET-3 presented, respectively, 3, 4, and 3 possible cleavage sites for dibasic endopeptidases (Itoh et al., 1988; Ohkubo et al., 1990; Bloch et al., 1989), the putative fragments big ET-1, big ET-2, and big ET-3 containing, respectively, 38, 37, and 41 amino acids and a second "endothelin like" sequence of 15 amino acids (Figure 1). It is noteworthy that recent studies suggest that the true big ET-2 is not composed of 37 residues but of 38 residues with an additional arginine in the carboxy-terminus (Kosaka et al., 1992; Yorimitsu et al., 1992).

Insights into the *in vivo* processing of human proET-1 were gained by using the *Xenopus* oocyte system that has proved to faithfully reproduce several steps involved in the synthesis of mammalian proteins (Foreman and Judah, 1987; Faust et al., 1987). Fabbrini et al. (1991b) demonstrated that upon microinjection of human preproET-1 synthetic RNA, *Xenopus* oocytes could generate peptides presenting immunological and biological characteristics of big ET-1 and ET-1, and that this process was sensitive to monensin, a drug blocking secretion proteins within the Golgi complex (Ledger and Tanzer, 1984; Anderson and Orci, 1988). Similar results were obtained using insect cells infected with a baculovirus vector carrying a human preproET-1 cDNA (Benatti et al., 1992). Such observations suggested that heterologous cells that normally do not secrete ET-1 possess, in their constitutive Golgi secretory pathway, the enzymatic machinery required for the posttranslational maturation of human preproET-1 and proET-1.

Two different types of dibasic endoproteases with prohormone processing activities have been reported to be Golgi-resident proteins. They are distinct gene products, KEX2 from the *kex2* gene (Fuller et al., 1989) and furin from the *fur* gene (Bresnahan et al., 1990), but belong to the same membrane-bound, calcium-dependent, serine endoprotease family, also called subtilisin-like proprotein convertases (for review see Barr, 1991; Steiner et al., 1992). The substrate requirement for efficient activity of such enzymes is known to be a basic amino acid four residues ahead of a dibasic doublet Lys-Arg or Arg-Arg (Bresnahan et al., 1990). All three putative dibasic sites observed in human proET-1 (Itoh et al., 1988) fit that requirement (Figure 1) and it was demonstrated, using a *Xenopus* oocyte system, that the processing of proET-1 was inhibited by a mutation at the dibasic site demarcating the C-terminus of big ET-1 (Fabbrini et al., 1993). This indicated that processing at this site is an obligatory event for the synthesis of ET-1 and that Golgi-located subtilisin-like proprotein convertases are most likely involved in the maturation of proET-1. The presence of furin mRNA was recently reported in endothelial cells (Laporte et al., 1993), thus supporting a possible role of this enzyme in the maturation of proET-1. Subtilisin-like enzymes have been shown to play a role in the constitutive processing of many precursor proteins including pro-albumin, pro-von Willebrand factor, pro-insulin B and C chains, and pro-renin (Barr, 1991). It remains to be demonstrated whether or not they are involved in the processing of proET-2 and proET-3. It is noteworthy that only the dibasic sites demarcating big ET-2 and big ET-3 fit with their substrate requirement (Figure 1). The involvement of other, still unidentified enzymatic activities, however, cannot be excluded.

TABLE 1

Sequences of the Precursor Forms of the Three Human Endothelins ET-1, ET-2, and ET-3

	N terminal 1 3 11 15 21 22 C terminal 38
Big ET-1	C S C S S L M D K E C V Y F C H L D I I W V N T P E H V V P Y G L G S P R S
Big ET-2	C S C S S <u>W</u> <u>L</u> D K E C V Y F C H L D I I W V N T P E <u>Q</u> <u>T</u> <u>A</u> P Y G L G <u>N</u> P <u>P(R)</u>
Big ET 3	C <u>T</u> C <u>F</u> <u>T</u> <u>Y</u> <u>K</u> D K E C V Y <u>Y</u> C H L D I I W <u>I</u> N T P E <u>Q</u> <u>T</u> V P Y G L <u>S</u> <u>N</u> <u>Y</u> <u>R</u> <u>G</u> S <u>F</u> <u>R</u>

Note: Disulfide bonds link cysteine residues in positions 1-15 and 3-11. The cleavage site for ECE is between the residues 21-22.The differences compared to big ET-1 are underlined.

Nothing is known about the biological significance of the non-big endothelin sequences of proendothelin peptides. It has been clearly demonstrated, using *Xenopus* oocytes, that the N-terminal sequence of proET-1 is not necessary for the synthesis and secretion of big ET-1 and ET-1 (Fabbrini et al., 1993). However, unknown endocrine functions for these fragments, especially for the endothelin-like domains, cannot be ruled out.

C. Maturation of Big Endothelins

Yanagisawa et al. (1988) suggested that the formation of porcine ET-1 from big ET-1 would require proteolytic cleavage between Trp73 and Val74 and, because of the absence of paired basic amino acids around that putative cleavage site, proposed the existence of an as-yet unknown specific endopeptidase, endothelin converting enzyme (ECE). The sequencing of human big ET-1, big ET-2, and big ET-3 (Itoh et al., 1988; Inoue et al., 1989; Ohkubo et al., 1990; Bloch et al., 1989) suggested that all 3 peptides be matured using similar enzymatic processing between their Trp^{21} and the residue 22, which is Val in the case of big ET-1 and big ET-2 and Ile in the case of big ET-3 (Table 1).

This hypothesis received support from studies showing that mature ET-1, big ET-1, and its C-terminal fragment could be detected in the supernatant of vascular endothelial cells (Saito et al., 1989; Emori et al., 1989; Sawamura et al., 1989; Wagner et al., 1992). Furthermore, both big ET-1 and ET-1 were detected in human plasma (Miyauchi et al., 1989; Koller et al., 1991) whereas, in animal models, intravascular injections of big ET-1 increased the ET-1 circulating level (D'Orleans-Juste et al., 1990; Hemsen et al., 1991). The physiological importance of the cleavage of big ET-1 into ET-1 was further substantiated by a 100-fold lower binding affinity of big ET-1 for endothelin receptors and of its 100- to 200-fold lower *in vitro* vasoconstrictor activity compared to that of ET-1 (Hirata et al., 1989; Neuser et al., 1991; Kimura et al., 1989). On the other hand, big ET-1, big ET-2, and big ET-3 revealed marked hemodynamic activities *in vivo* (mainly constrictor effects) (Kashiwabara et al., 1989; Pollock and Opgenorth 1991a; Gardiner et al., 1992a,b; Matsumara et al., 1993), suggesting that exogenous big endothelins could be matured from the extracellular side of the vascular system. With such results, the existence of ECEs, essential for the synthesis of mature endothelin peptides, has been unambiguously recognized.

Yanagisawa et al. (1988) initially supposed ECE to be a chymotrypsin-like endopeptidase possibly existing as a soluble intracellular enzyme, a membrane-bound enzyme (like the angiotensin converting enzyme), or a secreted enzyme active outside the cell. However, since Watanabe et al. (1991) reported the existence of an ECE activity in human whole blood but not in serum, it was believed that ET-1 maturation would be strictly a cell-mediated process and that ECE would probably not exist as a secreted form, at least under physiological conditions. Nevertheless,

TABLE 2

Enzymes Able to Generate Mature ET-1 From Big ET-1 *In Vitro*

Enzymatic Class	Cleavage Site	Optimum pH	Ref.
Serine protease			
Chymotrypsin	Trp^{21}-Val^{22}	8	McMahon et al., 1989
	Tyr^{31}-Gly^{32}		Takaoka et al., 1990a
	Tyr^{13}-Phe^{14}		Patterson et al., 1990
Chymase	Trp^{21}-Val^{22}	8.5	Wypij et al., 1992
Aspartyl protease			
Cathepsin D	Trp^{21}-Val^{22}		
	Asp^{18}-Ile^{19}	3.5	Sawamura et al., 1990a
			Takaoka et al., 1990b
			Lees et al., 1990
			Savage et al., 1993
Cathepsin E	Trp^{21}-Val^{22}	3.5	Lees et al., 1990
Pepsin	Trp^{21}-Val^{22}	2.3	Takaoka et al., 1990c
	Phe^{14}-Cys^{15}		Savage et al., 1993
Chymosin	Trp^{21}-Val^{22}	3.5	Savage et al., 1993.

Wypij et al. (1992) demonstrated that by stimulating the degranulation of mast cells in the perfused rat lung, one could induce the release of an ECE activity into the perfusate, suggesting that certain pathophysiological conditions could well be associated with an increase in circulating ECE. With the recent discovery of an ECE activity in a lipoprotein fraction of serum of healthy human donors (Ohwaki et al., 1993), it is now established that ECE indeed exists as a secreted form. The existence of cellular ECE has also been well established with the accumulation of evidence demonstrating that the synthesis of endothelins is not restricted to endothelial cells and that such a faculty is also shared by a large variety of cultured cells including epithelial cells (Marumo and Tomita, 1990), glial cells (MacCumber et al., 1990), glomerular mesangial cells (Sakamoto et al., 1990), smooth muscle cells (Resink et al., 1990), kidney cells (Kosaka et al., 1989), neurons (Krsmanovic et al., 1991), and some macrophages and leukocytes (Ehrenreich et al., 1990; Sessa et al., 1991a; Kaw et al., 1992). Therefore, it could be assumed that a similar, if not identical, ECE might exist in all these different cell types, as it might also exist in cells that do not normally produce endothelins, such as *Xenopus* oocytes (Fabbrini et al., 1991, 1993). Today, the nature and the cellular location of ECEs are still matters of debate.

III. *IN VITRO* IDENTIFICATION OF ENDOTHELIN-CONVERTING ACTIVITIES

The cleavage site of big ET-1 by ECE, located between two hydrophobic residues (Trp^{21}-Val^{22}) and qualified as "unusual" by Yanagisawa et al. (1988), in fact has no special feature that would help in the search of a specific proteolytic activity. This view is reinforced by the fact that common nonspecific serine and aspartyl proteases have demonstrated ECE activities *in vitro*, with variable degrees of selectivity (see Table 2), and that several ECE activities, representing the four classes of endopeptidases, have been identified in tissues or cultured cells. Furthermore, Fabbrini et al. (1993) demonstrated that in *Xenopus* oocytes, the conversion of big ET-1 into ET-1 does not require a Trp-Val sequence, suggesting that conformation of the processing site, rather than specific sequence, defines cleavage specificity by the ECE.

A. Serine Proteases with ECE-Like Activities

McMahon et al. (1989) demonstrated that chymotrypsin-treated big ET-1 could generate, among various fragments, native ET-1 with *in vitro* and *in vivo* constrictor activities. Such observations, supporting the previous hypothesis of Yanagisawa et al. (1988), were confirmed by Takaoka et al. (1990a) and Patterson et al. (1990) who also observed that chymotrypsin not only generated ET-1 by hydrolyzing big ET-1 at Tyr^{31}-Gly^{32} and then at Trp^{21}-Val^{22}, but could also degrade ET-1 by cleaving between Tyr^{13} and Phe^{14}. Sessa et al. (1991a,b) also described, in activated human polymorphonuclear leukocytes, a serine protease activity able to degrade ET-1.

Ohlstein et al. (1990) reported that the ability of cultured vascular endothelial cells to produce biologically active ET-1 was reduced by *N-p*-tosyl-L-lysine-chloromethylketone (TPCK), an irreversible inhibitor of chymotrypsin-like serine proteases, or by isatoic anhydride, an irreversible inactivator of chymotrypsin. The existence of an ECE with serine protease characteristics was proposed, although other general inhibitors of serine proteases such as chymostatin and phenylmethylsulfonyl fluorides (PMSF) were inactive in this preparation.

Another serine-protease activity demonstrating an ECE activity *in situ* was detected in rat lungs, associated with the granules of activated mast cells and copurified with a chymostatin-sensitive basic protease, chymase I (Wypij et al., 1992). It was also reported that chymase could generate ET-1 from big ET-1 *in vitro* without degrading ET-1. Such an activity could be released into the circulation by stimulating the degranulation of mast cells.

The ECE activity found by Ohwaki et al. (1993) in a human serum lipoprotein fraction was sensitive to chymostatin and could therefore be considered as a chymotrypsin-like enzyme.

An atypical serine protease activity, purified from the soluble fraction of porcine lungs, was reported to cleave big ET-1 between Val^{22}-Asn^{23}, at neutral pH and generate a peptide called ET-1-Val (Watanabe and Yokosowa, 1992). Similarly to big ET-1, the biological activities of ET-1-Val were trivial *in vitro* but comparable to ET-1 *in vivo*, thus suggesting that ET-1-Val could be an intermediate in the maturation of ET-1, and the involvement of a cascade of events initiated by a serine protease was proposed. Kaw et al. (1992) suggested that an elastase-like enzyme could initiate such a cascade in polymorphonuclear leukocytes.

B. Cysteine Proteases with ECE-Like Activities

The sole report concerning an ECE with cysteine protease characteristics was published by Deng et al. (1992). It describes the identification and partial purification, from endothelial cells of porcine aorta, of a neutral soluble proteolytic activity able to convert big ET-1 into an ET-1-like peptide, as determined by its characteristic vasoconstrictor activity and HPLC retention time.

Knap et al. (1993) mentioned a neutral ECE activity detected in the cytosolic extract of endothelial cells from porcine aortae that could well be similar to the thiol ECE activity previously described by Deng et al. (1992). However, no supporting data have been presented.

C. Aspartyl Proteases with ECE-Like Activities

Four aspartyl proteases (cathepsin D, cathepsin E, pepsin, and chymosin) have been reported as having ECE activities *in vitro* (see Table 2). Among them, only cathepsin D could also rapidly degrade ET-1.

TABLE 3

Enzymatic Preparations Revealing ECE Activities with Properties of Aspartyl Proteases

Source	Preparation	pH	Ref.
Endothelial cells			
Porcine aortae	Crude cytosolic extract	4.0	Matsumura et al., 1990a
	Crude cytosolic extract	4.0	Ikegawa et al., 1990
	Purified cell membranes	4.0	Knap et al., 1993
Bovine aortae	Crude homogenate	3.0	Ohnaka et al. 1990
	Crude cytosolic extract	3.5	Sawamura et al., 1990b
Smooth muscle cells			
Bovine carotid artery	Cytosolic extract	3–5	Hioki et al., 1991
Adrenal medulla			
Bovine	Purified chromaffin granules	3.5	Sawamura et al., 1990c
Lung			
Rat	Purified cell membranes	4.0	Wu-Wong et al., 1990
			Wu-Wong et al., 1991
			Shiosaki et al., 1993

The likelihood for cathepsin-like, pepsin-like, or chymosin-like enzymes being ECEs was emphasized by several studies describing ECE activities with the characteristics of aspartyl proteases in preparations from various tissues (endothelial cells, vascular smooth muscle cells, adrenal medulla, lung) (Table 3). All of them were indeed able to cleave big ET-1 into its C terminal fragment and ET-1 at a low pH, were highly sensitive to pepstatin A (a nonspecific inhibitor of aspartyl proteases), and resistant to inhibitors of other classes of endopeptidases.

The aspartyl ECE activities found in the cytosol or membrane extracts of vascular endothelial and smooth muscle cells (Table 3) were all crude preparations and have not been characterized further. Sawamura et al. (1990) purified the activity detected in the chromaffin granules of bovine adrenal medulla more than 3000-fold and found that, similarly to cathepsin D, it contained three major components of 45, 30, and 15 kDa. The ECE activity detected in the membrane extract from rat lung (Wu-Wong et al., 1990, 1991; Shiosaki et al., 1993) was purified about 190 times and appeared as a single protein of 90 kDa with the particularity of being activated (about 2.5-fold) by metal ions (especially Mn^{2+}) and cleaving big ET-1 in preference to big ET-3.

D. Metalloproteases with ECE-Like Activities

Ohnaka et al. (1990) were first to identify and characterize in a crude homogenate of cultured bovine aortic endothelial cells a neutral EDTA- and EGTA-sensitive big ET-1- to ET-1-converting activity that was unaffected by various inhibitors of other classes of endopeptidases. This suggested that ECE would be a metal-dependent protease. Shortly thereafter, Okada et al. (1990) located a similar activity in the membrane fraction of cultured bovine endothelial cells. They managed to solubilize it using the detergent Triton X-100, and described an activity of about 100 kDa presenting the characteristics of a metalloendopeptidase: sensitivity to metal chelators such as EDTA and *o*-phenanthroline; reactivation by divalent metal cations such as Zn^{2+}, Co^{2+}, Mn^{2+}, and Fe^{2+}; and sensitivity to the metalloprotease inhibitor, phosphoramidon. Since its substrate specificity was highly in favor of big ET-1, compared to big ET-2 and big ET-3, such an enzymatic activity was later named ECE-1 by Okada et al. (1991), who also demonstrated the importance of the C-terminal region of big ET-1 for its

TABLE 4

In vitro Preparations Revealing ECE Activities with Properties of Metalloproteases

Source	Preparation	pH	Ref.
Endothelial cells			
Bovine aortas	Crude homogenate	7.0	Ohnaka et al., 1990
	Purified cell membranes	6.6–7.6	Okada et al., 1990
	Cytosolic and membranal	6.5, 7.5	Okada et al., 1991, 1993 Takada et al., 1991
	Cytosolic and membranal	7.4	Warner et al., 1991
	Membranes	7.0	Ahn et al., 1992
	Microsomes	7.0	Ashizawa et al., 1994a
Porcine aortas	Cytosolic and membranal	6.5, 7.0	Matsumura et al., 1990b Matsumura et al., 1991a,b
	Membranes	7.4	Ohnaka et al., 1992, 1993
Human umbilical vein	Cytosolic and membranal	7.0	Ahn et al., 1992
Permanent human cell line (EAHY 926)	Membranes	7.0	Waxman et al., 1994
Smooth muscle cells			
Porcine aorta	Membranes	6.0–7.5	Matsumura et al., 1990c
Lung			
Rat	Membranes	6.5	Matsumura et al., 1992
	Microsomes		McMahon et al., 1991
	Microsomes	7.5	Takahashi et al., 1993
	Golgi apparatus	7.0	Gui et al., 1993
Bovine	Membranes	7.5	Kundu and Wilson, 1992 Kundu and Wilson, 1993
Brain			
Rat	Membranal and cytoskeletal	7.4	Warner et al., 1992a,b
Serum			
Human	Lipoprotein fraction	7.0	Ohwaki et al., 1993

recognition by the enzyme (Okada et al., 1993). A phosphoramidon-sensitive ECE-1 was also found in the cytosolic fraction of bovine endothelial cells (Okada et al., 1991).

Several metalloproteases with endothelin-converting activity have also been found in other tissue preparations (Table 4) and a lot of effort has been devoted to their purification using as the main criteria their resistance to inhibitors of cysteine, serine, and aspartic proteases and their sensitivity to metal chelators and phosphoramidon (K_i in the micromolar range) (see review by Turner and Murphy, 1996). In vascular endothelial cells, neutral metalloprotease-ECE activities have been localized both in membrane and cytosolic fractions (Table 4).

The membrane-bound activities characterized so far share the features of being sensitive to phosphoramidon; resistant to thiorphan and captopril; specific inhibitors of, respectively, enkephalinase (neutral endopeptidase; EC 3.4.24.11); and, angiotensin converting enzyme (EC 3.4.15.1) and similarly to ECE-1 (Okada et al., 1991, 1993), selective for big ET-1. An ECE activity presenting all these properties has been partially purified by Ohnaka et al. (1992, 1993) who recently achieved a 12,000-fold purification of a 120- to 131-kDa monomeric glycoprotein processing big ET-1 (K_m and V_{max} of about 3.3 μM and 0.41 μmol/min/mg, respectively) with a high selectivity over big ET-2 and big ET-3. The carboxy-terminal of big ET-1 appeared essential for its recognition by the enzyme. Very similar activities have been found in the membrane and cytoskeletal fractions from brain and in membrane fractions from lung (see Table 4 for references). Takahashi et al. (1993) recently attained, from a membrane fraction of rat lungs, a 13,000-fold purification of an enzyme sharing several properties with the endothelial enzyme: similar size, similar inhibitor profile, and similar substrate specificity. Gui et al. (1993) studied the intracellular localization of such an activity and found that it was mainly concentrated within vesicles of the Golgi apparatus. A very similar activity to those purified by Okada et al., Ohnaka et al., and Takahashi et al. was very recently purified 1000-fold by Waxman et al. (1994) from a permanent human endothelial cell line. Its molecular weight, however, appeared to be about twice as heavy (250 to 280 kDa). Okada et al. (1991) suggested the existence of enzymes other than ECE-1, with different substrate specificities, that could be involved in the processing of big ET-2 and big ET-3. Indeed, neutral, phosphoramidon-sensitive and thiorphan-resistant converting activities have been detected in membrane fractions of vascular endothelial cells and in renal adenocarcinoma cells with no selectivity for big ET-1 over big ET-3 (Matsumura et al., 1992) and a higher specificity for big ET-2 over big ET-1 and big ET-3 (Yorimitsu et al., 1992; Shinmi et al., 1993). The latter enzyme was called ECE-2.

Two types of EDTA- and *o*-phenanthroline-sensitive ECE activities have been detected in the cytosolic fractions of vascular endothelial cells: one phosphoramidon-sensitive (Takada et al., 1991; Warner et al., 1991) and one phosphoramidon-resistant (Matsumura et al., 1990, 1991; Ahn et al., 1992). The phosphoramidon-sensitive form, with a molecular weight five to six times higher than the phosphoramidon-sensitive membrane-bound enzyme (Takada et al., 1991), could easily be differentiated from the phosphoramidon-resistant form by the fact that only the phosphoramidon-resistant form was deactivated by *N*-ethylmaleimide, a sulfhydryl blocking reagent (Matsumura et al., 1991a). A thiol-dependent type metalloprotease has been proposed for an ECE activity such as has also been described in the membrane fraction obtained from cultured vascular smooth muscle cells (Matsumura et al., 1991).

The ECE activity found by Ohwaki et al. (1993) in a lipoprotein from human serum fraction exhibited the particularity of having both metalloprotease and serine protease properties.

IV. NATURE OF PHYSIOLOGICALLY RELEVANT ECE

As seen previously, a large variety of enzymes, or more or less purified enzymatic preparations, presenting *in vitro* ECE activities have been proposed as potential ECEs and, therefore, the identity of the physiologically or pathophysiologically relevant ECE(s) is still uncertain. Likely candidates have been detected both in the cellular compartment (cytoplasm, microsomal, and membranal fractions) and in the circulation, and it might well be possible that the maturation of ET-1 occurs both at the intra- and extracellular levels. This was already predicted by Yanagisawa et al. (1988) and has been clearly demonstrated by *in vitro* studies showing that cultured cells not only secrete both ET-1 and big ET-1, but also generate ET-1 from exogenously applied big ET-1 (Sawamura et al., 1989; Suzuki et al., 1989; Ikegawa et al., 1991). It has also been shown *in vivo* that both big ET-1 and ET-1 are present in plasma, and that intravascular injections of big ET-1 increase the ET-1 circulating level (D'Orleans-Juste et al., 1990). In other words, the *de novo* synthesized big ET-1 most probably is not exclusively processed intracellularly.

A. Intracellular ECEs

It has been clearly established that the secretion of ET-1 by cultured endothelial cells is not affected by the intracellular accumulation of pepstatin, a classical inhibitor of aspartyl proteases (Shields et al., 1991; Nichols et al., 1991). It is therefore most unlikely in such cells that an aspartyl ECE is responsible for the maturation of big ET-1. It is not known whether or not this conclusion can be extrapolated to other cell types and to the processing of big ET-2 or big ET-3.

On the other hand, the secretion of ET-1 by cultured endothelial cells is decreased by TPCK, an irreversible inhibitor of a variety of serine and cysteine proteases, and by isatoic anhydride, an inactivator of chymotrypsin (Ohlstein et al., 1989). While such observations are consistent with the involvement of a chymotrypsin-like ECE, it is not known to what extent they reflect a real inhibition of an ECE or the inhibition of other enzymes which process preproendothelin-1.

Sawamura et al. (1991) demonstrated that the incubation of cultured endothelial cells in the presence of phosphoramidon resulted in an inversion of the intracellular ratio ET-1/big ET-1 from 3.3/1 to 1/3.5, indicating an accumulation within the cells of big ET-1 and a reduction in *de novo* synthesized ET-1.

Taken together, such observations suggest that two types of ECEs might mainly be involved in the pathophysiological intracellular maturation of big ET-1: a chymotrypsin-like enzyme and a phosphoramidon-sensitive metalloprotease. Although the results available thus far do not support a biological role for intracellular aspartyl ECE activities, more information is certainly required.

B. Extracellular ECEs

Evidence is now accumulating to suggest that a neutral, membrane-bound phosphoramidon-sensitive metalloprotease is the most likely candidate for the conversion of extracellular big ET-1 (for review, see Turner and Murphy, 1996). This is mainly supported by the fact that in cultured endothelial cells extracellular phosphoramidon not only suppresses the secretion of ET-1 while causing a simultaneous increase in unprocessed big ET-1 (Ikegawa et al., 1990b), but also inhibits the conversion of exogenously applied big ET-1 (Okada et al., 1990; Ikegawa et al., 1991; Corder et al., 1993). The ability of cultured vascular smooth muscle cells and human polymorphonuclear cells to mature exogenous big ET-1 is also phosphoramidon sensitive (Ikegawa et al., 1991; Sessa et al., 1991a,b). Classical inhibitors of aspartyl, serine, and cysteine endopeptidases are all inactive in such preparations (Okada et al., 1990; Ikegawa et al., 1990a; Sessa et al., 1991a) as is thiorphan, suggesting that the ECE involved is a metalloprotease different from enkephalinase (neutral endopeptidase; EC 3.4.24.11).

Phosphoramidon has been shown not to affect [^{125}I]ET-1 binding in placental membranes (Pollock et al., 1993a). It was observed that it could also suppress both the constrictor effects and the simultaneous generation of ET-1 following big ET-1 administration in preparations of isolated vascular beds (Hisaki et al., 1991, 1993; Fukuroda et al., 1990; Balwierczak et al., 1993; Kaw et al., 1993; Verbeuren et al., 1993) or isolated lungs and kidneys (Ishikawa et al., 1992; Télémaque and D'Orléans-Juste, 1991). In these reports, classical selective inhibitors of aspartyl, serine, or cysteine endoproteases, cathepsin E (Bird et al., 1992), enkephalinase, and angiotensin converting enzyme failed to inhibit big ET-1 conversion. There are also several reports demonstrating that phosphoramidon could inhibit the exogenous big ET-1-mediated, but not the ET-1-mediated, hypertensor, systemic vasoconstrictor, and bronchoconstrictor effects observed in anesthetized or conscious animals (Fukuroda et al., 1990; Matsumura et al., 1990d; Le Monnier de Gouville and Cavero, 1991; McMahon et al., 1991a,b; Pollock and Opgenorth, 1991b; Gardiner et al., 1991; Nogushi et al., 1991, 1992b; Pons et al., 1992; Pollock et al., 1993b; Lawrence and Brain, 1993b; Verbeuren et al., 1993). However, once the effects of ET-1 or of big ET-1 have been allowed to develop, they are no longer sensitive to phosphoramidon (Pollock and Opgenorth, 1991b). Phosphoramidon was also shown to be effective at antagonizing both vasospasm and increases in ET-1 content in cerebrospinal fluid following intracisternal administrations of big ET-1 to anesthetized dogs (Shinyama et al., 1991). Big ET-1-induced sudden death and elevation of the plasma ET-1 level in mice were effectively blocked by phosphoramidon (Matsuura et al., 1992).

Taken together, all these observations indicate that if extracellular processing of big ET-1 is pathophysiologically relevant, then the ECE involved cannot be an aspartyl, serine, or cysteine endopeptidase. It is more likely a new type of membrane-bound phophoramidon-sensitive metalloendopeptidase capable of converting *in vitro, ex vivo,* and *in vivo* exogenous big ET-1 into ET-1. They also suggest that such an enzyme might be closely related, if not identical, to those purified from lung and endothelial cell membranes by Takahashi et al. (1993) and Ohnaka et al. (1993), respectively. Alternatively, the involvement of a circulating ECE could also be considered (Ohwaki et al., 1993).

The conversions of exogenous big ET-2 and big ET-3 are also phosphoramidon sensitive and thiorphan resistant in the anesthetized rat (Gardiner et al., 1992b; Lawrence and Brain, 1993; Pollock et al., 1993b; Mattera et al., 1993), suggesting that in the cardiovascular system exogenous big ET-1, big ET-2, and big ET-3 are processed by similar enzymatic activities. It is, however, not known whether or not different enzymes with different substrate specificities are involved.

Nevertheless, it may be misleading to extrapolate from the biological maturation of endogenous endothelin peptides based on the *in vivo* effects of exogenous big endothelins and their sensitivity to phosphoramidon. Indeed, phosphoramidon, a natural compound isolated by Umezawa et al. (1972), is a nonspecific inhibitor of several metalloproteases such as thermolysin (EC 3.4.24.4), enkephalinase (EC 3.4.24.11), angiotensin converting enzyme (EC 3.4.15.1), carboxypeptidase A (EC 3.4.17.1) and collagenase (EC 3.4.24.7)(Schwartz et al., 1985; Komiyama et al., 1975; Hudgin et al., 1981; Matthews, 1988; Holmquist and Vallee, 1979; Kam et al., 1979; Gomez-Monterrey et al., 1992; Kortylewicz and Galardy, 1990). Concerning enkephalinase, it has been recognized as responsible for the degradation of endothelins (Abassi et al., 1992, 1993) and of other vasoactive neurohormonal peptides including atrial natriuretic peptide, bradykinin, enkephalins, and substance P (Yamagushi et al., 1992; Erdös and Skidgel, 1989). Therefore, inhibition of the *in vivo* effects of exogenous big endothelins by phosphoramidon might not necessarily imply only the inhibition of an ECE but could result from a complex balance between maturation of endothelins and inhibition of the degradation of several other vasoactive peptides, including ET-1. This might be illustrated by a report from Modin et al. (1991), showing that in anesthetized rats and pigs, phosphoramidon was able to inhibit the vasoconstrictor effects of exogenous big ET-1, but not the elevation of plasma ET-1. Furthermore, McMahon et al. (1991a), Gardiner et al. (1992c), and Trapani et al. (1993) have reported that thiorphan or SQ 28,603, specific inhibitors of enkephalinase, were able to weakly diminish the vasoconstrictor effects of exogenous big ET-1 in both anesthetized and conscious rats.

Therefore, it is clear that experiments showing that only one given inhibitor can alter the conversion of exogenous big ET-1 *in vivo* cannot provide definitive proof of the existence of a pathophysiologically relevant phosphoramidon-sensitive metallo-ECE, whatever its localization. They should be complemented by demonstrating that the inhibitor is also able to alter the production of endogenous ET-1. On that point, research has focused on studying the effects of phosphoramidon.

C. Is Production of Endogenous ET-1 Phosphoramidon-Sensitive?

It has been well established *in vitro* (cultured cells and isolated organs) that the spontaneous (above) as well as the cyclosporin-stimulated (Yuan et al., 1992) generation of endogenous ET-1 could be inhibited by phosphoramidon. Vemulapalli et al. (1992) demonstrated in isolated perfused guinea-pig lungs that ischemia/hypoxia-induced endogenous ET-1 release was abolished by pretreating the animals with phosphoramidon.

Things appear more complex *in vivo* since protective actions of phosphoramidon were observed in experimental models of subarachnoid hemorrhage and ischemic acute renal failure (Matsumura et al., 1991b; Vemulapalli et al., 1993), where the endothelin-receptor antagonists BQ 123 (Mino et al., 1992; Clozel and Watanabe, 1993a), FR 139317 (Nirei et al., 1993) and Ro 46-2005 (Clozel et al., 1993b) demonstrated protective activities as well. On the other hand, Shinego et al. (1991) and Cosentino et al. (1993) did not observe any protective effects of either phosphoramidon or of BQ 123 in models of subarachnoid hemorrhage-induced vasospasm. Furthermore, phosphoramidon was reported to inhibit the increase in circulating ET-1 stimulated either by endotoxin (Pollock et al., 1993a), hemorrhage, cytokines, and hypoxia (Vemulapalli et al., 1994) in the anesthetized rat. It is therefore not clear to what extent phosphoramidon can actually interfere with the stimulation of endogenous ET-1 production *in vivo*.

Whatever the case, a nonselective inhibitor of metalloproteinases like phosphoramidon is probably not an appropriate pharmacological tool to interfere with the bioactivation of endothelin peptides. Potent and selective inhibitors of the various ECEs identified thus far are therefore required.

V. THE SEARCH FOR NEW SELECTIVE ECE INHIBITORS

The search for selective enzyme inhibitors has been made with some of the various enzymatic preparations described previously, using big ET-1 as the substrate and analyzing the generation of ET-1 mainly by HPLC or RIA techniques. The *in vitro* or *in vivo* inhibition of exogenous big ET-1-mediated effects (vasoconstriction, bronchoconstriction, hypertension) has also been used as criteria of efficacy and physiological relevance of potential ECE inhibitors.

Various approaches have been used for the discovery of ECE inhibitors such as the synthesis of substrate-based compounds, phosphoramidon or pepstatin analogues, and the screening of substances produced by microorganisms.

A. Substrate-Based ECE Inhibitors

Using phosphoramidon-sensitive neutral ECE purified from microsomal fractions of bovine endothelial cells and called ECE-1 because of its selectivity for big ET-1 over big ET-3, Okada et al. (1991) suggested that the carboxyl-terminal sequence at residues Gly32-Arg37 (Table 1) was important for the conversion of big ET-1, whereas the disulfide loop structure could interfere with access of big ET-1 to ECE. Later, the same authors (Okada et al., 1993) demonstrated that the minimum substrate for ECE-1 was the 19-34 fragment of big ET-1, of which the region His27-Val-Pro-Tyr-Gly-Leu-Gly34 was most important. It was proposed that such a fragment, five residues distant from the cleavage site, would be a recognition site for the enzyme. On the other hand, the deletion of the disulfide loop structure resulted in a fivefold increase in conversion rate, suggesting that the N-terminal fragment of big ET-1 might introduce some unfavorable steric hindrance in the enzymatic processing. Unfortunately, none of the big ET-1 fragments synthesized for these studies was tested for inhibitory activity. It is also unknown to what extent big ET-3, which is not converted by ECE-1, can compete with the conversion of big ET-1. Similar results were obtained by Ohnaka et al. (1993) using their phosphoramidon-sensitive membrane-bound ECE, 12,000-fold purified from porcine aortic endothelium. Among the various fragments synthesized they reported that the peptides big ET-2(1-37), big ET-2(18-34), big ET-3(1-41) and the substituted peptide (Phe^{21})big ET-1(18-34) showed weak inhibitory activities of big ET-1 cleavage, the hydrolysis rate being reduced to at most 70% of the original rate in the presence of these peptides at 100 μM. Taken together, these results suggest that the design of substrate-based compounds might lead to selective but weak ECE inhibitors.

This is supported by a recent report concerning a synthetic peptide, called EV-D, that could competitively inhibit *in vitro* and *in vivo* exogenous big ET-1-mediated effects, albeit at high concentrations (300 μM)(Inagaki et al., 1993). The structure of such a peptide was described as similar to that of big ET-1 but has not been published as yet. Also, the peptide Ac-Asp-Ile-Ile-Trp-Cys-NH_2, related to big ET-1(18-22), is claimed to be an inhibitor of ECE (Biohovsky et al., 1993).

B. Pepstatin Analogues as ECE Inhibitors

Only one major report has been published concerning the design of potent and selective ECE inhibitors using, as enzymatic assay, a pepstatin-sensitive ECE activity from the membranes of rat lungs (Shiosaki et al., 1993). This might be explained by the limited amount of evidence supporting an aspartyl protease nature for the pathophysiologically relevant ECE(s). Potent (nanomolar activities) and selective inhibitors were designed by incorporating the statine amino acid or a dihydroxyethylene building block into peptide sequences (Figure 2, structures 1 and 2). In general, parallel trends for inhibition between ECE and cathepsin D were observed; however, some compounds demonstrated significant selectivity (more than 50-fold) suggesting that the two enzymes are indeed distinct. No consistent result could be obtained concerning the ability of such compounds to inhibit the hypertensive effect of exogenous big ET-1 in the rat. This strengthens the idea that the ECE involved in the *in vivo* processing of exogenous big ET-1 is not an aspartyl protease. However, the action of an intracellular ECE could not be revealed by such experiments.

C. Phosphoramidon Analogues as ECE Inhibitors

The structures of phosphoramidon (*N*-[α-L-rhamnopyranosyloxy-hydroxyphosphinyl]-L-leucyl-L-tryptophan) and of various analogues are presented in Table 5.

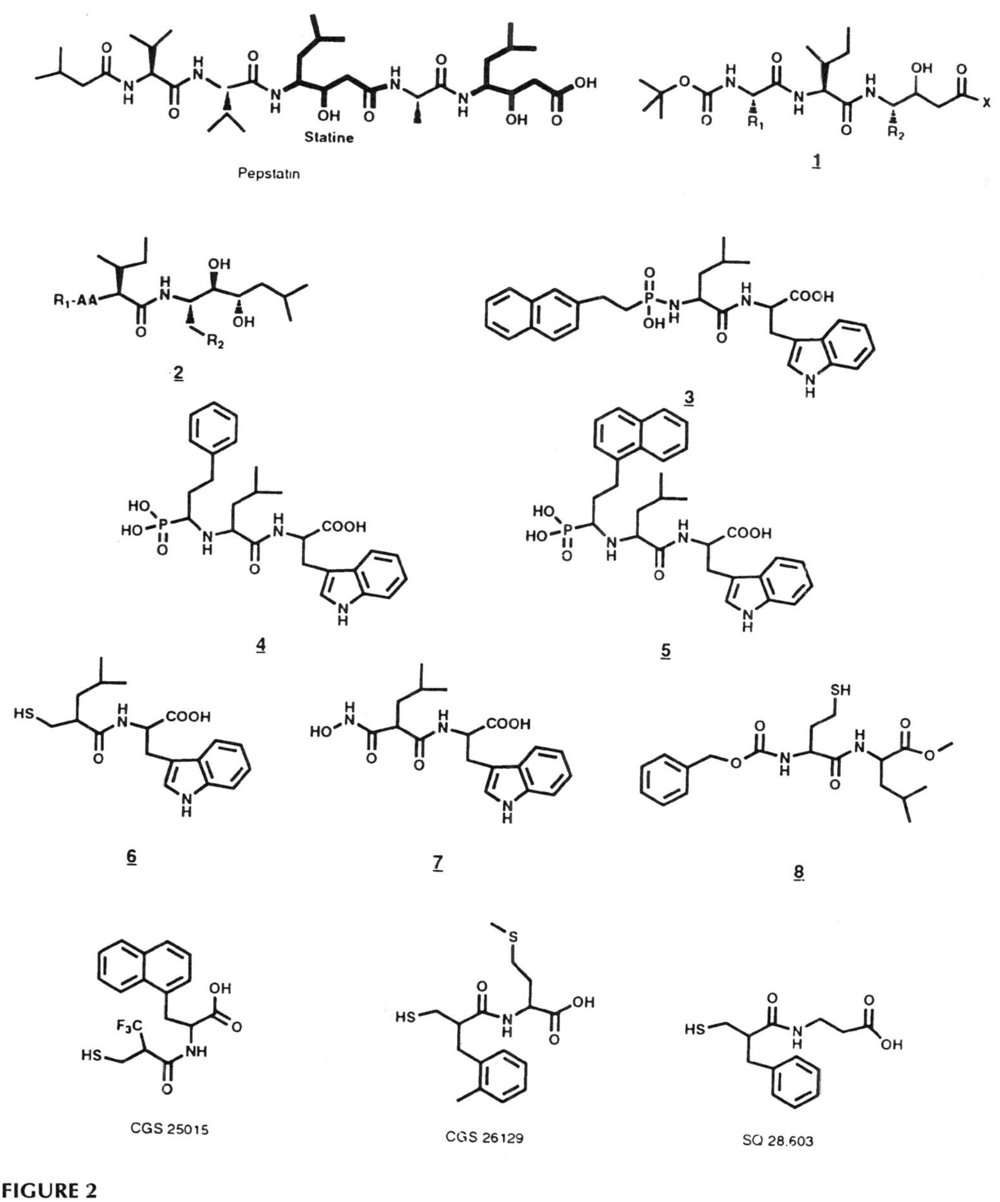

FIGURE 2
Chemical structures showing various degrees of ECE inhibitory activity. Structures 1 and 2 are statine based; structures 3 to 8 are phosphoramidon derivatives, and the last three structures are thiol based.

1. Phosphoramidate-Based Inhibitors

The molecular mechanism involved in the inhibition of the metalloproteases by phosphoramidon has been extensively studied in the case of thermolysin, where phosphoramidon binds with a single oxygen of the phosphoryl moiety to the catalytic Zn within the active site of the enzyme (Tronrud et al., 1986). On the other hand, the rhamnose sugar of the molecule appeared to be only weakly involved in the binding to thermolysin (Weaver et al., 1977) and its removal, resulting in *N*-phosphoryl-Leu-Trp, did not affect much the inhibitory activity of the molecule (Komiyama et al.,

TABLE 5

Influence of the Phosphorus Function of Various Phosphoramidon Analogues for ECE Inhibition *In Vitro* (Data from Bertenshaw et al., 1993a.)

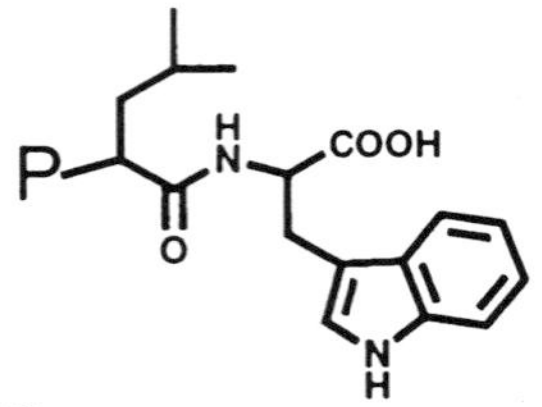

P / H / N / COOH / O / N / H

Entry	P	name	ECE inhibition IC_{50} (μM)
1	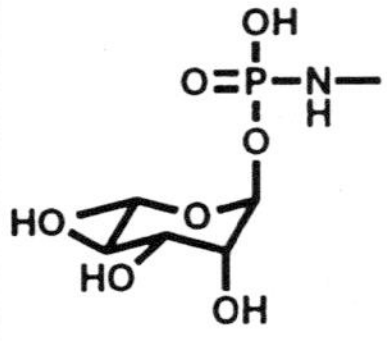 OH / O=P–N– / H / O / HO / O / HO / OH	Phosphoramidon	2
2	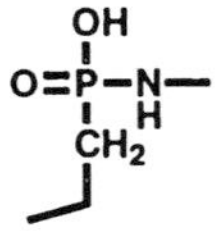 OH / O=P–N– / H / CH_2	Propyl Phosphonamidate	2
3	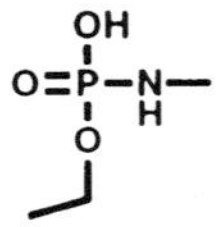 OH / O=P–N– / H / O	Ethyl Phosphoramidate	109
4	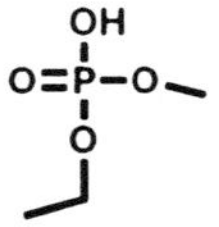 OH / O=P–O– / O	Ethyl Phosphate	>>100
5	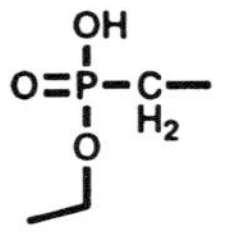 OH / O=P–C– / H_2 / O	Ethyl Phosphonate	>>100
6	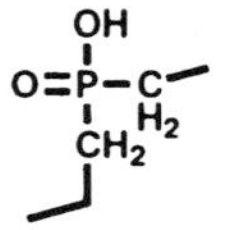 OH / O=P–C– / H_2 / CH_2	Propyl Phosphinate	>>100

1975). The absence of sugar also facilitates the chemistry, making these phosphoramidates readily available by synthesis.

Considering the high affinity of *N*-phosphoryl-Leu-Trp for thermolysin and enkephalinase (Alstein et al., 1982), it was not very surprising to observe that continuous infusion of *N*-phosphoryl-Leu-Trp into anesthetized rats could abolish big ET-1-mediated hypertension to the same extent as did phosphoramidon (Pollock et al., 1992). The *in vivo* ECE inhibitory activity of N-phosphoryl-Leu-Trp has been confirmed (Bigaud et al., 1994), however with a lower potency than phosphoramidon. In contrast to the former experiments, the inhibitor was administered as a unique bolus injection and therefore its activity was more dependent on its *in vivo* stability than when continuously infused. The lower ECE-inhibitory activity of phosphoryl-Leu-Trp compared to phosphoramidon could then be explained by a higher resistance of phosphoramidon to degradation *in vivo*. *N*-phosphoryldipeptides were indeed found to be rapidly hydrolyzed under neutral conditions, whereas phosphoramidon was found unaffected (Poncz et al., 1984). It could be suggested that the rhamnose ring, although not important for the activity of phosphoramidon, stabilizes the molecule. From a practical point of view, *N*-phosphoryldipeptide compounds decompose within a few days at room temperature, even in the solid state, and should be tested shortly after being synthesized. For example, by monitoring the hydrolysis of *N*-phosphoryl-Leu-Trp by ^{31}P NMR in D_2O at pH 8.5 and 37°C, a half life of 2.8 h was estimated (personal communication of Piriou, F., Marion Merrell Dow Analytical Chemistry Department, Strasbourg). This instability, however, did not preclude a small structure-activity relationship study of *in vivo* ECE inhibition, and it was observed that the tryptophan residue, but not leucine, was essential for the inhibitory activity of the molecule. The existence of an hydrophobic pocket, specific for the recognition of the tryptophan residue of phosphoramidon, was therefore suggested for the active site of ECE (Bigaud et al., 1994). Such an hypothesis received support from Fukami et al. (1994), who reported that *N*-phosphoryl-Leu-3-(1-naphthyl)-Ala elicits a good inhibition of ECE prepared from microsomal fractions of bovine cultured endothelial cells, whereas *N*-phosphoryl-Val-Asn is almost inactive.

Bertenshaw et al. (1993a) demonstrated that the replacement in the phosphoramidon molecule of the rhamnose ring by an ethoxy group (Table 5, entry 3) reduced ECE inhibitory activity by a factor of about 60 fold, when assayed on partially purified ECE from rabbit lung. Nevertheless, this ethyl phosphoramidate could also dose-dependently inhibit the pressor response of big ET-1, with a potency slightly less than that of phosphoramidon. Moreover, the corresponding cyclohexyl and phenyl phosphoramidates exhibited ECE inhibitory activities 20 fold lower than that of phosphoramidon. On the other hand, the corresponding ethyl phosphate (Table 5, entry 4) and propyl phosphinate (Table 5, entry 6) were found poorly active, suggesting an important contribution of the NH group for binding to the enzyme.

However, phosphoramidate-based inhibitors, although active against ECE, are not ideal tools to assess the role of endothelin peptides *in vivo* because of their instability.

2. *Phosphonamidate-Based Inhibitors*

Such compounds (Table 5), although also unstable, are good inhibitors of enkephalinase, collagenase, and angiotensin converting enzyme (Elliott et al., 1985; Mookhtiar et al., 1987). They also exhibit similar, or slightly better, ECE inhibitory activities than the corresponding phosphoramidates (Bertenshaw et al., 1993a). For example, modifying phosphoramidon into a propyl phosphonamide (Table 5, Entry 2) resulted in a fivefold increase in ECE inhibition *in vivo*. Mitsuhiro et al. (1992) reported that on a crude ECE preparation from pig lungs, the compound *N*-[2-(2-napththyl)ethylphosphonyl]-Leu-Trp (Figure 2, structure 3) presented an IC_{50} of about 0.1 μM compared to about 100 μM for phosphoramidon. Moreover, a structure-activity relationship study indicated a preference for the Leu-Trp sequence.

A rather obvious strategy to increase the stability of such compounds was to design phosphonate analogues of phosphoramidon.

3. *Phosphonate-Based Inhibitors*

N-phosphonoalkyl peptides are potent inhibitors of ACE (Flynn and Giroux, 1986), enkephalinase (De Lombaert et al., 1994), and collagenase (Bird et al., 1994). Ishikawa et al. (1993) claim that *N*-phosphonoaralkyl-Leu-Trp derivatives can also inhibit a partially purified ECE similarly to phosphoramidon. This also has been claimed by Biohovsky et al. (1993) but without showing biological data. For example, Ishikawa et al. (1993) observed that the compound *N*-[N-(3-phenyl-1-phosphonopropyl)-L-leucyl]-L-tryptophan (Figure 2, Structure 4) induced, at 10 μM, 88% ECE inhibition *in vitro* compared to 64% with phosphoramidon. Furthermore, the most active compound, *N*-{N-[3-(l-naphthyl)-1-phosphonopropyl]-L-leucyl}-L-tryptophan (Figure 2, Structure 5), was reported as ten times more potent than phosphoramidon and also had an improved selectivity for ECE over enkephalinase (Fukami et al., 1994).

Another attempt to design phosphonate analogues was to replace the NH group of the phosphoramidate by a methylene group. Unfortunately, such a compound (Table 5, Entry 5) was reportedly inactive *in vitro* (Bertenshaw et al., 1993a), stressing again the importance of the NH group for a good affinity for ECE.

4. *Thiol- or Hydroxamate-Based Inhibitors*

Other inhibitors have been designed by combining the metal-coordinating properties of either thiol or hydroxamate moieties with the peptide recognition elements found in phosphoramidon. The thiol and hydroxamate analogues of phosphoramidon (Figure 2, Structures 6 and 7) were, however, found to be slightly less active than phosphoramidon on a crude ECE preparation from rabbit lungs with a K_i of, respectively, 12 μM and 24 μM compared to 4 μM for phosphoramidon (Bertenshaw et al., 1993b).

D. Other Approaches for the Discovery of ECE Inhibitors

1. *Thiol-Based Inhibitors*

Some thiol inhibitors not related to phosphoramidon may also reveal good inhibitory activities against various ECE preparations. It is noteworthy that with such an approach, compounds with an acid carboxy-terminal were in some cases less active than the corresponding methyl esters (Bertenshaw et al., 1993b). For example, the compound Cbz-homo-Cys-Leu-OCH_3 (Figure 2, Structure 8) exhibited activity similar to that of phosphoramidon, whereas the corresponding Cbz-homo-Cys-Leu-COOH was significantly less potent.

Two thiol-dipeptides CGS 25015 and CGS 26129 (Figure 2)(Trapani et al., 1993) revealed weaker ECE-inhibitory activities than phosphoramidon both *in vitro* and *in vivo*.

A thiorphan analogue, SQ 28,603 (Figure 2), was found to significantly attenuate the effects of big ET-1 and prolong the effects of atrial natriuretic peptide in the conscious rat (Gardiner et al., 1992c).

2. *Metal Chelators as ECE Inhibitors*

Metalloproteases are generally inhibited *in vitro* by metal chelators such as EDTA and *o*-phenanthroline, and such compounds show ECE-inhibitory activities *in vitro* (Ashizawa et al., 1994a). However, because of their poor selectivity resulting in a

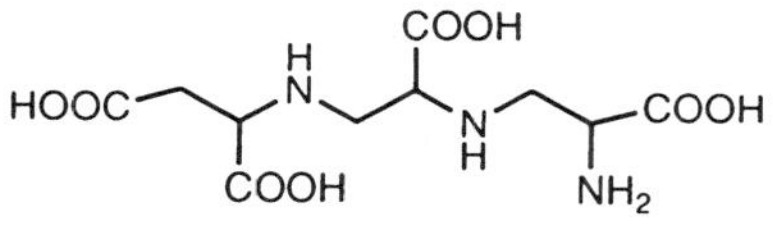

Aspergillomarasmine A

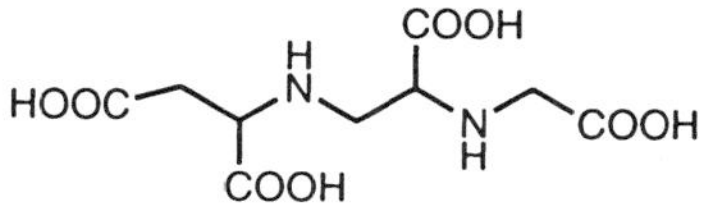

Aspergillomarasmine B

FIGURE 3
Chemical structures of metal chelators inhibiting the bioactivation of exogenous big ET-1 *in vitro* and *in vivo*.

rapid inactivation and toxicity, they exhibit rather weak ECE inhibition *in vivo* (Ashizawa et al., 1994b). Nevertheless, EDTA-related compounds, called aspergillomarasmines-A and -B, which are phytotoxins produced by fungi (Figure 3), have been shown to inhibit a partially purified ECE preparation *in vitro*, to abolish the pressor effect of exogenous big ET-1 in the anesthetized rat, and to significantly prolong the latency to sudden death induced by big ET-1 (Matsuura et al., 1993; Arai et al., 1993).

E. Concluding Remarks

Because of the experimental evidence supporting the phosphoramidon-sensitive metalloprotease nature of the pathophysiologically relevant ECE, the discovery of selective inhibitors has been mainly oriented towards the synthesis of phosphoramidon analogues. Their activities *in vitro* are usually similar to phosphoramidon (affinities within the micromolar range), except for some phosphonamidates which are one order of magnitude more potent.

Such a poor diversity may reflect the difficulties encountered thus far to establish an *in vitro* structure-activity relationship with a reasonably pure enzyme preparation. Most of the inhibition experiments were performed on crude preparations which may contain other metalloendopeptidases with similar substrate specificity. This could explain the discrepancies sometimes observed between the results obtained *in vivo* and those obtained *in vitro*. Furthermore, the relevance of the *in vivo* tests, using the effects of exogenous big ET-1 as a marker of enzymatic conversion, remains to be established since intracellular ECE is not involved in such a test.

Selective and stable inhibitors of ECE would be good tools to assess the pathophysiological roles of endothelin peptides. As already mentioned, phosphoramidon is not a selective inhibitor of ECE but has a broad spectrum of activity on metalloproteases. Therefore, phosphoramidon analogues may also be expected to be nonselective inhibitors of metalloproteases. There is some hope that the selectivity problem could be solved in the near future by using highly purified ECE activities, as recently described (Ohnaka et al., 1993; Takahashi et al., 1993; Waxman et al., 1994).

VI. THERAPEUTIC POTENTIAL FOR ECE INHIBITORS

With the recent discovery of potent and specific endothelin-receptor antagonists (Clozel et al., 1993b; for review see Huggins et al., 1993), evidence is now accumulating to strengthen the idea that endothelin peptides play important roles in the pathophysiology of various disease states such as hypertension, asthma, congestive heart failure, renal failure, and cerebral vasospasm (for reviews see Doherty, 1992; Masaki et al. 1992). However, all the functions of the known endothelin receptor subtypes (ET_A, ET_B, ET_C) have not yet been clearly identified and it is becoming evident that the family of endothelin receptors might not simply be restricted to these three subtypes (see Chapter 2). Therefore, despite all the unsolved problems encountered in attempts to isolate and purify pathophysiologically relevant ECE(s) and to design potent and selective inhibitors, it is still possible that ECE inhibitors might provide some advantages over endothelin-receptor antagonists to interfere with the multiple deleterious consequences provoked by elevated production of endothelins. An analogy could be made with the therapeutic usefulness of angiotensin-converting enzyme inhibitors.

NOTE ADDED IN PROOF

The molecular pharmacology of ECEs was recently reviewed (Turner and Murphy, 1996). Two metalloproteases ECE-l and ECE-2 were cloned, sequenced and expressed (Xu, et al., 1994; Emoto and Yanagisawa, 1995). Phosphinic acids analogs of phosphoramidon were reported to inhibit ECE with a high degree of selectivity against neutral endopeptidase (Lloyd et al., 1996; Chackalamannil et al., 1996). Novel analogs of phosphoramidon were synthesized (Keller, et al., 1996). Hydroxamates derived from malonyl amino acids were found to be extremely potent inhibitors of ECE in the subnanomolar range (Biohovsky et al., 1995). Optimization of retrothiorphan afforded a thiol derivative twofold more potent than phosphoramidon (Kukkola et al., 1996). A natural compound, FR901533, having a benzo[a]naphtacen quinone structure showed a potent and selective inhibitory activity (Tsurumi, et al., 1995).

REFERENCES

Abassi, Z.A., Tate, J.E., Golomb, E. and Keiser, H.R., Role of neutral endopeptidase in the metabolism of endothelin, *Hypertension*, 20, 89-95. 1992.

Abassi, Z.A., Golomb, E., Bridenbaugh, R. and Keiser, H.R., Metabolism of endothelin-1 by recombinant neutral endopeptidase EC.3.4.24.11., *Br. J. Pharmacol.*, 109, 1024-1028. 1993.

Ahn, K., Beningo, K., Olds, G. and Hupe, D., The endothelin-converting enzyme from human umbilical vein is a membrane-bound metalloprotease similar to that from bovine aortic endothelial cells, *Proc. Natl. Acad. Sci. U.S.A.*, 89, 8606-8610. 1992.

Alstein, M., Blumberg, S. and Vogel, Z., Phosphoryl-Leu-Phe: a potent inhibitor of the degradation of enkephalin by enkephalinase, *Eur. J. Pharmacol.*, 76, 299-300. 1982.

Anderson, R.G.W. and Orci, L., A view of acidic intracellular compartments, *J. Cell. Biol.*, 106, 539-543. 1988.

Arai, K., Ashikawa, N., Nakakita, Y., Matsuura, A., Ashikawa, N. and Munekata, M., Aspergillomarasmine A and B, potent microbial inhibitors of endothelin-converting enzyme, *Biosci. Biotech. Biochem.*, 57, 1944-1945. 1993.

Ashizawa, N., Okumura, H., Kobayashi, F., Aotsuka, T., Takahashi, M., Asakura, R., Arai, K. and Matsuura, A., Inhibitory activities of metal chelators on endothelin-converting enzyme. I. *In vitro* studies, *Biol. Pharm. Bull.*, 17, 207-211. 1994a.

Ashizawa, N., Okumura, H., Kobayashi, F., Aotsuka, M., Asakura, R., Arai, K., Ashikawa, N. and Matsuura, A., Inhibitory activities of metal chelators on endothelin-converting enzyme. II. *In vivo* studies, *Biol. Pharm. Bull.*, 17, 212-216. 1994b.

Balwierczak, J.L., Wong, M. and Jeng, A.Y., A simple assay for the measurement of conversion of big endothelin-1 to endothelin-1 by smooth muscle, *Eur. J. Pharmacol.*, 250, 379-384. 1993.

Barr, P.J., Mammalian subtilisins: the long-sought dibasic processing endoproteases, *Cell*, 66, 1-3. 1991.

Benatti, L., Cozz, L., Zamai, M., Tamburin, M., Vaghi, F., Caiolfa, V.R., Fabbrini, M.S. and Sarmientos, P., Human preproendothelin-1 is converted into active endothelin-1 by baculovirus-infected insect cells, *Biochem. Biophys. Res. Commun.*, 186, 753-759. 1992.

Bertenshaw, S.R., Rogers, R.S., Stern, M.K., Norman, B.H., Moore, W.M., Jerome, G.M., Branson, L.M., McDonald, J.F., McMahon, E.G. and Palomo, M.A., Phosphorus-containing inhibitors of endothelin converting enzyme: effects of the electronic nature of phosphorus on inhibitor potency, *J. Med. Chem.*, 36, 173-176. 1993a.

Bertenshaw, S.R., Talley, J.J., Rogers, R.S., Carter, J.S., Moore, W.M., Branson, L.M. and Koboldt, C.M., Thiol and hydroxamic acid containing inhibitors of endothelin converting enzyme, *Bioorg. Med. Chem. Lett.*, 3, 1953-1958. 1993b.

Bigaud, M., Hauss, B., Schalk, C., Jauch, M.F. and D'Orchymont, H., Structure activity relationship of phosphoramidon derivatives for *in vivo* endothelin-converting-enzyme inhibition, *Fundam. Clin. Pharmacol.*, 8, 155-161. 1994.

Biohovsky, R.H., Erhardt, P.W., Lampe, J.W., Mohan, R. and Shaw, K.J., Inhibitors of the Conversion of Big Endothelin to Endothelin, World Pat. Appl., WO9311154-A1. 1993.

Biohovsky, R., Levinson, B.L., Loewi, R.C., Erhardt, P.W. and Polokoff, M.A., Hydroxamic acids as potent inhibitors of endothelin-converting enzyme from human bronchiolar smooth muscle, *J. Med. Chem.*, 38, 2119-2129. 1995.

Bird, J.E., Waldron, T.L., Little, D.K., Asaad, M.M., Dorso, C.R., DiDonato, G. and Norman, J.A., The effects of novel cathepsin E inhibitors on the big endothelin pressor response in conscious rats, *Biochem. Biophys. Res. Commun.*, 182, 224-231. 1992.

Bird, J.C., De Mello, R., Harper, G.P., Hunter, D.J., Karran, E.H., Markwell, R.E., Miles-Williams, A.J., Rahman, S.S. and Ward, R.W., Synthesis of novel N-phosphonoalkyl dipeptide inhibitors of human collagenase, *J. Med. Chem.*, 37, 158-169. 1994.

Bloch, K.D., Eddy, R.L, Shows, T.B. and Quertermous, T., cDNA cloning and chromosomal assignment of the gene encoding endothelin-3, *J. Biol. Chem.*, 264, 18156-18161. 1989.

Bresnahan, P.A., Leduc, R., Thomas, L., Thorner, J., Gibson, H.L., Brake, A.J., Barr, P.J. and Thomas, G., Human *fur* gene encodes a yeast KEX2-like endoprotease that cleaves pro-β-NGF *in vivo*, *J. Cell. Biol.*, 111, 2851-2859. 1990.

Chackalamannil, S., Chung, S., Stamford, A.W., McKittrick, B.A., Wang, Y., Tsai, H., Cleven, R., Fawzi, A. and Czarniecki, M., Highly potent and selective inhibitors of endothelin converting enzyme, *Biorg. Med. Chem. Lett.*, 6, 1257-1260. 1996.

Clozel, M. and Watanabe, H., BQ 123, a peptidic endothelin ET_A receptor antagonist, prevents the early cerebral vasospasm following subarachnoid hemorrhage after intracisternal but not intravenous injection, *Life Sci.*, 52, 825-834. 1993a.

Clozel, M., Breu, V., Burri, K., Cassal, J.M., Fischli, W., Gray, G.A., Hirth, G., Löffler, B.M., Müller, M., Neidhart, W. and Ramuz, H., Pathophysiological role of endothelin revealed by the first orally active endothelin receptor antagonist, *Nature*, 365, 759-761. 1993b.

Corder, R., Harrison, V.J., Khan, N., Anggård, E.E. and Vane, J., Effects of phosphoramidon in endothelial cell cultures on the endogenous synthesis of endothelin-1 and on conversion of exogenous big endothelin-1 to endothelin-1, *J. Cardiovasc. Pharmacol.*, 22 (Suppl. 8), S73-S76, 1993.

Cosentino, F., McMahon, E.G, Carter, J.S. and Katusic, Effect of endothelin-A receptor antagonist BQ-123 and phosphoramidon on cerebral vasospasm, *J. Cardiovasc. Pharmacol.*, 22 (Suppl. 8), S332-S335. 1993.

De Lombaert, S., Erion, M.D., Tan, J., Blanchard, L., El-Chehabi, L., Ghai, R.D., Sakane, Y., Berry, C. and Trapani, A.J., N-Phosphonomethyl dipeptides and their phosphonate prodrugs, a new generation of neutral endopeptidase (NEP, EC 3.4.24.11) inhibitors, *J. Med. Chem.*, 37, 498-511. 1994.

Deng, Y., Savage, P., Shetty, S.S, Martin, L.L. and Jeng, A.Y., Identification and partial purification of a thiol endothelin converting enzyme from porcine aortic endothelial cells, *J. Biochem.*, 111, 346-351. 1992.

Doherty, A.M., Endothelin: a new challenge, *J. Med Chem.* 35,1493-1508. 1992.

D'Orleans-Juste, P., Lidbury, P.S., Warner, T.D. and Vane, J.R., Intravascular big endothelin increases circulating levels of endothelin-1 and prostanoids in the rabbit, *Biochem. Pharmacol.*, 39, R21-R22. 1990.

Ehrenreich, H., Anderson, R.W., Fox, C.H., Riekmann, P., Hoffman, G.S., Travis, W.D., Coligan, J.E., Kehrl, J.H. and Fauci, A.S., Endothelins, peptides with potent vasoactive properties, are produced by human macrophages, *J. Exp. Med.*, 172, 1741-1748. 1990.

Elliott, R.L., Marks, N., Berg, M.J. and Portoghese, P.S., Synthesis and biological evaluation of phosphonamidate peptide inhibitors of enkephalinase and angiotensin-converting enzyme, *J. Med. Chem.*, 28, 1208-1216. 1985.

Emori, T., Hirata, Y., Ohta, K., Shichiri, M., Shimokado, K. and Marumo, F., Concomitant secretion of big endothelin and its C-terminal fragment from human and bovine endothelial cells, *Biochem. Biophys. Res. Commun.*, 162, 217-223. 1989.

Emoto and Yanagisawa, Endothelin-coverting enzyme-2 is a membrane-bound, phosphormamidon-sensitive metalloprotease with acidic pH optimum, *J. Biol. Chem.*, 270, 15262-15268. 1995.

Erdös, E.G. and Skidgel, R.A., Neutral endopeptidase 24.11 (enkephalinase) and related regulators of peptides hormones, *FASEB J.*, 3, 145-151. 1989.

Fabbrini, M.S., Valsasina, B., Nitti, G., Benatti, L. and Vitale, A., The signal peptide of human preproendothelin-1, *FEBS Lett.*, 286, 91-94. 1991a.

Fabbrini, M.S., Vitale, A., Patrono, C., Zamai, M., Vaghi, F., Caiolfa, V., Monaco, L., and Benatti, L., Heterologous *in vivo* processing of human preproendothelin 1 into bioactive peptides, *Proc. Natl. Acad. Sci. U.S.A.*, 88, 8939-8943. 1991.

Fabbrini, M.S., Vitale, A., Pedrazzini, E., Nitti, G., Zamai, M., Tamburin, M., Caiolfa, V.R., Patrono, C. and Benatti, L., *In vivo* expression of mutant preproendothelins. Hierarchy of processing events but no strict requirement of Trp-Val at the processing site, *Proc. Natl. Acad. Sci.U.S.A.*, 90, 3923-3927. 1993.

Faust, P.L., Wall, D.A., Perara, E., Lingappa, V.R. and Kornfeld, S., Expression of human cathepsin D in *Xenopus* oocytes. Phosphorylation and intracellular targeting, *J. Cell. Biol.*, 105, 1937-1945. 1987.

Flynn, G.A. and Giroux, E.L., The synthesis of an aminophosphonic acid converting enzyme inhibitor, *Tetrahedron Lett.*, 27, 1575-1758. 1986.

Foreman, R.C. and Judah, J.D., The processing and secretion of rat serum albumin by oocytes from *Xenopus laevis*. *FEBS Lett.*, 219, 75-78. 1987.

Fuller, R.S., Brake, A.J. and Thorner, J., Intracellular targeting and structural conservation of a prohormone-processing endoprotease, *Science*, 246, 482-486. 1989.

Fukami, T., Hayama, T., Amano, Y., Nakamura, Y., Arai, Y., Matsuyama, K., Yano, M. and Ishikawa, K., Aminophosphonate endothelin converting enzyme inhibitors: potency-enhancing and selectivity improving modifications of phosphoramidon, *Bioorg. Med. Chem. Lett.*, 4, 1257-1262. 1994.

Fukuroda, T., Noguchi, K., Tsuchida, S., Nishikibe, M., Ikemoto, F., Okada, K. and Yano, M., Inhibition of biological actions of big endothelin-1 by phosphoramidon, *Biochem. Biophys. Res. Commun.*, 172, 390-395. 1990.

Gardiner, S.M., Compton, A.M., Kemp, P.A. and Bennett, T., The effects of phosphoramidon on the regional haemodynamic responses to human proendothelin [1-38] in conscious rats, *Br. J. Pharmacol.*, 130, 200-92015. 1991.

Gardiner, S.M., Kemp, P.A., Compton, A.M. and Bennett, T., Coeliac haemodynamic effects of endothelin-1, endothelin-3, proendothelin-1 [1-38] and proendothelin-3 [1-41] in conscious rats, *Br. J. Pharmacol.*, 106, 483-488. 1992a.

Gardiner, S.M., Kemp, P.A. and Bennett, T., Inhibition by phosphoramidon of the regional haemodynamic effects of pro-endothelin-2 and -3 in conscious rats, *Br. J. Pharmacol.*, 107, 584-590. 1992b.

Gardiner, S.M., Kemp, P.A. and Bennett, T., Effects of the neutral endopeptidase inhibitor, SQ 28,603, on regional haemodynamic responses to atrial natriuretic peptide or proendothelin-1 [1-38] in conscious rats, *Br. J. Pharmacol.*, 106, 180-186. 1992c.

Gomez-Monterrey, I., Muniz, R., Pérez-Martin, C., Lopéz de Ceballos, M., Del Rio, J. and Garcia-Lopez, M.T., Ketomethylene analogues of phosphoryl dipeptides related to phosphoramidon: synthesis and inhibition of proteases, *Arch. Pharm.*, 325, 261-265. 1992.

Gui, G., Xu, D., Emoto, N. and Yanagisawa, M., Intracellular localization of membrane-bound endothelin-converting enzyme from rat lung, *J. Cardiovasc. Pharmacol.*, 22 (Suppl. 8), S53-S56. 1993.

Hemsen, A., Pernow, J. and Lundberg, J.M., Regional extraction of endothelins and conversion of big endothelin to endothelin-1 in the pig, *Acta Physiol. Scand.*, 141, 325-334. 1991.

Hioki, Y., Okada, K., Ito, H., Matsuyama, K. and Yano, M., Endothelin converting enzyme of bovine carotid artery smooth muscles, *Biochem. Biophys. Res. Commun.*, 174, 446-451. 1991.

Hirata, Y., Kanno, K., Watanabe, T.X., Kumagaye, S.I., Nakajima, K., Kimura, T., Sakakibara, S. and Marumo, F., Receptor binding and vasoconstrictor activity of big endothelin, *Eur. J. Pharmacol.*, 176, 225-228. 1990.

Hisaki, K., Matsumura, Y., Ikegawa, R., Nishiguchi, S., Hayashi, K., Takaoka, M. and Morimoto, S., Evidence for phosphoramidon-sensitive conversion of big endothelin-1 to endothelin-1 in isolated rat mesenteric artery, *Biochem. Biophys. Res. Commun.*, 177, 1127-1132. 1991.

Hisaki, K., Matsumura, Y., Nishiguchi, S., Fujita, K., Takaoka, M. and Morimoto, S., Endothelium-independent pressor effect of big endothelin-1 and its inhibition by phosphoramidon in rat mesenteric artery, *Eur. J. Pharmacol.*, 241, 75-81. 1993.

Holmquist, B. and Vallee, B.L., Metal-coordinating substrate analogs as inhibitors of metalloenzymes, *Proc. Natl. Acad. Sci. U.S.A.*, 76, 6216-6220. 1979.

Hudgin, R.L., Charleson, S.E., Zimmerman, M., Mumford, R. and Wood, P.L., Enkephalinase: selective peptide inhibitors, *Life Sci.*, 29, 2593-2601. 1981.

Huggins, J.P., Pelton, J.T. and Miller, R.C., The structure and specificity of endothelin receptors: their importance in physiology and medicine, *Pharmacol. Ther.*, 59, 55-123. 1993.

Ikegawa, R., Matsumura, Y., Takaoka, M. and Morimoto, S., Evidence for pepstatin-sensitive conversion of porcine big endothelin-1 to endothelin-1 by the endothelial cell extract, *Biochem. Biophys. Res. Commun.*, 167, 860-866. 1990a.

Ikegawa, R., Matsumura, Y., Tsukahara, Y., Takaoka, M. and Morimoto, S., Phosphoramidon, a metalloproteinase inhibitor, suppresses the secretion of endothelin-1 from cultured endothelial cell by inhibiting a big endothelin-1 converting enzyme, *Biochem. Biophys. Res. Commun.*, 171, 669-675. 1990b.

Ikegawa, R., Matsumura, Y., Tsukahara, Y., Takaoka, M. and Morimoto, S., Phosphoramidon inhibits the generation of endothelin-1 from exogenously applied big-endothelin-1 in cultured vascular endothelial cells and smooth muscle cells, *FEBS Lett.*, 293, 45-48. 1991.

Inagaki, S., Ikeda, M., Tomita, T., Morita, A., Nomizu, M., Kumagai, T., Okitsu, M., Yokogoshi, H. and Roller, P.P., *In vitro* and *in vivo* effects of a new endothelin-converting enzyme inhibitor, EV-D, *Jpn. J. Pharmacol.*, 61 (Suppl. 1), 214P [Abstr.]. 1993.

Inoue, A., Yanagisawa, M., Kimura, S., Kasuya, Y., Miyaushi, T., Goto, K. and Masaki, T., The human endothelin family: three structurally and pharmacologically distinct isopeptides predicted by three separate genes, *Proc. Natl. Acad. Sci. U.S.A.*, 86, 2863-2867. 1989.

Ishikawa, S., Tsukada, H., Yuasa, H., Fukue, M., Wei, S., Onizuka, M., Miyauchi, T., Ishikawa, T., Mitsui, K., Goto, K. and Hori, M., Effects of endothelin-1 and conversion of big endothelin-1 in the isolated perfused rabbit lung, *J. Appl. Physiol.*, 72, 2387-2395. 1992.

Ishikawa, K., Fukami, T. and Hayama, S., New Aminophosphoric Acid Derivatives Inhibit Endothelin-Converting Enzyme for Treating Hypertension, Diabetes Mellitus, Bronchial Asthma, Atherosclerosis, Acute Renal Failure, Myocardial Infarction, Angina, Endotoxin Shock, Etc., Jpn. Pat., JP05148277-A1. 1993.

Itoh, Y., Yanagisawa, M., Ohkubo, S., Kimura, C., Kosaka, T., Inoue, A., Ishida, N., Mitsui, Y., Onda, H., Fujino, M. and Masaki, T., Cloning and sequence analysis of cDNA encoding the precursor of a human endothelium-derived vasoconstrictor peptide, endothelin: identity of human and porcine endothelin, *FEBS Lett.*, 231, 440-444. 1988.

Kam, C.M., Nishino, N. and Powers, J.C., Inhibition of thermolysin and carboxypeptidase A by phosphoramidates, *Biochemistry*,18, 3032-3038. 1979.

Kashiwabara, T., Inagaki, Y., Ohta, H., Iwamatsu, A., Nomizu, M., Morita, A. and Nishikori, K., Putative precursors of endothelin have less vasoconstrictor activity *in vitro* but a potent pressor effect *in vivo*, *FEBS Lett.*, 247, 73-76. 1989.

Kaw, S., Hecker, M. and Vane, J.R., The two-step conversion of big endothelin-1 to endothelin-1 and degradation of endothelin-1 by subcellular fractions from human polymorphonuclear leukocytes, *Proc. Natl. Acad. Sci. U.S.A.*, 89, 6886-6890. 1992.

Kaw, S., Warner, T.D. and Vane J.R., The metabolism of endothelin-1 and big endothelin-1 by the isolated perfused kidney of the rabbit, *J. Cardiovasc. Pharmacol.*, 22 (Suppl. 8), S65-S68. 1993.

Keller, P.M., Lee, C.-P., Fenwick, A.E., Atkinson, S.T., Elliott, J.D. and DeWolf, W.E., Endothelin-converting enzyme: substrate specificity and inhibition by novel analogs of phosphoramidon, *Biochem. Biophys. Res. Commun.*, 223, 371-378. 1996.

Kimura, S., Kasuya, Y., Sawamura, T., Shinmi, O., Sugita, Y., Yanagisawa, M., Goto, K. and Masaki, T., Conversion of big endothelin-1 to 21-residue endothelin-1 is essential for expression of full vasoconstrictor activity: structure-activity relationships of big endothelin-1, *J. Cardiovasc. Pharmacol.*, 13 (Suppl. 5), S5-S7. 1989.

Knap, A.K., Soriano, A., Savage, P., Del Grande, D. and Shetty, S.S., Identification of a novel aspartyl endothelin converting enzyme in porcine aortic endothelial cells, *Biochem. Mol. Biol. Int.*, 29, 739-745. 1993.

Koller, J., Mair, P., Wieser, C. and Pomaroli, A., Endothelin and big endothelin concentrations in injured patients, *N. Engl. J. Med.*, 325, 1518. 1991.

Komiyama, T., Suda, H., Aoyagi, T., Takeuchi, T. and Umezawa, H., Studies on inhibitory effect of phosphoramidon and its analogs on thermolysin, *Arch. Biochem. Biophys.*, 171, 727-731. 1975.

Kortylewicz, Z.P. and Galardy, R.E., Phosphoramidate peptide inhibitors of human skin fibroblast collagenase, *J. Med. Chem.*, 33, 263-273.1990.

Kosaka, T., Suzuki, N., Matsumoto, H., Itoh, Y., Yasuhara, T., Onda, H. and Fijino, M., Synthesis of the vasoconstrictor peptide endothelin in kidney cells, *FEBS Lett.*, 249, 42-46. 1989.

Kosaka, T., Ishibashi, Y., Suzuki, N., Matsumoto, H., Ohkubo, S., Kitada, C., Onda, H. and Fujino, M., The mechanism of human ET-2 secretion, in *Peptide Chemistry 1991*, Suzuki, A., Ed., Protein Research Foundation, Ozaka, Japan, 1992, 19-24.

Krsmanocic, L.Z., Stojilkovic, S.S., Ball, T., Al-Damluji, S., Weiner, R.I. and Catt, K.J., Receptors and neurosecretory actions of endothelin in hypothalamic neurons, *Proc. Natl. Acad. Sci. U.S.A.*, 88, 11124-11128. 1991.

Kukkola, P.J., Bilci, N.A., Kozak, W.Z., Savage, P. and Jeng, A.Y., Optimization of retro-thiorphan for inhibition of endothelin converting enzyme, *Biorg. Med. Chem. Lett.*, 6, 619-624. 1996.

Kundu, G.C. and Wilson, I.B., Identification of endothelin converting enzyme in bovine lung membranes using a new fluorogenic substrate, *Life Sci.*, 50, 965-970. 1992.

Kundu, G.C. and Wilson, I.B., Endothelin-converting enzyme: the binding of metal ions, *Int. J. Peptide Protein Res.*, 42, 64-67. 1993.

Laporte, S., Denault, J.B., D'Orléans-Juste, P. and Leduc, R., Presence of furin mRNA in culture bovine endothelial cells and possible involvement of furin in the processing of endothelin precursor, *J. Cardiovasc. Pharmacol.*, 22 (Suppl. 8), S7-S10. 1993.

Lawrence, E. and Brain, S.D., Big endothelin-1 and big endothelin-3 are constrictor agents in the microvasculature: evidence for the local phosphoramidon-sensitive conversion of big endothelin-1, *Eur. J. Pharmacol.*, 233, 243-250. 1993.

Ledger, P.W. and Tanzer, M.L., Monensin: a perturbant of cellular physiology, *Trends Biochem. Sci.*, 9, 313-314. 1984.

Lees, W.E., Kalinka, S., Meech, J., Capper, S.J., Cook, N.D. and Kay, J., Generation of endothelin by cathepsin E, *FEBS Lett.*, 273, 99-102. 1990.

Le Monnier de Gouville, A.C. and Cavero, I., Differential pharmacological profile of endothelin-1 and its precursor, big endothelin, *J. Cardiovasc. Pharmacol.*, 17 (Suppl. 7), S362-S365. 1991.

Lloyd J., Schmidt, J.B., Hunt, J.T., Barrish, J.C., Little, D.K. and Tymiak, A.A., Solid phase synthesis of phosphinic acid endothelin converting enzyme inhibitors, *Biorg. Med. Chem. Lett.*, 6, 1323-1326. 1996.

MacCumber, M.W., Ross, C.A. and Snyder, S.H., Endothelin in brain: receptors, mitogenesis and biosynthesis in glial cells, *Proc. Natl. Acad. Sci. U.S.A.*, 87, 2359-2363. 1990.

Marumo, F. and Tomita, K., Secretion of endothelin and related peptides from renal epithelial cell lines, *Jpn. J. Physiol.*, 40, S77. 1990.

Masaki, T., Yanagisawa, M. and Goto, K., Physiology and pharmacology of endothelins, *Med. Res. Rev.*, 12, 391-421. 1992.

Matsumura, Y., Ikegawa, R., Takaoka, M. and Morimoto, S., Conversion of porcine big endothelin by an extract from the porcine aortic endothelial cells, *Biochem. Biophys. Res. Commun.*, 167, 203-210. 1990a.

Matsumura, Y., Ikegawa, R., Tsukahara, Y., Takaoka, M. and Morimoto, S., Conversion of big endothelin-1 to endothelin-1 by two types of metalloproteinases derived from porcine aortic endothelial cells, *FEBS Lett.*, 272, 166-170. 1990b.

Matsumura, Y., Ikegawa, R., Tsukahara, Y., Takaoka, M. and Morimoto, S., Conversion of big endothelin-1 to endothelin-1 by two-types of metalloproteinases of cultured porcine vascular smooth muscle cells, *Biochem. Biophys. Res. Commun.*, 178, 899-905. 1990c.

Matsumura, Y., Hisaki, K., Takaoka, M. and Morimoto, S., Phosphoramidon, a metalloproteinase inhibitor, suppresses the hypertensive effect of big endothelin-1, *Eur. J. Pharmacol.*, 185, 103-106. 1990d.

Matsumura, Y., Ikegawa, R., Tsukahara, Y., Takaoka, M. and Morimoto, S., N-ethylmaleimide differentiates endothelin converting activity by two types of metalloproteinases derived from vascular endothelial cells, *Biochem. Biophys. Res. Commun.*, 178, 531-538. 1991a.

Matsumura, Y., Ikagawa, R., Suzuki, Y., Takaoka, M., Uchida, T., Kido, H., Shinyama, H., Hayashi, K., Watanabe, M. and Morimoto, S., Phosphoramidon prevents cerebral vasospasm following subarachnoid hemorrhage in dogs: the relationship to endothelin-1 levels in the cerbrospinal fluid, *Life Sci.*, 49, 841-848. 1991b.

Matsumura, Y., Umekawa, T., Kawamura, H., Takaoka, M., Robinson, P.S., Cook, N.D. and Morimoto, S., A simple method for the measurement of phosphoramidon-sensitive endothelin converting enzyme activity, *Life Sci.*, 51, 1603-1611. 1992.

Matsumura, Y., Tsukahara, T., Kuninibu, K., Takaoka, M. and Morimoto, S., Phosphoramidon-sensitive endothelin-converting enzyme in vascular endothelial cells converts big endothelin-1 and big endothelin-3 to their mature form, *FEBS Lett.*, 305, 86-90. 1992.

Matsumara, Y., Fujita, K., Takaoka, M. and Morimoto, S., Big endothelin-3-induced hypertension and its inhibition by phosphoramidon., *Eur. J. Pharmacol.*, 230, 89-93. 1993.

Matsuura, A., Okumura, H., Ashizawa, N. and Kobayashi, F., Big endothelin-1-induced sudden death is inhibited by phosphoramidon in mice, *Life Sci.*, 50, 1631-1638. 1992.

Matsuura, A., Okumura, H., Asakura, R., Ashizawa, N., Takahashi, M., Kobayashi, F., Ashikawa, N. and Arai, K., Pharmacological profiles of Aspergillomarasmines as endothelin converting enzyme inhibitors, *Jpn. J. Pharmacol.*, 63, 187-193. 1993.

Mattera, G.G., Eglezos, A., Renzetti, A.R. and Mizrahi, J., Comparison of the cardiovascular and neural activity of endothelin-1, -2, -3 and respective proendothelins: effects of phosphoramidon and thiorphan, *Br. J. Pharmacol.*, 110, 331-337. 1993.

Matthews, B.W., Structural basis of the action of thermolysin and related zinc peptidases, *Acc. Chem. Res.*, 21, 333-340. 1988.

McMahon, E.G., Fok, K.F., Moore, W.M., Smith, C.E., Siegel, N.R. and Trapani, A.J., *In vitro* and *in vivo* activity of chymotrypsin-activated big endothelin (porcine 1-40), *Biochem. Biophys. Res. Commun.*, 161, 406-413. 1989.

McMahon, E.G., Palomo, M.A., Moore, W.M., McDonald, J.F. and Stern, M.K., Phosphoramidon blocks the pressor activity of porcine big endothelin-1-(1-39) *in vivo* and conversion of big endothelin-1-(1-39) to endothelin-1-(1-21) *in vitro*, *Proc. Natl. Acad. Sci. U.S.A.*, 88, 703-707. 1991a.

McMahon, E.G., Palomo, M.A. and Moore, W.M., Phosphoramidon blocks the pressor activity of big endothelin[1-39] and lowers blood pressure in spontaneously hypertensive rats, *J. Cardiovasc. Pharmacol.*, 17 (Suppl. 7), S29-S33. 1991b.

Mino, N., Kobayashi, M., Nakajima, A., Amano, H., Shimamoto, K., Ishikawa, K., Watanabe, K., Nishikibe, M., Yano, M. and Ikemoto, F., Protective effect of a selective endothelin receptor antagonist, BQ-123, in ischemic acute renal failure in rats, *Eur. J. Pharmacol.*, 221, 77-83. 1992.

Mitsuhiro, W., Masaaki, M. and Akira, K., Phosphonic Acid Derivatives, Production and Use Thereof, Eur. Pat. Appl., 518299-A2. 1992.

Miyauchi, T., Yanagisawa, M., Tomizawa, T., Sugishita, Y., Suzuki, N., Funino, M., Ajisaka, R., Goto, K., and Masaki, T., Increased plasma concentration of endothelin-1 and big endothelin-1 in acute myocardial infarction, *Lancet*, ii, 53-54. 1989.

Modin, A., Pernow, J. and Lundberg, J.M., Phosphoramidon inhibits the vasoconstrictor effects evoked by big endothelin-1 but not the elevation of plasma endothelin-1 *in vivo*, *Life Sci.*, 49, 1619-1625. 1991.

Mookhtiar, K.A., Marlowe, C.K., Bartlett, P.A. and Van Wart, H.E., Phosphonamidate inhibitors of human neutrophil collagenase, *Biochemistry*, 26, 1962-1965. 1987.

Nakagawa, O., Nakao, K., Saito, Y., Shirakami, G., Jougasaki, M., Mukoyama, M., Hosoda, K., Suga, S.I., Ogawa, Y., Kishimoto, I., Komatsu, Y., Hama, N. and Imura, H., Isolation and characterization of porcine endothelin-like peptide, *J. Cardiovasc. Pharmacol.*, 17(Suppl. 7), S13-S16. 1991.

Neuser, D., Steinke, W., Dellweg, H., Kazda, S. and Stasch, J.P., ^{125}I-endothelin-1 and ^{125}I-big endothelin-1 in rat tissues: autoradiographic localization and receptor binding, *Histochemistry*, 95, 621-628. 1991.

Nirei, H., Hamada, K., Shoubo, M., Sogabe, K., Notsu, Y. and Ono, T., An endothelin ET_A receptor antagonist, FR139317, ameliorates cerebral vasospasm in dogs, *Life Sci.*, 52, 1869-1874. 1993.

Nogushi, K., Fukuroda, T., Ikeno, Y., Hirose, H., Tsukada, Y., Nishikibe, M., Ikemoto, F., Matsuyama, K. and Yano, M., Local formation and degradation of endothelin-1 in guinea-pig airway tissues, *Biochem. Biophys. Res. Commun.*, 179, 830-835. 1991.

Nichols, J.S., Berman, J., Wypij, D.M. and Wiseman, J.S., Evidence aginst a role for aspartyl proteases in intracellular processing of big endothelin, *J. Cardiovasc. Pharmacol.*, 17 (Suppl. 7), S10-S12. 1991.

Okada, K., Miyazaki, Y., Takada, J., Matsuyama, K., Yamaki, T. and Yano, M., Conversion of big endothelin-1 by membrane-bound metalloendopeptidase in cultured bovine endothelial cells, *Biochem. Biophys. Res. Commun.*, 171, 1192-1198. 1990.

Okada, K., Takada, J., Arai, Y., Matsuyama, K. and Yano, M., Importance of the C-terminal region of big-endothelin-1 for specific conversion by phosphoramidon-sensitive endothelin converting enzyme, *Biochem. Biophys. Res. Commun.*, 180, 1019-1023. 1991.

Okada, K., Arai, Y., Hata, M., Matsuyama, K. and Yano, M., Big endothelin-1 structure important for specific processing by endothelin-converting enzyme of bovine endothelial cells, *Eur. J. Biochem.*, 218, 493-498. 1993.

Ohkubo, S., Ogi, K., Hosoya, M., Matsumoto, H., Suzuki, N., Kimura, C., Onda, H. and Fujino, M., Specific expression of human endothelin-2 (ET-2) gene in a renal adenocarcinoma cell line. Molecular cloning of cDNA encoding the precursor of ET-2 and its characterization, *FEBS Lett.*, 274, 136-140. 1990.

Ohlstein, E.H., Arleth, A., Ezekiel, M., Horohonich, S., Ator, M.A., Caltabiano, M.M. and Sung, C.P., Biosynthesis and modulation of endothelin from bovine pulmonary arterial endothelial cells, *Life Sci.*, 46, 181-188. 1990.

Ohnaka, K., Takayanagi, R., Yamauchi, T., Okazaki, H., Ohashi, M., Umeda, F. and Nawata, H., Identification and characterization of endothelin converting activity in cultured bovine endothelial cells, *Biochem. Biophys. Res. Commun.*, 168 1128-1136. 1990.

Ohnaka, K., Nishikawa, M., Takayanagi, R., Haji, M. and Nawata, H., Partial purification of phosphoramidon-sensitive endothelin converting enzyme in porcine aortic endothelial cells: high affinity for *Ricinus communis* agglutinin, *Biochem. Biophys. Res. Commun.*, 185, 611-616. 1992.

Ohnaka, K., Takayanagi, R., Nishikawa, M., Haji, M. and Nawata, H., Purification and characterization of a phosphoramidon-sensitive endothelin-converting enzyme in porcine aortic endothelium, *J. Biol. Chem.*, 268, 26759-26766. 1993.

Ohwaki, T., Sakai, H. and Hirata, Y., Endothelin-converting enzyme activity in human serum lipoprotein fraction, *FEBS Lett.*, 320, 165-168. 1993.

Opgenorth, T.J., Wu-Wong, J.R. and Shiosaki, K., Endothelin-converting enzymes, *FASEB J.*, 6, 2653-2659. 1992.

Patterson, K., MacNaof, R., Rubanyi, G.M. and Parker-Botelho, L.H., Separation and biological activity of chymotrypsin and cathepsin G cleavage product of big endothelin [Abstract], *FASEB J.*, 4, 909. 1990.

Pollock, D.M. and Opgenorth, T.J., Comparison of the hemodynamic effects of endothelin-1 and big endothelin-1 in the rat, *Biochem. Biophys. Res. Commun.*, 179, 1122-1126. 1991a.

Pollock, D.M. and Opgenorth, T.J., Evidence for metalloprotease involvement in the *in vivo* effects of big endothelin 1, *Am. J. Physiol.*, 261, R257-R263. 1991b.

Pollock, D.M., Shiosaki, K., Sullivan, G.M. and Opgenorth, T.J., Rhamnose moiety of phosphoramidon is not required for *in vivo* inhibition of endothelin converting enzyme, *Biochem. Biophys. Res. Commun.*, 186, 1146-1150. 1992.

Pollock, D.M., Divish, B.J., Milicic, I., Novosad, E.I., Burres, N.S. and Opgenorth, T.J., *In vivo* characterization of a phosphoramidon-sensitive endothelin-converting enzyme in the rat, *Eur. J. Pharmacol.*, 231, 459-464. 1993a.

Pollock, D.M., Divish, B.J. and Opgenorth, T.J., Stimulation of endogenous endothelin release in the anesthetized rat, *J. Cardiovasc. Pharmacol.*, 22 (Suppl. 8), S295-S298. 1993b.

Poncz, L., Gerken, T.A., Dearborn, D.G., Grobelny, D. and Galardy, R.E., Inhibition of elastase of *Pseudomonas aeruginosa* by Nα-Phosphoryl dipeptides and kinetics of spontaneous hydrolysis of the inhibitors, *Biochemistry*, 23, 2766-2772. 1984.

Pons, F., Touvay, C., Lagente, V., Mencia-Huerta, J.M. and Braquet, P., Involvement of a phosphoramidon-sensitive endopeptidase in the processing of big endothelin-1 in the guinea pig, *Eur. J. Pharmacol.*, 217, 65-70. 1992.

Rang, H.P. and Dale, M.M., *Pharmacology*, Churchill Livingstone, Edinburgh, 1991, chap. 10.

Resink, T.J., Hahn, A.W.A., Scott-Burden, T., Powell, J., Weber, E. and Buhler, F.R., Inducible endothelin messenger RNA expression and peptide secretion in cultured human vascular smooth muscle cells, *Biochem. Biophys. Res. Commun.*, 168, 1303-1310. 1990.

Saito, Y., Nakao, K., Itoh, H., Yamada, T., Mukoyama, M., Arai, H., Hosoda, K., Shirakami, G., Suga, S., Jougasaki, M., Morichika, S. and Imura, H., Endothelin in human plasma and culture medium of aortic endothelial cells — detection and characterization with radioimmunoassay using monoclonal antibody, *Biochem. Biophys. Res. Commun.*, 161, 320-326. 1989.

Sakamoto, H., Sasaki, S., Hirata, Y., Imai, T., Ando, K., Ida, T., Sakurai, T., Yanagisawa, M., Masaki, T. and Marumo, F., Production of endothelin-1 by rat cultured mesangial cells, *Biochem. Biophys. Res. Commun.*, 169, 462-468. 1990.

Savage, P., Shetty, S.S., Martin, L.L. and Jeng, A.Y., Conversion of proendothelin-1 into endothelin-1 by aspartylprotease, *Int. Peptide Protein Res.*, 42, 227-232. 1993.

Sawamura, T., Kimura, S., Shinmi, O., Sugita, Y., Yanagisawa, M. and Masaki, T., Analysis of endothelin related peptides in culture supernatant of porcine aortic endothelial cells: evidence for biosynthetic pathway of endothelin-1, *Biochem. Biophys. Res. Commun.*, 162, 1287-1294. 1989.

Sawamura, T., Shinmi, O., Kishi, N., Sugita, Y., Yanagisawa, M., Goto, K., Masaki, T. and Kimura, S., Analysis of big endothelin-1 digestion by cathepsin D, *Biochem. Biophys. Res. Commun.*, 172, 883-889. 1990a.

Sawamura, T., Kimura, S., Shinmi, O., Sugita, Y., Kobayashi, M., Mitsui, Y., Yanagisawa, M., Goto, K. and Masaki, T., Characterization of endothelin converting enzyme activities in soluble fraction of bovine cultured endothelial cells, *Biochem. Biophys. Res. Commun.*, 169, 1138-1144. 1990b.

Sawamura, T., Kimura, S., Shinmi, O., Sugita, Y., Yanagisawa, M., Goto, K. and Masaki, T., Purification and characterization of putative endothelin converting enzyme in bovine adrenal medulla: evidence for a cathepsin D-like enzyme, *Biochem. Biophys. Res. Commun.*, 168, 1230-1236. 1990c.

Sawamura, T., Kasuya, Y., Matsushita, Y., Suzuki, N., Shinmi, O., Kishi, N., Sugita, Y., Yanagisawa, M., Goto, K., Masaki, T. and Kimura, S., Phosphoramidon inhibits the intracellular conversion of big endothelin-1 in cultured endothelial cells, *Biochem. Biophys. Res. Commun.*, 174, 779-784. 1991.

Schwartz, J.C., Costentin, J. and Lecomte, J.M., Pharmacology of enkephalinase inhibitors, *TIPS*, 6, 472-476. 1985.

Sessa, W.C., Kaw, S., Hecker, M. and Vane, J.R., The biosynthesis of endothelin-1 by human polymorphonuclear leukocytes, *Biochem. Biophys. Res. Commun.*, 174, 613-618. 1991a.

Sessa, W.C., Kaw, S., Zembowicz, A., Änggård, E., Hecker, M. and Vane, J.R., Human polymorphonuclear leukocytes generate and degrade endothelin-1 by two distinct neutral protease, *J. Cardiovasc. Pharmacol.*, 17(Suppl. 7), S34-S38. 1991b.

Shields, P.P., Gonzales, T.A., Charles, D., Gilligan, J.P. and Stern, W., Accumulation of pepstatin in cultured endothelial cells and its effect on endothelin processing, *Biochem. Biophys. Res. Commun.*, 177, 1006-1012. 1991.

Shinego, T., Mima, T., Yanagisawa, M., Saito, A., Fujimori, A., Shiba, R., Goto, K., Kimura, S., Yamashita, K., Yamasaki, Y., Masaki, T. and Takakura, K., Possible role of endothelin in the pathogenesis of cerebral vasospasm, *J. Cardiovasc. Pharmacol.*, 17 (Suppl. 7), S480-S483. 1991.

Shinmi, O., Yorimitsu, K., Moroi, K., Nishiyama, M., Sugita, Y., Saito, T., Inagaki, Y., Masaki, T. and Kimura, S., Endothelin-2-converting enzyme from human renal adenocarcinoma cells is a phosphoramidon-sensitive, membrane-bound metalloprotease, *J. Cardiovasc. Pharmacol.*, 22 (Suppl. 8), S61-S64. 1993.

Shinyama, H., Uchida, T., Kido, H., Hayashi, K., Watanabe, M., Matsumura, Y., Ikegawa, R., Takaoka, M. and Morimoto, S., Phosphoramidon inhibits the conversion of intracisternally administered big endothelin-1 to endothelin-1, *Biochem. Biophys. Res. Commun.*, 178, 24-30. 1991.

Shiosaki, K., Tasker, A.S., Sullivan, G.M., Sorensen, B.K., von Geldern, T.W., Wu-Wong, J.R., Marselle, C.A. and Opgenorth, T.J., Potent and selective inhibitors of an aspartyl protease-like endothelin converting enzyme identified in rat lung, *J. Med. Chem.*, 36, 468-478. 1993.

Steiner, D.F., Smeekens, S.P., Ohagi, S. and Chan, S.J., The new enzymology of precursor processing endoproteases, *J. Biol. Chem.*, 267, 23435-23438. 1992.

Suzuki, N., Matsumoto, H., Kitada, C., Kimura, S. and Fujino, M., Production of endothelin-1 and big-endothelin-1 by tumor cells with epithelial-like morphology, *J. Biochem. (Tokyo)*, 106, 736-741. 1989.

Takada, J., Okada, K., Ikenaga, T., Matsuyama, K. and Yano, M., Phosphoramidon-sensitive endothelin-converting enzyme in the cytosol of cultured bovine endothelial cells, *Biochem. Biophys. Res. Commun.*, 176, 860-865. 1991.

Takahashi, M., Matsushita, Y., Iijima, Y. and Tanzawa, K., Purification and characterization of endothelin-coverting enzyme from rat lung, *J. Biol. Chem.*, 268, 21394-21398. 1993.

Takaoka, M., Miyata, Y., Takenobu, Y., Ikegawa, R., Matsumura, Y. and Morimoto, S., Mode of cleavage of pig endothelin-1 by chymotrypsin. Production and degradation of mature endothelin-1, *Biochem. J.*, 270, 541-544. 1990a.

Takaoka, M., Hukumori, Y., Shiragami, K., Ikegawa, R., Matsumura, Y. and Morimoto, S., Proteolytic processing of porcine big endothelin-1 catalyzed by cathepsin D, *Biochem. Biophys. Res. Commun.*, 173, 1218-1223. 1990b.

Takaoka, M., Takenobu, Y., Miyata, Y., Ikegawa, R., Matsumura, Y. and Morimoto, S., Pepsin, an aspartic protease, converts porcine big endothelin to 21-residue endothelin, *Biochem. Biophys. Res. Commun.*, 166, 436-442. 1990c.

Télémaque, S. and D'Orléans-Juste, P., Presence of a phosphoramidon-sensitive endothelin-converting enzyme which converts big-endothelin-1, but not big-endothelin-3, in the rat vas deferens, *Naunyn-Schmiedeberg's Arch. Pharmacol.*, 344, 505-507. 1991.

Trapani, A.J., Balwierczak, J.L., Lappe, R.W., Stanton, J.L., Graybill, S.C., Hopkins, M.F., Savage, P., Sperbeck, D.M. and Jeng, A.Y., Effects of metalloprotease inhibitors on the conversion of proendothelin-1 to endothelin-1, *Biochem. Mol. Biol. Int.*, 31, 861-867. 1993.

Tronrud, D.E., Monzingo, A.F. and Matthews, B.W., Crystallographic structural analysis of phosphoramidates as inhibitors and transition-state analogs of thermolysin, *Eur. J. Biochem.*, 157, 261-268. 1986.

Tsurumi, Y., Fujie, K., Nishikawa, M., Kiyoto, S. and Okuhara, M., Biological and pharmacological properties of highly selective new endothelin converting enzyme inhibitor WS79089B isolated from Streptosporangium roseum No. 79089, *J. Antibiotics*, 48, 169-174. 1995.

Turner, A.J. and Murphy, L.J., Molecular pharmacology of endothelin converting enzymes, *Biochem. Pharmacol.*, 51, 91-102. 1996.

Umezawa, S., Tatsuta, K., Izawa, O. and Tsuchiya, T., A new microbial metabolite phosphoramidon (isolation and structure), *Tetrahedron Lett.*, 97-100. 1972.

Vemulapalli, S., Rivelli, M., Chiu, P.J.S., del Prado, M. and Hey, J.A., Phosphoramidon abolishes the increases in endothelin-1 release induced by ischemia-hypoxia in isolated perfused guinea-pig lungs, *J. Pharmacol. Exp. Ther.*, 262, 1062-1069. 1992.

Vemulapalli, S., Chiu, P.J.S., Chintal, M. and Bernardino, V., Attenuation of ischemic acute renal failure by phosphoramidon in rats, *Pharmacology*, 47, 188-193. 1993.

Vemulapalli, S., Chiu, P.J.S., Gristi, K., Brown, A., Kurowski, S. and Sybertz, E.J., Phosphoramidon does not inhibit endogenous endothelin-1 release stimulated by hemorrhage, cytokines and hypoxia in rats, *Eur. J. Pharmacol.*, 257, 95-102. 1994.

Verbeuren, T.J., Mennecier, P., Merceron, D., Simonet, S., de Nanteuil, G., Vincent, M. and Laubie, M., Phosphoramidon inhibits the conversion of big ET-1 into ET-1 in the pithed rat and in isolated perfused rat kidneys, *J. Cardiovasc. Pharmacol.*, 22 (Suppl. 8), S81-S84. 1993.

Walter, P. and Blobel, G., Translation of proteins across the endoplasmic reticulum. II. Signal recognition protein (SRP) mediates the selective binding to microsomal membranes of *in vitro*-assembled polysomes synthesizing secretory protein, *J. Cell. Biol.*, 91, 551-556. 1981.

Wagner, O.F., Christ, G., Wojta, J., Vierhapper, H., Parzer, S., Nowotny, P.J., Schneider, B., Waldhäusl, W. and Binder, B.R., Polar secretion of endothelin-1 by cultured endothelial cells, *J. Biol. Chem.*, 267, 16066-16068. 1992.

Warner, T.D., Mitchell, J.A., D'Orleans-Juste, P., Ishii, K., Forstermann, U. and Murad, F., Characterization of endothelin-converting enzyme from endothelial cells and rat brain: detection of the formation of biologically active endothelin-1 by rapid bioassay, *Mol. Pharmacol.*, 41, 399-403. 1992a.

Warner, T.D., Budzik, G.P., Matsumoto, T., Mitchell, J.A., Forstermann, U. and Murad, F., Regional differences in endothelin converting enzyme activity in rat brain: inhibition by phosphoramidon and EDTA, *Br. J. Pharmacol.*, 106, 948-952. 1992.

Watanabe, Y., Naruse, M., Monzen, C., Naruse, K., Ohsumi, K., Horiuchi, J., Yoshihara, I., Kato, Y., Nakamura, N., Kato, M., Sugino, N. and Demura, H., Is big endothelin converted to endothelin-1 in circulating blood, *J. Cardiovasc. Pharmacol.*, 17(Suppl.7), S503-S505. 1991.

Watanabe, T. and Yokosawa, H., The generation of big-endothelin (1-22) (endothelin-valine) from big-endothelin in the soluble fraction of porcine lung, *Biochem. Int.*, 27, 1-8. 1992.

Waxman, L., Doshi, K.P., Gaul, S.L., Wang, S., Bednar, R.A. and Stern, A.M., Identification and characterization of endothelin converting activity from EAHY 926 cells: evidence for the physiologically relevant human enzyme, *Arch. Biochem. Biophys.*, 308, 240-253. 1994.

Weaver, L.H., Kester, W.R. and Matthews, B.W., A crystallographic study of the complex of phosphoramidon with thermolysin. A model for the presumed catalytic transition state and the binding of extended substrates, *J. Mol. Biol.*, 114, 119-132. 1977.

Wu-Wong, J.R., Budzik, G.P., Devine, E.M. and Opgenorth, T.J., Characterization of endothelin converting enzyme in rat lung, *Biochem. Biophys. Res. Commun.*, 171, 1291-1296. 1990.

Wu-Wong, J.R., Devine, E.M., Budzik, G.P. and Opgenorth, T.J., Characterization and partial purification of endothelin-converting enzyme in rat lung, *J. Cardiovasc. Pharmacol.*, 17 (Suppl. 7), S20-S25. 1991.

Wypij, D.M., Nichols, J.S., Novak, P.J., Stacy, D.L., Berman, J. and Wiseman, J.S., Role of mast cell chymase in the extracellular processing of big-endothelin-1 to endothelin-1 in the perfused rat lung, *Biochem. Pharmacol.*, 43, 845-853. 1992.

Xu, D., Emoto, N., Giaid, A., Slaughter, C., Kaw, S., deWit, D. and Yanagisawa, M., ECE-1: A membrane bound metalloprotease that catalyzes the proteolytic activation of big endothelin-1, *Cell*, 78, 473-485. 1994.

Yanagisawa, M., Kurihara, H., Kimura, S., Tomobe, Y., Kobayashi, M., Mitsui, Y., Yazaki, Y., Goto, K. and Masaki, T., A novel potent vasoconstrictor peptide produced by vascular endothelial cells, *Nature*, 332, 411-415. 1988.

Yamagushi, T., Kido, H. and Katunuma, N., A membrane-bound metallo-endopeptidase from rat kidney. Characteristics of its hydrolysis of peptide hormones and neuropeptides, *Eur. J. Biochem.*, 204, 547-552. 1992.

Yorimitsu, K., Shinmi, O., Nishiyama, M., Moroi, K., Sugita, Y., Saito, T., Inagaki, Y., Masaki, T. and Kimura, S., Effect of phosphoramidon on big endothelin-2 conversion into endothelin-2 in human renal adenocarcinoma (ACHN) cells. Analysis of endothelin-2 biosynthetic pathway, *FEBS Lett.*, 314, 395-398. 1992.

Yuan, C.M., Kiandoli, L. and Moore, J., Phosphoramidon inhibits endothelin production by endothelial cells exposed to cyclosporine, *Clin. Res.*, 40, 316A. 1992.

Chapter 7

ENDOTHELIN METABOLISM

Semiko Kaw and Masashi Yanagisawa

CONTENTS

I. OVERVIEW

Although the synthesis of endothelins (ETs) has been the focus of intense research, comparatively little attention has been paid to the mechanisms responsible for the inactivation of this potent pressor peptide. Over 60% of an intravenous bolus of [^{125}I]ET-1 injected into rats disappears from the bloodstream within the first minute with the highest uptake occurring in the lung, followed by the kidney and liver

0-8493-6975-4/97/$0.00+$.50

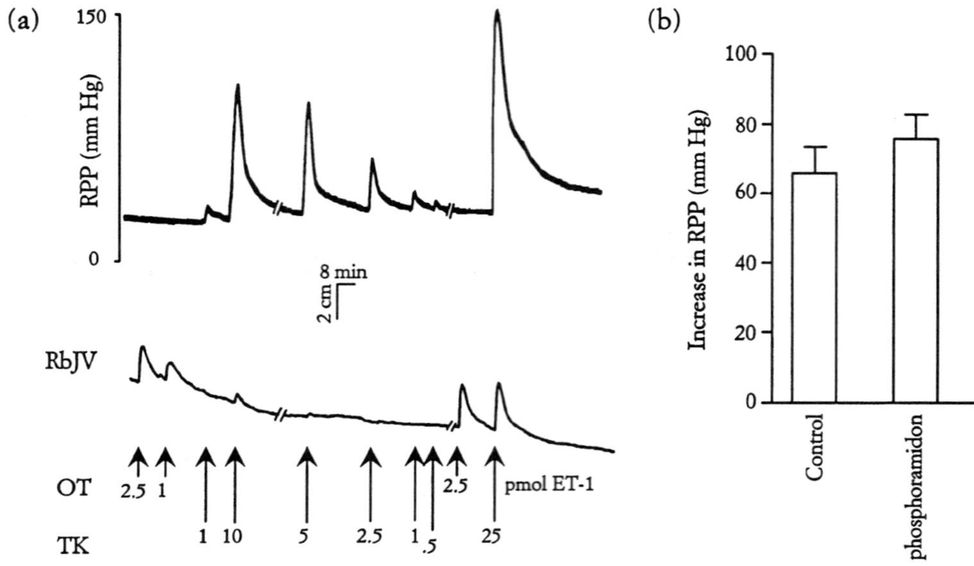

FIGURE 1
Inactivation of ET-1 by the isolated perfused kidney of the rabbit. A helically cut strip of rabbit jugular vein (RbJV) was superfused with the effluent from an isolated perfused kidney of the rabbit. (a) The representative bioassay trace shows the contraction of the RbJV when ET-1 (1 or 2.5 pmol) was injected over the tissue (OT), and the dose-dependent increases in renal perfusion pressure (RPP) when ET-1 (0.5 to 25 pmol) was administered through the kidney (TK). At least 10 times the dose of ET-1 given OT was required TK to produce an equivalent contraction of the RbJV, suggesting that >90% of ET-1 is inactivated or removed by the rabbit kidney. (b) The bar graph depicts that an infusion of phosphoramidon (10 μM) does not significantly alter the increase in RPP caused by ET-1 (control, 10 pmol; n = 12).

(Ånggård et al., 1989; Shiba et al., 1989; Sirviö et al., 1990). Two thirds of the biological activity of ET-1 disappears in a single passage through guinea pig isolated perfused lungs (de Nucci et al., 1988), confirming that the pulmonary circulation is largely responsible for the disappearance of ET-1 from the circulation. However, the capacity of the lungs for ET-1 uptake is saturable and under these conditions the kidney or liver may participate in the clearance and/or removal of ET-1. Indeed, in the rat pretreatment with nonlabeled ET-1 shifts the distribution of subsequently injected [^{125}I]ET-1 from the lung to the kidney (Sirviö et al., 1990). Furthermore, 90% of a bolus dose of ET-1 is removed by the isolated perfused kidney of the rabbit (Kaw et al., 1993., Figure 1a) and the renal vascular bed also participates in the clearance of exogenous ET-1 in humans (Weitzberg et al., 1991; Gasic et al., 1992) or pigs (Pernow et al., 1989b; Hemsen et al., 1991). The plasma concentration of immunoreactive endothelin is elevated in pathological conditions that precipitate or contribute to renal injury such as acute (Tomita et al., 1989) or chronic renal failure (Koyama et al., 1989), cyclosporine toxicity (Kon et al., 1990; Perico et al., 1990), endotoxemia (Sugira et al., 1989; Pernow et al., 1989a; Morel et al., 1989), and hypertension (Saito et al., 1990; Kohno et al., 1990), reflecting an increased synthesis or reduced metabolism and/or clearance of the peptide. Moreover, bilateral nephrectomy in rats causes a prolongation of the pressor response to ET-1 which is associated with a delayed disappearance of the peptide from the circulation (Kohno et al., 1989).

Isolated cells have also been shown to metabolize ET-1. For instance, primary cultures of rat hepatocytes or Kupffer cells readily degrade [^{125}I]ET-1 (Gandhi et al., 1993), supporting a role for the liver in the metabolism of ETs. Activated human neutrophils inactivate ET-1 rapidly (Figure 4; Sessa et al., 1991a,b; Patrignani et al.,

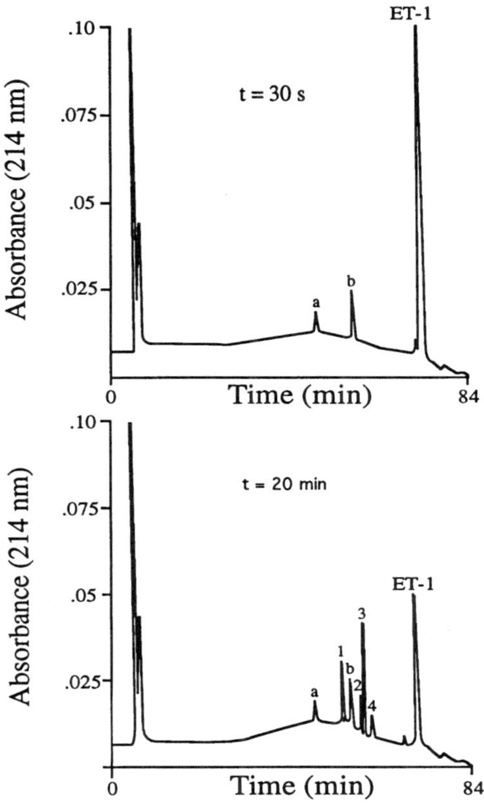

FIGURE 2
Metabolism of ET-1 by rat kidney NEP 24.11. ET-1 (24 μM) was incubated with 135 ng purified rat kidney NEP and the resulting metabolites separated by HPLC. The figure depicts representative chromatograms 30 s (top panel) or 20 min (bottom panel) into the reaction. The peaks are numbered in order of their appearance on the chromatograms: 1 = Two-chain ET-$1_{1\text{-}16}$, 2 = ET-$1_{1\text{-}16}$, 3 = C-terminal tripeptide, and 4 = ET-$1_{1\text{-}18}$; *a*, a contaminant in the solvent, and *b*, added bestatin. (Modified from Vijayaraghavan, J., Scili, A. G., and Carretero, O. A., et al., *J. Biol. Chem.*, 265, 14150, 1990. With permission.)

1991; Fagny et al., 1992) suggesting that these cells may be involved in the metabolism of ETs in pathological conditions associated with neutrophil activation. *In vitro* ETs are substrates for a variety of proteolytic enzymes including aspartic, serine, or neutral metalloproteinases and, like other peptides, ETs may be degraded extracellularly by nonspecific proteinases or be metabolized intracellularly by lysosomal enzymes following receptor-mediated endocytosis. The kidneys, lungs, liver, and neutrophils contain an abundance of peptidases, and enzymatic degradation of ET-1 by these tissues or cells may play a vital role in regulating ET-1 levels *in vivo*.

II. METABOLISM OF ENDOTHELINS *IN VITRO*

A. Membrane-Bound Metalloendopeptidases

Membrane-bound metalloendopeptidases are key enzymes involved in the generation and inactivation of biologically active peptides (Turner, 1993). At least three metalloproteinase activities have been found to metabolize ETs *in vitro*.

1. *Neutral Endopeptidase (NEP) 24.11*

NEP 24.11 was the first peptidase shown to possess endothelin-metabolizing activity. This 93-kDa transmembrane metalloendopeptidase, with zinc at its active center, cleaves peptide bonds on the amino side of hydrophobic amino acids (see Erdös and Skidgel, 1989; Roques et al., 1993). Originally recognized as an inactivator of enkephalins (Malfroy et al., 1978; Schwartz et al., 1981), NEP 24.11 is now known to metabolize a number of peptides *in vitro* and *in vivo* including bradykinin (Gafford et al., 1983; Ura et al., 1987; Skidgel et al., 1991), substance P (Matsas et al., 1983; Skidgel et al., 1991), the C-terminal peptide of cholecystokinin (Najdovski et al., 1985), *N*-formyl-L-methionyl-L-leucyl-L-phenylalanine (fMLP) (Connelly et al., 1985), gastrin (Deschodt-Lanckman et al., 1988), and atrial natriuretic peptide (ANP) (Stephenson and Kenny, 1987; Kenny and Stephenson, 1988).

NEP 24.11 was discovered in, and isolated from, the brush border membranes of the rabbit kidney (Kerr and Kenny, 1974a,b). Indeed, the isolated perfused kidney of the rabbit removes 90% of a bolus dose of ET-1 (Figure 1; Kaw et al., 1993), which could be a direct consequence of its degradation by NEP 24.11. The involvement of NEP 24.11 in the metabolism of ET-1, however, was first suspected when membrane fractions from the rat kidney were shown to metabolize ET-1, and this activity was inhibited by phosphoramidon, a selective inhibitor of NEP 24.11 (Scili et al., 1989). It was subsequently demonstrated that purified bovine (Sokolovsky et al., 1990), rat (Vijayaraghavan et al., 1990), or human (Fagny et al., 1991) kidney NEP 24.11 degraded ET-1, resulting in a loss of its biochemical (Sokolovsky et al., 1990) or biological (Fagny et al., 1991) activity.

a. Endothelin Metabolism by Renal NEP 24.11

Amino acid analysis of rat kidney NEP 24.11-derived metabolites of ET-1 reveals the formation of a two-chain ET-$1_{1\text{-}16}$ molecule, cleaved at the Ser^5-Leu^6 and His^{16}-Leu^{17} bonds, with intact disulfide bridges (Table 1; Figure 2; Vijayaraghavan et al., 1990). ET-$1_{1\text{-}16}$, ET-$1_{1\text{-}18}$, and the C-terminal tripeptide are also detected (Vijayaraghavan et al., 1990). The formation of the two chain ET-$1_{1\text{-}16}$ molecule has been proposed to occur via initial hydrolysis of the Ser^5-Leu^6 bond, rapidly followed by sequential cleavage first at Asp^{18}-Ile^{19} and then at the His^{16}-Leu^{17} bond, whereas slower hydrolysis of intact ET-1 at the His^{16}-Leu^{17} and Asp^{18}-Ile^{19} bonds generates ET-$1_{1\text{-}16}$ and ET-$1_{1\text{-}18}$ and its corresponding C-terminal tripeptide, respectively (Vijayaraghavan et al., 1990). The K_m value of rat NEP 24.11-catalyzed hydrolysis of ET-1 is 2.3 μM (Vijayaraghavan et al., 1990).

Bovine kidney NEP 24.11 generates a metabolite from ET-1 whose amino acid composition is consistent with cleavage between Ser^5-Leu^6 as well as Asp^{18}-Ile^{19} bonds, with a small amount of material additionally cleaved between the Cys^{11}-Val^{12} bond (Sokolovsky et al., 1990). The C-terminal tripeptide, indicative of the Asp^{18}-Ile^{19} cleavage, however, was not detected (Sokolovsky et al., 1990). Moreover, in contrast to ET-1 hydrolysis by rat kidney NEP 24.11, there is no indication of further cleavage of this metabolite at the His^{16}-Leu^{17} bond, nor is there any evidence of hydrolysis exclusively at the Asp^{18}-Ile^{19} or His^{16}-Leu^{17} bonds. Interestingly, the hydrolysis of ET-1 by bovine kidney NEP 24.11 yielded a K_m value 30 ± 10 μM (Sokolovsky et al., 1990), higher than that for rat kidney NEP 24.11. Compared to the intact peptide, the bovine kidney NEP 24.11-derived metabolite of ET-1 does not (1) bind to rat heart myocytes, (2) induce hydrolysis of inositol phospholipids, and (3) is at least 50 times less lethal than ET-1 following intravenous administration into mice (Sokolovsky et al., 1990).

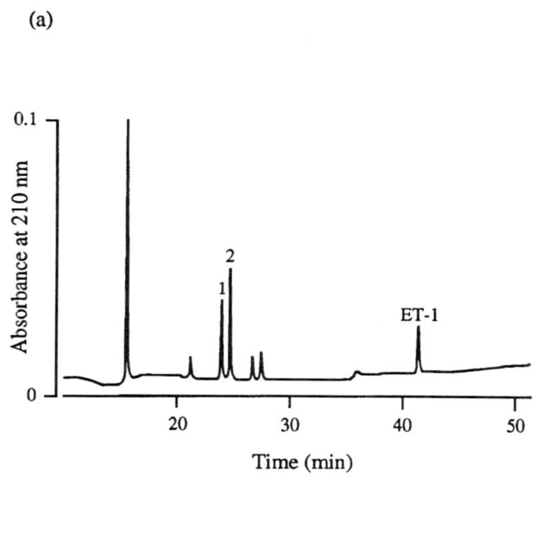

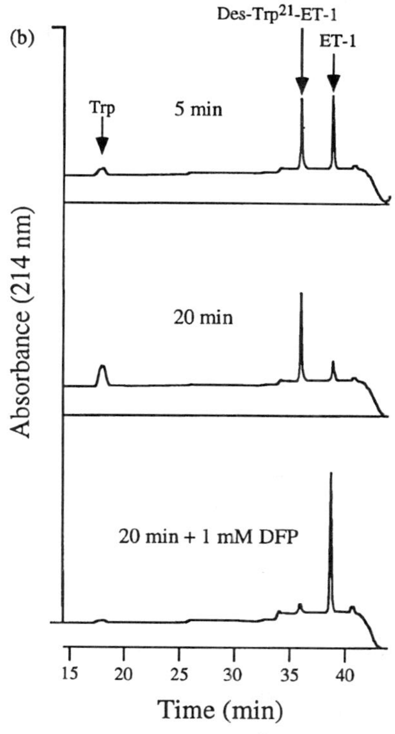

FIGURE 3
Metabolism of ET-1 by (a) cathepsin D or (b) deamidase. ET-1 (1 nmol) was incubated with 0.1 μg cathepsin D for 24 h and the metabolites analyzed by HPLC. Two major products ET-$1_{1\text{-}18}$ (1) and the C-terminal tripeptide (2) were identified, consistent with hydrolysis of the Asp^{18}-Ile^{19} bond. Purified deamidase (14.4 ng) removes the C-terminal tryptophan of ET-1 (35 μM) resulting in the formation of Des-Trp^{21}-ET-1. This carboxypeptidase activity is prevented by co-incubation with 1 mM DFP. (Modified from Sawamura, T., Shinmi, O., Kishi, N., et al., *Biochem. Biophys. Res. Commun.*, 172, 883, 1990c (a) and Jackman, H.L. et al., *J. Biol. Chem.*, 267, 2872, 1992 (b). With permission.)

Incubation of ET-1 with human renal NEP 24.11 mainly leads to the formation of ET-$1_{1\text{-}16}$, ET-$1_{1\text{-}18}$, and the C-terminal tri, penta, or hexa peptides, indicating hydrolysis of the His^{16}-Leu^{17}, Asp^{18}-Ile^{19}, or Cys^{15}-His^{16} bonds (Fagny et al., 1991). In contrast to the intact peptide, ET-$1_{1\text{-}16}$ does not elicit a contractile response from the rat aorta (Fagny et al., 1991). Incubations of ET-1 for more than 24 h with human NEP 24.11 result in the generation of ET-$1_{1\text{-}3}$ and the 6-16 fragment of ET-1, indicating cleavage of the Cys^{3}-Ser^{4} and Ser^{5}-Leu^{6} bonds, respectively (Fagny et al., 1991).

The metabolism of ET-1 by bovine, rat, or human kidney NEP 24.11 is completely inhibited by phosphoramidon (Sokolovsky et al., 1990; Vijayaraghavan et al., 1990; Fagny et al., 1991).

ET-1 is not the only member of the ET family to be metabolized by NEP 24.11. Indeed, ET-2 and ET-3 are also substrates for NEP 24.11 (Sokolovsky et al., 1990; Vijayaraghavan et al., 1990). However, as both ET-2 and ET-3 contain amino acid residues that differ from ET-1 at position 6, no cleavage between positions 5 and 6 of these isoforms is observed. Instead ET-2 is hydrolyzed at the His^{16}-Leu^{17} and Asp^{18}-Ile^{19} bonds producing ET-$2_{1\text{-}16}$, ET-$2_{1\text{-}18}$, and the C-terminal tripeptide, whereas cleavage of the Cys^{3}-Phe^{4}, Cys^{11}-Val^{12}, and His^{16}-Leu^{17} bonds of ET-3 results in the formation of three-chain ET-$3_{1\text{-}16}$ (Vijayaraghavan et al., 1990). Two-chain ET-$3_{1\text{-}16}$ and ET-$3_{1\text{-}18}$ and the corresponding C-terminal tripeptide were also identified as NEP

24.11-derived metabolites of ET-3 resulting from cleavage of the Cys^{11}-Val^{12} and His^{16}-Leu^{17} and Asp^{18}-Ile^{19} bonds, respectively (Vijayaraghavan et al., 1990).

In contrast to ETs, sarafotoxins (STXs) are relatively resistant to hydrolysis by bovine NEP 24.11. Only 40 to 50% of STX-b is degraded after a 4-h incubation with bovine NEP 24.11, compared to 50% of ET-1 after 1 h, whereas STX-c remains intact for up to 2 h following incubation with this metalloproteinase (Sokolovsky et al., 1990).

b. Extrarenal NEP 24.11 Activity

Although not ubiquitous, NEP 24.11 is widely distributed in the body (see Erdös and Skidgel, 1989). NEP 24.11 could, therefore, also be responsible for the extrarenal degradation of ETs. Indeed, in addition to the kidney, membrane fractions from rat lung were also found to inactivate ET-1 in a phosphoramidon-sensitive manner (Scili et al., 1989). Moreover, membrane fractions from human choroid plexus also contain phosphoramidon-sensitive NEP 24.11-like endothelin-degrading activity (Fagny et al., 1991). Furthermore, aortic smooth muscle cells from the rat fetus (A10 cells) actively metabolize [^{125}I]ET-1 in a manner comparable to NEP 24.11 purified from rat kidney brush borders (Dickinson et al., 1991).

ETs induce potent contractions of tracheal and bronchial preparations from various species *in vitro* (see Filep, 1993), and these effects are potentiated by removal of the airway epithelium (Candenas et al., 1992). As phosphoramidon also augments the contractions induced by ETs in airway smooth muscle, the protective role of the epithelium is attributed to degradation of these peptides by epithelial NEP 24.11. (Noguchi et al., 1991; Candenas et al., 1992; Di-Maria et al., 1992; Yamaguchi et al., 1992c). Activated human neutrophils metabolize ET-1 *in vitro* (Sessa et al., 1991a, b; Patrignani et al., 1991; Fagny et al., 1992) and NEP 24.11 is also present on the neutrophil cell surface (Connelly et al., 1985). However, phosphoramidon does not inhibit the degradation of ET-1 by these cells indicating that NEP 24.11 does not play a major role in the neutrophil-induced inactivation of ET-1.

2. Thermolysin

Thermolysin, a metalloproteinase derived from *Bacillus thermoproteolyticus* (Matsubara, 1970), also hydrolyzes bonds on the amino side of hydrophobic amino acids and has a substrate specificity similar to that of NEP 24.11 (Hersh and Morihara, 1986). Indeed, thermolysin metabolizes ET-1 to two-chain $ET\text{-}1_{1\text{-}16}$, $ET\text{-}1_{1\text{-}16}$, $ET\text{-}1_{1\text{-}18}$ and the C-terminal tripeptide, similar to the hydrolysis of ET-1 by rat NEP 24.11 (Vijayaraghavan et al., 1990). However, an additional cleavage between the Cys^{11}-Val^{12} bond, leading to the formation of three-chain $ET\text{-}1_{1\text{-}16}$ and the C-terminal pentapeptide (Leu-Asp-Ile-Ile-Trp) also occurs (Table 1; Vijayaraghavan et al., 1990). The liberation of the C-terminal pentapeptide by thermolysin indicates that, in contrast to rat NEP 24.11, initial cleavage of ET-1 by this bacterial endopeptidase occurs in the C-terminal tail region followed by hydrolysis within the ring structure (Vijayaraghavan et al., 1990). The thermolysin-catalyzed hydrolysis of ET-1 exhibits a K_m of 40 μM (Vijayaraghavan et al., 1990).

3. Phosphoramidon-Insensitive Metalloendopeptidase Activity

A recent investigation of ET-1 hydrolysis by rat kidney membranes using reverse-phase HPLC and gas-phase protein sequencing revealed the formation of three phosphoramidon-sensitive fragments and one phosphoramidon-insensitive metabolite (Yamaguchi et al., 1992a). Whereas all the phosphoramidon-sensitive

TABLE 1

In Vitro Degradation of ET-1 by Various Proteolytic Enzymes — Major Cleavage Sites and Metabolites Are Listed

Proteinase	Cleavage Site(s)	Metabolite(s)
Rat kidney NEP 24.11	Ser^5-Leu^6, His^{16}-Leu^{17}	Two-chain ET-$1_{1\text{-}16}$
	Asp^{18}-Ile^{19}	ET-$1_{1\text{-}18}$, C-terminal tripeptide
	His^{16}-Leu^{17}	ET-$1_{1\text{-}16}$
Thermolysin	Ser^5-Leu^6, Cys^{11}-Val^{12}, His^{16}-Leu^{17}	Three chain ET-$1_{1\text{-}16}$
	Ser^5-Leu^6, His^{16}-Leu^{17}	Two chain ET-$1_{1\text{-}16}$
	Asp^{18}-Ile^{19}	ET-$1_{1\text{-}18}$, C-terminal tripeptide
	His^{16}-Leu^{17}	ET-$1_{1\text{-}16}$, C-terminal pentapeptide
Bovine spleen cathepsin D	Asp^{18}-Ile^{19}	ET-$1_{1\text{-}18}$, C-terminal tripeptide
Human leukocyte cathepsin G	His^{16}-Leu^{17}	ET-$1_{1\text{-}16}$, C-terminal pentapeptide
Bovine pancreatic chymotrypsin	Tyr^{13}-Phe^{14}	Tyr^{13} nicked ET-$1_{1\text{-}21}$
Mast cell chymase	Tyr^{13}-Phe^{14}	Tyr^{13} nicked ET-$1_{1\text{-}21}$
	Phe^{14}-Cys^{15}	Phe^{14} nicked ET-$1_{1\text{-}21}$
Platelet deamidase	Ile^{20}-Trp^{21}	Des-Trp^{21}-ET-1

metabolites were generated by cleavage at the Ser^5-Leu^6 bond of ET-1, in a manner similar to that observed with purified rat NEP 24.11, the phosphoramidon-insensitive fragment was formed via cleavage of the Leu^{17}-Asp^{18} bond. This metabolism is due to an 88-kDa membrane-bound metalloproteinase that has been purified from rat kidney (Yamaguchi et al., 1991). This enzyme has a pH optimum of 8.0 to 8.5 and is inhibited by EDTA, EGTA, *o*-phenanthroline, but not by phosphoramidon (Yamaguchi et al., 1991). Initially shown to degrade human parathyroid hormone, this metalloendopeptidase is now known to hydrolyze a number of substrates including ET-1, cleaving peptide bonds flanked by hydrophilic residues (Yamaguchi et al., 1991, 1992b). Kinetics of ET-1 hydrolysis by the isolated proteinase yielded a K_m value of 71.5 μM (Yamaguchi et al., 1992a), higher than that for NEP 24.11, indicating that NEP 24.11 metabolizes ET-1 more efficiently than the more recently purified metalloproteinase from rat kidney.

B. Aspartic Proteinases

Aspartic proteinases such as pepsin, cathepsin D, cathepsin E, and renin are produced by many mammalian cells and tissues. Many are secreted into the extracellular space, whereas others exert their effects within the cell of origin. Most of these proteinases are enzymatically active at an acidic pH (Barrett, 1977).

1. Cathepsin D

Cathepsin D is a ubiquitous mammalian lysosomal enzyme which is mainly involved in the intracellular degradation of endocytosed proteins (Barrett, 1977; Diment and Stahl, 1985). This 42-kDa aspartic proteinase has an acidic pH optimum and usually cleaves peptide bonds that involve markedly hydrophobic residues, at least one of which is aromatic (Barrett, 1977). In addition to its general function as a lysosomal enzyme, cathepsin D has been implicated in the generation of γ endorphin from β endorphin (Gráf et al., 1979) and in the conversion of procollagen to collagen (Helseth and Veis, 1984).

Recently, cathepsin D was also shown to convert big ET-1 to ET-1 *in vitro*, and cathepsin D-like endothelin converting enzyme (ECE) activities were purified from bovine adrenal medulla (Sawamura et al., 1990b) or endothelial cells (Sawamura et al., 1990a). In the adrenal medulla this ECE was found to have a molecular weight similar to cathepsin D (Sawamura et al., 1990a). Furthermore, this ECE activity was precipitated by an antibody raised against bovine spleen cathepsin D (Sawamura et al., 1990a). Interestingly, however, the time course of immunoreactive ET-1 generation from big ET-1, using different amounts of cathepsin D, demonstrated a marked reduction in ET-1 production following an increase of either the incubation time or cathepsin D concentration, suggesting that mature ET-1 may be degraded further (Sawamura et al. 1990c). Indeed reverse-phase HPLC analysis of cathepsin D-derived metabolites of big ET-1 showed that cleavage of the Trp^{21}-Val^{22} bond was followed by hydrolysis of the Asp^{18}-Ile^{19} bond of ET-1 (Sawamura et al., 1990c; Takaoka et al., 1990a) resulting in the formation of ET-$1_{1\text{-}18}$ and the C-terminal tripeptide (Table 1) in addition to the ET-1 (Sawamura et al., 1990c; Takaoka et al., 1990a). Sawamura et al. (1990c) confirmed this further by analyzing the metabolism of ET-1 by cathepsin D using reverse-phase HPLC which also showed the generation of ET-$1_{1\text{-}18}$ and the C-terminal tripeptide from the mature peptide (Figure 3a).

Currently, the endothelin-degrading potential of other aspartic proteinases is unknown.

C. Serine Proteinases

Serine proteinases represent the largest, best characterized, and most versatile class of proteinases. The mammalian serine proteinases such as trypsin, chymotrypsin, thrombin, elastase, cathepsin G, and kallikrein are implicated in diverse physiological processes including digestion, blood coagulation/fibrinolysis, complement activation, and release of bioactive peptides.

Human neutrophils are a rich source of serine proteinases which are released following cell activation. Interestingly, incubation of ET-1 with neutrophils activated by the chemotactic peptide fMLP results in a loss of its contractile activity, and this effect is prevented by the general serine proteinase inhibitor phenyl methyl sulfonyl fluoride (PMSF; Figure 4a) (Sessa et al., 1991a, b). PMSF also prevents the metabolism of $[^{125}I]$ET-1 by fMLP-activated neutrophils, indicating that serine proteinases released from these cells are responsible for degrading ET-1 (Figure 4b and c). These findings were confirmed by two other studies showing that either fMLP (Patrignani et al., 1991) or phorbol-12-myristate-13-acetate (PMA)-activated neutrophils (Fagny et al., 1992) metabolized ET-1, and that this degradation was prevented by the serine proteinase inhibitors eglin C (Patrignani et al., 1991) and soy bean trypsin inhibitor (Fagny et al., 1992), respectively. Cathepsin G and elastase are the two major serine proteinases present in and secreted by human neutrophils (Travis, 1988).

1. Human Neutrophil Cathepsin G

Human neutrophil cathepsin G is a chymotrypsin-like serine proteinase and a glycoprotein with a molecular weight ranging from 23 to 25 kDa depending upon its carbohydrate content (Travis et al., 1979). Cathepsin G effectively degrades proteoglycans, fibronectin, and collagen, and together with elastase has been implicated in the development of pulmonary emphysema (Travis, 1988). Although its physiological function *in vivo* remains unknown, cathepsin G has been shown to produce angiotensin II by two distinct pathways, one involving the activation of pro-renin to renin (Dzau et al., 1987) and another converting angiotensinogen or angiotensin I directly to angiotensin II (Wintroub et al., 1984). Cathepsin G is also a strong platelet

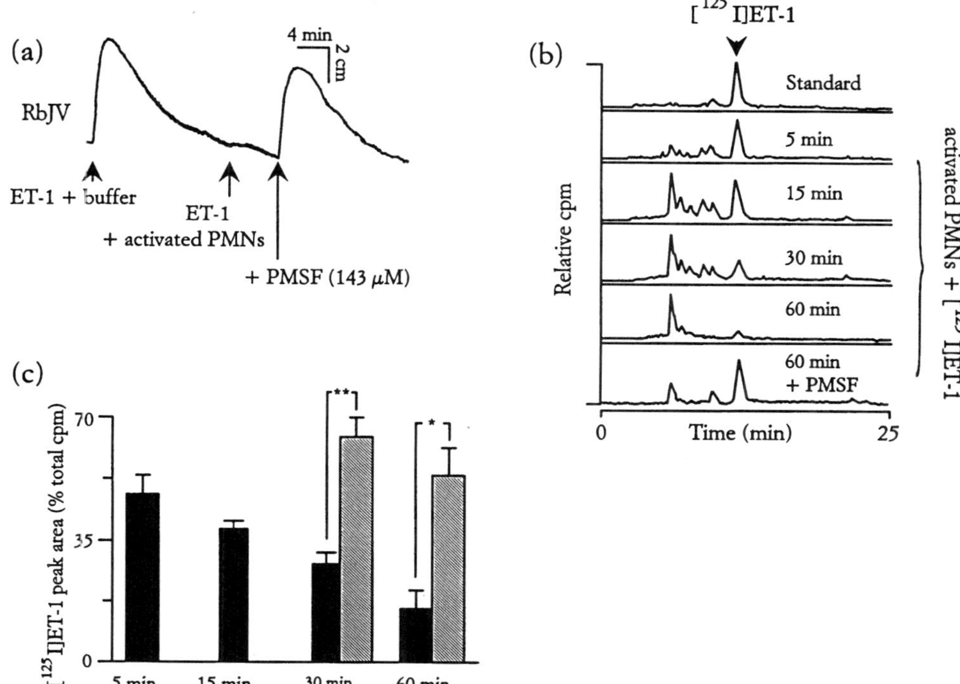

FIGURE 4
Metabolism of ET-1 by activated neutrophils. ET-1 (0.4 μM) was incubated with 4×10^6 neutrophils (PMNs) activated with fMLP (100 nM) for 30 min at 37°C and aliquots of the incubation mixture were (a) assayed for contractile activity on the rabbit jugular vein (RbJV), or (b) analyzed by reverse-phase HPLC. Activated PMNs abrogate the contractile activity of ET-1 and this inactivation is prevented by the serine proteinase inhibitor PMSF. The representative chromatograms show that PMSF also inhibits the time-dependent degradation of [^{125}I]ET-1 by activated neutrophils, and (c) the bar graph sums up the HPLC data in terms of [^{125}I]ET-1 recovered at 5 (n = 3), 15 (n = 3), 30 (n = 5), or 60 min (n = 3; filled columns). PMSF (hatched columns) inhibited the neutrophil-induced metabolism of [^{125}I]ET-1 at 30 (n = 4) or 60 min (n = 3); * $p <$ .05, ** $p <$.01, Bonferroni's *t*-test.

aggregating agent (Selak et al., 1988) and possesses microbicidal properties unrelated to its proteolytic activity (Travis, 1988).

Incubation of ET-1 with cathepsin G virtually abolishes the contractile activity of the peptide and this effect is reversed by PMSF (Figure 5a). Reverse-phase HPLC analysis of the cathepsin G-induced metabolism of [^{125}I]ET-1 (Figure 5b and c) reveals that this serine proteinase degrades the labeled peptide rapidly and in a manner similar to that observed with activated neutrophils (Figure 4b; Sessa et al., 1991b). Amino acid analysis of the cathepsin G-derived metabolites of nonlabeled ET-1 led Fagny et al. (1992) to identify the ET-$1_{1\text{-}16}$ sequence and the C-terminal pentapeptide, consistent with hydrolyis of the His^{16}-Leu^{17} bond (Table 1).

Elastase is colocalized with cathepsin G in the azurophilic granules of neutrophils. This enzyme is a glycoprotein with a molecular weight ranging from 29 to 31 kDa and catalyzes the hydrolysis of elastin, collagen, proteoglycan, and fibronectin (Travis et al., 1979). Surprisingly, in contrast to cathepsin G treatment, incubation of ET-1 with elastase does not dramatically reduce its contractile activity (Figure 5a), a finding confirmed by Fagny et al. (1992) who also found cathepsin G to be more effective than elastase as an ET-degrading serine proteinase. The small reduction in

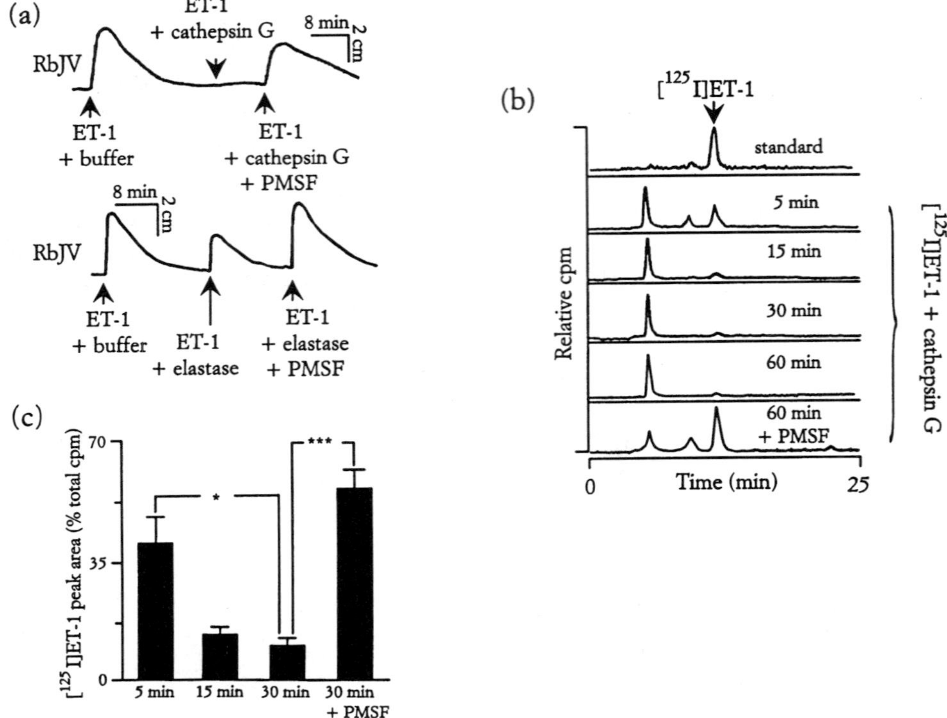

FIGURE 5
Degradation of ET-1 by human neutrophil cathepsin G. ET-1 (0.4 μM) was incubated with cathepsin G or elastase (5 μg/ml each) for 30 min at 37°C and the incubation mixture was either (a) assayed for biological activity on strips of RbJV or (b) subjected to HPLC analysis. Cathepsin G virtually abolished the contractile activity of ET-1, in contrast to elastase which produced only a minor reduction in the activity of ET-1. However, the degradation of ET-1 by either serine proteinase is inhibited by PMSF (143 μM). The representative chromatograms illustrate the rapid, time-dependent, and PMSF-sensitive degradation of [^{125}I]ET-1 by cathepsin G, generating a large peak similar to the one produced from [^{125}I]ET-1 by activated neutrophils. The bar graph (c) sums up the recovery of [^{125}I]ET-1 following incubation with cathepsin G for 5 (n = 3), 15 (n = 3), or 30 min (n = 4) which is significantly inhibited by PMSF; * $p < .05$, *** $p < 0.01$, Bonferroni's *t*-test.

the contractile activity of ET-1 following elastase treatment, however, is also inhibited by PMSF (Figure 5a).

2. *Chymotrypsin*

ET-1 is generated from big ET-1 by cleavage of the Trp^{21}-Val^{22} bond (Yanagisawa et al., 1988). Because chymotrypsin cleaves peptide bonds on the carboxy side of hydrophobic amino acids, it was postulated that an enzyme with a chymotrypsin-like specificity could be responsible for the conversion of big ET-1 to ET-1 (Yanagisawa et al., 1988). Indeed, chymotrypsin-treated big ET-1 (1-40) produced an ET-1-like contraction of rat aorta *in vitro* or pressor effect *in vivo* (McMahon et al., 1989). Interestingly, reverse-phase HPLC and sequence analysis of the chymotrypsin-derived metabolites of porcine big ET-1 (1-39) revealed that big ET-1 was converted to ET-1 via the formation of an intermediate big ET-1$_{1\text{-}31}$ (Takaoka et al., 1990b). Moreover, the ET-1 formed was cleaved further between the Tyr^{13}-Phe^{14} bond (Table 1). This metabolite was designated Tyr^{13} nicked ET-1 (Takaoka et al., 1990b). However, the effect of this cleavage on the biological activity of ET-1 remains unknown.

3. *Chymase*

Mast cell chymase also generates ET-1 from big ET-1 (Wypij et al., 1992). However, the newly formed ET-1 is degraded at approximately the same rate as its formation by cleavage at the Tyr^{13}-Phe^{14} and Phe^{14}-Cys^{15} bond, to generate their corresponding internally nicked ET-1 products (Table 1; Wypij et al., 1992). Sequence analysis reveals that both metabolites contain the amino as well as carboxy terminal sequences, indicating the presence of an intact disulfide bond(s) (Wypij et al., 1992).

D. Miscellaneous

1. *Carboxypeptidases*

a. Renal

Incubation of ET-1 with the soluble fraction from the rat kidney homogenate results in a loss of its contractile activity as compared to ET-1 incubated with buffer alone (Deng et al., 1992). This activity, distinct from carboxypeptidases A or B, was designated endothelin-degrading enzyme (EDE) and purified partially by anion exchange and concanavalin A-Sepharose column chromatography (Deng et al., 1992). EDE cleaves the C-terminal tryptophan residue of ET-1 resulting in a peptide with a significantly reduced constrictor activity as compared to the intact peptide, and also removes the C-terminal tyrosine of ANP, although the biological activity of the resulting peptide remains unchanged (Deng et al., 1992). The C-terminal tryptophan of ET-3 is also cleaved by EDE; however, neither big ET-1 nor angiotensin II are substrates for this enzyme (Deng et al., 1992).

b. Extrarenal

A cathepsin A-like proteinase, having deamidase as well as carboxypeptidase activity, has been purified from human platelets (Jackman et al., 1990). With peptide substrates deamidation is the major reaction at neutral pH whereas carboxypeptidase activity prevails at acidic pH (Jackman et al., 1990). As removal of the C-terminal tryptophan attenuates the contractile activity of ET-1, Jackman et al. (1992) hypothesized that this enzyme may play a role in the inactivation of ET-1.

Incubation of ET-1 with purified deamidase resulted in a time-dependent degradation of the peptide, which was prevented by co-incubation with diisopropyl fluorophosphate (DFP; Figure 3b; Jackman et al., 1992). The deamidase-treated ET-1 was inactive as a constrictor of the guinea-pig ileum (Jackman et al., 1992). HPLC analysis revealed that this inactivation of ET-1 was due to the removal of its C-terminal tryptophan by purified deamidase, resulting in the formation of Des-Trp^{21}-ET-1 (Figure 3b; Jackman et al., 1992). The K_m value determined for the deamidase-catalyzed hydrolysis of ET-1 was 22 μM (Jackman et al., 1992). Deamidase-like ET-degrading activity was also identified in cultured rat vascular smooth muscle cell homogenates (Jackman et al., 1992).

2. *Aminopeptidase-Like Endothelin-Degrading Activity*

Primary cultures of rat hepatocytes or Kupffer cells rapidly internalize and metabolize [^{125}I]ET-1 (Gandhi et al., 1993). This metabolism is prevented by the aminopeptidase inhibitor bacitracin, with a concomitant increase in the intracellular content of [^{125}I]ET-1 indicating that aminopeptidase activity in the liver is responsible for the observed degradation of ET-1 (Gandhi et al., 1993).

III. ROLE OF ENDOTHELIN-DEGRADING ENZYMES IN REGULATING ENDOTHELIN LEVELS *IN VIVO*

Despite the fact that ETs are substrates for several proteinase activities *in vitro*, relatively little is known about the enzymatic degradation of ET-1 *in vivo*. The lung, kidney, and liver (Änggård et al., 1989; Shiba et al., 1989; Sirviö et al., 1990) contribute to the rapid clearance of ET-1 *in vivo*, indicating that these tissues are the major sites of ET metabolism.

Both the kidney and lung are sources of NEP 24.11, and as the K_m for NEP 24.11-catalyzed hydrolysis of ETs is amongst the lowest reported for this enzyme, ETs may be important physiological substrates for NEP 24.11 (Vijayaraghavan et al., 1990). Furthermore, Vijayaraghavan et al. (1990) also proposed that the fivefold increase in plasma ET-1 concentration observed in patients with acute renal failure (Tomita et al., 1989) may be explained by decreased NEP 24.11 activity. However, whereas phosphoramidon prevents the metabolism of ET-1 by rat lung homogenates (Scili et al., 1989) and potentiates ET-1-induced bronchoconstriction in airway smooth muscle (Noguchi et al., 1991; Candenas et al., 1992; Di-Maria et al., 1992; Yamaguchi et al., 1992c), this NEP 24.11 inhibitor does not enhance the renal vasoconstrictor effects of ET-1 in the rabbit kidney (Figure 1b), a finding confirmed by Télémaque et al. (1992). A possible explanation for this lack of effect may be that the site of ET-1 degradation is distal to its site of action in the isolated perfused kidney of the rabbit. However, there is also conflicting evidence with regard to the effects of NEP 24.11 inhibition on ET metabolism *in vivo*, thus complicating the issue further. Neither phosphoramidon nor thiorphan enhances the pressor effects of ET-1 (Matsumura et al., 1990; McMahon et al., 1991), nor does selective NEP 24.11 inhibition with SQ 28,603 alter endogenous plasma ET-1 levels or the clearance of exogenous ET-1 in rats at doses that increase plasma ANP and cyclic GMP concentrations (Asaad et al., 1993), suggesting that NEP 24.11 is not responsible for the degradation of ET-1 *in vivo*. On the other hand, Abassi et al. (1992) reported a small increase in plasma ET-1 and a marked increase in the urinary excretion of the intact peptide in rats using SQ 29,072, another selective NEP 24.11 inhibitor. In the same study the NEP 24.11 inhibitor SQ 28,603 increased urinary excretion of [^{125}I]ET-1 and accumulation of intact ET-1 in the lungs and kidney, suggesting a positive role for NEP 24.11 in the metabolism of ET-1 (Abassi et al., 1992). Furthermore, Salvati et al. (1992) also showed that phosphoramidon caused a small potentiation of the renal vasoconstrictor effects of ET-1 in the rat kidney. The reason for these discrepancies is not clear, but may be related to the differences in the concentrations of inhibitors used. Nevertheless, the lack of a clear-cut involvement of NEP 24.11 in the metabolism of endogenous ET-1 indicates that other proteinases may be more important in inactivating ET-1 *in vivo*.

Like other peptides, ETs may undergo receptor-mediated endocytosis followed by intracellular digestion within lysosomes (Gammmeltoft, 1991). The lysosomal localization of cathepsin D or deamidase, which degrade ET-1 *in vitro*, indicates that these enzymes are ideal candidates for the intracellular metabolism of ETs. Furthermore, activities of lysosomal enzymes are increased in aortic endothelial cells isolated from spontaneously or renal hypertensive rats (Sasahara et al., 1988), and under these circumstances such enzymes may serve a homeostatic function to regulate endogenous levels of this potent pressor peptide.

Degradation of ET-1 by neutrophil-derived serine proteinases could also provide a homeostatic mechanism for regulating ET-1 production in diseases where these cells are activated. Indeed, co-incubation of primary cultures of human umbilical vein endothelial cells with serum from pre-eclamptic women suppressed the production of ET-1 by these cells (Ware Branch et al., 1991). The authors suggested that endopeptidase activity in pre-eclamptic serum could be degrading the ET-1 released. However, as neutrophils are activated in pre-eclampsia (Greer et al., 1989) the release of serine proteinases such as cathepsin G is more likely to be responsible for this effect.

Conflicting reports exist in the literature as to the plasma levels of ET-1 measured in essential hypertension. Saito et al. (1990) found elevated plasma ET-1 in 20 patients with essential hypertension, but Davenport et al. (1990) found no difference between untreated hypertensives and age-matched controls, and a negative correlation between ET-1 and systolic blood pressure in hypertensive patients. Plasma concentrations of ET-1 and big ET-1 are lower in stroke-prone spontaneously hypertensive rats compared to Wistar Kyoto rats (Suzuki et al., 1990). As an increase in neutrophil counts and activation has been demonstrated in spontaneously hypertensive rats (Schmid-Schönbein et al., 1991), and these cells have also been implicated in the conversion of big ET-1 to ET-1 (Sessa et al., 1991a,b), it is tempting to speculate that the variability in the capacity of neutrophils to metabolize big ET-1 or ET-1 may account for the observed differences in the ET-1 levels measured.

The circulating levels of ET-1 are too low to exert systemic effects and possibly represent only a "spillover" of the peptide into the bloodstream. Indeed, conversion experiments in the rabbit kidney suggest that the concentrations of ET-1 at its site of synthesis are considerably higher and that the newly synthesized peptide acts locally to produce a constrictor response, followed by its prompt removal or inactivation (Kaw et al., 1993). Local formation and degradation of ET-1 has also been demonstrated in guinea-pig airway tissues where phosphoramidon (100 to 1000 μM) suppresses big ET-1-induced bronchoconstriction by inhibiting ECE activity, whereas lower doses of phosphoramidon (0.01 to 0.1 μM) potentiate the bronchoconstrictor response to big ET-1 by preventing the concomitant NEP 24.11-induced degradation of the ET-1 generated from its precursor (Noguchi et al., 1991). Local metabolism, therefore, could also occur in other tissues where ETs are synthesized and/or exert their effects. The sites of syntheses or actions of ETs, in turn, are likely to determine which enzyme activity is relevant to the metabolism of these bioactive peptides *in vivo.*

REFERENCES

Abassi, Z. A., Tate, J. E., Golomb, E. and Keiser, H. R. (1992). Role of neutral endopeptidase in the metabolism of endothelin. *Hypertension*, 20, 89-95.

Änggård, E., Galton, S., Rae, G., Thomas, R., McLoughlin, L., de Nucci, G. and Vane, J. R. (1989). The fate of radioiodinated endothelin-1 and endothelin-3 in the rat. *J. Cardiovasc. Pharmacol.*, 13(Suppl. 5), S46-S49.

Asaad, M. M., Dorso, C. R. and Moreland, S. M. (1993). Effects of neutral endopeptidase inhibition on the clearance of exogenously administered endothelin in Sprague Dawley rats. *J. Cardiovasc. Pharmacol.*, 21, 633-636.

Barrett, A. J. (1977). Cathepsin D and other carboxyl proteinases, in *Proteinases in Mammalian Cells and Tissues*, Barrett, A. J., Ed., Elsevier/North Holland, Amsterdam, 209-248.

Candenas, M.-L., Naline, E., Sarria, B. and Advenier, C. (1992). Effect of epithelium removal and of enkephalin inhibition on the bronchoconstrictor response to three endothelins of the human isolated bronchus. *Eur. J. Pharmacol.*, 210, 291-297.

Connelly, J. C., Skidgel, R. A., Schulz, W. W., Johnson, A. R. and Erdös, E. G. (1985). Neutral endopeptidase 24.11 in human neutrophils: cleavage of chemotactic peptide. *Proc. Natl. Acad. Sci. U.S.A.*, 82, 8737-8741.

Davenport, A. P., Ashby, M. J., Easton, P., Ella, S., Bedford, J., Dickerson, C., Nunez, D. J., Capper, S. J. and Brown, M. J. (1990). A sensitive radioimmunoassay measuring endothelin-like immunoreactivity in human plasma: comparison of levels in patients with essential hypertension and normotensive control subjects. *Clin. Sci.*, 78, 261-264.

De Nucci, G., Thomas, R., D'Orleans-Juste, P., Antunes, E., Walder, C., Warner, T. D. and Vane, J. R. (1988). Pressor effects of circulating endothelin are limited by its removal in the pulmonary circulation and by the release of prostacyclin and endothelium derived relaxing factor. *Proc. Natl. Acad. Sci. U.S.A.*, 85, 9797-9800.

Deng, Y., Martin, L. L., Del Grande, D. and Jeng, A. Y. (1992). A soluble proteinase identified from rat kidney degrades endothelin-1 but not proendothelin-1. *J. Biochem.*, 112, 168-172.

Deschodt-Lanckman, M., Pauwels, S., Najdovski, T., Dimaline, R. and Dockray, G. J. (1988). *In vitro* and *in vivo* degradation of human gastrin by endopeptidase 24.11. *Gastroenterology*, 94, 712-721.

Dickinson, K. E. J., Tymiak, A. A., Cohen, R. B., Liu, E. C.-K., Webb, M. L. and Hedberg, A. (1991). Vascular A10 cell membranes contain an endothelin metabolising neutral endopeptidase. *Biochem. Biophys. Res. Commun.*, 176, 423-430.

Di-Maria, G. U., Katayama, M., Borson, D. B. and Nadel, J. A. (1992). Neutral endopeptidase modulates endothelin-1-induced airway smooth muscle contraction in guinea-pig trachea. *Regul. Pep.* 39, 137-145.

Diment, S. and Stahl, P. (1985). Macrophage endosomes contain proteases which degrade endocytosed protein ligands. *J. Biol. Chem.*, 260, 15311-15317.

Dzau, V. J., Gonzalez, D., Kaempfer, C., Dubin, D. and Wintroub, B. U. (1987). Human neutrophils release serine proteases capable of activating pro-renin. *Circ. Res.*, 60, 595-601.

Erdös, E. G. and Skidgel, R. A. (1989). Neutral endopeptidase 24.11 (enkephalinase) and related regulators of peptide hormones. *FASEB. J.*, 3, 145-151.

Fagny, C., Michel, A., Leonard, I., Berkenboom, G., Fontaine, J. and Deschodt-Lanckman, M. (1991). *In vitro* degradation of endothelin-1 by endopeptidase 24.11 (enkephalinase). *Peptides*, 12, 773-778.

Fagny, C., Michel, A., Nortier, J. and Deschodt-Lanckman, M. (1992). Enzymatic degradation of endothelin-1 by activated human polymorphonuclear neutrophils. *Regul. Pep.*, 42, 27-37.

Filep, J. G. (1993) Endothelin peptides: biological actions and pathophysiological significance in the lung. *Life Sci.*, 52, 119-133.

Gafford, J. T., Skidgel, R. A., Erdös, E. G. and Hersh, L. B. (1983). Human kidney "enkephalinase" a neutral metalloendopeptidase that cleaves active peptides. *Biochemistry*, 22, 3265-3271.

Gammeltoft, S. (1991). Receptor mediated endocytosis and degradation of polypeptide hormones, growth factors, and neuropeptides, in *Degradation Of Bioactive Substances*, Henriksen, J. H., Ed., CRC Press, Boca Raton, FL, 81-111.

Gandhi, C. R., Harvey, S. A. K. and Olson, M. S. (1993). Hepatic effects of endothelin: metabolism of [^{125}I]endothelin-1 by liver-derived cells. *Arch. Biochem. Biophys.*, 305, 38-46.

Gasic, S., Wagner, O. F., Vierhapper, H., Nowotny, P. and Waldhäusl, W. (1992). Regional haemodynamic effects and clearance of endothelin-1 in humans: renal and peripheral tissues may contribute to the overall disposal of the peptide. *J. Cardiovasc. Pharmacol.*, 19, 176-180.

Gráf, L., Kenessey, A., Patthy, A., Grynbaum, A., Marks, N. and Lajtha, A. (1979). Cathepsin D generates γ-endorphin from β endorphin. *Arch. Biochem. Biophys.*, 193, 101-109.

Greer, I. A., Haddad, N. G., Dawes, J., Johnstone, F. D. and Calder, A. A. (1989). Neutrophil activation in pregnancy induced hypertension. *Br. J. Obstet Gynaecol.*, 96, 978-982.

Helseth, D. L. and Veis, A. (1984). Cathepsin D-mediated processing of procollagen: lysosomal enzyme involvement in secretory processing of procollagen. *Proc. Natl. Acad. Sci. U.S.A.*, 81, 3302-3306.

Hemsen, A., Pernow, J. and Lundberg, J. M. (1991). Regional extraction of endothelins and conversion of big endothelin to endothelin-1 in the pig. *Acta Physiol. Scand.*, 141, 325-334.

Hersh, L. B. and Morihara, K. (1986). Comparison of subsite specificity of mammalian neutral endopeptidase 24.11 (enkephalinase) to the bacterial neutral endopeptidase thermolysin. *J. Biol. Chem.*, 261, 6433-6437.

Jackman, H. L., Morris, P. W., Deddish, P. A., Skidgel, R. A. and Erdös, E. G. (1992). Inactivation of endothelin 1 by deamidase (lysosomal protective protein). *J. Biol. Chem.*, 267, 2872-2875.

Jackman, H. L., Tan, F., Tamei, H., Beurling-Harbury, C., Li, X.-Y., Skidgel, R. A. and Erdös, E. G. (1990). A peptidase in human platelets that deamidates tachykinins. *J. Biol. Chem.*, 265, 11265-11272.

Kaw, S., Warner, T. D. and Vane, J. R. (1993). The metabolism of endothelin-1 and big endothelin-1 by the isolated perfused kidney of the rabbit. *J. Cardiovasc. Pharmacol.*, 22(Suppl. 8), S65-S68.

Kerr, M. A. and Kenny, A. J. (1974a). The purification and specificity of a neutral endopeptidase from rabbit kidney brush border. *Biochem. J.*, 137, 477-488.

Kerr, M. A. and Kenny, A. J. (1974b). The molecular weight and properties of a neutral metallo-endopeptidase from rabbit kidney brush border. *Biochem. J.*, 137, 489-495.

Kenny, A. J. and Stephenson, S. L. (1988). Role of endopeptidase 24.11 in the inactivation of atrial natriuretic peptide. *FEBS Lett.*, 232, 1-8

Kohno, M., Murakawa, K., Yasunari, K., Yokokawa, K., Horio, T., Kurihara, N. and Takeda, T. (1989). Prolonged blood pressure elevation after endothelin administration in bilaterally nephrectomised rats. *Metabolism*, 38, 712-713.

Kohno, M., Yasunari, K., Murakawa, K.-I., Yokokawa, K., Horio, T., Fukui, T. and Takeda, T. (1990). Plasma immunoreactive endothelin in essential hypertension. *Am. J. Med.*, 88, 614-618.

Kon, V., Sugira, M., Inagami, T., Harvie, B. R., Ichikawa, I. and Hoover, R. L. (1990). Role of endothelin in cyclosporine-induced glomerular dysfunction. *Kidney Int.*, 37, 1487-1491.

Koyama, H., Tabata, T., Nishizawa, Y., Inoue, T., Morii, H. and Yamaji, T. (1989). Plasma endothelin levels in patients with uraemia. *Lancet*, 333, 991-992.

Malfroy, B., Swerts, J. P., Guyon, A., Roques, B. P. and Schwartz, J.-C. (1978). High affinity enkephalin-degrading peptidase in mouse brain and its enhanced activity following morphine. *Nature*, 276, 523-526.

Matsas, R., Fulcher, I. S., Kenny, A. J. and Turner, A. J. (1983). Substance P and Leu-enkephalin are hydrolysed by an enzyme in pig caudate synaptic membranes that is identical with the endopeptidase of kidney microvilli. *Proc. Natl. Acad. Sci. U.S.A.*, 80, 3111-3114.

Matsubara, H. (1970). Purification and assay of thermolysin. *Methods. Enzymol.*, 19, 642-651.

Matsumura, Y., Hisaki, K., Takaoka, M. and Morimoto, S. (1990). Phosphoramidon, a metalloprotease inhibitor, suppresses the hypertensive effect of big endothelin-1. *Eur. J. Pharmacol.*, 185, 103-106.

McMahon, E. G., Fok, K. F., Moore, W. M., Smith, C. E., Siegel, N. R. and Trapani, A. J. (1989). *In vitro* and *in vivo* activity of chymotrypsin-activated big endothelin (porcine 1-40). *Biochem. Biophys. Res. Commun.*, 161, 406-413.

McMahon, E. G., Palomo, M. A. and Moore, W. M. (1991). Phosphoramidon blocks the pressor activity of big endothelin (1-39) and lowers blood pressure in spontaneously hypertensive rats. *J. Cardiovasc. Pharmacol.*, 17(Suppl. 7), S29-S33.

Morel, D. R., Silvain Lacroix, J. S., Hemsen, A., Steinig, D. A., Pittet, J. F. and Lundberg, J. M. (1989). Increased plasma and pulmonary lymph levels of endothelin during endotoxic shock. *Eur. J. Pharmacol.*, 167, 427-428.

Najdovski, T., Collette, N. and Deschodt-Lanckman, M. (1985). Hydrolysis of the C-terminal octapeptide of cholecystokinin by rat kidney membranes: characterization of the cleavage by solubilized-endopeptidase 24.11. *Life Sci.*, 37, 827-834.

Noguchi, K., Fukuroda, T., Ikeno, Y., Hirose, H., Tsukada, Y., Nishikibe, M., Ikemoto, F., Matsuyama, K. and Yano, M. (1991). Local formation and degradation of endothelin-1 in guinea-pig airway tissues. *Biochem. Biophys. Res. Commun.*, 179, 830-835.

Patrignani, P., Del Maschio, A., Bazzoni, G., Daffonchio, L., Hernandez, A., Modica, R., Montesanti, L., Volpi, D., Patrono, C. and Dejana, E. (1991). Inactivation of endothelin by polymorphonuclear leukocyte-derived lytic enzymes. *Blood*, 78, 2715-2720.

Perico, N., Dadan, J. and Remuzzi, G. (1990). Endothelin mediates the renal vasoconstriction induced by cyclosporine in the rat. *J. Am. Soc. Nephrol.*, 1, 76-83.

Pernow, J., Hemsen, A. and Lundberg, J. M. (1989a). Increased plasma levels of endothelin-like immunoreactivity during endotoxin administration in the pig. *Acta Physiol. Scand.*, 137, 317-318.

Pernow, J., Hemsen, A. and Lundberg, J. M. (1989b). Tissue specific distribution and vascular effects of endothelin in the pig. *Biochem. Biophys. Res. Commun.*, 161, 647-653.

Roques, B. P., Noble, F., Dauge, V., Fournie-Zaluski, M.-C. and Beaumont, A. (1993). Neutral endopeptidase 24.11: structure, inhibition, and experimental and clinical pharmacology. *Pharmacol. Rev.*, 45, 87-145.

Saito, Y., Nakao, K., Mukoyama, M. and Imura, H. (1990). Increased plasma endothelin level in patients with essential hypertension. *N. Engl. J. Med.*, 332, 205.

Salvati, P., Dho, L., Calabresi, M., Rosa, B. and Patrono, C. (1992). Evidence for a direct vasoconstrictor effect of big endothelin-1 in the rat kidney. *Eur. J. Pharmacol.*, 221, 267-273.

Sasahara, M., Hazama, F., Amano, S., Hayase, Y., Yukioka, N., Kawai, J. and Kataoka, H. (1988). Effect of hypertension on lysosomal enzyme activities in aortic endothelial cells. *Atherosclerosis*, 70, 53-62.

Sawamura, T., Kimura, S., Shinmi, O., Sugita, Y., Kobayashi, M., Mitsui, M., Yanagisawa, M., Goto, K. and Masaki, T. (1990a). Characterization of endothelin converting enzyme activities in the soluble fraction of bovine cultured endothelial cells. *Biochem. Biophys. Res. Commun.*, 169, 1138-1144.

Sawamura, T., Kimura, S., Shinmi O., Sugita, Y., Yanagisawa, M., Goto, K. and Masaki, T. (1990b). Purification and characterization of putative endothelin converting enzyme in bovine adrenal medulla: evidence for a cathepsin D-like enzyme. *Biochem. Biophys. Res. Commun.*, 168, 1230-1236.

Sawamura, T., Shinmi, O., Kishi, N., Sugita, Y., Yanagisawa, M., Goto, K., Masaki, T. and Kimura, S. (1990c). Analysis of big endothelin-1 digestion by cathepsin D. *Biochem. Biophys. Res. Commun.*, 172, 883-889.

Schmid-Schönbein, G. W., Sieffge, D., DeLano, F. A., Shen, K. and Zweifach, W. (1991). Leukocyte counts and activation in spontaneously hypertensive and normotensive rats. *Hypertension*, 17, 323-330.

Schwartz, J. C., Malfroy, B. and De La Baume, S. (1981). Biological inactivation of enkephalins: the role of enkephalin-dipeptidyl-carboxypeptidase "enkephalinase" as a neuropeptidase. *Life Sci.*, 29, 1715-1740.

Scili, A. G., Vijayaraghavan, J., Hersh, L. and Carretero, O. (1989). Neutral endopeptidase 24.11 (NEP) inactivates endothelin. *Hypertension*, 14, 353.

Selak, M. A., Chignard, M. and Smith, J. B. (1988). Cathepsin G is a strong platelet agonist released by neutrophils. *Biochem. J.*, 251, 293-299.

Sessa, W. C., Kaw, S., Hecker, M. and Vane, J. R. (1991a). The biosynthesis of endothelin-1 by human polymorphonuclear leukocytes. *Biochem. Biophys. Res. Commun.*, 174, 613-618.

Sessa, W. C., Kaw, S., Zembowicz, A., Änggård, E., Hecker, M. and Vane, J. R. (1991b). Human polymorphonuclear leukocytes generate and degrade endothelin-1 by two distinct neutral proteases. *J. Cardiovasc. Pharmacol.*, 17(Suppl. 7), S34-S38.

Shiba, R., Yanagisawa, M., Miyauchi, T., Ishii, Y., Kimura, S., Uchiyama, Y., Masaki, T. and Goto, K. (1989). Elimination of intravenously injected endothelin-1 from the circulation of the rat. *J. Cardiovasc. Pharmacol.*, 13(Suppl. 5), S98-S101.

Sirviö, M.-L., Metsärinne, K., Saijonmaa, O. and Fyhrquist, F. (1990). Tissue distribution and half life of ^{125}I endothelin in the rat: importance of pulmonary clearance. *Biochem. Biophys. Res. Commun.*, 167, 1191-1195.

Skidgel, R. A., Jackman, H. L. and Erdös, E. G. (1991). Metabolism of substance P and bradykinin by human neutrophils. *Biochem. Pharmacol.*, 41, 1335-1344.

Sokolovsky, M., Galron, R., Kloog, Y., Bdolah, A., Indig, F. E., Blumberg, S. and Fleminger, G. (1990). Endothelins are more sensitive than sarafotoxins to neutral endopeptidase. Possible physiological significance. *Proc. Natl. Acad. Sci. U.S.A.*, 87, 4702-4706.

Stephenson, S. L. and Kenny, A. J. (1987). The hydrolysis of α human atrial natriuretic peptide by pig kidney microvillar membranes is initiated by endopeptidase 24.11. *Biochem. J.*, 243, 183-187.

Sugira, M., Inagami, T. and Kon, V. (1989). Endotoxin stimulates endothelin release *in vivo* and *in vitro* as determined by radioimmunoassay. *Biochem. Biophys. Res. Commun.*, 161, 1220-1227.

Suzuki, N., Miyauchi, T., Tomobe, Y., Matsumoto, H., Goto, K., Masaki, T. and Fujino, M. (1990). Plasma concentrations of endothelin-1 in spontaneously hypertensive rats and DOCA-salt hypertensive rats. *Biochem. Biophys. Res. Commun.*, 167, 941-947.

Takaoka, M., Hukumori, Y., Shiragami, K., Ikegawa, R., Matsumura, Y. and Morimoto, S. (1990a). Proteolytic processing of porcine big endothelin-1 catalysed by cathepsin D. *Biochem. Biophys. Res. Commun.*, 173, 1218-1223.

Takaoka, M., Miyata, Y., Takenobu, Y., Ikegawa, R., Matsumura, Y. and Morimoto, S. (1990b). Mode of cleavage of pig big endothelin-1 by chymotrypsin. Production and degradation of mature endothelin-1. *Biochem. J.*, 270, 541-544.

Télémaque, S., Lemaire, D., Claing, A. and D'Orleans-Juste, P. (1992). Phosphoramidon-sensitive effects of big endothelins in the perfused rabbit kidney. *Hypertension*, 20, 518-523.

Tomita, K., Ujiie, K., Nakanishi, T., Tomura, S., Matsuda, O., Ando, K., Shichiri, M., Hirata, Y. and Marumo, F. (1989). Plasma endothelin levels in patients with acute renal failure. *N. Engl. J. Med.*, 321, 1127

Travis, J. (1988). Structure, function and control of neutrophil proteinases. *Am. J. Med.*, 84(Suppl. 6A), 37-42.

Travis, J., Giles, P. J., Porcelli, L., Reilly, C. F., Baugh, R. and Powers, J. (1979). Human leukocyte elastase and cathepsin G: structural and functional characteristics. *Excerpta Med.* (Ciba Found. Symp.), 75, 51-68.

Turner, A. J. (1993). Endothelin-converting enzymes and other families of metallo-endopeptidases. *Biochem. Soc. Trans.*, 21, 697-701.

Ura, N., Carretero, O. A. and Erdös, E. G. (1987). Role of renal endopeptidase 24.11 in kinin metabolism *in vitro* and *in vivo*. *Kidney Int.*, 32, 507-513.

Vijayaraghavan, J., Scili, A. G., Carretero, O. A., Slaughter, C., Moomaw, C. and Hersh, L. B. (1990). The hydrolysis of endothelins by neutral endopeptidase 24.11 (enkephalinase). *J. Biol. Chem.*, 265, 14150-14155.

Ware Branch, D., Dudley, D. J. and Mitchell, M. D. (1991). Preliminary evidence for a homeostatic mechanism regulating endothelin production in pre-eclampsia. *Lancet*, 337, 943-945.

Weitzberg, E., Ahlborg, G. and Lundberg, J. M. (1991). Long-lasting vasoconstriction and efficient regional extraction of endothelin-1 in human splanchnic and renal tissues. *Biochem. Biophys. Res. Commun.*, 180, 1298-1303.

Wintroub, B. U., Klickstein, L. B., Dzau, V. J. and Watt, K. W. K. (1984). Granulocyte-angiotensin system: identification of angiotensinogen as the plasma protein substrate of leukocyte cathepsin G. *Biochemistry*, 23, 227-232.

Wypij, D. M., Nichols, J. S., Novak, P. J., Lowell Stacy, D., Berman, J. and Wiseman, J. S. (1992). Role of mast cell chymase in the extracellular processing of big-endothelin-1 to endothelin-1 in the perfused rat lung. *Biochem. Pharmacol.*, 43, 845-853.

Yamaguchi, T., Fukase, M., Arao, M., Sugimoto, T. and Chihara, K. (1992a). Endothelin-1 hydrolysis by rat kidney membranes. *FEBS Lett.*, 309, 303-306.

Yamaguchi, T., Kido, H., Fukase, M., Fujita, T. and Katunuma, N. (1991). A membrane-bound metalloendopeptidase from rat kidney hydrolyzing parathyroid hormone. Purification and characterization. *Eur. J. Biochem.*, 200, 563-571.

Yamaguchi, T., Kido, H. and Katunuma, N. (1992b). A membrane-bound metalloendopeptidase from rat kidney. Characteristics of its hydrolysis of peptide hormones and neuropeptides. *Eur. J. Biochem.*, 204, 547-552.

Yamaguchi, T., Kohrogi, H., Kawano, O., Ando, M. and Araki, S. (1992c). Neutral endopeptidase inhibitor potentiates endothelin-1-induced airway smooth muscle contraction. *J. Appl. Physiol.*, 73, 1108-1113.

Yanagisawa, M., Kurihara, H., Kimura, S., Tomobe, Y., Kobayashi, M., Mitsui, Y., Yazaki, Y., Goto, K. and Masaki, T. (1988). A novel potent vasoconstrictor peptide produced by vascular endothelial cells. *Nature*, 332, 411-415.

Part III

Endothelin Pathophysiology

Chapter 8

Cardiovascular Effects of Endothelin

Bulent Gumusel, Qingzhong Hao, Jaw-Kang Chang, Albert Hyman, and Howard Lippton

CONTENTS

I. INTRODUCTION

In 1985, Hickey et al. showed that cultured endothelial cells produced a peptidergic contractile factor. This factor was isolated, sequenced, cloned, and named "Endothelin" by Yanagisawa in 1988. The endothelin (ET) isolated from supernatant of porcine aortic endothelial cells was a peptide of 21 amino acid residues with 2 disulfide bridges linking Cys^{1}-Cys^{15} and Cys^{3}-Cys^{11}. Analysis of human genomic sequences showed that there are three distinct genes for ET which encode three distinct ET peptides and are named ET-1, ET-2, and ET-3 (Inoue et al., 1989a, b). The gene sequences predict that all ETs are derived from "preproendothelin" precursors. These large precursors are initially subjected to intermediate processing by endopeptidases to yield proendothelins (called Big ETs). Subsequent peptide cleavage of a Trp-Val (ET-1 and ET-2) or Trp-Ile (ET-3) bond appears to depend on processing by a recently described "endothelin converting enzyme-1 (ECE)" (Inoue et al., 1989b),

0-8493-6975-4/97/$0.00+$.50

which has been reported to be expressed in both intracellular (Golgi bodies) and extracellular forms. Autoradiographic studies show that following intravenous injection into mammals, ET was largely localized in lungs, kidneys, liver, and to a lesser extent in spleen, heart, stomach, skin, intestine, and pancreas.

II. CARDIOVASCULAR EFFECTS OF ENDOTHELIN

A. Heart

ET is produced by cultured endocardial cells and circulates in blood (Lewis and Sham, 1994). The secreted peptide may act on myocytes, the conduction system of the heart, or coronary vessels. ETs have a direct effect on the myocardium (Hu et al., 1988; Ishikawa et al., 1988); it possesses positive inotropic activity on the isolated perfused rat heart (Baydoun et al., 1989; Firth et al., 1990), on isolated human, guinea pig, and rat atria (Davenport et al., 1989), in both ferret (Shah et al., 1989) and rabbit papillary muscles (Watanabe et al., 1989a), and in isolated adult rat and rabbit ventricular cells (Kelly et al., 1990). ET-1 produces a dose-dependent increase in heart rate in isolated spontaneously beating atrial preparations (Ishikawa et al., 1988). The positive inotropic response is accompanied by prolongation of the cardiac action potential (Watanabe et al., 1989a,b). In contrast to the positive inotropic and chronotropic effects seen in isolated heart studies, a decrease in cardiac output and heart rate is observed with i.v. infusion of ET in the intact animal (Lerman et al., 1991, 1992). This bradycardia is likely due to both direct and indirect effects of ET-1 on the heart (Karwatowska Prokopczuk and Wennmalm, 1990). Although ET-1 may directly influence the cardiac conduction system *in vivo*, it may also indirectly decrease heart rate by coronary vasoconstriction (Hom et al., 1992), and reflex response to systemic hypertension as well as to changes in neural outflow from the central nervous system. The decrease in cardiac output in response to ET-1 in animals is due to a combination of actions, including a decreasing heart rate, decreasing coronary perfusion, and a reduction in cardiac performance. These actions of ET-1 appear to counteract the potential increase in cardiac output due to increased preload as a result of the marked contractile effect of ET-1 on large veins. The divergent properties of ET-1 on cardiac tissue *in vitro* and *in vivo* may relate to release of additional mediators and alterations in receptor number and function. Although the chronotropic effect of ET may be species dependent, the positive inotropic activity of ET appears universal (Karwatowska Prokopczuk and Wennmalm, 1990). The positive inotropic response is accompanied by prolongation of the cardiac action potential (Watanabe et al., 1989a,b).

B. Pulmonary Vascular Bed

Since the lung both releases and removes ET-1, studies were undertaken to determine the effects of ET-1 in the pulmonary vascular bed *in vivo*. In light of the difficulties in attempting to study the direct effects of a substance on the pulmonary vascular bed *in vivo*, a novel technique was employed to investigate the effects of ET isopeptides on the pulmonary bed in the closed-chest, spontaneously beating animal. In these experiments, a specially designed 6-F triple-lumen balloon perfusion catheter was passed from the left external jugular vein into the pulmonary artery and then into the left lower lobe, using fluoroscopy. The left lower lobe artery was isolated by distention of the balloon cuff on the perfusion catheter. The left lower

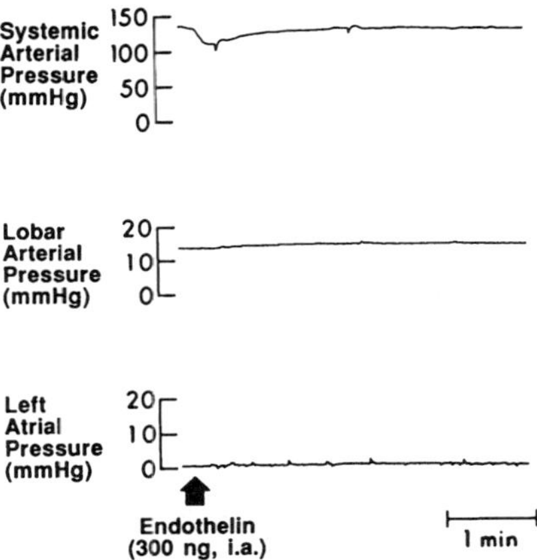

FIGURE 1

Effects of intralobar injections of endothelin on lobar arterial pressure under conditions of resting and elevated pulmonary vasoconstrictor tone in cats. (From Lippton, H. L., Hauth, T. A., Summer, W. R., and Hyman, A. L., *J. Appl. Physiol.*, 66(2), 1008, 1989. With permission.)

lobe was then perfused using a peristaltic pump and blood from a femoral. Pulmonary blood flow and left atrial pressure were held constant so changes in lobar arterial pressure directly reflected changes in pulmonary vascular resistance (Lippton et al., 1993). Rapid bolus injections of ET into the lobar artery in doses of 100, 300, and 1000 ng produced small increases in arterial pressure without altering left atrial pressure. The pulmonary vasoconstrictor response to endothelin appeared within 1 min and lasted from 2 to 8 min.

In contrast to pulmonary vasoconstriction, intralobar injections of endothelin produced concomitant, dose-related reductions in systemic arterial pressure (Lippton et al., 1989a) (Figure 1). Under conditions of resting pulmonary vasomotor tone in the intact rabbit, intralobar arterial bolus injections of ET-1 (0.3 to 1 μg) and ET-3 (3 μg) significantly increased lobar arterial pressure (Figure 2). The increases in lobar arterial pressure in response to ET-1 as well as to ET-3 were inhibited by nitrendepine, a dihydropyridine calcium channel blocker (Figure 3) but were not altered by indomethacin, suggesting that ET-1 and ET-3 promote calcium influx via the L-type calcium channel to promote constriction of pulmonary resistance vessels *in vivo*. Moreover, the pulmonary hemodynamic effects of ET-1 and ET-3 in the rabbit do not depend on cyclooxygenase products (Lippton et al., 1991a). In the anesthetized, closed-chest cat, *in vivo* pulmonary vasomotor tone was elevated by intralobar infusions of U-46619, a thromboxane mimic; single intralobar bolus injections of ET-1, ET-2, and ET-3 significantly decreased lobar arterial pressure without altering left atrial pressure. The pulmonary vasodilator response, but not the systemic vasodilator response to each ET isopeptide was blocked by glybenclamide, an inhibitor of ATP-dependent potassium channel activation. The present study suggests that, in the pulmonary vascular bed, activation of potassium channels mediates the vasodilator response to ET-1, ET-2, and ET-3 (Lippton et al., 1991b) (Table 1).

In another study in intact cats under similar elevated pulmonary vasomotor tone conditions, the first intralobar arterial bolus injections of ET-1 significantly decreased lobar and systemic arterial pressure. This pulmonary vasodilator response abruptly disappeared after two subsequent injections of ET-1 and did not reappear during the following 4 h. The pulmonary vasoconstrictor response to ET-1 remained unchanged with all ET-1 injections and had a duration of 2 to 8 min, whereas this systemic vasodepressor response was decreased by the third ET-1 injection and

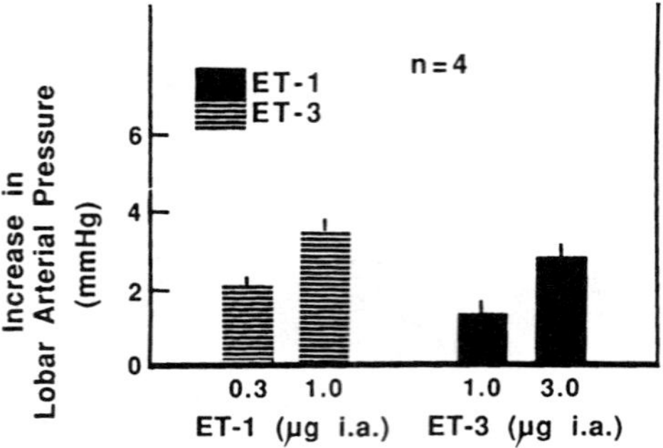

FIGURE 2
Comparison of effects of intralobar injections of ET-1 and ET-3 in rabbit pulmonary vascular bed. (From Lippton, H. L., Ohlstein, E. H., Summer, W. R., and Hyman, A. L., *J. Appl. Physiol.*, 70(1), 331, 1991. With permission.)

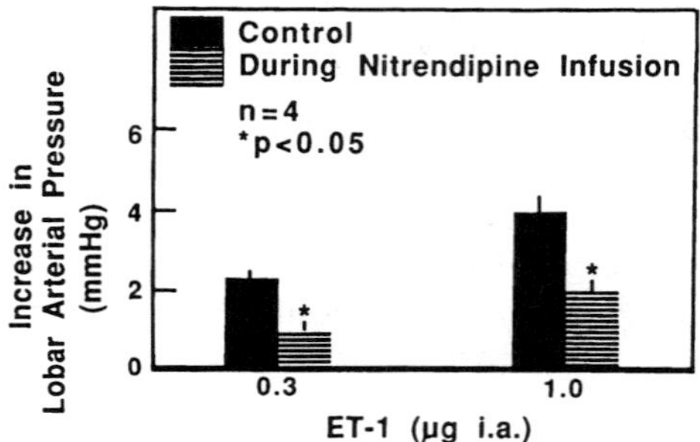

FIGURE 3
Influence of nitrendipine on pulmonary vasoconstrictor responses to ET-1. Responses to intralobar injections of ET-1 were obtained before (control) and 5 min after start of intralobar infusion of nitrendipine (5 µg/min). (From Lippton, H. L., Ohlstein, E. H., Summer, W. R., and Hyman, A. L., *J. Appl. Physiol.*, 70(1), 331, 1991. With permission.)

disappeared by the fourth ET-1 injection. The fourth and subsequent ET-1 injections produced similar systemic vasoconstrictor responses (Figure 4). Tachyphylaxis to the pulmonary vasodilator response and systemic vasodilator response to ET-2 also developed, but more slowly than the tachyphylaxis to ET-1. Repeated intralobar arterial bolus injections of ET-3 produced significant pulmonary vasodilation. The pulmonary vasodilator response to ET-3 injection did not show tachyphylaxis. In contrast, the systemic vasodilator response to ET-3 was lost by the fourth injection, and subsequent injections of ET-3 produced systemic vasoconstriction (Figure 5). In cats that had received only seven injections of ET-1, initial injection of ET-2 caused vasoconstriction in both pulmonary and systemic vascular beds; intralobar arterial injections of ET-3 into ET-1-treated cats produced pulmonary vasodilator responses that were similar to those seen with the initial ET-3 injections. In contrast, the systemic vasodilator response to ET-3 was no longer present in these cats (Figure 5). After seven injections of ET-3 in the third group of cats, pulmonary vasodilator response to single injections of ET-1 and ET-2 was absent, but the pulmonary vasodilator response to a single injection of ET-3 remained. Moreover, the systemic vasodilator responses to ET-1, ET-2, STXb, and ET-3 were absent in cats that received seven prior injections of ET-3 (Figure 6).

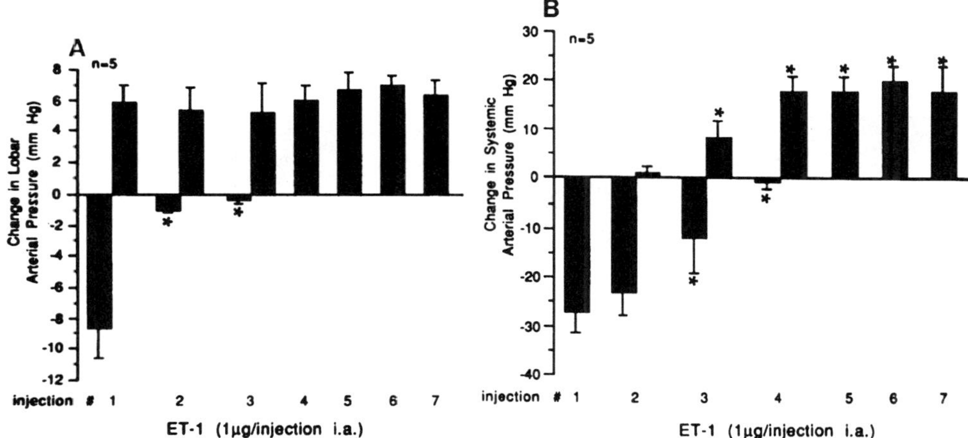

FIGURE 4
Effects of repeated intralobar bolus injections of ET-1 on lobar (A) and systemic (B) arterial pressures in anesthetized cats under conditions of elevated pulmonary vasomotor tone. Total of seven injections of ET-1 (1 µg/injection) were administered. (From Lippton, H. L., Hauth, T. A., Cohen, G. A., and Hyman, A. L., *J. Appl. Physiol.*, 75(1), 38, 1993. With permission.)

Results of these experiments demonstrate that ET-1, ET-2, ET-3, and STXb dilate the pulmonary vascular bed *in vivo* under conditions of elevated pulmonary vasomotor tone. The present data extend previous studies by demonstrating that the pulmonary, similar to the systemic, vasodilator response to both ET-1 and ET-2 undergoes tachyphylaxis (Lippton et al., 1993). The tachyphylactic nature of the pulmonary vasodilator response to ET-1 was first reported in cat (Le Monnier de Gouville et al., 1990a). The development of tachyphylaxis to the pulmonary vascular response to ET-1 *in vivo* does not appear to be limited to this species, since the vasodilator response to ET-1 in the fetal lamb lung *in situ* was subsequently reported to be decreased and reversed after injections of increasing doses of the peptide. Tachyphylaxis and desensitization of a biological response to a variety of endogenous peptides such as angiotensin II and opioids are well known.

Tachyphylaxis to the contractile response to ET-1 has been reported with isolated guinea-pig femoral artery (Wiklund et al., 1988) and porcine aorta (Ishikawa et al., 1988). The present study indicates that ET-1 and ET-2 isopeptides also undergo tachyphylaxis. The cross-tachyphylaxis of the pulmonary and systemic vasodilator response to ET-1, ET-2, and STXb indicates these isopeptides may share a common population of receptors. However, heterologous desensitization of ET receptors in systemic blood vessels remains a possibility.

ET-1 and STXb have been shown *in vitro* to interact at a high-affinity ET receptor in the aorta. The reason for the lack of tachyphylaxis to the pulmonary vasodilation to ET-3 is unclear, but suggests the existence of a heterogeneous population of ET receptors mediating vasodilation in the pulmonary vascular bed. This conclusion is supported by the demonstration of at least two different forms of ET receptors in the rat (Masuda et al., 1989; Sakurai et al., 1990), and guinea pig (Cardell et al., 1992) lung. ET-1 is formed by proteolytic metabolism of its precursor, big ET-1. Moreover, a specific protease, termed endothelin converting enzyme (ECE), has been reported to specifically cleave the Trp^{21}-Val^{22} bond that leads to generation of ET-1. In the pulmonary vascular bed of the intact cat, intralobar bolus injections of porcine big ET-1 and ET-1 increased lobar arterial pressure in a dose-dependent manner. Under

TABLE 1

Effects of Blocking Agents on Pulmonary and Systemic Vascular Responses to Intralobar Bolus Injections of ET-1, ET-2, and ET-3 Under Conditions of Elevated Pulmonary Vasomotor Tone

	Lobar Arterial Pressure (mmHg)	Left Atrial Pressure (mmHg)	Systemic Arterial Pressure (mmHg)	Cardiac Output (ml/min)	Systemic Vascular Resistance (mmHg·ml^{-1}·min)
Control					
Baseline	34 ± 2	3 ± 1	135 ± 6	585 ± 41	0.231 ± 0.021
ET-1 (1 µg ia)	25 ± 1*	3 ± 1	102 ± 5*	691 ± 54*	0.148 ± 0.016*
Baseline	35 ± 1	3 ± 1	133 ± 7*	682 ± 49	0.202 ± 0.025
ET-2 (1 µg ia)	26 ± 1*	3 ± 1	103 ± 6*	774 ± 52*	0.136 ± 0.017*
Baseline	34 ± 1	3 ± 1	127 ± 3	610 ± 47	0.220 ± 0.022
ET-3 (3 µg ia)	24 ± 1*	3 ± 1	91 ± 5*	738 ± 59*	0.132 ± 0.016*
After glybenclamide (5 mg/kg ia)					
Baseline	35 ± 2	4 ± 1	149 ± 5	561 ± 51	0.266 ± 0.021
ET-1 (1 µg ia)	32 ± 2	4 ± 1	114 ± 5*	700 ± 61*	0.162 ± 0.018*
Baseline	35 ± 1	3 ± 1	140 ± 8	522 ± 51	0.268 ± 0.043
ET-2 (1 µg ia)	32 ± 1	3 ± 1	109 ± 4*	634 ± 56*	0.172 ± 0.039*
Baseline	35 ± 1	4 ± 1	137 ± 4	491 ± 51	0.311 ± 0.042
ET-3 (3 µg ia)	31 ± 1*	4 ± 1	98 ± 4*	643 ± 65*	0.152 ± 0.036*
After indomethacin (2.5 mg/kg iv)					
Baseline	35 ± 1	3 ± 1	131 ± 6	589 ± 61	0.222 ± 0.024
ET-1 (1 µg ia)	25 ± 1*	3 ± 1	98 ± 5*	700 ± 58*	0.140 ± 0.015*
Baseline	34 ± 2	4 ± 1	129 ± 5	546 ± 49	0.236 ± 0.021
ET-2 (1 µg ia)	25 ± 1*	4 ± 1	97 ± 6*	651 ± 52*	0.149 ± 0.013*
Baseline	35 ± 2	4 ± 1	130 ± 6	610 ± 51	0.213 ± 0.024
ET-3 (3 µg ia)	24 ± 1*	4 ± 1	93 ± 5*	742 ± 69*	0.125 ± 0.014*
After atropine (1 mg/kg iv)					
Baseline	34 ± 1	3 ± 1	131 ± 6	589 ± 51	0.222 ± 0.019
ET-1 (1 µg ia)	24 ± 2*	3 ± 1	97 ± 5*	695 ± 59*	0.140 ± 0.011*
Baseline	35 ± 1	4 ± 1	129 ± 5	553 ± 48	0.233 ± 0.020
ET-2 (1 µg ia)	26 ± 1*	4 ± 1	96 ± 6*	671 ± 61*	0.143 ± 0.015*
Baseline	34 ± 1	3 ± 1	133 ± 6	608 ± 54	0.219 ± 0.019
ET-3 (3 µg ia)	24 ± 1*	3 ± 1	95 ± 6*	748 ± 610*	0.127 ± 0.013*
After ICI 118511 (1 mg/kg iv)					
Baseline	35 ± 2	3 ± 1	131 ± 5	581 ± 49	0.225 ± 0.022
ET-1 (1 µg ia)	25 ± 2*	3 ± 1	98 ± 6*	610 ± 51*	0.161 ± 0.015*
Baseline	34 ± 2	4 ± 1	134 ± 8	535 ± 41	0.250 ± 0.019
ET-2 (1 µg ia)	25 ± 1*	4 ± 1	97 ± 8*	649 ± 48*	0.149 ± 0.015*
Baseline	35 ± 1	3 ± 1	128 ± 5	551 ± 46	0.232 ± 0.022
ET-3 (3 µg ia)	25 ± 1*	3 ± 1	89 ± 5*	691 ± 53*	0.129 ± 0.013*

Note: Values are means ± SE; n = 5 cats; $p \leq .05$ compared with corresponding control (paired *t* test). From Lippton, H. L., Cohen, G. A., McMurtry, I. F., and Hyman, A. L., *J. Appl. Physiol.*, 70(2), 947, 1991. With permission.

conditions of actively increased pulmonary vasomotor tone, intralobar bolus administration of porcine big ET-1 produced pulmonary vasoconstriction, whereas ET-1 produced pulmonary vasodilation. The pulmonary vasoconstrictor responses to porcine big ET-1 under conditions of resting and elevated pulmonary vasomotor tone

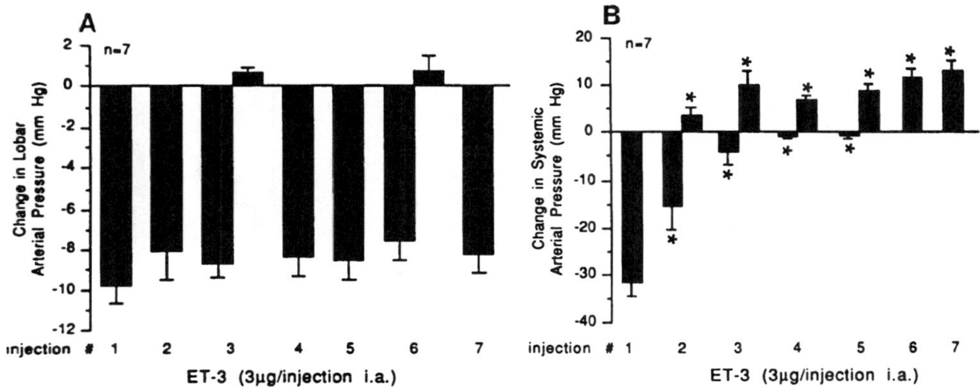

FIGURE 5
Effects of repeated intralobar bolus injections of ET-3 on lobar (A) and systemic (B) arterial pressures in anesthetized cats under conditions of elevated pulmonary vasomotor tone. Total of seven injections of ET-1 (1 μg/injection) were administered. (From Lippton, H. L., Hauth, T. A., Cohen, G. A., and Hyman, A. L., *J. Appl. Physiol.*, 75(1), 38, 1993. With permission.)

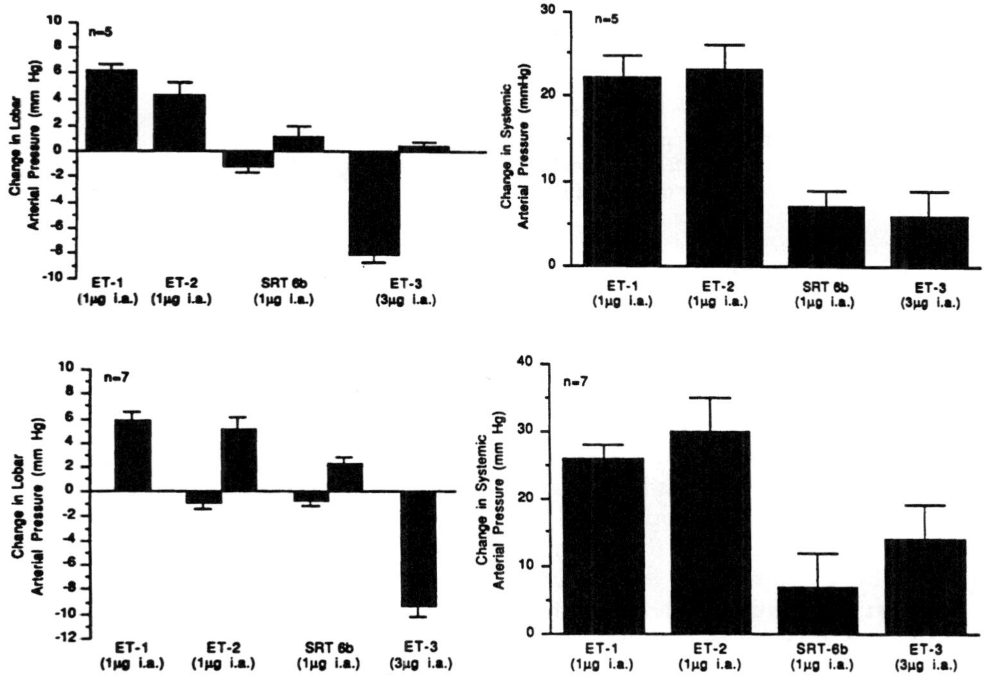

FIGURE 6
Influence of seven intralobar injections of ET-1 and ET-3 on lobar (A) and systemic (B) vascular responses to ET-1, ET-2, ET-3, and SRT-6b in anesthetized cats under conditions of elevated PVT. (From Lippton, H. L., Hauth, T. A., Cohen, G. A., and Hyman, A. L., *J. Appl. Physiol.*, 75(1), 38, 1993. With permission.)

are not altered by phosphoramidon, which has been shown to inhibit the conversion of big ET-1 to ET-1.

These data suggest that the ability of porcine big ET-1 to constrict the pulmonary vascular bed *in vivo* does not depend on formation of ET-1. Although porcine big

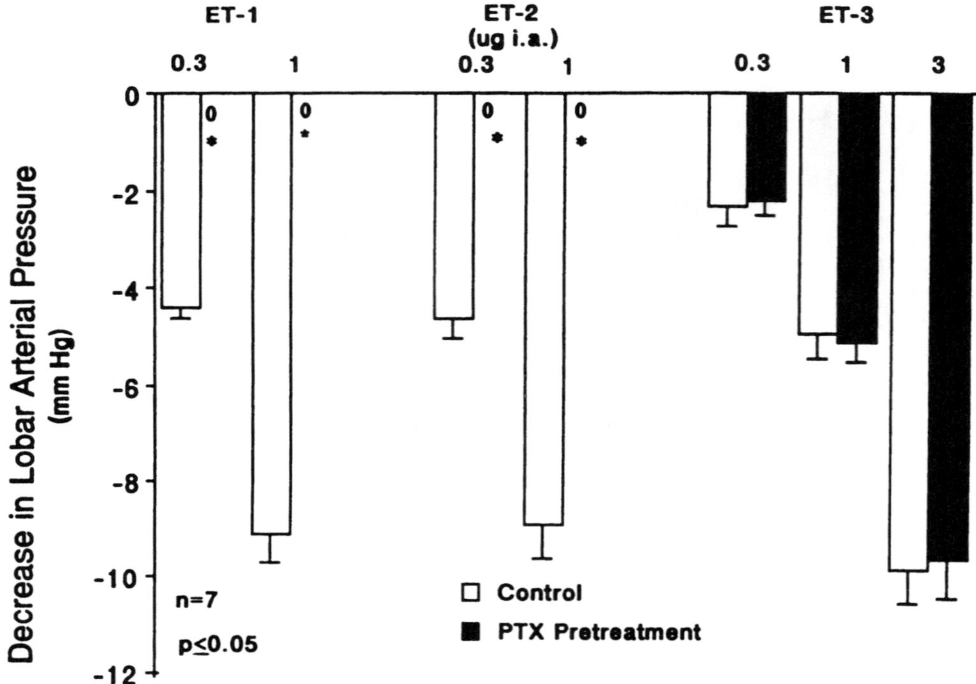

FIGURE 7
Effects of PTX pretreatment on decreases in lobar arterial pressure in response to intralobar arterial bolus injections of ET-1 (0.1 to 1 µg), ET-2 (0.1 to 1 µg), and ET-3 (0.3 to 3 µg) under conditions of elevated pulmonary vasomotor tone. (From Lippton, H. L., Hao, Q., Erdemli, O., and Hyman, A., *J. Appl. Physiol.*, 78(6), 2062, 1995. With permission.)

ET-1 constricts the pulmonary and systemic vascular bed *in vivo*, only the systemic vasoconstrictor response to porcine big ET-1 is blocked by phosphoramidon. This study provides evidence suggesting that the mediation of the pulmonary and systemic vascular responses to porcine big ET-1 differ. In intact cats under conditions of constant pulmonary blood flow and left atrial pressure, when pulmonary vasomotor tone was actively increased by an intralobar arterial infusion of U-46619, the vasodilator responses to ET-1 and ET-2 were abolished by pertussis toxin (PTX), which possesses the enzymatic activity that transfers the ADP-ribose moiety of NAD to the α-subunit of a subpopulation of G proteins (Cortina and Barieri, 1991); PTX pretreatment did not alter the pulmonary vasodilator response to ET-3 (Figures 7 and 8; Table 2). To confirm *in vivo* that PTX-sensitive G proteins were functionally uncoupled, pulmonary vasoconstrictor responses to prostaglandin $F_{2\alpha}$ ($PGF_{2\alpha}$), methoxamine (an α_1-adrenoceptor agonist) and Uk–14304 (an α_2-adrenoceptor agonist) were compared in animals with and without PTX treatment. In animals pretreated with PTX, pulmonary vasoconstrictor response to Uk–14304 was abolished, whereas the pulmonary vasoconstrictor responses to $PGF_{2\alpha}$ and methoxamine were not altered. Selective blockade of the pressor response mediated by α_2-adrenoceptors by PTX affords evidence *in vivo* of selective uncoupling of PTX-sensitive G proteins (Nichols et al., 1988; Nichols et al., 1989) (Table 3). The discordant influence of PTX on the pulmonary vasodilator responses to ET isopeptides in the present study suggests that the signal transduction mechanisms for ET receptor-mediated vasodilation in the lung differ even though the ET receptors mediating vasodilation in the lung share a common effector mechanism. Based on the present data and recent work

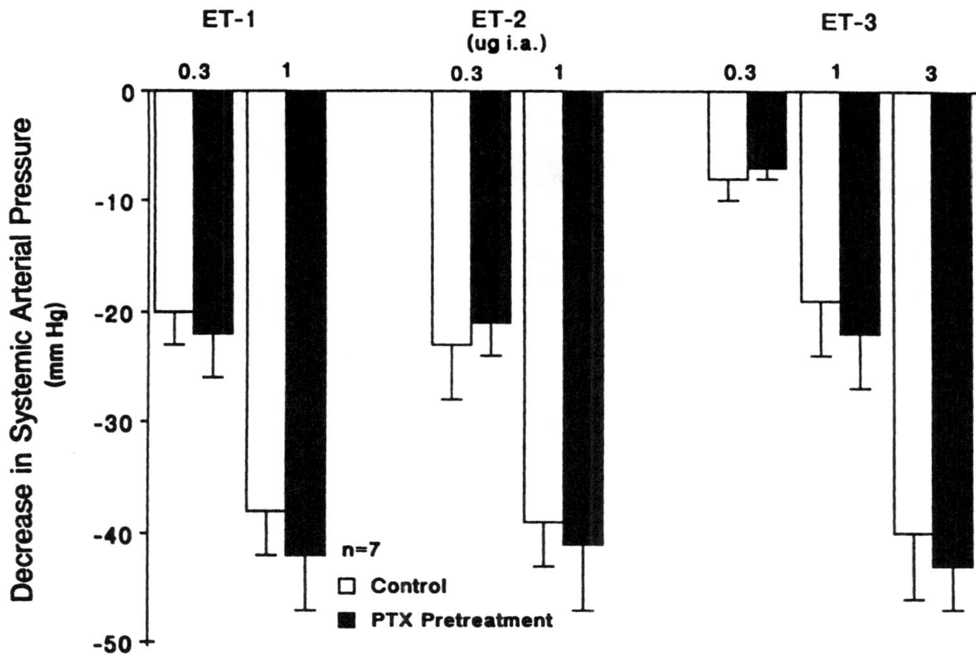

FIGURE 8
Effects of PTX treatment on decreases in systemic arterial pressure in response to bolus injections of ET-1, ET-2, and ET-3 under conditions of elevated pulmonary vasomotor tone. (From Lippton, H. L., Hao, Q., Erdemli, O., and Hyman, A., *J. Appl. Physiol.*, 78(6), 2062, 1995. With permission.)

TABLE 2
Influence of PTX (50 μg/kg, i.v. 72 h Before Study) Pretreatment on Systemic Hemodynamics Responses to Intralobar Arterial Injections of ET Isopeptides

	Increase in Lobar Arterial Pressure (mmHg)	
	Control	PTX Pretreatment
$PGF_{2\alpha}$		
10 ng ia	4 ± 0.3	4 ± 0.6
30 ng ia	8 ± 0.9	8 ± 0.3
Methoxamine		
100 μg ia	4 ± 0.4	4 ± 0.7
300 μg ia	9 ± 0.9	9 ± 1.2
UK-14304		
100 μg ia	4 ± 0.7	0
300 μg ia	8 ± 0.9	0

Note: Values are means ± SE; n = 6 animals.
From Lippton, H. L., Hao, Q., Erdemli, O., and Hyman, A., *J. Appl. Physiol.*, 78(6), 2062, 1995. With permission.

from this laboratory (Lippton et al., 1993), it appears that ET_A-like receptors mediating vasodilation in the cat lung are coupled to PTX-sensitive G proteins, whereas ET_C-like receptors mediating vasodilation in the feline pulmonary vascular bed are not coupled to PTX-sensitive G proteins (Lippton et al., 1995). Thus, the transduction mechanism for the presumed ET_A-like and ET_C-like receptors in the adult cat lung

TABLE 3

Effects of PT Pretreatment on Pulmonary Vasoconstrictor Responses

	Systemic Vascular Resistance	
	Control	After Pretreatment
ET-1 (1 μg ia)	40 ± 3	36 ± 4
ET-2 (1 μg ia)	42 ± 4	37 ± 4
ET-3 (3 μg ia)	44 ± 3	39 ± 5

Note: Values are means ± SE of percent decrease; n = 7 animals.

From Lippton, H. L., Hao, Q., Erdemli, O., and Hyman, A., *J. Appl. Physiol.*, 78(6), 2062, 1995. With permssion.

differ; however, activation of both these ET-like receptor subtypes dilates the pulmonary vascular bed through cellular hyperpolarization via K_{ATP} channels (Lippton et al., 1991b, 1993).

C. Hemodynamic Actions

The complex hemodynamic response to ET in the intact animal was initially interpreted to be monophasic, consisting of prolonged systemic hypertension and systemic vasoconstriction (Yanagisawa et al., 1988). Data from this and other laboratories have provided the basis for the following generalizations regarding the systemic hemodynamic response to exogenously administrated ET-1 to adult mammals *in vivo:* specifically (1) the initial degree of vasomotor tone, (2) rate of administration of the peptide, (3) route of administration, (4) the total dose administered acutely, and (5) the vascular bed under study.

The systemic hemodynamic response to ET-1 occurs independent of species and the animal's state of consciousness (Lippton et al., 1989b). In the initial study by Yanagisawa et al. (1988), ET-1 produced transient mild systemic hypotension followed by a prolonged systemic pressor response in the anesthetized rats that were chemically sympathectamized. In pithed rats, i.v. bolus injections of ET-1 produced dose-related increases in mean arterial blood pressure which reached a maximum within the first 2 to 3 min after the injection and waned during the subsequent 20 to 50 min. All doses produced a significant increase in systemic vascular resistance. The effects of the higher doses were accompanied by significant dose-related decreases in cardiac output (Le Monnier de Gouville et al., 1990b)(Figure 9).

Tone-dependent responses in normotensive cats showed decreased systemic arterial pressure following an i.v. bolus injection of ET-1 (300 ng), which was rapid in onset, lasted 20 to 60 s, was reversible and reproducible over time, significantly decreased systemic vascular resistance, and increased cardiac output (Lippton et al., 1988)(Figure 10). In anesthetized cats, i.v. bolus injections of ET-1 (1 μg) and ET-3 (3 μg) produced a rapidly appearing but short-lasting fall in aortic blood pressure, followed in the case of ET-1 only by a small pressor response. When these peptides were administrated repeatedly at 10- to 12-min intervals, there was a gradual attenuation of the hypotension that was replaced by a monophasic pressor response by the fourth injection (Le Monnier de Gouville et al., 1990a) (Figure 11). Since the systemic hypotensive responses to ET-1 in anesthetized dogs (Given et al., 1989) rabbit, cat, and rats (Le Monnier de Gouville and Cavero, 1989; Wright and Frozard, 1988) and in conscious dogs (Figure 12) and rats are similar, the ability of ET to

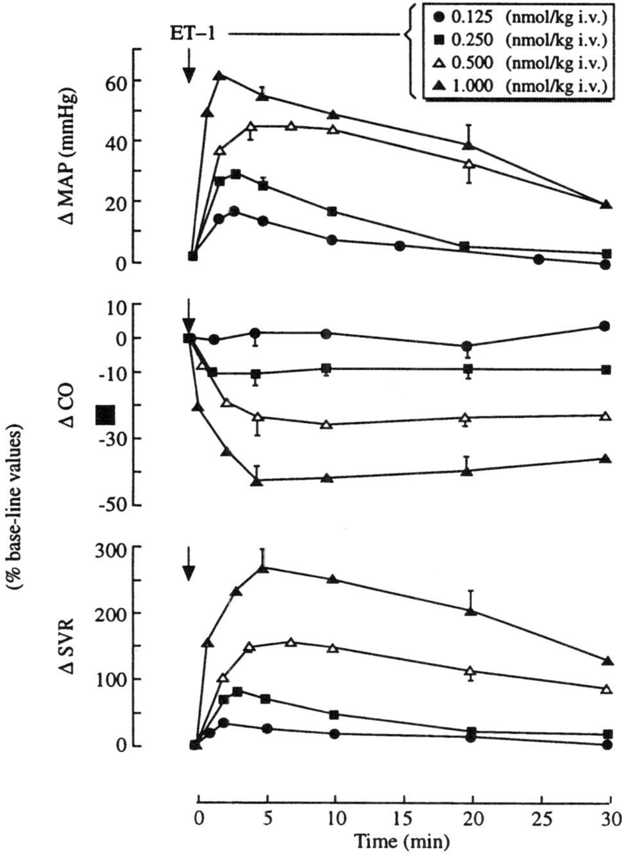

FIGURE 9
Time course effects of i.v. bolus injections of increasing doses of ET-1 on ΔMAP, systemic blood flow (CO) and systemic vascular resistance (SVR) in pithed rats. (From Le Monnier de Gouville, A. C., Mondot, S., Lippton, H., et al., *J. Pharmacol. Exp. Ther.*, 252, 300, 1990. With permission.)

decrease systemic arterial pressure does not appear to depend on an animal's species and preparation.

The decrease in systemic arterial pressure in response to ET-1 is the result of the systemic vasodilator properties of this peptide since cardiac output in rats, cats, and dogs (Lippton et al., 1988; Le Monnier de Gouville et al., 1990a; Cairo et al., 1989) is not decreased or slightly increased. Although ET-1 has been reported to release PGI_2 *in vivo*, the systemic vasodilator response to ET-1 is not altered by indomethacin, suggesting PGI_2 and other vasodilator PGs do not contribute in large measure to the systemic vasodilator response to ET-1 (DeNucci et al., 1988).

Some studies showed that L-N^G-nitro-L-arginine methyl ester can inhibit ET-1-induced vasodilatation, but other studies showed no effect (Todd and Cassin, 1992). In intact cat, glybenclamide, an inhibitor of ATP-dependent potassium channel activation, blocks the pulmonary but not the systemic vasodilator response to ET-1, ET-2, and ET-3. These pulmonary and systemic vasodilator responses to each ET isopeptide were not altered by atropine, ICI-118551, and indomethacin. This study demonstrates that the systemic vasodilator response to ETs does not depend on activation of muscarinic β_2-adrenergic receptors and formation of cyclooxygenase products (Lippton et al., 1991b). The nature of the initial depressor response is still uncertain and

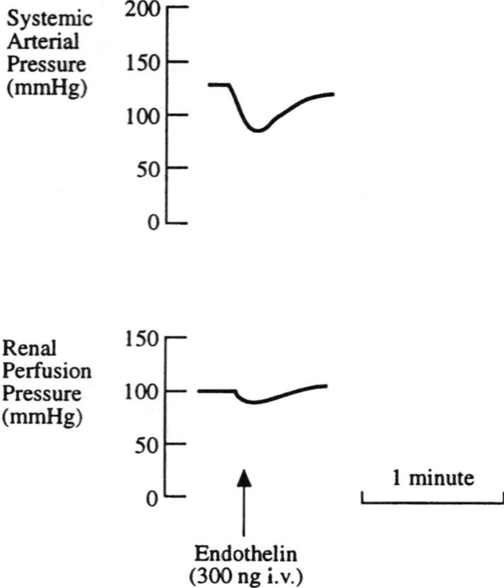

FIGURE 10
Typical response to i.v. administration of endothelin (300 ng) on systemic arterial pressure and renal perfusion pressure. (From Lippton, H., Goff, J., and Hyman, A., *Eur. J. Pharmacol.*, 155, 197, 1988. With permission.)

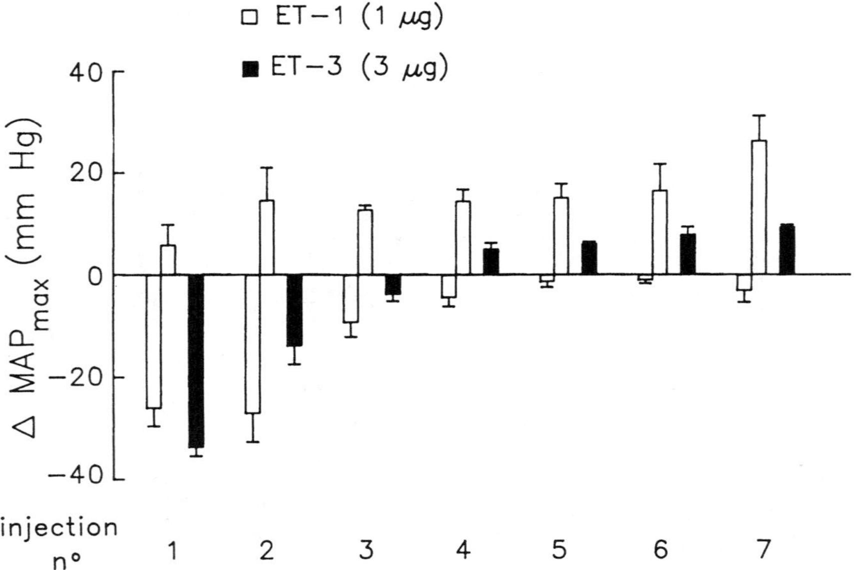

FIGURE 11
Effects of seven repeated bolus i.v. injections of ET-1 and ET-3; 10 min elapsed between two continuous doses. (From Le Monnier de Gouville, A. C., Lippton, H., Cohen, G., et al., *J. Pharmacol. Exp. Ther.*, 253(3), 1024, 1990. With permission.)

remains to be determined. The pressor response is due to direct activation of vascular smooth muscle contraction mediated predominantly by the ET_A receptor subtype. However, in some vascular beds activation of ET_B receptors also causes vasoconstriction.

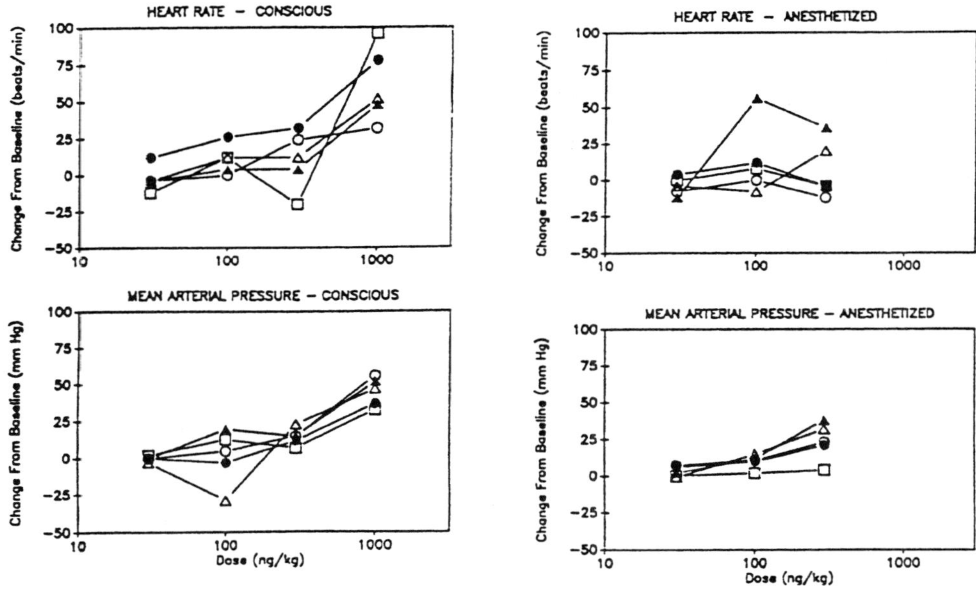

FIGURE 12
Effects of ET-1 on heart rate and mean arterial blood pressure in conscious and anesthetized dogs. (From Given, M. B., Lowe, R. F., Lippton, H., et al., *Peptides*, 10, 41, 1989. With permission.)

D. Renal Vascular Bed

ET is secreted at several sites in the kidney and may act in a paracrine or autocrine manner on target cells possessing ET receptors. Systemic infusion of ET-1 increases renal vascular resistance and decreases renal blood flow (Badr et al., 1989; Cao and Banks, 1990; Chou et al., 1990). Similar to the systemic effect of ET-1 on arterial pressure, renal vasoconstriction is often preceded by a transient renal vasodilation. ET-1 reduces both cortical and medullar blood flow, although the cortical vasculature is more sensitive to ET-1 (Tsuchiya et al., 1989). ET-1 and ET-3 can equipotently induce renal vasoconstriction in anesthetized rats (Cristol et al., 1993), suggesting that the ET_B receptor subtype mediates the response. In addition to contractile ET_A receptors, ET_{B1} (dilator) and ET_{B2} (constrictor) receptors were reported to play an important role in mediating the renal hemodynamic effects. Indeed, ET-1-induced renal vasoconstriction could not be attenuated by ET_A receptor antagonist BQ-123 in rats *in vivo* (Pollock and Opgenorth, 1993). However, the contribution of ET_{B2} receptors to mediate renal vasoconstriction may be species dependent, since in the dog and rabbit the increase in renal vascular tone appears to be predominantly ET_A mediated (Baldi and Dunn, 1991). The direct intrarenal actions of ETs were investigated in the isolated perfused kidney. These experiments suggested that ET-1 is indeed a potent renal vasoconstrictor in isolated perfused rat and rabbit kidneys *in vitro* (Loutzenhiser et al., 1989). Intravenous bolus infusion of ET-1 in rats and dogs produces a dose-dependent, sustained decrease in glomerular filtration rate (Badr et al., 1989; Cao and Banks, 1990). Systemic infusion of ET-1 decreases sodium excretion because of the reduction in filtered load (Miller et al., 1989) and also decreased diuresis indirectly due to a reduction in renal blood flow and glomerular filtration rate (Goetz et al., 1988).

E. Hindquarters Vascular Bed

Since ETs are released from endothelial cells, a series of *in vivo* experiments were performed in the hindquarters vascular bed of the cat, to better understand the regional vascular response to these peptides. Since hindquarters blood flow was kept constant, changes in hindquarters perfusion pressure directly reflected changes in hindquarters vascular resistance. Intraarterial injections of ET-1 decreased hindquarters perfusion pressure in a dose-related fashion. However, further increases in the dose of ET-1 produced a biphasic response with a secondary increase in hindquarters perfusion pressure. The vasodilator response to ET-1 lasted 0.5 to 3 min and the vasoconstrictor response to ET-1 lasted 4 to 18 min. Intraarterial bolus injections of ET-2 or ET-3 in the hindquarters altered the perfusion pressure similar to ET-1. However, the secondary vasoconstrictor responses to the higher doses of ET-2 were significantly greater than to the same doses of ET-1. Although all three ET isoforms had dilator efficacy in the hindquarters vascular bed, the dose of ET-3 required to produce a vasodilator response similar to ET-1 or ET-2 was approximately three times greater at the lower dose range.

The present data are consistent with previous work demonstrating that the ETs have a similar qualitative pattern of effects in the hindquarters vascular bed. However, the present data provides new information by showing that relative to the other regional circulation, the hindquarters vascular bed is exquisitely sensitive to the vasodilator properties of ETs. In the hindquarters vascular bed, the vasodilator response to ET undergoes tachyphylaxis. Upon repeated exposure to ET-1 or ET-3 the hindquarters vascular bed becomes refractory to the vasodilator effects of either peptide. Upon loss of the vasodilatory response, ET-1 and ET-3 constrict the hindquarters vascular bed. Since the vasoconstrictor activity to repeated injections of ET-1 remains essentially unchanged despite the loss of the vasodilator component, increased vasoconstrictor activity alone cannot account for the tachyphylactic nature of the hindquarters vasodilator response to ET-1. Moreover, tachyphylaxis to the hindquarters vasodilator response to ET-3 occurred without the development of secondary vasoconstriction (Lippton et al., 1996) (Figure 13). It is important to point out that with respect to ET evoking an initial vasodilator response, the systemic as well as regional vascular beds exist on a continuum. Specifically, the hindquarters vascular bed is exquisitely sensitive to the vasodilator properties of ETs. ETs are approximately 1000-fold more potent in their vasodilator activity than isoproterenol in the hindquarters vascular bed of the cat (Figure 14). In fact, ET-1 dilates the feline hindquarters vascular bed at a dose as little as 300 fmol. In contrast, visceral vascular beds are much less responsive to the vasodilator properties of ET-1, and in the cutaneous circulation ETs appear to produce only vasoconstriction.

F. Other Vascular Beds

When injected as an intraarterial bolus, ET-1 has a qualitatively similar hemodynamic effect in the hindquarters, mesenteric, celiac, and renal vascular beds. However, the hindquarter vascular bed, in light of its marked sensitivity to the vasodilator action of ET-1, could be considered as a bioassay for the presence of ET-1. Although the vasodilator activity of ET-1 in the renal, mesenteric, and celiac vascular beds *in vivo* is small, ET-1 (13 to 130 pmol/kg, i.a.) has markedly greater vasoconstrictor activity in these vascular beds when compared to the hindquarter region. ET-1 appears to have the greatest vasoconstrictor efficacy in the mesenteric vascular beds (Le Monnier et al., 1995) and bone vascular bed (Brinker et al., 1990).

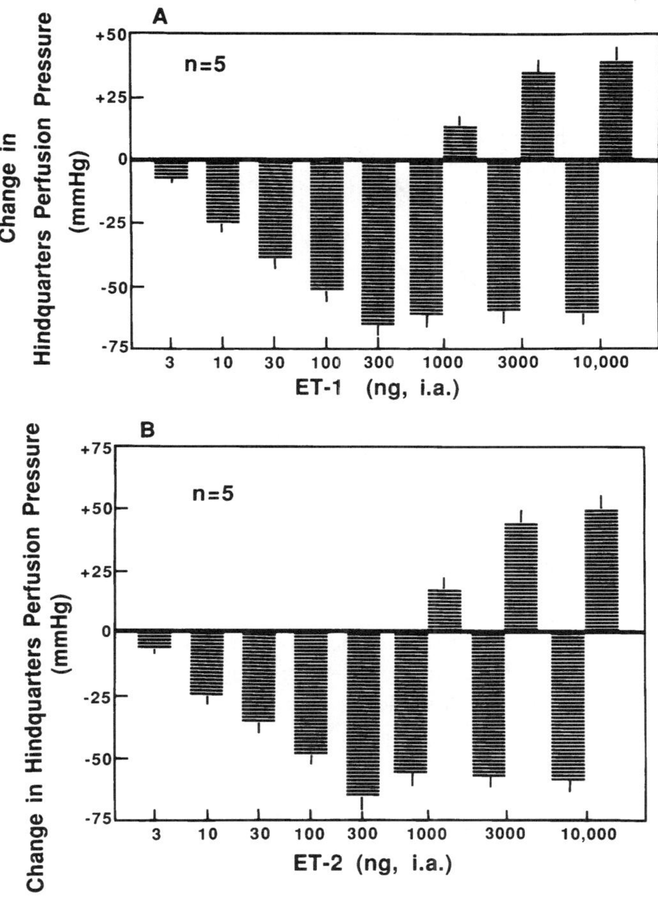

FIGURE 13
Effects of ET-1, ET-2, ET-3, and SRT 6b in the hindquarters vascular bed of the cat.

G. Blood Vessels

ET-1 has potent contractile activity on a variety of blood vessels from a number of species including rat (Tamobe et al., 1988; Augent et al., 1988), rabbit (Sugiura et al., 1989; Marsden et al., 1989), dog (Withrington et al., 1989), and human (Power et al., 1989). ET-1 elicits contractile effects which are slow in onset, long in duration, and resistant to wash (Yanagisawa et al., 1988; Sugiura et al., 1989). In general, these contractions are initiated by the binding of ET isopeptides to ET_A receptors on vascular smooth muscle (Arai et al., 1990; Masaki et al., 1994). However, recent studies demonstrated that, in addition to ET_A, activation of the ET_B receptor subtype can also mediate vasoconstriction in some vascular preparations (Webb and Lappe, 1993; Harrison et al., 1992; Naguchi et al., 1993). *In vivo* studies confirmed these *in vitro* observations showing ET_B receptor-mediated vasoconstriction in coronary arteries of isolated rat hearts (Balvierczak, 1993) and in the pulmonary circulation of the rabbit (Clozel et al., 1992). In our laboratories, we compared the vascular responses of ET isoforms on bovine pulmonary arterial and venous rings in a cumulative concentration-response fashion. The relative contractile potency of the three peptides was ET-1 > ET-2 > ET-3. The pulmonary venous rings were more sensitive to the

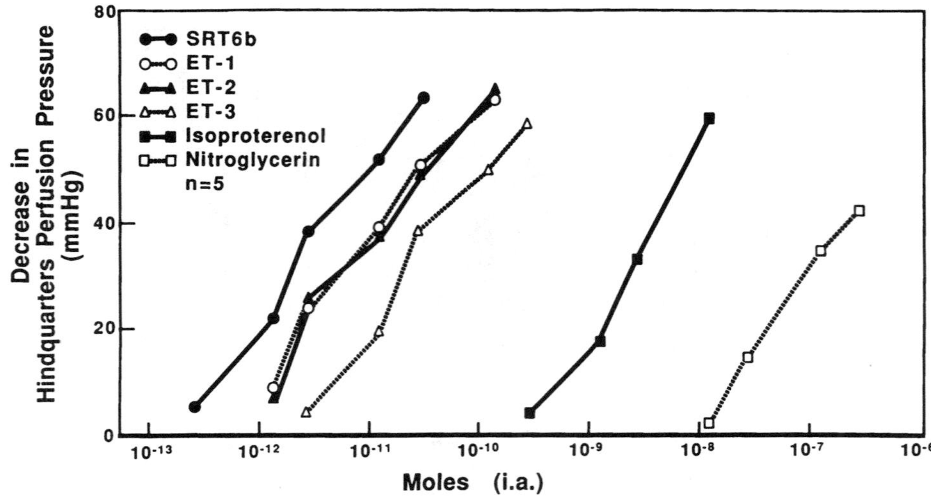

FIGURE 14
Dose-response curves comparing decreases in hindquarters perfusion pressure in response to ET-1, ET-2, ET-3, STX-b, isoproterenol, and nitroglycerin.

peptides than were the pulmonary arterial rings. Removal of the endothelium in pulmonary vessels did not enhance the sensitivity of arterial and venous smooth muscle to ET. In addition, the contractile potency of big ET-1 was less than that of ET-1. Phosphoramidon, an inhibitor of ECE, markedly inhibited the contractile activities of big ET-1 by blocking the conversion of big ET-1 to ET-1. Since contractile activity of big ET-1 was much lower than that of ET-1, the conversion of big ET-1 to ET-1 appears to be an essential step for the ET to evoke its full agonist activity as a contractile substance.

Accumulating data suggest that ET-1 has a greater contractile activity on dog coronary, mesenteric, femoral, renal, and internal mammary veins than on arteries. These *in vitro* observations were also confirmed *in vivo*. Infusion of ET-1 (5 pmol/min) into human forearm caused significantly more constriction of dorsal hand veins than arteries (Haynes and Webb, 1991; Haynes et al., 1992).

The contractile response to ET was initially demonstrated using arterial segments with a damaged endothelial cell layer (Yanagisawa et al., 1988). The contractile response to ET in rat aorta, dog femoral vein, and cat cerebral artery is significantly enhanced after removal of the endothelium (Miller et al., 1989; Saito et al., 1989). ETs are potent coronary vasoconstrictors. In anesthetized dogs, bolus intracoronary injections of ET-1 (30 pmol/kg) reduce coronary blood flow by approximately 90% (Kurihara et al., 1989a, b). Coronary veins are more sensitive than are coronary arteries; the effects are endothelium independent (Cocks et al., 1989a, b).

ET-1 and ET-3 were reported to stimulate the release of endothelium-derived relaxing factor (EDRF) (NO) from bovine native endothelial cells (Warner et al., 1992) isolated from perfused rat and rabbit blood vessels (Warner et al., 1989a, b) and from several other vascular tissues (Bottling and Vane, 1990; Namiki et al., 1992). ET-1 and ET-3 also stimulate the release of PGI_2 (prostacyclin) from bovine and human vascular endothelial cells (Emori et al., 1991; Filep et al., 1991), from isolated rat mesenteric arteries (Rakugi et al., 1989), from perfused rat lung (D'Orleans-Juste et al., 1992), and from perfused rabbit heart (Karwatowska Prokopczuk and Wennmalm, 1990). ETs can also stimulate PGI_2 release *in vivo* in rats (DeNucci et al., 1988) and rabbits (Thiemermann et al., 1989).

EDRF and PGI_2 release were suggested to be responsible for the hypotensive (at least in part) effect of ET-1 and ET-3 *in vivo*, however, data to the contrary have been reported *in vivo* (Lippton et al., 1991b).

REFERENCES

Arai, H., Hori, S., Aramori, I., Ohkuba, H., and Nakanishi, S.: Cloning and expression of cDNA encoding an endothelin receptor. *Nature (London).* 348:730-732, 1990.

Auguet, N., Delaflotle, S., Chabrier, I. E., Pirotzky, E., Clostre, R., and Braquet, P.: Endothelin and Ca^{+2} agonist Bay K 8664. Different vasoconstrictive properties. *Biochem. Biophys. Res. Commun.*, 156:186-192, 1988.

Badr, K., Murray, J. J., Breyer, M. D., Takahaski, K., Inagami, T., and Harris, R. C.: Mesangial cell, glomerular and renal vascular responses to endothelin in the rat kidney. *J. Clin. Invest.*, 83:336-342, 1989.

Baldi, E. and Dunn, M. J.: Endothelin binding and receptor downregulation in rat glomerular mesangial cells. *J. Pharmacol. Exp. Ther.*, 256:581-586, 1991.

Balvierczak, J. L.: Two subtypes of the ET-1 receptor (ET_A and ET_B) mediate the coronary vasoconstrictor effect of ET-1. In Proc. 3rd Int. Conf. Endothelin, p. 4, 1993, Houston, TX, February, 1993.

Baydoun, A. R., Peers, S. H., Cirina, G., and Woodward, B.: Effects of endothelin-1 on the rat isolated heart. *J. Cardiovasc. Pharmacol.*, 13(Suppl. 5):S193-196, 1989.

Bottling, R. M. and Vane, J. R.: Endothelins: potent releasers of prostacyclin and EDRF. *Pul. J. Pharmacol. Pharm.*, 42:203-218, 1990.

Brinker, M. R., Lippton, H. L., Cook, S. D., and Hyman, H. L.: Pharmacological regulation of the bone circulation. *J. Bone Joint Surg.*, 72-A: 964-975, 1990.

Cairo, J., Pellet, A., Lippton, H., Summer, W., Hyman, A., and Levitsky, A.: In vivo effects of endothelin on vascular dynamics. *FASEB J.*, 3 A878, 1989.

Cao, L. Q. and Banks, R. O.: Cardiovascular and renal actions of endothelin: effect of calcium channel blockers. *Am. J. Physiol.*, 258:254-258, 1990.

Cardell, L. O., Uddman, R., and Edvinsson, L.: Evidence for multiple endothelin receptors in the guinea-pig pulmonary artery and trachea. *Br. J. Pharmacol.* 105:376-380, 1992.

Chou, S. Y., Dahhan, A., and Porush, J. G.: Renal actions of endothelin: interaction with prostacylin. *Am. J. Physiol.*, 259:F645-652, 1990.

Clozel, M., Gray, G. A., Breu, V., Loffler, B. M., and Osterwalter, R.: The endothelin ET_B receptor mediates both vasodilation and vasoconstriction *in vivo*. *Biochem. Biophys. Res. Commun.*, 186:867-873, 1992.

Cocks, T. M., Broughton, A., Dih, M., Sudhir, K., and Angus, J. A.: Endothelin is blood vessel selective: studies on a variety of human and dog vessels *in vitro* and on regional blood flow in the conscious rabbit. *Clin. Exp. Pharmacol. Physiol.*, 16:243-246, 1989a.

Cocks, T. M., Faulkner, N. L., Sudhir, K., and Angus, J. A.: Reactivity of ET-1 on human and canine large veins compared with large arteries *in vitro*. *Eur. J. Pharmacol.*, 171:17-24, 1989b.

Cortina, G. and Barieri, J. T.: Localization of a region of the S1 subunit of pertussis-toxin required for efficient ADP-ribosyl-transferase activity. *J. Biol. Chem.*, 266:3022-3030, 1991.

Cristol, J. P., Warner, T. D., Thiemermann, C., and Vane, J. R.: Mediation via different receptors of the vasoconstrictor effects of endothelins and sarafotoxins in the systemic circulation and renal vasculature of the anesthetized rat. *Br. J. Pharmacol.*, 108:776-779, 1993.

D'Orleans-Juste, P., Telemaque, S., Claing, A., Ihara, M., and Yano, M.: Human big ET-1 and ET-1 release prostacyclin via the activation of ET-1 receptors in the rat perfused lung. *Br. J. Pharmacol.*, 105:773-775, 1992.

Davenport, A. P., Nunez, D. J., and Brown, M. J.: Binding sites for ^{125}I labeled endothelin-1 in the kidneys. *Clin. Sci.*, 77:129-131, 1989.

DeNucci, G., Thomas, R., D'Orleans-Juste, P., Antunes, E., Walder, C., Warner, T. D., and Vane, J. R.: Pressor effects of circulating endothelin are limited by its removal in the pulmonary circulation and by the release of prostacyclin and EDRF. *Proc. Natl. Acad. Sci. U.S.A.*, 85:9797-9800, 1988b.

Emori, T., Hirata, Y., and Maruma, F.: ET-3 stimulates prostacyclin production in cultured bovine endothelial cells. *J. Cardiovasc. Pharmacol.*, 17(Suppl. 7):S140-142, 1991.

Filep, J. G., Battistini, B., Cote, Y. P., Beaudoin, A. R., and Sirois, P.: ET-1 induces prostacyclin release from bovine aortic endothelial cells. *Biochem. Biophys. Res. Commun.*, 177:171-176, 1991.

Firth, J. D., Roberts, A. F., and Raine, A. E.: Effects of endothelin on the function of the isolated perfused working rat heart. *Clin. Sci.*, 79:221-226, 1990.

Given, M. B., Lowe, R. F., Lippton, H., Hyman, A. L., Sander., G. E., and Giles, J. D.: Hemodynamic actions of endothelin in conscious and anesthetized dogs. *Peptides*, 10:41-44, 1989.

Goetz, K. L., Wang, B. C., Madwed, J. B., Zhu, J. L., and Leadley, R. Y.: Cardiovascular, renal and endocrine responses to intravenous endothelin in conscious dogs. *Am. J. Physiol.*, 255:R1064-1068, 1988.

Harrison, V. J., Randriantsoa, A., and Schoeffter, P.: Heterogeneity of endothelin sarafatoxin receptors mediating contraction of pig coronary artery. *Br. J. Pharmacol.*, 105:511-513, 1992.

Haynes, W. G. and Webb, D. J.: Administration of ET-1 in humans. *Circulation*, 83:1121-1124, 1991.

Haynes, W. G., Clarke, J. G., Cockcroft, J. R., and Webb, D. J.: Pharmacology of Endothelin-1 *in vivo* in humans. *J. Cardiovasc. Pharmacol.*, 17(Suppl. 7): S284-286, 1992.

Hickey, K. A., Rubanyi, G. M., Paul, R. J., and Highsmith, R. F.: Characterization of a coronary vasoconstriction produced by cultured endothelial cells. *Am. J. Physiol.*, 248:C550-556, 1985.

Hom, G. J., Touhey, B., and Rubanyi, G. M.: Effects of intracoronary administration of endothelin in anesthetized dogs: comparison with BAY K 8664 and U46619. *J. Cardiovasc. Pharmacol.*, 19:194-200, 1992.

Hu, J. R., Von Harsdorf, R., and Lang, R. E.: Endothelin has potent inotropic effects in rat atria. *Eur. J. Pharmacol.*, 158:275-278, 1988.

Inoue, A., Yanagisawa, M., Kimura, S., Kasuya, Y., Miyauchi, T., Goto, K., and Masaki, T.: The human endothelin family: three structurally and pharmacologically distinct isopeptides predicted by three separate genes. *Proc. Natl. Acad. Sci. U. S. A.*, 86:2863-2867, 1989a.

Inoue, A., Yanagisawa, M., Takuwa, Y., Mitsui, Y., Kobayashi, M., and Masaki, T.: The human preproendothelin 1 gene. Complete nucleotide sequence and regulation of expression. *J. Biol. Chem.*, 264:14954-14959, 1989b.

Ishikawa, T. M., Yanagisawa, M., Kimura, S., Goto, K., and Masaki, T.: Positive inotropic action of novel vasoconstrictor peptide endothelin on guinea pig atria. *Pfluegers Arch.*, 413:108-110, 1988.

Karwatowska Prokopczuk, E. and Wennmalm, A.: Effects of endothelin on coronary flow, mechanical performance, oxygen uptake, and formation of purines and on outflow of PGI_2 in the isolated rabbit heart. *Circ. Res.*, 66:46-54, 1990.

Kelly, R. A., Eid, H., Kramer, B. K., O'Neill, M., Liang, B. T., Reers, M., and Smith, T. W.: Endothelin enhances the contractile responsiveness of adult rat ventricular myocytes to calcium by a pertussis toxin sensitive pathway. *J. Clin. Invest.*, 86:1164-1171, 1990.

Kurihara, H., Yamaoki, K., Nagai, R., Yoshizumi, M., Takaku, F., Satoh, H., Inui, J., and Yazaki, Y.: Endothelin: a potent vasoconstrictor associated with coronary vasospasm. *Life Sci.*, 44:1937-1943, 1989a.

Kurihara, H., Yoshizumi, M., Sugiyama, T., Yamaoki, K., Nagai, R., Takaku, F., et al.: The possible role of endothelin 1 in the pathogenesis of coronary vasospasm. *J. Cardiovasc. Pharmacol.*, 13(Suppl. 5):S132-137, 1989b.

Le Monnier de Gouville, A. C. and Cavero, I.: Haemodynamic and pharmacological profile of endothelin in pithed rats. *Br. J. Pharmacol.*, 96:95P, 1989.

Le Monnier de Gouville, A. C., Lippton, H., Cohen, G., Cavero, I., and Hyman, A.: Vasodilator activity of ET-1 and ET-3. Rapid development of cross-tachyphylaxis and dependence on the rate of endothelin administration. *J. Pharmacol. Exp. Ther.*, 253(3):1024-1027, 1990a.

Le Monnier de Gouville, A. C., Mondot, S., Lippton, H., Hyman, A., and Cavero, I.: Hemodynamic and pharmacological evaluation of the vasodilator and vasoconstrictor effects of ET-1 in rats. *J. Pharmacol. and Exp. Ther.*, 252:300-311, 1990b.

Lerman, A., Hildebrand, F. L., Aarhus, L. L., and Burnett, J. C.: Endothelin has biological actions at pathophysiological concentration. *Circulation*, 83; 1808-1814, 1991.

Lerman, A., Sandok, E. K., Hilderbrand, F. L., and Burnett, J. C.: Inhibition of endothelium derived relaxing factor enhances endothelin mediated vasoconstriction. *Circulation*, 85:1894-1898, 1992.

Lewis, M. J. and Sham, A. M.: Endothelial modulation of myocardial contraction. *Endothelium*, in press, 1994.

Lippton, H. L., Hao, Q., Erdemli, O., and Hyman, A.: Role of G proteins in the vasodilator response to endothelin isopeptides *in vivo*. *J. Appl. Physiol.*, 78(6):2062-2069, 1995.

Lippton, H., Goff, J., and Hyman, A.: Effects of endothelin in systemic and renal vascular beds *in vivo*. *Eur. J. Pharmacol.*, 155:197-199, 1988.

Lippton, H., Hao, H., Gumusel, B., and Hyman, A.: Effects of endothelin isopeptides in hindquarters vascular bed of the cat (in press).

Lippton, H. L., Cohen, G. A., McMurtry, I. F., and Hyman, A. L.: Pulmonary vasodilation to endothelin isopeptides *in vivo* is mediated by potassium channel activation. *J. Appl. Physiol.*, 70(2):947-952, 1991b.

Lippton, H. L., Hauth, T. A., Cohen, G. A., and Hyman, A. L.: Functional evidence for different endothelin receptors in the lung. *J. Appl. Physiol.*, 75(1):38-48, 1993.

Lippton, H. L., Hauth, T. A., Summer, W. R., and Hyman, A. L.: Endothelin produces pulmonary vasoconstriction and systemic vasodilation. *J. Appl. Physiol.*, 66(2):1008-1012, 1989a.

Lippton, H. L., Ohlstein, E. H., Summer, W. R., and Hyman, A. L.: Analysis of responses to endothelins in the rabbit pulmonary and systemic vascular beds. *J. Appl. Physiol.*, 70(1):331-341, 1991a.

Lippton, H. L., Pellett, A., Cairo, J., Summer, W., Lowe, R., Sander, G. E., Giles, T., Cohen, G., and Levitzky, M. G.: Endothelin produces systemic vasodilation independent of the state of consciousness. *Peptides*, 10(5):939-943, 1989b.

Loutzenhiser, R., Hayaski, K., and Epstein, M.: Direct visualization of effects of endothelin on the renal microvasculature. *Am. J. Physiol.*, 258 (1, Pt2): F61-68, 1989.

Marsden, P. A., Danthuluri, N. R., Brenner, B. M., Ballermawn, B. J., and Brock, T. A.: Endothelin action on vascular smooth muscle involves inositol triphosphate and calcium mobilization. *Biochem. Biophys. Res. Commun.*, 158:86-93, 1989.

Masaki, T., Vane, J. R., and Vanhoutte, P. M.: V. International Union of Pharmacology Nomenclature of Endothelin Receptors. *Pharmacol. Rev.*, 46: 127-142, 1994.

Masuda, Y., Miyazaki, H., Kondoh, M., Watanabe, H., Yanagisawa, M., Masaki, T., and Murakami, K.: Two different forms of endothelin receptors in rat lung. *FEBS Lett.*, 257:208-213, 1989.

Miller, W. L., Redfield, M. M., and Burnett, J. C.: Integrated cardiac, renal and endocrine actions of endothelin. *J. Clin. Invest.*, 83:317-320, 1989.

Miller, V. M., Komori, K., Burnett, J. C., and Vanhoutte, P. M.: Differential sensitivity to endothelin in canine arteries and veins. *Am. J. Physiol.*, 257 (4,Pt2): H1127-1131, 1989.

Naguchi, K., Naguchi, Y., Hirose, H., Nishikibrie, M., Ihara, M., Ishikawa, K., and Yano, M.: Role of endothelin ET_B receptors in bronchoconstrictor and vasoconstrictor responses in guinea pig. *Eur. J. Pharmacol.*, 233:47-51, 1993.

Namiki, A., Hirata, Y., Fukazawa, M., Ishikawa, M., Moroi, M., Aikawa, J., Yabuki, S., and Machii, K.: ET-1 and ET-3 induced vasorelaxation via EDRF. *Jpn. J. Pharmacol.*, 58 (Suppl. 2): 326p, 1992.

Nichols, A. J., Motley, E. D., and Ruffalo, R. R.: Different effect of pertussis toxin on pre-and postjunctional α_2-adrenoceptors in cardiovascular system of pithed rat. *Eur. J. Pharmacol.*, 145:345-349, 1988.

Nichols, A. J., Motley, E. D., and Ruffalo, R. R.: Effect of pertussis-toxin treatment on postjunctional α_1 and α_2 adrenoceptor function in the cardiovascular system of the pithed rat. *J. Pharmacol. Exp. Ther.*, 248:203-209, 1989.

Pollock, D. M. and Opgenorth, T. J.: Evidence for endothelin induced renal vasoconstriction independent of ET_A receptor activation. *Am. J. Physiol.*, 264:R222-226, 1993.

Power, R. F., Wharton, J., Salas, S., Kanse, S., Ghater, N., Bloom, S. R., and Polak, J. M.: Autoradiographic localization of endothelin binding sites in human and porcine coronary arteries. *Eur. J. Pharmacol.*, 160:199-200, 1989.

Rakugi, H., Nakamanu, M., Tabuchi, Y., Nagano, M., Mikami, H., and Ogihara, T.: Endothelin stimulates the release of prostacyclin from rat mesenteric arteries. *Biochem. Biophys. Res. Commun.*, 160:924-928, 1989.

Saito, A., Shiba, R., Kimura, S., Yanagisawa, M., Goto, K., and Masahi, T.: Vasoconstrictor response of large cerebral arteries of cats to endothelins, an endothelium-derived vasoactive peptide. *Eur. J. Pharmacol.*, 162:353-358, 1989.

Sakurai, T., Yanagisawa, M., Tukuwa, Y., Miyazaki, H., Kimura, S., Goto, K., and Masaki, T.: Cloning of a cDNA encoding a non-isopeptide-selective subtype of the endothelin receptors. *Nature (London)*, 348:732-735, 1990.

Shah, A. M., Lewis, M. J., and Henderson, A. H.: Inotropic effects of endothelin in ferret ventricular myocardium. *Eur. J. Pharmacol.*, 63:365-367, 1989.

Sugiura, T., Inagami, T., Hare, G. M. T., and Jones, J. A.: Endothelin action: Inhibition by a protein kinase C inhibitor and involvement of phosphoinositols. *Biochem. Biophys. Res. Commun.*, 158:170-176, 1989.

Tamobe, Y., Miyauchi, T., Saito, A., Yanagisawa, N., Kimura, S., Gato, K., and Masaki, T.: Effects of endothelin on the renal artery from spontaneously hypertensive and Wistar Kyoto rats. *Eur. J. Pharmacol.*, 152:373-376, 1988.

Thiemermann, C., Lidbury, P. S., Thomas, G. R., and Vane, J. R.: Endothelin 1 releases prostacyclin and inhibits *ex vivo* platelet aggregation in the anesthetized rabbit. *J. Cardiovasc. Pharmacol.*, 13 (Suppl. 5):S138-141, 1989.

Todd, M. L. and Cassin, S.: ET-1-induced pulmonary arterial dilation is reduced by N^W- nitro-L-arginine in fetal lambs. *J. Appl. Physiol.*, 72:1730-1734, 1992.

Tsuchiya, K., Naruse, M., Sanaka, T., Naruse, K., Niha, K., Demura, H., and Sugino, H.: Effects of endothelin on renal regional blood flow in dogs. *Eur. J. Pharmacol.*, 166:541-543, 1989.

Warner, T. D., DeNucci, G., and Vane, J. R.: Rat endothelin is a vasodilator in the isolated perfused mesentery of the rat. *Eur. J. Pharmacol.*, 159:325-326, 1989a.

Warner, T. D., Mitchells, J. A., DeNucci, G., and Vane, J. R.: ET-1 and 3 release EDRF from isolated perfused arterial vessels of the rat and rabbit. *J Cardiovasc. Pharmacol.*, 13 (Suppl. 5):S85-88, 1989b.

Warner, T. D., Schmidt, H. H., and Murad, F.: Interactions of endothelins and EDRF in bovine native endothelial cells: selective effects of ET-3. *Am. J. Physiol.*, 262:H1600-1605, 1992.

Watanabe, H., Miyazaki, H., Kandoh, M., Masuda, Y., Yanagisawa, M., Masaki, T., and Murakami, K.: Two distinct types of endothelin receptors are present on chick cardiac membranes. *Biochem. Biophys. Res. Commun.*, 161:1252-1259, 1989b.

Watanabe, T., Kusumoto, K., Kitayoshi, T., and Shimamoto, N.: Positive inotropic and vasoconstrictive effects of endothelin-1 in *in vivo* and *in vitro* experiments: characteristics and the role of L-type calcium channels. *J. Cardiovasc. Pharmacol.*, 13(Suppl. 5):S108-111, 1989a.

Webb, R. L. and Lappe, R. W. ; Hemodynamic effects of an endothelin (ET_B) selective agonist, IRL 1620 in WKY and SHR. In Proc. 3rd Int. Conf. Endothelin, p. 137, 1993, Houston, TX, February, 1993.

Wiklund, N. P., Ohlen, A., and Cederquist, B.: Inhibition of adrenergic neuroeffector transmission by endothelin in guinea-pig femoral artery. *Acta Physiol. Scand.*, 134:311-312, 1988.

Withrington, P. G., DeNucci, G., and Gand Vane, J. R.: ET-1 causes vasoconstriction and vasodilation in the blood perfused liver of the dog. *J. Cardiovasc. Pharmacol.*, 13: S209-210, 1989.

Wright, C. E. and Frozard, J. R.: Regional vasodilation is a prominent feature of the haemodynamic response to endothelin in anesthetized, spontaneously hypertensive rats. *Eur. J. Pharmacol.*, 155:201-203, 1988.

Yanagisawa, M., Kurihara, H., Kimura, S., Tomobe, Y., Kobayashi, M., Mitsui, Y., Yazaki, Y., Goto, K., and Masaki, T. ; A novel potent vasoconstrictor peptide produced by vascular endothelial cells. *Nature (London)*, 332:411-415, 1988.

Chapter 9

EFFECTS OF ENDOTHELIN IN THE KIDNEY

Ponnal Nambi and David P. Brooks

CONTENTS

I. INTRODUCTION

Endothelin-1, a 21-amino-acid peptide initially isolated from the culture medium of porcine aortic endothelial cells by Yanagisawa et al. (1988) is the most potent renal vasoconstrictor yet identified. ET-1 is also synthesized and released from a number of other cell types such as mesangial cells, renal tubular cells, smooth muscle cells,

0-8493-6975-4/97/$0.00+$.50

and neuronal cells (Yanagisawa et al., 1988; Yanagisawa and Masaki, 1989a; Lemonnier Degouville et al., 1989; Simonson and Dunn, 1990a). Endothelin has been shown to mediate its effects by binding to specific cell surface receptors that are located on many tissues and cells (for review, see Rubanyi and Parker Botelho, 1991; Yanagisawa and Masaki, 1989b; Sokolovsky, 1992; Thomas et al., 1992; Huggins et al., 1993). Based on the physiological responses mediated by ET-1, it is proposed that this peptide can act in an autocrine and paracrine fashion, indicating that it can be released at or near the site of action. The receptors for endothelin have been classified as ET_A and ET_B based on the binding profile of ET-1 and related agonists such as ET-3 and STX-c (a snake venom peptide that shares a high degree of homology with ET-1). Both ET_A and ET_B receptors have been cloned and expressed from many species including humans (Arai et al., 1990; Sakurai et al., 1990; Elshourbagy et al., 1993; Elshourbagy et al., 1992). ET_A receptors display high affinity for ET-1 and ET-2, whereas ET_B receptors display equal affinity for ET-1, ET-2, ET-3, and STX-c (Arai et al., 1990; Sakurai et al., 1990; Elshourbagy et al., 1992, 1993). ET-3 and S6c are 100 and 1000 times, respectively, less potent than ET-1 and ET-2 for ET_A receptors. In addition, two more subtypes of ET receptors, ET_C and ET_{AX}, have been cloned and expressed from *Xenopus* melanophores and heart, respectively (Karne et al., 1993; Kumar et al., 1994).

II. ENDOTHELIN EXPRESSION IN THE KIDNEY

Molecular biological techniques such as Northern blot and ribonuclease protection analyses of total RNA from rat kidney displayed messenger RNAs for both prepro ET-1 and prepro ET-3 (Maccumber et al., 1989; Nunez et al., 1991; Sakurai et al., 1991; Firth et al., 1992; Ujiie et al., 1992). The expression of these two messenger RNAs appears to be species specific because prepro ET-1 expression was predominant in pig kidney, whereas prepro ET-3 expression was predominant in rat kidney (Firth et al., 1992; Ujiie et al., 1992; Nunez et al., 1990). *In situ* hybridization studies have also been performed to localize ET-1 in the kidney and the data indicate that prepro ET-1 was predominantly expressed in the vasa recta bundles of medulla (Maccumber et al., 1989) and vascular sites of cortex and medulla (Nunez et al., 1990). Using reverse transcriptase and polymerase chain reaction, Ujiie et al. (1992) and Uchida et al. (1992) have localized ET-1 mRNA to rat nephrons where it was found primarily in the glomerulus and inner medullary collecting duct. ET-1 mRNA was also shown to be present in the proximal convoluted and straight tubules (Chen et al., 1993). Using immunological techniques, Yoshimi et al. (1989) showed that rat kidney is a rich source of ET-1. ET-1-like immunoreactivity was observed in renal cortex, medulla, and papilla, although the vasa recta of distal nephron segments in the renal papilla showed the predominant localization of this peptide (Wilkes et al., 1991c).

The synthesis and/or release of ET-1 in the kidney has been shown to be increased by a number of factors. Many renal cells in culture secrete ET-1, and this secretion is potentiated by various factors such as thrombin, bradykinin, ATP, platelet activating factor, transforming growth factor β (TGFβ), tumor necrosis factor, interleukin-1β (IL-1β), and cyclosporine A. These effects have been demonstrated in many renal epithelial cell lines such as MDCK, LLCPK, NRK-52E, BHK-21, and Rk–13 (Kosaka et al., 1989; Shichiri et al., 1989; Ohta et al., 1990). In addition to epithelial cells, human mesangial cells have also been shown to secrete ETs in response to TGFβ, arginine vasopressin, and thrombin (Bakris et al., 1991; Zoja et al., 1991; Sakamoto et al., 1990). On the other hand, atrial natriuretic peptide (ANP) (Kohno et al.,

1991, 1993; Uemasu et al., 1993) and heparin (Kohno et al., 1994) have been shown to decrease ET-1 synthesis and release. ANP also causes a decrease in plasma ET-1 concentrations. In addition to synthesis and release, the kidney also has the capability of degrading ET-1 (Yamaguchi et al., 1992; Janas et al., 1994; Edano et al., 1994; Deng et al., 1994), most probably by the action of enzymes such as neutral endopeptidase. The observation that immunoreactive ET in urine is six times higher than that present in plasma of normal volunteers indicates that the kidney also plays a major role in ET synthesis as well as metabolism (Ando et al., 1991; Berbinschi and Ketelslegers, 1989). The observation that plasma ET levels were significantly increased in patients undergoing hemodialysis, renal transplantation, or peritoneal dialysis further supports a role for ET in renal function (Koyama et al., 1989; Predel et al., 1990; Totsune et al., 1989; Warrens et al., 1990). The synthesis of ETs by the kidney is also consistent with the hypothesis that ET peptides mediate their effects in an autocrine or paracrine fashion because the kidney has been shown to possess both ET_A and ET_B receptors, as explained below.

A. Renal Endothelin Receptors

Binding sites for ET-1 in kidneys from a number of species were described as early as 1989 using [^{125}I]ET-1 binding as well as autoradiographic studies (Jones et al., 1989; Kohzuki et al., 1989; Gauquelin et al., 1989). Although these studies did not distinguish receptor subtypes, a very high density of ET binding sites was identified in mammalian kidney in the glomeruli as well as inner medulla and inner strip of the outer medulla corresponding to the vasa recta (Kohzuki et al., 1989). Proximal tubules displayed moderate binding sites, whereas no binding was observed in the outer medulla (Jones et al., 1989; Kohzuki et al., 1989). Later studies by a number of investigators demonstrated the presence of subtypes of ET receptors based on the binding profile of the subtype-selective ligands ET-3 and STX-c (Martin et al., 1990; Kohzuki et al., 1991; Nambi et al., 1992a,b; Kohan et al., 1992a). In addition, discovery of BQ-123 as an ET_A-selective antagonist enabled the quantitation of the subtypes of ET receptors in various regions of the kidney (Nambi et al., 1992a,b; Karet et al., 1993). ET receptors have also been shown to be present in microdissected rat nephron segments (Takemoto et al., 1993). In addition to the traditional radioligand binding studies, sophisticated molecular biological techniques such as reverse transcriptase polymerase chain reaction have been used to detect large signals for the ET_B receptor subtype in the initial and terminal inner medullary collecting duct and glomerulus, whereas small signals were found in the cortical collecting duct, outer medullary collecting duct, vasa recta, and arcuate artery. Although ET_A receptor mRNA was shown to be present in the glomerulus, binding studies using [^{125}I]ET-1 or [^{125}I]ET-3 indicated that the predominant subtype was ET_B receptors (Edwards et al., 1992). ET receptors are differentially expressed along the nephron and some segments such as the inner medullary collecting duct displayed both ET_A and ET_B subtypes (Kohan et al., 1992). Studies performed with cells isolated and cultured from various regions of the kidney indicate the presence of high levels of ET_B receptors in glomerular, epithelial, and endothelial cells of the thin segments of Henle's loop (Hori et al., 1992). Studies reported on mesangial cell ET receptors are conflicting. Takeda et al. (1992) and Albrightson et al. (1995) have presented data on the presence of ET_A receptors on mesangial cells, whereas Martin et al. (1990) and Sugiura et al. (1989) have shown the presence of both subtypes of ET receptors on mesangial cells. The reason for these differences is unclear.

B. Endothelin Receptor Signal Transduction in the Kidney

Both ET_A and ET_B receptors belong to the superfamily of seven-transmembrane G-protein-coupled receptors which mediate their biological actions by coupling through a variety of GTP binding proteins and specific effector molecules. Binding of ET to its receptors results in the formation of a number of second messengers including inositol phosphates, diacylglycerol, Ca^{2+}, cyclic AMP, cyclic GMP, and arachidonic acid. Depending on the coupling of the receptor to the effector system, different second messenger(s) are generated in different tissues. ET mediates both short-term (contraction) and long-term (mitogenesis) responses. ET-1 stimulates phospholipase C activity in mesangial cells, resulting in the generation of inositol trisphosphate and diacylglycerol. This activation is mediated by a pertussis toxin-sensitive G protein (Simonson et al., 1989). While inisitol bisphosphate stimulates the release of intracellular Ca^{2+} from the endoplasmic reticulum, diacylglycerol activates protein kinase C (PKC) which results in the release of arachidonic acid. In mesangial cells, this pathway appears to be very critical since arachidonic acid metabolism results in the synthesis of prostaglandins which regulate a number of functions in the kidney including natriuresis, diuresis, vascular tone, and renin secretion (Simonson and Dunn, 1991). ET has also been shown to stimulate phospholipase D in mesangial cells through a PKC-dependent pathway (Kester et al., 1992). The activation of phospholipase D results in the hydrolysis of phospholipids like phosphatidylcholine and phosphatidylethanolamine, producing phosphatidic acid and choline or ethanolamine, depending upon the phospholipids hydrolyzed. ET-mediated activation of phospholipase D in mesangial cells can be inhibited by the PKC inhibitor, sangiavamycin, or by downregulation of PKC (Kester et al., 1992). In rat aorta, however, ET-mediated activation of phospholipase D was insensitive to PKC inhibition (Liu et al., 1992), suggesting that both PKC-dependent as well as -independent pathways are involved in ET-mediated activation of phospholipase D. In addition to phospholipase C and D activation, ET also activates phospholipase A_2 in mesangial cells, leading to the formation of arachidonic acid which is metabolized to specific eicosanoids, depending on the species (Resink et al., 1990; DeNucci et al., 1988; Reynolds and Mok, 1990). Studies performed with cloned and expressed ET_A and ET_B receptors indicate that ET peptides stimulated phospholipase A_2 by interacting with ET_A as well as ET_B receptors (Aramori and Nakanishi, 1992).

In contrast to the observation that addition of ET to most cells and tissues resulted in the activation of phospholipase C and hydrolysis of phosphoinositides, studies performed to evaluate the involvement of ET in the activation of cyclic nucleotide pathways have provided different results. Generally, ET has not been shown to modulate cyclic AMP levels (Ohlstein et al., 1989), however, Abdel-Latif et al. (1991) have demonstrated an ET-mediated increase in cyclic AMP accumulation in the iris sphincter of different species including cat, dog, monkey, cow, and human. There have been suggestions that ET-mediated cyclic AMP accumulation may not be due to direct activation of adenylate cyclase, but indirectly through the stimulation of phospholipase C. Indeed, studies in rat nephron segments have shown that ET-1 inhibits vasopressin-mediated cyclic AMP accumulation by a PKC-dependent pathway (Tomita et al., 1990), whereas in rat mesangial cells ET-1 has been shown to stimulate cyclic AMP accumulation through a PGE_2-dependent pathway (Simonson and Dunn, 1990b). ETs have also been shown to inhibit isoproterenol- and forskolin-stimulated adenylate cyclase activity in membrane preparations through a pertussis toxin-sensitive G protein, indicating direct coupling of ET receptors to adenylate cyclase (Hilal-Dandan et al., 1992).

Exposure of NG 108-15 cells (mouse neuroblastoma/rat glioma hybrid cells) to ET-1 resulted in a Ca^{2+}-dependent increase in the accumulation of cyclic GMP that was inhibited by nitric oxide synthase inhibitors and hemoglobin (Reiser, 1990). Similar observations have been reported using rat glomeruli and the increase in cyclic GMP was shown to be mediated by ET_B receptors (Edwards et al., 1992). Thus, ET-induced increase in cyclic GMP appears to be mediated by the nitric oxide synthase pathway.

In addition to activating various second messenger pathways, ETs have also been shown to activate nuclear signal transduction pathways resulting in the expression of transcription factors such as *c-fos*, *c-jun* and *c-myc* (Simonson et al., 1989, 1992a). Involvement of these transcription factors has been implicated in the mitogenic property of ETs observed in many cells (Brown and Littlewood, 1989; Dubin et al., 1989; Jaffer et al., 1990; Macnulty et al., 1990; Takagi et al., 1990). The pattern of *c-fos* and *c-jun* expression is differentially regulated by ET receptor subtypes (Takagi et al., 1990). Studies have shown increased expression of collagenase (Simonson et al., 1992c), platelet-derived growth factor A and B chains (Simonson et al., 1992c), and prostaglandin endoperoxidase synthase (Simonson et al., 1991b) in response to ET-1. Activation of protein tyrosine kinases such as MAPK has been suggested to be involved in ET-1-induced activation of nuclear events (Simonson et al., 1992b; Simonson, 1993). These processes are reviewed in Chapter 5.

C. Endothelins and Mesangium

Exposure of mesangial cells to ET has been shown to result in increased cell number and [^{3}H]thymidine incorporation, which are indicative of a proliferative response (Sugiura et al., 1989; Jaffer et al., 1990). ET-1 has also been shown to increase mesangial DNA topoisomerase I activity and this activation was inhibited by pretreatment with pertussis toxin (Nambi et al., 1990b). Although the mitogenic signals mediated by ET-1 are not well understood, it is possible that an ET-1-induced increase in platelet-derived growth factor A and B chains might be involved in this process (Simonson et al., 1992a). In addition, the induction of protooncogenes such as *c-fos*, *c-jun*, and *c-myc* might contribute to ET-1-mediated mitogenic response in mesangial cells.

ET-mediated mesangial cell contraction has been shown as an increase in the intensity and number of tension-generated wrinkles of the silicon rubber substratum on which mesangial cells were grown (Badr et al., 1989) as well as a decrease in mesangial cell planar surface area (Tomita et al., 1990; López-Farré et al., 1991a.). An ET-1-induced increase in intracellular Ca^{2+} has been shown to be responsible for ET-mediated contraction in mesangial cells. Since the contractile state of mesangium is a key factor regulating ultrafiltration coefficient (K_f) (Mené et al., 1989), it is possible that ET peptides regulate K_f. Indeed, ET-1 has been shown to cause a dramatic increase in afferent and efferent arteriolar resistance, and a decline in glomerular capillary flow rate (Q_A) as well as K_f, which resulted in a decrease in glomerular filtration rate (GFR) (Badr et al., 1989; Kon et al., 1989; King and Brenner, 1991). Prostaglandins have been shown to be involved in the regulation of mesangial cell mitogenesis, as well as contraction, and exposure of mesangial cells to ETs results in the release of arachidonic acid and formation of PGE_2 (Tomita et al., 1990, Badr et al., 1989). In an elegant study in which mesangial cells and endothelial cells were cocultured, Uchida and Ballermann (1992) have shown increased formation of PGE_2 in mesangial cells through an ET-dependent mechanism. This suggests a paracrine role for ET which is released from the endothelial cells and acts on mesangial cells.

D. Endothelins and Tubular Function

ETs have been shown to cause natriuresis at low concentrations. A pressure natriuresis as well as inhibition of tubular sodium reabsorption are thought to be involved in this response since removal of the renal capsule or maintenance of renal arterial perfusion pressure with an aortic clamp can attenuate ET-mediated natriuresis (Harris et al., 1991; Takabatake et al., 1992; Uzuner and Banks, 1993). Perico et al. (1991), however, have shown an ET-mediated diuresis and natriuresis with no change in renal or systemic hemodynamics. ETs have been shown to inhibit tubular sodium and bicarbonate transport as well as Na^+/K^+ ATPase activity (Garvin and Sanders, 1991; Zeidel et al., 1989). This supports the observation made in dogs (Brooks et al., 1994b) and rats (Perico et al., 1991), where administration of ET-1 at a dose which causes vasoconstriction and a reduction in total sodium excretion, also causes an increase in fractional sodium excretion. ET-mediated inhibition of Na^+/K^+ ATPase appears to be mediated by prostaglandins because this response was abolished by ibuprofen (Zeidel et al., 1989). Evaluation of the effect of systemic infusions of ET-1 on sodium handling has provided variable data. In some studies, systemic infusion of ETs has been shown to cause a decrease in sodium excretion (Goetz et al., 1988; Hirata et al., 1989; Miller et al., 1989), whereas in other studies, ETs caused a natriuresis in spite of a decrease in renal blood flow (Garcia et al., 1990; King et al., 1989) and GFR (Perico et al., 1991). Studies performed with isolated perfused kidney also demonstrate ET-mediated increase in Na^+ excretion, in spite of a decrease in GFR (Ferrario et al., 1989; Nitta et al., 1990), which is different from whole-animal studies where intrarenal infusion of low doses of ET-1 was without effect on Na^+ excretion, whereas higher doses caused a decrease in Na^+ excretion (Katoh et al., 1990; Stacy et al., 1990). ETs have also been shown to increase plasma levels of ANP (Goetz et al., 1988; Miller et al., 1989), which is another important regulator of renal sodium concentration. ANP has been shown to increase sodium excretion by increasing GFR and decreasing sodium reabsorption at the collecting duct as well as by inhibiting the renin-angiotensin system (Stacy et al., 1990). Co-infusion of ANP along with ET-1 has resulted in inhibition of ET-mediated renal effects (Jones et al., 1989; Hirata et al., 1989; Ballermann et al., 1991; Ota et al., 1992). In addition, ANP is a potent vasodilator and attenuates ET-mediated proliferation in vascular smooth muscle cells (Neuser et al., 1990).

In addition to being natriuretic, ET has also been shown to have some water diuretic activity, in spite of its inhibitory effects on GFR and renal blood flow (Badr et al., 1989; Goetz et al., 1988), indicating that it has an effect on water reabsorption. The water diuretic effect of ET-1 may be species specific. This diuretic effect of ET has been suggested to be mediated through its inhibitory effect on arginine vasopressin (AVP)-induced cyclic AMP accumulation and independent of its effect on Na^+ reabsorption and renal hemodynamics. ETs inhibited AVP-induced cyclic AMP accumulation in cortical collecting duct, outer medullary collecting duct, and inner medullary collecting duct (Tomita et al., 1990), whereas it did not affect AVP-mediated cyclic AMP accumulation in other nephron segments. ET-mediated inhibitory effect on AVP-induced cyclic AMP accumulation is dependent on PKC activity (Tomita et al., 1990). ETs have no effect on isoproterenol-stimulated cyclic AMP accumulation. In addition to its effects on the kidney, ETs may act on the posterior pituitary to alter vasopressin receptors, and there are suggestions that ETs might interact with AVP at the hypothalamic-pituitary axis (Ritz et al., 1992).

The renin-angiotensin system is the key regulator of extracellular fluid volume and ETs appear to have multiple effects on renin release, depending on whether they act in an endocrine or autocrine/paracrine fashion. Systemic infusions of ETs resulted in dramatic increases in renin activity (Goetz et al., 1988; Miller et al., 1989), possibly through the intrarenal baroreceptor and macula-densa-mediated pathways. On the other hand, *in vitro* experiments performed with cortical slices, glomeruli, or juxtaglomerular cells demonstrated an inhibition of renin release by ET-1 (Moe et al., 1991; Matsumura et al., 1989; Rakugi et al., 1988; Takagi et al., 1988). The inhibitory effect of ET-1 was dependent on both extra- as well as intracellular Ca^{2+} release because removal of extracellular Ca^{2+} or addition of Ca^{2+} channel antagonists attenuated ET-mediated inhibition of renin release (Takagi et al., 1989; Bell and Henrich, 1990; Churchill, 1985; Hallam and Rink, 1989). In similar studies, ET-1 has also been shown to inhibit both basal as well as isoproterenol-stimulated renin release (Moe et al., 1991; Matsumura et al., 1989).

E. Endothelins and Renal Vasculature

Infusion of ET-1 causes differential renal effects depending on the mode of infusion. Systemic infusions of ET-1 cause a significant increase in renal vascular resistance and a concomitant decrease in renal blood flow (Badr et al., 1989; King and Brenner, 1991; Hirata et al., 1989; Miller et al., 1989; King et al., 1989; Cao and Banks, 1990; Chou et al., 1990; Denton and Anderson, 1990; Goetz et al., 1989; King and Brenner, 1992; Kon and Badr, 1991; Miura et al., 1990; Tsuchiya et al., 1990; Wilkes et al., 1991b). ET-3 was much weaker than ET-1 in increasing renal vascular resistance (Zimmerman et al., 1990) and, in addition, caused a transient vasorelaxation (Yamashita et al., 1991), although the magnitude of this vasodilation was less than that observed in other vascular beds (Yanagisawa et al., 1988; Lemonnier Degouville et al., 1989; Lerman et al., 1990; Vane et al., 1989). Involvement of PGE_2 or PGI_2 in ET-mediated vascular effects in the kidney has been implicated because the vasodilatory response mediated by ETs was attenuated by indomethacin, whereas the vasoconstrictor response was augmented, suggesting that the prostaglandins might serve to counteract the effects of ETs on renal vasoconstriction (Chou et al., 1990; Miura et al., 1991). Micropuncture studies suggest that ET-1 can contract both afferent and efferent arterioles (Badr et al., 1989; Kon et al., 1989). King et al. (1989) showed a preferential increase in efferent arteriolar resistance resulting in an increase in glomerular capillary pressure, whereas Loutzenhiser et al. (1990) demonstrated a preferential effect of ET-1 on the afferent arteriole using direct visualization of the renal microvasculature in isolated perfused hydronephrotic kidneys. The signal transduction pathways involved in ET-mediated renal vasoconstriction involve both influx of extracellular Ca^{2+} as well as mobilization of intracellular Ca^{2+}. This conclusion is based upon the observations that Ca^{2+} channel antagonists such as verapamil and nicardipine attenuated ET-mediated afferent arteriolar response only, whereas the efferent arteriolar responses appear to be mediated by the release of intracellular Ca^{2+} (Simonson and Dunn 1990a; Loutzenhiser et al., 1990; Edwards et al., 1990; Masaki et al., 1990).

Direct infusion of ET-1 into the renal artery of rats or dogs resulted in a transient increase followed by a sustained decrease of renal blood flow (Katoh et al., 1990; Stacy et al., 1990; Banks, 1990). Experiments performed with isolated kidneys of rat and rabbit also demonstrated an ET-mediated increase in vascular resistance which was partially dependent on the presence of extracellular Ca^{2+} (Ferrario et al., 1989; Nitta et al., 1990; Cairns et al., 1989; Perico et al., 1990b). The involvement of prostaglandins or thromboxane on ET-mediated renal effects is controversial and needs further investigation (Ferrario et al., 1989; Perico et al., 1990b). The subtypes of ET receptors involved in mediating renal vasoconstriction have been shown to be different between species. In the rat, ET-1 mediates vasoconstriction by activating ET_B receptors because both ET-3 and STX-c, ET_B-selective agonists, also caused potent vasoconstriction (Stier et al., 1992; Cristol et al., 1993; Warner et al., 1993a). In addition, BQ-123, an ET_A-selective antagonist, was ineffective in blocking the decrease in renal blood flow mediated by ET-1, ET-3 or STX-c (Cristol et al., 1993; Pollock and Opgenorth, 1993).

Gellai et al. (1994a) and Wellings et al. (1994) have shown that renal vasoconstriction in rats is mediated predominantly by ET_B receptors. If this is the case, then it is possible that there are two subtypes of ET_B receptors, one mediating vasodilation and one mediating vasoconstriction. The vasodilatory response may be mediated by nitric oxide synthase pathway, whereas the vasoconstrictor pathway may involve the release of intracellular Ca^{2+} as well as the influx of extracellular Ca^{2+}. It is very likely that both subtypes of ET_B receptors may be present in the kidney because both responses (transient vasodilation followed by sustained vasoconstriction) have been shown in rat kidney in response to ET_B-selective ligands such as STX-c. Also, in rat glomeruli ETs stimulate cyclic GMP accumulation via an L-arginine-dependent pathway and the receptors involved in this process belong to the ET_B subtype (Edwards et al., 1992). Radioligand binding studies performed in membranes prepared from various regions of rat kidney indicate that the ratio of ET_A to ET_B receptors in cortex, medulla, and papilla were 50:50, 30:70, and 10:90, respectively (Nambi et al., 1992b; Gellai et al., 1994a). In addition, recent studies performed with Wistar Kyoto and spontaneously hypertensive rats indicated an increased proportion of ET_B receptors and increased renal vasoconstriction mediated by these receptors in spontaneously hypertensive rats (Gellai et al., 1994a).

The subtype of ET receptors involved in the renal vasoconstriction in dogs may be different than in rats and appears to be ET_A. This conclusion is based on the observations that ET-1-mediated vasoconstriction could be blocked by the ET_A-selective antagonist BQ-123, and STX-c, an ET_B-selective agonist, did not have any effect on renal blood flow or glomerular filtration rate (Brooks et al., 1994b). ET-1-mediated renal vasoconstriction was also blocked by SB 209,670, the mixed nonpeptide ET_A/ET_B antagonist (Brooks et al., 1994a). In dogs, as in rats, bolus injections of ET-1 caused a transient vasodilation followed by a sustained decrease in renal blood flow (Tsuchiya et al., 1989). It is not clear whether this transient effect is mediated by ET_B receptors because in one study ET-3 infusion caused a renal vasodilation in dogs (Yamashita et al., 1991), whereas in another study infusion of STXc was without effect (Brooks et al., 1994b). As far as the distribution of the subtypes of ET receptors is concerned, both cortex and medulla display predominantly ET_B receptors, whereas papilla displays ET_A and ET_B in equal proportion.

III. ENDOTHELINS IN RENAL DISEASE

There is increasing evidence showing the involvement of ET-1 in a number of renal diseases such as ischemia-induced acute renal failure, chronic renal failure, and cyclosporine A- and radiocontrast-induced nephrotoxicity (Kon et al., 1991; Kohan 1993b; Marsen et al., 1994). Evidence accumulated so far, indicating a role of ET in the above diseases is very compelling because of the beneficial effects demonstrated by neutralizing antibodies to ET or ET receptor antagonists in these disease models. An involvement of ET-1 has also been implicated in other less extensively studied renal disease models including sepsis-induced renal failure (Morise et al., 1994), hepatorenal syndrome (Moore et al., 1992), cisplatin-induced nephrotoxicity (Ohta et al., 1991), amphotericin B-induced nephrotoxicity (Heyman et al., 1992a), and ureteral obstruction-induced nephrotoxicity (Kelleher et al., 1992).

A. Ischemia-Induced Acute Renal Failure

Since hypoxia is a potent stimulus of ET secretion, it is possible that ET might play an important role in ischemia-induced acute renal failure (Firth and Ratcliffe, 1992; Kon et al., 1989; Firth et al., 1988; López-Farré et al., 1991b; Sandok et al., 1992; Shibouta et al., 1990; Wilkes et al., 1991a). Indeed, the renal expression of prepro ET-1 mRNA as well as ET-1 mRNA was upregulated in rats 6 h after ischemia (Firth and Ratcliffe, 1992) and plasma ET-1 levels were elevated in patients with acute renal failure (Tomita et al., 1989). While ET-1 mRNA was upregulated in postischemic kidneys, ET-3 mRNA was downregulated following ischemia (Firth and Ratcliffe, 1992). In addition to the increased production of ET-1, ET-1 binding in the kidney was also increased following ischemia (Clozel et al., 1991; Nambi et al., 1993). This increase in binding was shown to be due to an increase in affinity of the receptor for the ligand rather than an increase in receptor density (Clozel et al., 1991; Nambi et al., 1993). On the other hand, the increase in ET binding observed after the induction of renal failure by glycerol appears to be due to an increase in density rather than affinity (Roubert et al., 1993). Since no change in the binding of ET to glomeruli was observed after ischemia-induced renal failure (Wilkes et al., 1991a), the increased affinity observed (Clozel et al., 1991; Nambi et al., 1993) could be due to the receptors present in cortical regions other than glomeruli.

More direct evidence for the involvement of ETs in acute renal failure comes from data obtained with ET antibodies or ET receptor antagonists in this model. Antibodies to ET-1 or ET receptor antagonists such as BQ-123, SB 209670, and Ro 46-2005 have been shown to protect against ischemia-induced acute renal failure (Kon et al., 1989; López-Farré et al., 1991b; Shibouta et al., 1990; Mino et al., 1992; Chan et al., 1994; Gellai et al., 1994b,c; Clozel et al., 1993). Indeed, in a very elegant study, Gellai et al. (1994b,c) have demonstrated that BQ-123 as well as SB 209670 can reverse ischemia-induced acute renal failure in rats. Since, in addition to SB 209670 and Ro 46-2005 which are nonselective antagonists, BQ-123 (an ET_A-selective antagonist) was also beneficial in ischemia-induced acute renal failure in rats, it appears that the receptor involved in this process might be ET_A. In contrast, in the dog, BQ-123 was ineffective in preventing the reduction in glomerular filtration rate induced by aortic cross-clamping, although it could attenuate the increase in systemic and renal vasoconstriction associated with aortic cross-clamping (Brooks et al., 1994a; Stingo et al., 1993). Following renal artery occlusion, however, the nonselective ET receptor antagonist, SB 209670, was beneficial in attenuating the decrease in glomerular filtration rate and the increase in fractional sodium excretion (Brooks et al., 1994a). These data indicate that in the dog, it is the ET_B receptor antagonist activity of SB 209670 that was beneficial, further emphasizing the species differences between rat and dog.

Based on the observations that in rat and dog renal vasoconstriction is mediated by ET_B and ET_A receptors, respectively, and that BQ-123, an ET_A-selective antagonist, was effective in rat but not in dog, mechanisms other than ET-induced renal vasoconstriction (e.g., tubular effects) have been suggested by some investigators to be involved in ischemia-induced acute renal failure (Brooks et al., 1994a; Gellai et al.,

1994b). Support for the tubular involvement in acute renal failure comes from the observations that acute moderate ischemia in dogs resulted in significant increases in urinary ET excretion, urine flow, and sodium excretion, with no change in renal blood flow or glomerular filtration rate (Nir et al., 1994); and that increased ET staining in the proximal and distal tubules following hypoxia was demonstrated by immunohistochemical techniques (Nir et al., 1994).

B. Chronic Renal Disease

Although the exact mechanisms involved in the development and progression of chronic renal failure are not known, the primary causes of chronic renal failure are hypertension and Type I and Type II diabetes. It is very likely that glomerular hypertension and/or hyperfiltration and mesangial cell proliferation and/or matrix production contribute to the development of chronic renal failure, and ET could be involved in all of these processes. ETs act as mitogens for mesangial cells and increase glomerular capillary pressure. They also increase Type I, III, and IV collagen as well as laminin expression in rat mesangial cells (Ishimura et al., 1991). Many of the proinflammatory mediators such as TGFβ, thrombin, and platelet-derived growth factor are potent mitogens in addition to their effects on increased matrix formation (Mené et al., 1989; Border et al., 1990; Johnson et al., 1990; Klahr et al., 1988) and these mediators have been shown to stimulate ET production in glomerular endothelial and mesangial cells (Zoja et al., 1991; Sakamoto et al., 1990; Marsden et al., 1991). ETs cause short-term as well as long-term effects in mesangial cells including contraction, proliferation, matrix formation, and expression of immediate early response genes (Simonson and Dunn, 1990a; Simonson et al., 1992a,b).

Angiotensin II, another important mediator of chronic renal failure, also causes mesangial cell proliferation which is mediated in part by ETs (Bakris and Re, 1993). Angiotensin converting enzyme inhibitors attenuated diabetic nephropathy as well as diabetes-induced increases in glomerular ET-1 messenger RNA (Fukui et al., 1994). Furthermore, the renal hemodynamics of rats with mesangial proliferative nephritis were much more sensitive to ET-1 and angiotensin II than control rats (Kanai et al., 1993). In rats with 5/6 nephrectomy-induced chronic renal failure, there was a significant increase in renal ET levels as well as urinary ET excretion which correlated well with the degree of proteinuria and hypertension (Brooks et al., 1991a; Benigni et al., 1991). ET gene expression in this model of renal failure correlated well with disease progression (Orisio et al., 1993). ET gene expression was also increased in the glomeruli of streptozotocin-induced diabetic rats (Fukui et al., 1993). In addition, in cpk/cpk mice which develop polycystic kidney disease, there was a good correlation between the disease progression and expression of ET-1 as well as ET_A and ET_B receptor messenger RNAs (Nakamura et al., 1993).

A number of clinical studies conducted on patients with chronic renal disease indicate increased urinary ET excretion or plasma levels (Koyama et al., 1989; Predel et al., 1990; Warrens et al., 1990; Stockenhuber et al., 1992; Collier et al., 1992; Lee et al., 1994; Saito et al., 1991). Stockenhuber et al. (1992) have shown a significant correlation between plasma ET and plasma creatinine levels in patients with chronic renal disease. In a very elegant study, Benigni et al. (1993) have demonstrated the beneficial effect of ET_A receptor antagonist FR 139,317 by limiting glomerular injury and preventing deterioration of renal function following 5/6 nephrectomy in rats.

There are some indications that ETs may be contributing factors to the problems associated with the subsequent treatment of patients with end-stage renal disease.

ETs may be factors in the hypertension observed in some patients who have undergone chronic dialysis and require recombinant erythropoietin for treatment of renal disease-induced anemia (Takahashi et al., 1993). This is supported by the observation that erythropoietin leads to an increase in ET-release from endothelial cells (Carlini et al., 1992), plasma levels of ETs are elevated in patients treated with recombinant erythropoietin and who are hypertensive (Carlini et al., 1992), and following intravenous administration of erythropoietin there is a significant increase in plasma ET levels that correlate well with the increase in blood pressure (Carlini et al., 1993).

C. Cyclosporine A-Induced Nephrotoxicity

There is increasing evidence to demonstrate the involvement of ETs in cyclosporine A-induced nephrotoxicity. This is based on the observations that cyclosporine A treatment causes (1) an increase in ET release from endothelial (Bunchman et al., 1991), renal, and mesangial cells (Nakahama et al., 1990; Moutabarrik et al., 1991); (2) an increase in urinary ET excretion (Brooks et al., 1991b; Benigni et al., 1991b); (3) an increase in plasma ET levels (Edwards et al., 1991; Grieff et al., 1993); (4) upregulation of renal and cardiac ET receptors (Nambi et al., 1990a; Nayler et al., 1989; Awazu et al., 1991); and (5) an increase in ET_B receptor mRNA in mesangial cells (Takeda et al., 1994). In addition, treatment with nifedipine inhibited cyclosporine A-mediated renal dysfunction as well as increased urinary ET excretion (Brooks et al., 1991b). Confirmative data for the involvement of ET in cyclosporine A-induced nephrotoxicity have come from studies performed using ET antibodies and ET receptor antagonists, which have been shown to attenuate or abolish cyclosporine A-induced renal dysfunction (Kon et al., 1990; Bloom et al., 1993; Fogo et al.,1992; Perico et al., 1990a). Furthermore, an ET receptor antagonist inhibited cyclosporine A- and ET-induced myosin light chain phosphorylation which plays an obligatory role in contraction (Takeda et al., 1992). Cyclosporine A-induced vasoconstriction of renal microvessels is mediated by ET in the afferent but not the efferent arteriole (Lanese and Conger, 1993). Brooks et al. (1995) have demonstrated that intravenous administration of SB 209670, a nonpeptide antagonist of ET_A/ET_B receptors, prevented the acute reductions in glomerular filtration rate and renal blood flow induced by cyclosporine A in the rat.

D. Radiocontrast-Induced Nephrotoxicity

Although radiocontrast media is of great use as a diagnostic tool, use of these agents has been shown to result in acute renal failure, a result of intense renal vasoconstriction (Sherwood and Lavender, 1969; Forrest et al., 1981). While increased renal prostaglandin production has been suggested to mediate radiocontrast-induced vasodilation (Cantley et al., 1993), the mechanism of radiocontrast-induced vasoconstriction is not known. Since changes in osmolality can alter ET-mRNA expression as well as ET-1 production in kidney cells or the inner medullary collecting duct *in vitro* (Kohan et al., 1993a; Schramek et al., 1993; Yang et al., 1993), and plasma as well as urinary ET levels are increased by radiocontrast media (Margulies et al., 1991; Heyman et al., 1992b), investigations have focused on the potential role of ET-1 in radiocontrast-induced nephrotoxicity. The tripeptide ET receptor antagonist, CP 170,687, blocked iatholamate-induced renal vasoconstriction in indomethacin-treated rats (Cantley et al., 1993) and the nonpeptide ET receptor antagonist, SB 209670, blocked hypaque-induced renal vasoconstriction in dogs (Brooks et al., 1994b).

IV. SUMMARY

Although innumerable research efforts are ongoing in the field of ETs and the kidney, many unanswered questions remain regarding the mechanism of ET action, the subtypes of ET receptors involved in various renal effects, and possible species differences. The potential of cloning and characterizing novel ET receptor subtypes and the development of potent subtype-selective nonpeptide ET receptor antagonists, along with techniques such as *in situ* hybridization to localize the subtypes of ET receptors, will help address these questions.

ACKNOWLEDGMENT

The authors would like to thank Sue Tirri for expert secretarial assistance.

REFERENCES

Abdel-Latif, A. A. and Zhang, Y., Species differences in effects of endothelin-1 on myoinositol trisphosphate accumulation, cyclic AMP formation and contraction of isolated iris sphincter of rabbit and other mammalian species, *Invest. Ophthalmol. Vis. Sci.*, 32, 2432, 1991.

Albrightson, C. R., Pullen, M., Wu, H.-L., Dytko, G., Hersh, L. B., and Nambi, P., Thrombin-mediated downregulation of endothelin receptors in mesangial cells coincides with the downregulation of neutral endopeptidase activity (submitted for publication), 1995.

Ando, K., Hirata, Y., Takei, Y., Kawakami, M., and Marumo, F., Endothelin-1-like immunoreactivity in human urine, *Nephron*, 57, 36, 1991.

Arai, H., Hori, S., Aramori, I., Ohkubo, H., and Nakanishi, S., Cloning and expression of a cDNA encoding an endothelin receptor, *Nature*, 348, 730, 1990.

Aramori, I. and Nakanishi, S., Coupling of two endothelin receptor subtypes to differing signal transduction in transfected Chinese hamster ovary cells, *J. Biol. Chem.*, 267, 12468, 1992.

Awazu, M., Sugiura, M., Inagami, T., Ichikawa, I., and Kon, V., Cyclosporine promotes glomerular ET binding *in vivo*, *J. Am. Soc. Nephrol.*, 1, 1253, 1991.

Badr, K. F., Murray, J. J., Breyer, M. D., Takahashi, K., Inagami, T., and Harris, R. C., Mesangial cell, glomerular and renal vascular responses to ET in the rat kidney, *J. Clin. Invest.*, 83, 336, 1989.

Bakris, G. L. and Re, R. N., Endothelin modulates angiotensin II-induced mitogenesis of human mesangial cells, *Am. J. Physiol.*, 264, F937, 1993.

Bakris, G. L., Fairbanks, R., and Traish, A. M., Arginine vasopressin stimulates human mesangial cell production of ET, *J. Clin. Invest.*, 87, 1158, 1991.

Ballermann, B. J., Zeidel, M. L., Gunning, M. E., and Brenner, B. M., Vasoactive peptides and the kidney, In: *The Kidney*, ed. B. M. Brenner and F. J. Rector, Philadelphia, PA, Saunders, 510, 1991.

Banks, R. O., Effects of endothelin on renal function in dogs and rats, *Am. J. Physiol.*, 258, F775, 1990.

Benigni, A., Perico, N., Gaspari, F., Zoja, C., Bellizzi, L., Gabanelli, M., and Remuzzi, G., Increased renal ET production in rats with reduced renal mass, *Am. J. Physiol.*, 260, F331, 1991.

Benigni, A., Perico, N., Ladny, J. R., Imberti, O., Bellizzi, L., and Remuzzi, G., Increased urinary excretion of ET-1 and its precursor, big-ET-1, in rats chronically treated with cyclosporine, *Transplantation*, 52, 175, 1991b.

Benigni, A., Zoja, C., Corna, D., Orisio, S., Longaretti, L., Bertani, T., and Remuzzi, G., A specific ET subtype A receptor antagonist protects against injury in renal disease progression, *Kidney Int.*, 44, 440, 1993.

Berbinschi, A. and Ketelslegers, J. M., Endothelin in urine, *Lancet*, 2, 46, 1989.

Bloom, I. T. M., Bentley, F. R., and Garrison, R. N., Acute cyclosporine-induced renal vasoconstriction is mediated by ET-1, *Surgery*, 114, 480, 1993.

Border, W. A., Okuda, S., Languino, L. R., Sporn, M. B., and Ruoslahti, E., Suppression of experimental glomerulonephritis by antiserum against transforming growth factor β1, *Nature London*, 346, 371, 1990.

Brooks, D. P. and Contino, L. C., Prevention of cyclosporine A-induced renal vasoconstriction by the endothelin receptor antagonist SB 209670, *Eur. J. Pharmacol.*, 294, 571, 1995.

Brooks, D. P. and DePalma, P. D., Blockade of radiocontrast-induced renal vasoconstriction by the ET receptor antagonist, SB 209670, *Nephron*, 72, 629, 1996.

Brooks, D. P., Contino, L. C., Storer, B., and Ohlstein, E. H., Increased ET excretion in rats with renal failure induced by partial nephrectomy, *Br. J. Pharmacol.*, 104, 987, 1991a.

Brooks, D. P., DePalma, P. D., Gellai, M., Nambi, P., Ohlstein, E. H., Elliott, J. D., Gleason, J., and Ruffolo, R. R., Jr., Nonpeptide ET receptor antagonists. III. Effect of SB 209670 and BQ123 on acute renal failure in anesthetized dogs, *J. Pharmacol., Exp. Ther.* 271, 769, 1994a.

Brooks, D. P., DePalma, P. D., Pullen, M., and Nambi, P., Characterization of canine renal ET receptor subtypes and their function, *J. Pharmacol. Exp. Ther.*, 268, 1091, 1994b.

Brooks, D. P., Ohlstein, E. H., Contino, L. C., Storer, B., Pullen, M., Caltabiano, M., and Nambi, P., Effect of nifedipine on cyclosporine A-induced nephrotoxicity, urinary ET excretion and renal ET receptor number, *Eur. J. Pharmacol.*, 194, 115, 1991b.

Brown, K. D. and Littlewood, C. J., Endothelin stimulates DNA synthesis in Swiss 3T3 cells. Synergy with polypeptide growth factors, *Biochem. J.*, 263, 977, 1989.

Bunchman, T. E. and Brookshire, C. A., Cyclosporine-induced synthesis of ET by cultured human endothelial cells, *J. Clin. Invest.*, 88, 310, 1991.

Cairns, H. S., Rogerson, M. E., Fairbanks, L. D., Nield, G. H., and Westwick, J., Endothelin induces an increase in renal vascular resistance and a fall in glomerular filtration rate in the rabbit isolated perfused kidney, *Br. J. Pharmacol.*, 98, 155, 1989.

Campbell, W. B. and Henrich, W. L., Endothelial factors in the regulation of renin release, *Kidney Int.*, 38, 612, 1990.

Cantley, L. G., Spokes, K., Clark, B., McMahon, E. G., Carter, J., and Epstein, F. H., Role of ET and prostaglandins in radiocontrast-induced renal artery constriction, *Kidney Int.*, 44, 1217, 1993.

Cao, L. and Banks, R. O., Cardiovascular and renal actions of endothelin: effects of calcium-channel blockers, *Am. J. Physiol.*, 258, F254, 1990.

Carlini, R., Dusso, A., Obialo, C., and Rothstein, M., Erythropoietin increases ET-1 production and cell proliferation in bovine pulmonary arterial endothelial cells, *J. Am. Soc. Nephrol.*, 3, 423 (Abstr.), 1992.

Carlini, R., Obialo, C. I., and Rothstein, M., Intravenous erythropoietin (rHuEPO) administration increases plasma ET and blood pressure in hemodialysis patients, *Am. J. Hypertens.*, 6, 103, 1993.

Chan, L., Chittinandana, A., Shapiro, J. I., Shanley, P. F., and Schrier, R. W., Effect of an ET receptor antagonist on ischemic acute renal failure, *Am. J. Physiol.*, 266, F135, 1994.

Chen, M., Todd-Turla, K., Wang, W.-H., Cao, X., Smart, A., Brosius, F. C., Killen, P. D., Keiser, J. A., Briggs, J. P., and Schnermann, J., Endothelin-1 mRNA in glomerular and epithelial cells of kidney, *Am. J. Physiol.*, 265, F542, 1993.

Chou, S. Y., Dahhan, A., and Porush, J. G., Renal actions of endothelin: interaction with prostacyclin, *Am. J. Physiol.*, 259, F645, 1990.

Churchill, P. C., Second messengers in renin secretion, *Am. J. Physiol.*, 249, F175, 1985.

Clozel, M., Breu, V., Burri, K., Cassal, J.-M., Fischli, W., Gray, G. A., Hirth, G., Löffler, B.-M., Müller, M., Neidhart, W., and Ramuz, H., Pathophysiological role of ET revealed by the first orally active ET receptor antagonist, *Nature*, 365, 759, 1993.

Clozel, M., Löffler, B. M., and Gloor, H., Relative preservation of the responsiveness to ET-1 during reperfusion following renal ischemia in the rat, *J. Cardiovasc. Pharmacol.*, 17(Suppl. 7), S313, 1991.

Collier, A., Leach, J. P., McLellan, A., Jardine, A., Morton, J. J., and Small, M., Plasma ET-like immunoreactivity levels in IDDM patients with microalbuminuria, *Diabetic Care*, 15, 1038, 1992.

Cristol, J.-P., Warner, T. D., Thiemermann, C., and Vane, J. R., Mediation via different receptors of the vasoconstrictor effects of ETs and sarafotoxins in the systemic circulation and renal vasculature of the anesthetized rat, *Br. J. Pharmacol.*, 108, 776, 1993.

Deng, A. Y., Martin, L. L., Balwierczak, J. L., and Jeng, A. Y., Purification and characterization of an ET degradation enzyme from rat kidney, *J. Biochem.*, 115, 120, 1994.

Denton, K. M. and Anderson, W. P., Vascular actions of endothelin in the rabbit kidney, *Clin. Exp. Pharmacol. Physiol.*, 17, 861, 1990.

DeNucci, G., Thomas, R., D'Orleans-Juste, P., Antunes, E., Walder, C., Warner, T. D., and Vane, J. R., Pressor effects of circulating endothelin are limited by its removal in the pulmonary circulation and by the release of prostacyclin and endothelium-derived relaxing factor, *Proc. Natl. Acad. Sci. U.S.A.*, 85, 9797, 1988.

Dubin, D., Pratt, R. E., Cooke, J. P., and Dzau, V. J., Endothelin, a potent vasoconstrictor, is a vascular smooth muscle mitogen, *J. Vasc. Med. Biol.*, 1, 150, 1989.

Edano, T., Arai, K., Ohshima, T., Koshi, T., Hirata, M., Ohkuchi, M., and Okabe, T., Purification and characterization of ET-1 degradation activity from porcine kidney, *Biol. Pharm. Bull.*, 17, 379, 1994.

Edwards, B. S., Hunt, S. A., Fowler, M. B., Valentine, H. A., Anderson, L. M., and Lerman, A., Effect of cyclosporine on plasma ET levels in humans after cardiac transplantation, *Am. J. Cardiol.*, 67, 782, 1991.

Edwards, R. M., Pullen, M., and Nambi, P., Activation of ET ET_B receptors increases glomerular cyclic GMP via an L-arginine-dependent pathway, *Am. J. Physiol.*, 263, F1020, 1992.

Edwards, R. M., Trizna, W., and Ohlstein, E. H., Renal microvascular effects of ET, *Am. J. Physiol.*, 259, F217, 1990.

Elshourbagy, N. A., Korman, D. R., Wu, H.-L., Sylvester, D. R., Lee, J. A., Nuthulaganti, P., Bergsma, D. J., Kumar, C. S., and Nambi, P., Molecular characterization and regulation of the human endothelin receptors, *J. Biol. Chem.*, 268, 3873, 1993.

Elshourbagy, N. A., Lee, J. A., Korman, D. R., Nuthulaganti, P., Sylvester, D. R., Dilella, A. G., Sutiphong, J. A., and Kumar, C. S., Molecular cloning and characterization of the major endothelin receptor subtype in porcine cerebellum, *Mol. Pharmacol.*, 41, 465, 1992.

Ferrario, R. G., Foulkes, R., Salvati, P., and Patrono, C., Hemodynamic and tubular effects of endothelin and thromboxane in the isolated perfused rat kidney, *Eur. J. Pharmacol.*, 171, 127, 1989.

Firth, J. D. and Ratcliffe, P. J., Organ distribution of the three rat ET messenger RNAs and the effects of ischemia on renal gene expression, *J. Clin. Invest.*, 90, 1023, 1992.

Firth, J. D., Raine, A. E., Ratcliffe, P. J., and Ledingham, J. G., Endothelin: an important factor in acute renal failure, *Lancet*, 19, 1179, 1988.

Fogo, A., Hellings, S. E., Inagami, T., and Kon, V., Endothelin receptor antagonism is protective in *in vivo* acute cyclosporine toxicity, *Kidney Int.*, 42, 770, 1992.

Forrest, J. B., Howards, S. S., and Gillenwater, J. Y., Osmotic effects of intravenous contrast agents on renal function, *J. Urol.*, 125, 147, 1981.

Fukui, M., Nakamura, T., Ebihara, I., Makita, Y., Osada, S., Tomino, Y., and Koide, H., Effects of enalapril on ET-1 and growth factor gene expression in diabetic rat glomeruli, *J. Lab. Clin. Med.*, 123, 763, 1994.

Fukui, M., Nakamura, T., Ebihara, I., Osada, S., Tomino, Y., Masaki, T., Goto, K., Furuichi, Y., and Koide, H., Gene expression for ETs and their receptors in glomeruli of diabetic rats, *J. Lab. Clin. Med.*, 122, 149, 1993.

Garcia, R., Lachance, D., and Thibault, G., Positive inotropic action, natriuresis and atrial natriuretic factor release induced by endothelin in the conscious rat, *J. Hypertens.*, 8, 725, 1990.

Garvin, J. and Sanders, K., Endothelin inhibits fluid and bicarbonate transport in part by reducing Na^+/K^+ ATPase activity in the rat proximal straight tubule, *J. Am. Soc. Nephrol.*, 2, 976, 1991.

Gauquelin, G., Thibault, G., and Garcia, R., Characterization of renal glomerular ET receptors in the rat, *Biochem. Biophys. Res. Commun.*, 164, 54, 1989.

Gellai, M., DeWolf, R., Pullen, M., and Nambi, P., Distribution and functional role of renal ET receptor subtypes in normotensive and hypertensive rats, *Kidney Int.*, 46, 1287, 1994a.

Gellai, M., Jugus, M., Fletcher, T. A., DeWolf, R., and Nambi, P., Reversal of postischemic acute renal failure with a selective ET_A receptor antagonist in the rat, *J. Clin. Invest.*, 93, 900, 1994b.

Gellai, M., Jugus, M., Fletcher, T. A., Nambi, P., Brooks, D. P., Ohlstein, E. H., Elliott, J. D., Gleason, J., and Ruffolo, R. R., Jr., The ET receptor antagonist, (±)-SB 209670, reverses ischemia-induced acute renal failure (ARF) in the rat, *FASEB J.*, 8, A260 (Abstr.), 1994c.

Goetz, K. L., Wang, B. C., Madwed, J. B., Zhu, J. L., and Leadley, R. J., Cardiovascular, renal and endocrine responses to intravenous endothelin in conscious dogs, *Am. J. Physiol.*, 255, R1064, 1988.

Goetz, K., Wang, B. C., Leadley, R., Jr., Zhu, J. L., Madwed, J., and Bie, P., Endothelin and sarafotoxin produce dissimilar effects on renal blood flow, but both block the antidiuretic effects of vasopressin, *Proc. Soc. Exp. Biol. Med.*, 191, 425, 1989.

Grieff, M., Loertscher, R., Shohaib, S. A., and Stewart, D. J., Cyclosporine-induced elevation in circulating ET-1 in patients with solid-organ transplants, *Transplantation*, 56, 880, 1993.

Hallam, T. J. and Rink, T. J., Receptor-mediated Ca^{2+} entry: diversity of function and mechanism, *Trends Pharmacol. Sci.*, 10, 8, 1989.

Harris, P. J., Zhuo, J., Mendelsohn, F. A. O., and Skinner, S. L., Haemodynamic and renal tubular effects of low doses of ET in anesthetized rats, *J. Physiol.*, 433, 25, 1991.

Heyman, S. N., Clark, B. A., Kaiser, N., Epstein, F. H., Spokes, K., Rosen, S., and Brezis, M., *In vivo* and *in vitro* studies on the effect of amphotericin B on ET release, *J. Antimicrob. Chemother.*, 29, 69, 1992a.

Heyman, S. N., Clark, B. A., Kaiser, N., Spokes, K., Rosen, S., Brezis, M., and Epstein, F. H., Radiocontrast agents induce ET release *in vivo* and *in vitro*, *J. Am. Soc. Nephrol.*, 3, 58, 1992b.

Hilal-Dandan, R., Urasawa, K., and Brunton, L. L., Endothelin inhibits adenylate cyclase and stimulates phosphoinositide hydrolysis in adult cardiac monocytes, *J. Biol. Chem.*, 267, 10620, 1992.

Hirata, Y., Matsuoka, H., Kimura, K., Fukui, K., Hayakawa, H., Suzuki, E., Sugimoto, T., Yanagisawa, M., and Masaki, T., Renal vasoconstriction by the endothelial cell-derived peptide endothelin in spontaneously hypertensive rats, *Circ. Res.*, 65, 1370, 1989.

Hori, S., Komatsu, Y., Shigemoto, R., Mizuno, N., and Nakanishi, S., Distinct tissue distribution and cellular localization of two messenger ribonucleic acids encoding different subtypes of rat endothelin receptors, *Endocrinology*, 130, 1885, 1992.

Huggins, J. P., Pelton, J. T., and Miller, R. C., The structure and specificity of endothelin receptors: their importance in physiology and medicine, *Pharm. Ther.*, 59, 95, 1993.

Ishimura, E., Shouji, S., Nishizawa, Y., Morii, H., and Kashgarian, M., Regulation of mRNA expression for extracellular matrix (ECM) by cultured rat mesangial cells (MCs), *J. Am. Soc. Nephrol.*, 2, 546 (Abstr.), 1991.

Jaffer, F. E., Knauss, T. C., Poptic, E., and Abboud, H. E., Endothelin stimulates PDGF secretion in cultured human mesangial cells, *Kidney Int.*, 38, 1193, 1990.

Janas, J., Sitkiewicz, D., Pulawska, M. F., Warnawin, K., and Janas, R. M., Purification of ET-1 inactivating peptidase from the rat kidney, *J. Hypertens.*, 12, 375, 1994.

Johnson, R. L., Garcia, R. L., Pritzi, P., and Alpers, C. E., Platelets mediate glomerular cell proliferation in immune complex nephritis induced by anti-mesangial cells antibodies in the rat, *Am. J. Pathol.*, 136, 369, 1990.

Jones, C. R., Hiley, C. R., Pelton, J. T., and Miller, R. C., Autoradiographic localization of ET binding sites in kidney, *Eur. J. Pharmacol.*, 163, 379, 1989.

Kanai, H., Okuda, S., Kiyama, S., Tomooka, S., Hirakata, H., and Fujishima, M., Effects of ET and angiotensin II on renal hemodynamics in experimental mesangial proliferative nephritis, *Nephron*, 64, 609, 1993.

Karet, F. E., Kuc, R. E., and Davenport, A. P., Novel ligands BQ123 and BQ3020 characterize ET receptor subtypes ET_A and ET_B in human kidney, *Kidney Int.*, 44, 36, 1993.

Karne, S., Jayawickreme, C. K., and Lerner, M. R., Cloning and characterization of an endothelin-3 specific receptor (ET_C receptor) from *Xenopus laevis* dermal melanophores, *J. Biol. Chem.*, 268, 19126, 1993.

Katoh, T., Chang, H., Uchida, S., Okuda, T., and Kurokawa, K., Direct effects of endothelin in the rat kidney, *Am. J. Physiol.*, 258, F397, 1990.

Kelleher, J. P., Shah, V., Godley, M. L., Wakefield, A. J., Gordon, I., Ransley, P. G., Snell, M. E., and Risdon, R. A., Urinary ET (ET1) in complete ureteric obstruction in the miniature pig, *Urol. Res.*, 20, 63, 1992.

Kester, M., Simonson, M. S., McDermott, R. G., Baldi, E., and Dunn, M. J., Endothelin stimulates phosphatidic acid formation in cultured rat mesangial cells: role of a protein kinase C-dependent phospholipase D, *J. Cell. Physiol.*, 150, 578, 1992.

King, A. and Brenner, B. M., Renal and systemic hemodynamic actions of endothelin. In: *Endothelin*, Rubanyi, G., Ed., Oxford University Press, New York, 158, 1992.

King, A. J. and Brenner, B. M., Endothelium-derived vasoactive factors and the renal vasculature. *Am. J. Physiol.*, 260, R653, 1991.

King, A. J., Brenner, B. M., and Anderson, S., Endothelin: a potent renal and systemic vasoconstrictor peptide, *Am. J. Physiol.*, 256, F1051, 1989.

Klahr, S., Schreiner, G., and Ichikawa, I., The progression of renal disease, *N. Engl. J. Med.*, 318, 1657, 1988.

Kohan, D. E. and Padilla, E., Osmolar regulation of ET-1 production by rat inner medullary collecting duct, *J. Clin. Invest.*, 91, 1235, 1993a.

Kohan, D. E., Endothelins in the kidney: physiology and pathophysiology, *Am. J. Kidney Dis.*, 22, 493, 1993b.

Kohan, D. E., Hughes, A. K., and Perkins, S. L., Characterization of ET receptors in the inner medullary collecting duct of the rat, *J. Biol. Chem.*, 267, 12336, 1992.

Kohno, M., Horio, T., Ikeda, M., Yokokawa, K., Fukui, T., Yasunari, K., Murakawa, K.-I., Kurihara, N., and Takeda, T., Natriuretic peptides inhibit mesangial cell production of ET induced by arginine vasopressin, *Am. J. Physiol.*, 264, F678, 1993.

Kohno, M., Yasunari, K., Yokokawa, K., Murakawa, K., Horio, T., and Takeda, T., Inhibition by atrial and brain natriuretic peptides of endothelin-1 secretion after stimulation with angiotensin II and thrombin of cultured human endothelial cells, *J. Clin. Invest.*, 87, 1999, 1991.

Kohno, M., Yokokawa, K., Horio, T., Ikeda, M., Kurihara, N., Mandal, A. K., and Takeda, T., Heparin inhibits ET-1 production in cultured rat mesangial cells, *Kidney Int.*, 45, 137, 1994.

Kohzuki, M., Johnston, C. I., Abe, K., Chai, S. Y., Casley, D. J., Yasujima, M., Yoshinaga, K., and Mendelsohn, F. A. O., *In vitro* autoradiographic ET-1 binding sites and sarafotoxin S6B binding sites in rat tissues, *Clin. Exp. Pharmacol. Physiol.*, 18, 509, 1991.

Kohzuki, M., Johnston, C. I., Chai, S. Y., Casley, D. J., and Mendelsohn, F. A. O., Localization of ET receptors in rat kidney, *Eur. J. Pharmacol.*, 160, 193, 1989.

Kon, V. and Badr, K. F., Biological actions and pathophysiologic significance of ET in the kidney, *Kidney Int.*, 40, 1, 1991.

Kon, V., Sugiura, M., Inagami, T., Harvie, B. R., Ichikawa, I., and Hoover, R. L., Role of ET in cyclosporine-induced glomerular dysfunction, *Kidney Int.*, 37, 1487, 1990.

Kon, V., Yoshioka, T., Fogo, A., and Ichikawa, I., Glomerular actions of ET *in vivo*, *J. Clin. Invest.*, 83, 1762, 1989.

Kosaka, T., Suzuki, N., Matsumoto, H., Itoh, Y., Yasuhara, T., Onda, H., and Fujino, M., Synthesis of the vasoconstrictor peptide endothelin in kidney cells, *FEBS Lett.*, 249, 42, 1989.

Koyama, H., Nishizawa, Y., Morii, E. H., Tabata, T., Inoue, T., and Yamaji, T., Plasma endothelin levels in patients with uremia, *Lancet*, 1, 991, 1989.

Kumar, C., Mwangi, V., Nuthulaganti, P., Wu, H.-L., Pullen, M., Brun, K., Aiyar, H., Morris, R. A., Naughton, R., and Nambi, P., Cloning and characterization of a novel endothelin receptor from *Xenopus* heart, *J. Biol. Chem.*, 269, 13414, 1994.

Lanese, D. M. and Conger, J. D., Effects of ET receptor antagonist on cyclosporine-induced vasoconstriction in isolated rat renal arterioles, *J. Clin. Invest.*, 91, 2144, 1993.

Lee, Y.-J., Shin, S.-J., and Tsai, J.-H., Increased urinary ET-1-like immunoreactivity excretion in NIDDM patients with albuminuria, *Diabetic Care*, 17, 263, 1994.

Lemonnier Degouville, A.-C., Lippton, H. L., Cavero, I., Summer, W. R., and Hyman, A. L., Endothelin — a new family of endothelium-derived peptides with widespread biological properties, *Life Sci.*, 45, 1499, 1989.

Lerman, A., Hildebrand, F. L., Margulies, K. B., O'Murchu, B., Perrella, M. A., Heublien, D. M., Schwab, T. R., and Burnett, J. C., Endothelin: a new cardiovascular regulatory peptide, *Mayo Clin. Proc.*, 65, 1441, 1990.

Liu, Y., Geisbuhler, B., and Jones, A. W., Activation of multiple mechanisms including phospholipase D by endothelin-1 in rat aorta, *Am. J. Physiol.*, 262, C941, 1992.

López-Farré, A., Gómez-Garre, D., Bernabeu, F., and López-Novoa, J., A role for ET in the maintenance of post-ischemic renal failure in the rat, *J. Physiol.*, 444, 513, 1991b.

López-Farré, A., Gómez-Garre, D., Bernabeu, F., Montañés, I., Millás, I., and López-Novoa, J., Renal effects and mesangial cell contraction induced by ET are mediated by PAF, *Kidney Int.*, 39, 624, 1991a.

Loutzenhiser, R., Epstein, M., Hayashi, K., and Horton, C., Direct visualization of effects of ET on the renal microvasculature, *Am. J. Physiol.*, 258, F61, 1990.

Maccumber, M. W., Ross, C. A., Glaser, B. M., and Synder, S. H., Endothelin: visualization of mRNAs by *in situ* hybridization provides evidence for local action, *Proc. Natl. Acad. Sci. U.S.A.*, 86, 7285, 1989.

Macnulty, E. E., Plevin, R., and Wakelam, M. J., Stimulation of the hydrolysis of phosphatidylinositol 4,5-bisphosphate and phosphatidylcholine by endothelin, a complete mitogen for Rat-1 fibroblasts, *Biochem. J.*, 272, 761, 1990.

Margulies, K. B., Hildebrand, F. L., Heublein, D. M., and Burnett, J. C., Jr., Radiocontrast increases plasma and urinary ET, *J. Am. Soc. Nephrol.*, 2, 1041, 1991.

Marsden, P. A., Dorfman, D. M., Collins, T., Brenner, B. M., Orkin, S. H., and Ballermann, B. J., Regulated expression of ET 1 in glomerular capillary endothelial cells, *Am. J. Physiol.*, 261, F117, 1991.

Marsen, T. A., Schramek, H., and Dunn, M. J., Renal actions of ET: linking cellular signaling pathways to kidney disease, *Kidney Int.*, 45, 336, 1994.

Martin, E. R., Brenner, B. M., and Ballermann, B. J., Heterogeneity of cell surface ET receptors, *J. Biol. Chem.*, 265, 14044, 1990.

Masaki, T., Yanagisawa, M., Takuwa, Y., Kasuya, Y., Kimura, S., and Goto, K., Cellular mechanism of vasoconstriction induced by endothelin, *Adv. Second Messenger Phospho. Res.*, 24, 425, 1990.

Matsumura, Y., Nakase, K., Ikegawa, R., Hayashi, K., Ohyama, T., and Morimoto, S., The endothelium-derived vasoconstrictor peptide endothelin inhibits renin release *in vitro*, *Life Sci.*, 44, 149, 1989.

Mené, P., Simonson, M. S., and Dunn, M. J., Physiology of the mesangial cell, *Physiol. Rev.*, 69, 1347, 1989.

Miller, W. L., Redfield, M. M., and Burnett, J. C., Integrated cardiac, renal and endocrine actions of endothelin, *J. Clin. Invest.*, 83, 317, 1989.

Mino, N., Kobayashi, M., Nakajima, A., Amano, H., Shimamoto, K., Ishikawa, K., Watanabe, K., Nishikebe, M., Yano, M., and Ikemoto, F., Protective effect of a selective ET receptor antagonist, BQ123, in ischemic acute renal failure in rats, *Eur. J. Pharmacol.*, 221, 77, 1992.

Miura, K., Yukimura, T., Yamashita, Y., Shichino, K., Shimmen, T., Saito, M., Okumura, M., Imanishi, M., Yamanaka, S., and Yamamoto, K., Effects of endothelin on renal hemodynamics and renal function in anesthetized dogs, *Am. J. Hypertens.*, 3, 632, 1990.

Miura, K., Yukimura, T., Yamashita, Y., Shimmen, T., Okumura, M., Yamanaka, S., Imanishi, M., and Yamamoto, K., Renal and femoral vascular responses to endothelin-1 in dogs: role of prostaglandins, *J. Pharmacol. Exp. Ther.*, 256, 11, 1991.

Moe, O., Tejedor, A., Campbell, W. B., Alpern, R. J., and Henrich, W. L., Effects of ET on *in vitro* renin secretion, *Am. J. Physiol.*, 260, E521, 1991.

Moore, K., Wendon, J., Frazer, M., Karani, J., Williams, R., and Badr, K., Plasma ET immunoreactivity in liver disease and the hepatorenal syndrome, *N. Engl. J. Med.*, 327, 1774, 1992.

Morise, Z., Ueda, M., Aiura, K., Endo, M., and Kitajima, M., Pathophysiologic role of ET-1 in renal function in rats with endotoxin shock, *Surgery*, 115, 199, 1994.

Moutabarrik, A., Ishibashi, M., Fukunaga, M., Kameoka, H., Takano, Y., Kokado, Y., Takahar, S., Jiang, H., Sonoda, T., and Okuyama, A., FK506 mechanism of nephrotoxicity: stimulatory effect on ET secretion by cultured kidney cells, *Transplantation Proc.*, 23, 3133, 1991.

Nakahama, H., Stimulatory effect of cyclosporine A on ET secretion by a cultured renal epithelial cell line, LLC-PK1 cells, *Eur. J. Pharmacol.*, 180, 191, 1990.

Nakamura, T., Ebihara, I., Fukui, M., Osada, S., Tomino, Y., Masaki, T., Goto, K., Furuichi, Y., and Koide, H., Increased ET and ET receptor mRNA expression in polycystic kidneys of cpk mice, *J. Am. Soc. Nephrol.*, 4, 1064, 1993.

Nambi, P., Pullen, M., Contino, L. C., and Brooks, D. P., Upregulation of renal ET receptors in rats with cyclosporine A-induced nephrotoxicity, *Eur. J. Pharmacol.*, 187, 113, 1990a.

Nambi, P., Pullen, M., Jugus, M., and Gellai, M., Rat kidney ET receptors in ischemia-induced acute renal failure, *J. Pharmacol. Exp. Ther.*, 264, 345, 1993.

Nambi, P., Pullen, M., Wu, H.-L., Aiyar, N., Ohlstein, E. H., and Edwards, R.M., Identification of ET receptor subtypes in human renal cortex and medulla using subtype-selective ligands, *Endocrinology*, 131, 1081, 1992a.

Nambi, P., Wu, H-L., Pullen, M., Aiyar, N., Bryan, H., and Elliott, J., Identification of ET receptor subtypes in rat kidney cortex using subtype-selective ligands, *Mol. Pharmacol.*, 42, 336, 1992b.

Nambi, P., Wu, H.-L., Woessner, R. D., and Mattern, M. R., Inhibition of endothelin-mediated topoisomerase I activation by pertussis toxin, *FEBS Lett.*, 276, 17, 1990b.

Nayler, W. G., Gu, X. H., Casley, D. J., Panagiotopoulos, S., Liu, J., and Mottram, P. L., Cyclosporine increases ET-1 binding site density in cardiac cell membranes, *Biochem. Biophys. Res. Commun.*, 163, 1270, 1989.

Neuser, D., Knorr, A., Stasch, J. P., and Kazda, S., Mitogenic activity of endothelin-1 and -3 on vascular smooth muscle cells is inhibited by atrial natriuretic peptides, *Artery*, 17, 311, 1990.

Nir, A., Clavell, A. L., Heublein, D., Aarhus, L. L., and Burnett, J. C., Jr., Acute hypoxia and endogenous renal ET, *J. Am. Soc. Nephrol.*, 4, 1920, 1994.

Nitta, K., Naruse, M., Sanaka, T., Tsuchiya, K., Naruse, K., Zeng, Z., Demura, H., and Sugino, N., Natriuretic and diuretic effects of endothelin in isolated perfused rat kidney, *Endocrinol. Jpn.*, 36, 887, 1990.

Nunez, D. J., Brown, M. J., Davenport, A. P., Neylon, C. B., Schofield, J. P., and Wyse, R. K., Endothelin-1 mRNA is widely expressed in porcine and human tissues, *J. Clin. Invest.*, 85, 1537, 1990.

Nunez, D. J., Taylor, E. A., Oh, V. M., Schofield, J. P., and Brown, M. J., Endothelin-1 mRNA expression in the rat kidney, *Biochem. J.*, 275, 817, 1991.

Ohlstein, E. H., Horohonich, S., and Hay, D. W., Cellular mechanisms of endothelin in rabbit aorta, *J. Pharmacol. Exp. Ther.*, 250, 548, 1989.

Ohta, K., Hirata, Y., Imai, T., Kanno, K., Emori, T., Shichiri, M., and Marumo, F., Cytokine-induced release of ET-1 from porcine renal epithelial cell line, *Biochem. Biophys. Res. Commun.*, 169, 578, 1990.

Ohta, K., Hirata, Y., Shichiri, M., Ichioka, M., Kubota, T., and Marumo, F., Cisplatin-induced urinary ET excretion, *J. Am. Med. Assoc.*, 265, 1391, 1991.

Orisio, S., Benigni, A., Bruzzi, I., Corna, D., Perico, N., Zoja, C., Benatti, L., and Remuzzi, G., Renal ET gene expression is increased in remnant kidney and correlates with disease progression, *Kidney Int.*, 43, 354, 1993.

Ota, K., Kimura, T., Shoji, M., Inoue, M., Sato, K., Ohta, M., Yamamoto, T., Tsunoda, K., Abe, K., and Yoshinaga, K., Interaction of ANP with ET on cardiovascular, renal and endocrine function, *Am. J. Physiol.*, 262, E135, 1992.

Perico, N., Cornejo, R. P., Benigni, A., Malanchini, B., Ladny, J. R., and Remuzzi, G., Endothelin induces diuresis and natriuresis in the rat by acting on proximal tubular cells through a mechanism mediated by lipoxygenase products, *J. Am. Soc. Nephrol.*, 2, 57, 1991.

Perico, N., Dadan, J., and Remuzzi, G., Endothelin mediates the renal vasoconstriction induced by cyclosporine in the rat, *J. Am. Soc. Nephrol.*, 1, 76, 1990a.

Perico, N., Dadan, J., Gabanelli, M., and Remuzzi, G., Cyclooxygenase products and atrial natriuretic peptide modulate renal response to endothelin, *J. Pharmacol. Exp. Ther.*, 252, 1213, 1990b.

Pollock, D. M. and Opgenorth, T. J., Evidence for ET-induced renal vasoconstriction independent of ET_A receptor activation, *Am. J. Physiol.*, 264, R222, 1993.

Predel, H. G., Meyer-Lehnert, Backer, A., Stelkens, H., and Kramer, H. J., Plasma concentrations of endothelin in patients with abnormal vascular reactivity. Effects of ergometric exercise and acute saline loading, *Life Sci.*, 47, 1837, 1990.

Rakugi, H., Nakamaru, M., Saito, H., Higaki, J., and Ogihara, T., Endothelin inhibits renin release from isolated rat glomeruli, *Biochem. Biophys. Res. Commun.*, 155, 1244, 1988.

Reiser, G., Endothelin and a Ca^{2+} ionophore raise cyclic GMP levels in a neuronal cell line via formation of nitric oxide, *Br. J. Pharmacol.*, 101, 722, 1990.

Resink, T. J., Scott-Burden, T., and Buhler, F. R., Activation of multiple signal transduction pathways by endothelin in cultured human vascular smooth muscle cells, *Eur. J. Biochem.*, 189, 415, 1990.

Reynolds, E. E. and Mok, L. L., Role of thromboxane A_2/prostaglandin H_2 receptor in the vasoconstrictor response of rat aorta to endothelin, *J. Pharmacol. Exp. Ther.*, 252, 915, 1990.

Ritz, M. F., Stuenkel, E. L., Dayanithi, G., Jones, R., and Nordmann, J. J., Endothelin regulation of neuropeptide release from nerve endings of the posterior pituitary, *Proc. Natl. Acad. Sci. U.S.A.*, 89, 8371, 1992.

Roubert, P., Gillard-Roubert, V., Pourmarin, L., Cornet, S., Guilmard, C., Plas, P., Pirotzky, E., Chabrier, P. E., and Braquet, P., Endothelin receptor subtypes A and B are up-regulated in an experimental model of acute renal failure, *Mol. Pharmacol.*, 45, 182, 1993.

Rubanyi, G. M. and Parker Botelho, L. H., Endothelins, *FASEB J.*, 5, 2713, 1991.

Saito, Y., Kazuwa, N., Shirakami, G., Mukoyama, M., Arai, H., Hosoda, K., Suga, S., Ogawa, Y., and Imura, H., Endothelin in patients with chronic renal failure, *J. Cardiovasc. Pharmacol.*, 17(Suppl. 7), S437, 1991.

Sakamoto, H., Sasaki, S., Hirata, Y., Imai, T., Ando, K., Ida, T., Sakurai, T., Yanagisawa, M., Masaki, T., and Marumo, F., Production of endothelin-1 by rat cultured mesangial cells, *Biochem. Biophys. Res. Commun.*, 169, 462, 1990.

Sakurai, T., Yanagisawa, M., Inoue, A., Ryan, U.S., Kimura, S., Mitsui, Y., Goto, K., and Masaki, T., cDNA cloning, sequence analysis and tissue distribution of rat preproendothelin-1 mRNA, *Biochem. Biophys. Res. Commun.*, 175, 44, 1991.

Sakurai, T., Yanagisawa, M., Takuma, Y., Miyazaki, H., Kimura, S., Goto, K., and Masaki, T., Cloning of a cDNA encoding a nonisopeptide-selective subtype of an endothelin receptor, *Nature*, 348, 732, 1990.

Sandok, E. K., Lerman, A., Stingo, A. J., Perrella, M. A., Gloviczki, P., and Burnett, J. C., Endothelin in a model of acute ischemic renal dysfunction: modulating action of atrial natriuretic factor, *J. Am. Soc. Nephrol.*, 3, 196, 1992.

Schramek, H., Gstraunthaier, G., Willinger, C. C., and Pfaller, W., Hyperosmolality regulates ET release by Madin-Darby canine kidney cells, *J. Am. Soc. Nephrol.*, 4, 206, 1993.

Sherwood, T. and Lavender, J. P., Does renal blood flow rise or fall in response to diatrizoate?, *Invest. Radiol.*, 4, 327, 1969.

Shibouta, Y., Suzuki, N., Shino, A., Matsumoto, H., Terashita, Z.-I., Kondo, K., and Nishikawa, K., Pathophysiological role of endothelin in acute renal failure, *Life Sci.*, 46, 1611, 1990.

Shichiri, M., Hirata, Y., Emori, T., Ohta, K., Nakajima, T., Sato, K., Sato, A., and Marumo, F., Secretion of endothelin and related peptides from renal epithelial cell lines, *FEBS Lett.*, 253, 203, 1989.

Simonson, M. S. and Dunn, M. J., Cellular signaling by peptides of the endothelin gene family, *FASEB J.*, 4, 2989, 1990.

Simonson, M. S. and Dunn, M. J., Endothelin peptides: a possible role in glomerular inflammation, *Lab. Invest.*, 64, 1, 1991a.

Simonson, M. S. and Dunn, M. J., Endothelin-1 stimulates contraction of rat glomerular mesangial cells and potentiates β-adrenergic-mediated cyclic adenosine monophosphate accumulation, *J. Clin. Invest.*, 85, 790, 1990b.

Simonson, M. S., Endothelins: multifunctional renal peptides, *Physiol. Rev.*, 73, 375, 1993.

Simonson, M. S., Jones, J. M., and Dunn, M. J., Cytosolic and nuclear signaling by endothelin peptides: implications for the mesangial response to glomerular injury, *Kidney Int.*, 41, 542, 1992a.

Simonson, M. S., Jones, J. M., and Dunn, M. J., Differential regulation of *fos* and *jun* gene expression and AP-1 *cis*-element activity by ET isopeptides. Possible implications for mitogenic signaling by ET, *J. Biol. Chem.*, 267, 8643, 1992b.

Simonson, M. S., Wang, Y., and Dunn, M. J., Cellular signaling by endothelin peptides: pathways to the nucleus, *J. Am. Soc. Nephrol.*, 2, S116, 1992c.

Simonson, M. S., Wann, S., Mené, P., Dubyak, G. R., Kester, M., Nakazato, Y., Sedor, J. R., and Dunn, M. J., Endothelin stimulates phospholipase C, Na^+/H^+ exchange, *c-fos* expression, and mitogenesis in rat mesangial cells, *J. Clin. Invest.*, 83, 708, 1989.

Simonson, M. S., Wolfe, J. A., Konieczkowski, M., Sedor, J. R., and Dunn, M. J., Regulation of prostaglandin endoperoxidase synthase gene expression in cultured rat mesangial cells: induction by serum via a protein kinase C-dependent mechanism, *Mol. Endocrinol.*, 5, 441, 1991b.

Sokolovsky, M., Endothelins and sarafotoxins: physiological regulation, receptor subtypes and transmembrane signaling, *Pharm. Ther.*, 54, 129, 1992.

Stacy, D. L., Scott, J. W., and Granger, J. P., Control of renal function during intrarenal infusion of endothelin, *Am. J. Physiol.*, 258, F1232, 1990.

Stier, C. T., Jr., Quilley, C. P., and McGiff, J. C., Endothelin-3 effects on renal function and prostanoid release in the rat isolated kidney, *J. Pharmacol. Exp. Ther.*, 262, 252, 1992.

Stingo, A. J., Clavell, A. L., Aarhus, L. L., and Burnett, J. C., Jr., Biological role for the ET-A receptor in aortic cross-clamping, *Hypertension*, 22, 62, 1993.

Stockenhuber, F., Gottsauner-Wolf, M., Marosi, L., Liebisch, B., Kurz, R. W., and Balcke, P., Plasma levels of ET in chronic renal failure and after renal transplantation: impact on hypertension and cyclosporin A-associated nephrotoxicity, *Clin. Sci.*, 82, 255, 1992.

Sugiura, M., Snadjar, R. M., Schwartzberg, M., Badr, K. F., and Inagami, T., Identification of two types of specific ET receptors in rat mesangial cell, *Biochem. Biophys. Res. Commun.*, 162, 1396, 1989.

Takabatake, T., Ise, T., Ohta, K., and Kobayashi, K.-I., Effects of ET on renal hemodynamics and tubuloglomerular feedback, *Am. J. Physiol.*, 263, F103, 1992.

Takagi, M., Matsuoka, H., Atarash K., and Yagi, S., Endothelin: a new inhibitor of renin release, *Biochem. Biophys. Res. Commun.*, 157, 1164, 1988.

Takagi, N., Fukase, M., Takata, S., Yoshimi, H., Tokunaga, O., and Fujita, T., Autocrine effect of endothelin on DNA synthesis in human vascular endothelial cells, *Biochem. Biophys. Res. Commun.*, 168, 537, 1990.

Takagi, Y., Tsudada, H., Matsuoka, H., and Yagi, S., Inhibitory effect of endothelin on renin release *in vitro*, *Am. J. Physiol.*, 257, E833, 1989.

Takahashi, K., Totsune, K., Imai, Y., Sone, M., Nozuki, M., Murakami, O., Sekino, H., and Mouri, T., Plasma concentrations of immunoreactive-ET in patients with chronic renal failure treated with recombinant human erythropoietin, *Clin. Sci.*, 84, 47, 1993.

Takeda, M., Breyer, M. D., Noland, T. D., Homma, T., Hoover, R. L., Inagami, T., and Kon, V., Endothelin-1 receptor antagonist: effects on ET- and cyclosporine-treated mesangial cells, *Kidney Int.*, 42, 1713, 1992.

Takeda, M., Iwasaki, S., Hellings, S. E., Yoshida, H., Homma, T., and Kon, V., Divergent expression of ET_A and ET_B receptors in response to cyclosporine in mesangial cells, *Am. J. Pathol.*, 144, 473, 1994.

Takemoto, F., Uchida, S., Ogata, E., and Kurokawa, K., Endothelin-1 and ET-3 binding to rat nephrons, *Am. J. Physiol.*, 264, F827, 1993.

Thomas, C.P., Simonson, M. S., and Dunn, M. J., Endothelin: receptor and transmembrane signals, *NIPS*, 207, 1992.

Tomita, K., Nonoguchi, H., and Marumo, F., Effects of ET on peptide-dependent cyclic adenosine monophosphate accumulation along the nephron segments of the rat, *J. Clin. Invest.*, 85, 2014, 1990.

Tomita, K., Ujiie, K., Nakanishi, T., Tomura, S., Matsuda, O., Ando, K., Shichiri, M., Hirata, Y., and Marumo, F., Plasma ET levels in patients with acute renal failure, *Eur. J. Med.*, 32, 1127, 1989.

Totsune, K., Mouri, T., Takahashi, K., Ohneda, M., Sone, M., Saito, T., and Yoshinaga, K., Detection of immunoreactive endothelin in plasma of hemodialysis patients, *FEBS Lett.*, 249, 239, 1989.

Tsuchiya, K., Naruse, M., Sanaka, T., Naruse, K., Nitta, K., Demura, H., and Sugino, N., Effects of ET on renal regional blood flow in dogs, *Eur. J. Pharmacol.*, 166, 541, 1989.

Tsuchiya, K., Naruse, M., Sanaka, T., Naruse, K., Zeng, Z. P., Nitta, K., Demura, H., Shizume, K., and Sugino, N., Renal and hemodynamic effects of endothelin in anesthetized dogs, *Am. J. Hypertens.*, 3, 792, 1990.

Uchida, K. and Ballermann, B. J., Sustained activation of PGE_2 synthesis in mesangial cells cocultured with glomerular endothelial cells, *Am. J Physiol.*, 263, C200, 1992.

Uchida, S., Takemoto, F., Ogata, E., and Kurokawa, K., Detection of ET-1 mRNA by RT-PCR in isolated rat renal tubules, *Biochem. Biophys. Res. Commun.*, 188, 108, 1992.

Uemasu, J., Matsumoto, H., Kitano, M., and Kawasaki, H., Suppression of plasma ET-1 level by alpha-human atrial natriuretic peptide and angiotensin converting enzyme inhibition in normal men, *Life Sci.*, 53, 969, 1993.

Ujiie, K., Terada, Y., Nonoguchi, H., Shinohara, M., Tomita, K., and Marumo, F., Messenger RNA expression and synthesis of endothelin-1 along rat nephron segments, *J. Clin. Invest.*, 90, 1043, 1992.

Uzuner, K. and Banks, R. O., Endothelin-induced natriuresis and diuresis are pressure-dependent events in the rat, *Am. J. Physiol.*, 265, R90, 1993.

Vane, J. R., Botting, R., and Masaki, T., Endothelin, *J. Cardiovasc. Pharmacol.*, 13, S1, 1989.

Warner, T. D., Battistini, B., Allcock, G. H., and Vane, J. R., Endothelin ET_A and ET_B receptors mediate vasoconstriction and prostanoid release in the isolated kidney of the rat, *Eur. J. Pharmacol.*, 250, 447, 1993a.

Warrens, A. N., Cassidy, M. J., Takahashi, K., Ghatei, M. A., and Bloom, S. R., Endothelin in renal failure, *Nephrol. Dial. Transplant.*, 5, 418, 1990.

Wellings, R. P., Corder, R., Warner, T. D., Cristol, J.-P., Thiemermann, C., and Vane, J. R., Evidence from receptor antagonists of an important role for ET_B receptor-mediated vasoconstrictor effects of ET-1 in the rat kidney, *Br. J. Pharmacol.*, 111, 515, 1994.

Wilkes, B. M., Pearl, A. R., Mento, P. F., Maita, M. E., Macica, C. M., and Girardi, E. P., Glomerular ET receptors during initiation and maintenance of ischemic acute renal failure in rats, *Am. J. Physiol.*, 260, F110, 1991a.

Wilkes, B. M., Ruston, A. S., Mento, P., Girardi, E., Hart, D., Vander Molen, M., Barnett, R., and Nord, E. P., Characterization of ET 1 receptor and signal transduction mechanisms in rat medullary interstitial cells, *Am. J. Physiol.*, 260, F579, 1991b.

Wilkes, B. M., Susin, M., Mento, P. F., Macica, C. M., Girardi, E. P., Boss, E., and Nord, E. P., Localization of ET-like immunoreactivity in rat kidneys, *Am. J. Physiol.*, 260, F913, 1991c.

Yamaguchi, T., Fukase, M., Arao, M., Sugimoto, T., and Chihara, K., Endothelin 1 hydrolysis by rat kidney membranes, *FEBS Lett.*, 309, 303, 1992.

Yamashita, Y., Yukimura, T., Miura, K., Okumura, M., and Yamamoto, K., Effects of ET-3 on renal functions, *J. Pharmacol. Exp. Ther.*, 259, 1256, 1991.

Yanagisawa, M. and Masaki, T., Endothelin, a novel endothelium-derived peptide, *Biochem. Pharmacol.*, 38, 1877, 1989a.

Yanagisawa, M. and Masaki, T., Molecular biology and biochemistry of the endothelins, *Trends Pharmacol. Sci.*, 10, 374, 1989b.

Yanagisawa, M., Kurihara, H., Kimura, S., Tomobe, Y., Kobayashi, M., Mitsui, Y., Yazaki, Y., Goto, K., and Masaki, T., A novel potent vasoconstrictor peptide produced by vascular endothelial cells, *Nature (London)*, 332, 411, 1988.

Yang, T., Terada, Y., Nonoguchi, H., Ujiie, K., Tomita, K., and Marumo, F., Effect of hyperosmolality on production and mRNA expression of ET-1 in inner medullary collecting duct, *Am. J. Physiol.*, 264, F684, 1993.

Yoshimi, H., Hirata, Y., Fukuda, Y., Kawano, Y., Emori, T., Kuramochi, M., Omae, T., and Marumo, F., Regional distribution of immunoreactive ET in rats, *Peptides*, 10, 805, 1989.

Zeidel, M. L., Brady, H. R., Kone, B. C., Gullans, S. R., and Brenner, B. M., Endothelin, a peptide inhibitor of Na^+-K^+-ATPase in intact renal tubular epithelial cells, *Am. J. Physiol.*, 257, C1101, 1989.

Zimmerman, R. S., Martinez, A. J., Macphee, A. A., and Barbee, R. W., Cardio-renal effects of endothelin-3 in the rat, *Life Sci.*, 47, 2323, 1990.

Zoja, C., Orisio, S., Perico, N., Benigni, A., Morigi, M., Benatti, L., Rambaldi, A., and Remuzzi, G., Constitutive expression of ET gene in cultured human mesangial cells and its modulation by transforming growth factor-β, thrombin, and a thromboxane A_2 analog, *Lab. Invest.*, 64, 16, 1991.

Chapter **10**

Endothelin-Induced Responses in the Pulmonary System

Bruno Battistini and Timothy D. Warner

CONTENTS

0-8493-6975-4/97/$0.00+$.50

I. INTRODUCTION

Within months of the publication of the discovery of ET-1 (Yanagisawa et al., 1988) the first paper demonstrating its potent bronchoconstrictor activity followed (Uchida et al., 1988). ET-1 was shown to induce slowly developing and long-lasting contractions of the tracheal smooth muscle of the guinea pig similar to those observed in numerous isolated vascular preparations. Interestingly, even though ET-1 was less potent in airways than in the vasculature, it was still as potent as histamine, neurokinin A, or leukotriene (LT) D_4 at contracting these nonvascular smooth muscles. This observation, and a paper by DeNucci and colleagues (1988) showing that ET-1-induced the release of prostacyclin and thromboxane (Tx) from isolated perfused lungs of the rat or the guinea pig, were the forerunners of a formidable number of publications suggesting important roles for the ET isopeptides (ET-1, ET-2, and ET-3) as multifunctional pulmonary peptides.

II. PHARMACOKINETICS OF ET IN THE LUNG

A. Localization and Biosynthesis

Northern blot analysis and *in situ* hybridization have shown ET mRNA in the lungs of many species. Within rat lungs, ET mRNA is associated with epithelial cells of bronchioles and with blood vessels (MacCumber et al., 1989). In human lungs, it is associated with endothelial cells and the inner layer of small bronchiolar and alveolar blood vessels (Giaid et al., 1993a-c). Both ET-1 and ET-3 are abundantly expressed in fetal rat lungs (MacCumber et al., 1989) and human lungs (Bloch et al., 1989). Indeed, in both the human (Doherty, 1992) and rat (MacCumber et al., 1989) fetus, the greatest amount of ET-1 is found within the lungs. Similarly, immunoreactive-ET (ir-ET) is present in adult porcine (Kitamura et al., 1989; Hemsen et al., 1991), rat (Kitamura et al., 1990a), guinea pig (Franco-Cereceda et al., 1990), and human (Hemsen et al., 1990) lungs and ir-big ET-1-(1-39) is also abundant in porcine lungs (Kitamura et al., 1990b). In addition, ET-1 and ET-3 are also expressed in human lung macrophages (Ehreinreich et al., 1990), which may be of interest as ET-1 is a potent stimulator of superoxide production in these cells *in vitro* (Haller et al., 1991).

As might be expected, ET-1 is released by cultured endothelial cells isolated from bovine pulmonary arteries (Naruse et al., 1989; Ohlstein et al., 1990) and this release is increased by thrombin (Ohlstein et al., 1990). This may be of relevance, for infusion of thrombin to guinea pig perfused lungs stimulates the release of ET that subsequently causes an elevation in pulmonary perfusion pressure (PPP) (Moon et al., 1989). Possibly more interesting, ET-1 is also released by canine, porcine (Black et al.,

1989), and guinea pig (Ninomiya et al., 1991) isolated tracheal epithelial cells as well as human bronchial (Mattoli et al., 1990) and rat pleural mesothelial cells in culture (Kuwahara et al., 1992), suggesting that ET-1 may be a paracrine hormone in the airways. Interestingly, the release of ir-ET from guinea pig isolated tracheal epithelial cells (Ninomyia et al., 1991), from human isolated bronchial epithelial cells (Nakano et al., 1994), and from human isolated macrophages (Ehreinreich et al., 1990) is stimulated by endotoxin. The expression and release of ET-1 from human isolated bronchial epithelial cells is also stimulated by proinflammatory cytokines such as interleukin-1 or tumor necrosis factor (Nakano et al., 1994).

The conversion of big ET-1 to mature ET-1, which is essential for the full expression of its biological activity (Kimura et al., 1989), is dependent on a putative ET-converting enzyme (ECE). Most evidence strongly suggests that the physiologically relevant ECE is a phosphoramidon-sensitive metalloendopeptidase (Opgenorth et al., 1992) and, interestingly, big ET-1 is converted to ET-1 by a phosphoramidon-sensitive ECE in the rat lung (Wypij et al., 1992). Similarly, a phosphoramidon-sensitive ECE has been demonstrated in bovine (Kundu and Wilson, 1992) and porcine (Sawamura et al., 1993) lung membranes, in the pulmonary vascular bed of the guinea pig (D'Orléans-Juste et al., 1991), the rabbit (Ishikawa et al., 1992), the newborn piglet (Liben et al., 1993) and the rat (Hisaki et al., 1994), and in isolated bronchi of the guinea pig (Noguchi et al., 1991). Interestingly, the ECE activity is specific for big ET-1 and does not convert big ET-3 to ET-3 (D'Orléans-Juste et al., 1991). However, the lung contains numerous other enzymes that have been found to convert big ET-1 to ET-1. These include, in the rat lung, a mast cell chymase which can be inhibited by chymostatin (Wypij et al., 1992) and a metal ion-related aspartic protease which acts at pH 4.0 and which can be inhibited by pepstatin A (Wu-Wong et al., 1990, 1991; Shiosaki et al., 1993).

B. Metabolism and Removal From the Circulation

ET-1 is removed by 60% in a single passage when infused into the isolated perfused lungs of the guinea pig or rat (Antunes et al., 1988; DeNucci et al., 1988; Anggard et al., 1989) and is rapidly taken up into the lung when injected intravenously to the rat (Anggard et al., 1989). Similarly, there is a single-passage extraction of 31% in the canine lungs (Dupuis et al., 1994) and of 49% in the rabbit lungs (Rimar and Gillis, 1992). Conversely, it has been suggested that ET-1 does not undergo significant first-pass pulmonary metabolism in the cat (Lippton et al., 1989) or in human (Ray et al., 1992), even though there may be some concern regarding the methodology used in the human study. Thus, there may be species differences in the metabolism and removal of ET-1 within the lung; but we have to keep in mind that the lungs are an important site for not only the uptake, but also the release of ET-1.

III. ET RECEPTORS IN THE PULMONARY SYSTEM

A. Receptor Localization by Immunostaining and Autoradiography

ET receptors have been localized and characterized throughout the airways and pulmonary vasculature (see Table 1). Initial immunohistochemical studies showed the presence of ET receptors in bovine pulmonary artery endothelial cells (Naruse et al., 1989) and on mucous, serous, and Clara cells and in a few type II alveolar pneumocytes, although not on most basal and ciliated cells in rat and mouse bronchoalveolar epithelium (Rozengurt et al., 1990). Similarly, autoradiography has demonstrated ET binding

TABLE 1

Localization of High Affinity ET-Binding Sites in the Airways of Various Species

Sites/Species	Mouse	Rat	Guinea Pig	Bovine	Human
Airway SM	++	++	++++		++
Alveolar fibroblasts					
Alveolar wall/septa		++	++		++++
Basal cells	–				
Cartilage		–			±
Ciliated cells	–				
Clara cells	+				
Connective tissue		–			±
Endothelium				++++	++++
Epithelium	++	++	++		±
Intrapulm. blood vessels					++++
Mucous cells	++	++			
Nerve trunks		++	–		++
Serous cells	++	++			
Submucosal layer			++		±
Type I/II pneumocytes	++				
Vascular SM					++++

Note: Empty spaces = not determined.

sites throughout pulmonary tissues, with particularly high density in human (McKay et al., 1991b) and guinea pig (Cardell et al., 1991) pulmonary artery endothelium, in human airways (Power et al., 1989; Neuser et al., 1989; 1991; Marciniak et al., 1992), and rat trachea (Power et al., 1989) smooth muscle, in human pulmonary nerve trunks, and in human alveolar septa (Power et al., 1989). Autoradiographical studies have also shown high affinity binding sites for ET-1 in the smooth muscle, submucosa, and epithelium of guinea pig trachea (Tschirhart et al., 1991).

B. Competition Binding Studies

Binding studies in pulmonary tissues from a wide range of species have clearly confirmed the presence of at least one high-affinity ET receptor in the airways (see Table 2).

C. Purification of ET Receptors in Lungs

Cross-linking experiments have suggested proteins of different molecular weights to be ET receptors in the lung. For instance, in rat lung, a 44-kDa protein has been reported to have a higher affinity for ET-1 than ET-3, whereas another 32-kDa protein preferentially interacted with ET-3 (Masuda et al., 1989). In addition, in bovine lung two proteins of 34 and 52 kDa (Hagiwara et al., 1990; Kozuka et al., 1991), in rat and bovine lungs a 34-kDa protein (Kundu and Misono, 1991), and in human bronchial smooth muscle cells, a 70-kDa protein (Mattoli et al., 1991b) have been suggested as being ET receptors. The exact nature of such proteins, however,

* Abbreviations used in the tables in this chapter: SM: smooth muscle; SMC: SM cell; intrapulm.: intrapulmonary; concn: concentration or dose; EpC: epithelial cells; 5-HT: serotonin; Ach: acetylcholine; NKA: neurokinin-A; PVR: total pulmonary vascular resistance; Ppc: pulmonary capillary pressure; PIP: pulmonary inflation pressure; pulm.: pulmonary; aer.: aerosol; Em: maximal effect; BAL: bronchial alveolar lavage. For other abbreviations, refer to the text.

TABLE 2

Equilibrium Dissociation Constant and Maximal Density of Binding Sites for [^{125}I]ET-1 in Isolated Membranes From Pulmonary Tissues

Tissues	Species	K_d (nM)	B_{max}	Ref.
Alveolar septa	Rat	0.20	400 amol/mm^2	Power et al., 1989
Bronchial SMC	Human	0.113	22.1 fmol/10^6 cells	Mattoli et al., 1990
Lung alveolar cells	Rat	—	9.2 amol/mm^2	Davenport et al., 1989
Lung parenchyma	Bovine	0.15	2.7 pmol/mg	Kundu and Misono, 1991
	Guinea pig	0.051	2.31 pmol/mg	Pons et al., 1991a
		0.27	14.9 pmol/mg	Bolger et al., 1990
	Human	0.153	6.06 fmol/mg	Hemsen et al., 1990
	Pig	0.163	210 fmol/mg	Hemsen et al., 1991
	Rat	0.16	1.65 pmol/mg	Kanse et al., 1989
		0.22	6.1 pmol/mg	Kundu and Misono, 1991
		0.32	10 pmol/mg	Bolger et al., 1990
Pulmonary artery	Rat	—	2.5 amol/mm^2	Davenport et al., 1989
Tracheal EpC	Cat	0.03/0.21	15/35 fmol/10^7 cells	Wu et al., 1993
	Guinea pig	1.41	44 fmol/mg	Tschirhart et al., 1991
Tracheal SMC	Guinea pig	0.11	8.3 amol/mm^2	Henry et al., 1990
		2.76	279 fmol/mg	Tschirhart et al., 1991
	Human	—	42.7 amol/mm^2	Henry et al., 1990
	Lamb	0.40	104 fmol/10^6 cells	Glassberg et al., 1994
	Mouse	—	28.7 amol/mm^2	Henry et al., 1990
	Rat	0.12	210 amol/mm^2	Power et al., 1989
		0.13	1.2 fmol/mm^2	Turner et al., 1989b
		0.43	69 amol/mm^2	Henry et al., 1990
		0.59	23.4 pmol/mg	Bolger et al., 1990
		—	23 pmol/mg	Hemsen et al., 1990
Tracheal submucosa	Guinea pig	0.88	693 fmol/mg	Tschirhart et al., 1991

is unclear as both ET_A- and ET_B-receptors have been cloned and found to have molecular weights of around 50 kDa.

D. Cloning of ET Receptors in Lungs

It is worthy of note that the initial clonings of both ET_A- (427 residues) and ET_B- (441 residues) receptor subtypes, which were published simultaneously by two groups, were achieved using bovine and rat lungs, respectively (Arai et al., 1990; Sakurai et al., 1990). Since then, the ET_A-receptor subtype has also been cloned from rat (426 residues; Hori et al., 1992) and human (427 residues; Haendler et al., 1992) lungs while the ET_B-receptor subtype has been cloned again from rat (442 residues; Hori et al., 1992), bovine (441 residues; Saito et al., 1991), and human (442 residues; Haendler et al., 1992) lungs.

IV. BRONCHOPULMONARY RESPONSES INDUCED BY ETs

A. Effects of ETs *In Vitro*

1. *Isolated Tissues*

a. *Nonvascular Preparations*

The ET/STX peptides are potent contractors of smooth muscle taken from the airways of the mouse, rat, rabbit, guinea pig, ferret, and human (see Table 3). Furthermore, ET-1 is more potent, on a molar basis, in producing these effects than many other agonists, including 5-HT, histamine, and leukotriene (LT) D_4 (see Table 4).

TABLE 3

Contractions Induced by ET/STX Peptides in Isolated Airways

Tissues	Species	ET/STX	Conc. (nM)	Ref.
Bronchus	Ferret	ET-1	0.1–100	Lee et al., 1990
	Guinea pig	ET-1	1–100	Battistini et al., 1990a, b
		ET-1	1–100	Filep et al., 1990; 1991a
		ET-1, big $ET-1_{1-38}$	10–100	Noguchi et al., 1991
		ET-1, ET-2, ET-3, STX-c	0.01–300	Maggi et al., 1989a, b
		ET-1, ET-3, $ET-1_{16-21}$	0.1–100	Franco-Cereceda et al., 1990
		ET-1, ET-3, STX-b	0.01–300	Maggi et al., 1990
		ET-1, IRL 1620	1–300	Battistini et al., 1994a, b
		$ET-1_{16-21}$	1–30 000	Maggi et al., 1989a, b
		$ET-1_{16-21}$	10–10000	Rovero et al., 1990
	Human	ET-1, ET-2	0.001–300	McKay et al., 1991a, b
		ET-1, ET-2, ET-3	0.001–10	Hemsen et al., 1990
		ET-1, ET-2, ET-3	0.001–300	Advenier et al., 1990
Parenchyma	Guinea pig	ET-1	1–100	Filep et al., 1990; 1991a
		ET-1, IRL 1620	1–300	Battistini et al., 1994a, b
Trachea	Dog	ET-1	30	Sakata et al., 1989
	Ferret	ET-1	0.1–30	Lee et al., 1990
		ET-1	0.1–100	Yurdakos and Webber, 1991
	Guinea pig	ET-1	0.001–30	Uchida et al., 1988
		ET-1	0.001–32	Eglen et al., 1989
		ET-1	0.001–100	Sarria et al., 1990
		ET-1	0.1–30	Nagai et al., 1991
		ET-1	0.1–100	Cardell et al., 1990
		ET-1	0.1–100	O'Donnell et al., 1991
		ET-1	1–30	Schumacher et al., 1990
		ET-1	1–30	Schumacher et al., 1990
		ET-1	1–100	Battistini et al., 1990a, b
		ET-1	1–100	Filep et al., 1990; 1991a
		ET-1	1–100	Hay, 1989
		ET-1	1–300	Battistini et al., 1993b
		ET-1	1–300	Germain et al., 1993
		ET-1	1–300	Maggi et al., 1989a, b
		ET-1	10–300	Henry et al., 1990
		ET-1, ET-2, ET-3, STX-b	0.1–100	Tschirhart et al., 1991
		ET-1, IRL 1620	1–300	Battistini et al., 1994a, b
		$ET-1_{16-21}$	0.1–100	Hay, 1992
		$ET-1_{16-21}$	1–300	Maggi et al., 1989a, b
		$ET-1_{16-21}$, $ET-1_{16-21}-NH_2$	0.1–100	Tschirhart et al., 1991
	Human	ET-1	3–300	Henry et al., 1990
	Mouse	ET-1	1–300	Henry et al., 1990
Trachea	Rabbit	ET-1	0.1–30	Secrest and Cohen, 1989
		ET-1	0.1–1000	Grunstein et al., 1991
	Rat	ET-1	0.01–100	Borges et al., 1989
		ET-1	0.1–100	Secrest and Cohen, 1989
		ET-1	3–300	Henry et al., 1990
		ET-1	10–10,000	Turner et al., 1989a
		ET-1	10–1000	Chand et al., 1990

Contractions induced by ET-1 are slow developing (10 to 20 min), stable, and only slowly reversible even with repeated washing. In addition, in the guinea pig isolated trachea, they are not reversed by either $NiCl_2$ or ω-conotoxin, suggesting neither T- or N-type voltage-dependent Ca^{2+} channels are involved (Maggi et al., 1989b). In contrast, established contractions induced by ET-1 in the rat (Borges et al., 1989) or guinea pig (Maggi et al., 1989b) isolated trachea are reversed by nifedipine.

TABLE 4

Comparative Affinities of Several Agonists in Isolated Airway Preparations

Potency	Species	Tissues	Ref.
ET-1 ≫ 5-HT	Rat	Trachea	Chand et al., 1990
ET-1 ≫ ACh	Human	Bronchus	Advenier et al., 1990
	Rat	Trachea	Chand et al., 1990
ET-1 ≫ $CaCl_2$	Rat	Trachea	Chand et al., 1990
ET-1 ≫ Histamine	Guinea pig	Trachea, bronchus	Battistini et al., 1990a
	Guinea pig	Trachea, bronchus, parenchyma	Filep et al., 1990
	Human	Bronchus	Uchida et al., 1988
	Human	Bronchus	Uchida et al., 1988
	Human	Bronchus	Advenier et al., 1990
ET-1 ≫ KCl	Rat	Trachea	Chand et al., 1990
ET-1 ≫ LTD_4	Ferret	Trachea	Lee et al., 1990
	Guinea pig	Trachea	Uchida et al., 1988
	Human	Bronchus	Advenier et al., 1990
ET-1 ≫ NKA	Guinea pig	Trachea	Uchida et al., 1988
	Human	Bronchus	Uchida et al., 1988
ET-1 ≫ $PGF_{2\alpha}$	Guinea pig	Pulm. artery	O'Donnell et al., 1991
	Guinea pig	Trachea	O'Donnell et al., 1991
NKA ≫ ET-1	Guinea pig	Bronchus	Franco-Cereceda et al., 1990
	Human	Bronchus	Advenier et al., 1990
	Human	Bronchus	Hemsen et al., 1990

Dexamethasone (Filep et al., 1993b), isoprenaline, or EDTA (Maggi et al., 1989b) also reverse contractions induced by ET-1 in the guinea pig isolated trachea.

Interestingly, ET-1 can also relax rabbit (Grunstein et al., 1991) and guinea pig (Filep et al., 1993a; Saotome et al., 1991; Uchida et al., 1991) isolated trachea precontracted with acetylcholine (Grunstein et al., 1991), carbachol (Filep et al., 1993a) or following an anaphylactic-induced contraction (Saotome et al., 1991; Uchida et al., 1991). In the rabbit trachea, the relaxation is diminished by removal of the epithelium or incubation with indomethacin (Grunstein et al., 1991). In the guinea pig trachea, the relaxation is abolished by removal of the epithelium, by addition of an inhibitor of NO synthase such as L-NMMA, addition of oxyhemoglobin, or application of methylene blue (Filep et al., 1993a). In the guinea pig sensitized trachea, the relaxation is not diminished by indomethacin, FPL 55,712, or mechanical denudation of the tracheal epithelium, but is decreased by nordihyroguiaiaretic acid (NDGA), a lipoxygenase inhibitor, and AA 861 (Saotome et al., 1991; Uchida et al., 1991). Thus, airway responsiveness to ET-1 may be modulated by the release of epithelium-derived relaxing factors (e.g., NO, PGE_2, PGI_2, and/or hydroperoxides of arachidonic acid) from the epithelium and/or other sites (Grunstein et al., 1991; Saotome et al., 1991). The net effect of these relaxing and contracting activities in preparations under passive stretch can be a biphasic response to ET-1, characterized as an initial relaxation followed by a sustained contraction (Battistini et al., 1994a).

b. Vascular Preparations

ET/STX peptides are also very potent constrictors of isolated portions of the pulmonary vasculature (see Table 5) and, as in other blood vessels, these responses are modulated by the release of NO and prostanoids from the endothelium and vascular smooth muscle. For instance, vasorelaxation induced by ET-1 or ET-3 of porcine isolated pulmonary arteries precontracted with noradrenaline is blocked by N^G-nitro-L-arginine or methylene blue (Namiki et al., 1992). ET-3 also elicited a concentration-dependent relaxation of rat pulmonary arteries, an effect inhibited by

TABLE 5

Contractions Induced by ET/STX Peptides in Isolated Pulmonary Vascular Preparations

Tissues	Species	ET/STX	Conc. (nM)	Ref.
Pulmonary artery	Guinea pig	ET-1	0.1–100	O'Donnell et al., 1991
		ET-1, ET-2, ET-3	0.1–300	Cardell et al., 1991
		ET-1, ET-3, ET-$1_{16\text{-}21}$	1–100	Franco-Cereceda et al., 1990
		ET-1	3–100	Cardell et al., 1990
	Human	ET-1, ET-2	0.001–300	McKay et al., 1991a, b
		ET-1, ET-2, ET-3	1–100	Hemsen et al., 1990
	Lamb	ET-1	≤10	Chatfield et al., 1990
	Rabbit	ET-1, ET-2, ET-3, STX-b	0.1–1000	Maggi et al., 1989a
		ET-$1_{16\text{-}21}$	30–30,000	Maggi et al., 1989a
		STX-b	0.1–100	Maggi et al., 1989a
	Rat	ET-1	0.01–5	Hasunuma et al., 1990
		ET-1	0.1–100	Wanstall and O'Donnell, 1990
		ET-1	0.1–100	Wanstall and O'Donnell, 1991
		Porcine ET	0.01–10	Rodman et al., 1989
		Rat ET	1–1000	Rodman et al., 1989
Pulmonary vein	Guinea pig	ET-1	3–100	Cardell et al., 1990
	Rabbit	ET-1	0.1–300	Steffan and Russell, 1990

L-N^G-monomethyl-L-arginine (Crawley et al., 1992). For their part, prostanoids are also believed to be involved in mediating ET-1-induced vasoconstrictions of isolated veins from the sheep lung (Toga et al., 1992).

2. *Isolated Perfused Lungs*

In isolated perfused lungs of rats, guinea pigs, rabbits, cats, dogs, or lambs, intraarterial (i.a.) administration of the ET/STX peptides causes pulmonary vasoconstriction and so increases the pulmonary perfusion pressure (PPP) (see Table 6). In producing this effect, ET-1 is more potent on a molar basis, for instance, than substance P, α-thrombin, or arachidonic acid in guinea pig isolated perfused lungs (Horgan et al., 1991). Big ET-1 also causes concentration-related increases in perfusion pressure in rat, rabbit, and guinea pig isolated perfused lungs (see Table 6). The i.a. injection of ET-1 also induces an increase in pulmonary inflation pressure (PIP) in guinea pig isolated perfused lungs (Pons et al., 1991c).

In agreement with experiments using isolated preparations, ET-1 may also dilate the intact pulmonary vasculature. For instance, in isolated perfused lungs of newborn pigs, ET-1 produces a rapid decrease in PPP which is reduced by hemoglobin but not by indomethacin or glibenclamide, suggesting that this effect is mediated, at least in part, by NO (Pinheiro et al., 1990; Perreault and DeMarte, 1991). In the isolated blood-perfused lower left lobe of the fetal sheep lung, i.a. administration of ET-1 also causes a dose-related decrease in PPP that is mediated, in part, by NO (Cassin et al., 1991; Tod and Cassin, 1992). Similarly, addition of methylene blue, pyrogallol, or L-NMMA potentiates the increase in PPP induced by i.a. administration of ET-1 in rat isolated perfused lungs (Raffestin et al., 1991). ET-1 (Hasunuma et al., 1990) and ET-3 (Crawley et al., 1992) also induce vasodilatation in hypoxic rat isolated perfused lungs.

The i.a. administration of either ET-1, -2, or -3 increases edema formation in isolated perfused lungs of guinea pigs (Tschirhart et al., 1991; Pons et al., 1991a; Horgan et al., 1991), rats (Raffestin et al., 1991), and dogs (Barnard et al., 1991). This pro-edematous effect is thought to be due to the potent venoconstriction ET-1 produces

TABLE 6

Effects Induced by ETs in Isolated Perfused Lungs

Species	ET	Conc.	Effects	Ref.
Cat	ET-1,-2,-3	1–3 μg	↓PPP	Lippton et al., 1991a, b
Dog	ET-1	10 nM	↑PPP, Ppc	Barnard et al., 1991
Guinea pig	Big ET-3	100 nM	No effect	D'Orléans-Juste et al., 1991
	Big ET-1	100 nM	↑TxB_2, PGI_2	D'Orléans-Juste et al., 1991
	ET-1	0.01–10 nM	↑PVR	Horgan et al., 1991
	ET-1	0.1–10 nM	↑PPP	Spinella et al., 1991
	ET-1	3 μg/ml 2 min	↑PPP	Boichot et al., 1991b
	ET-1	3, 6, 10 μg/ml 2 min	↑PIP	Pons et al., 1991a-c
	ET-1	40, 120, 400 pmol	↑PPP, PIP	Touvay et al., 1990
	ET-1	40, 120, 400 pmol	↑PPP, PIP, Edema, TxB_2	Pons et al., 1991a-c
	ET-1,-3	2.5 nM	↑TxB_2, PGI_2	D'Orléans-Juste et al., 1991
	ET-3	1 nM	↑PPP	Spinella et al., 1991
Lamb	ET-1	50–500 ng/kg	↑PPP	Toga et al., 1991
Lamb fetus	ET-1	100–1000 ng/kg	↓PPP	Cassin et al., 1991
Newborn pig	ET-1	0.1 nM	↓PPP	Pinheiro et al., 1990
	ET-1	0.3–300 M/g dry lung	↑PPP	Perreault and DeMarte, 1991
Rabbit	Big ET-1	1 nmol	↑PPP	Ishikawa et al., 1992
	ET-1	1.2–8 nM	↑PPP	Mann et al., 1991
	ET-1	10 nM	↑PPP, Ppc	Barnard et al., 1991
Rat (hypoxic)	ET-1	0.5, 5 nM	Vasodilations	Hasunuma et al., 1990
Rat	Big ET-1	0.5-50 nM	↑PPP	Hisaki et al., 1994
	Big ET-1	1–10 μg	↑PPP, PIP	Feng et al., 1990
	Big ET-1	100 nM	↑PGI_2	D'Orléans-Juste et al., 1992
	ET-1	0.1–1 nM	↑PPP, edema	Raffestin et al., 1991
	ET-1	0.3–3 μg	↑PPP, PIP	Feng et al., 1990
	ET-1	5 nM	↑PGI_2	D'Orléans-Juste et al., 1992
	ET-1	10 nM	↑PPP	Barnard et al., 1991
	$ET\text{-}1_{16\text{-}21}$	1–10 μg	No effect	Feng et al., 1990

and the resultant pulmonary capillary hypertension (Horgan et al., 1991; Barnard et al., 1991). ET-1 also increases microvascular permeability in rat isolated blood-perfused lungs (Helset et al., 1993). However, ET-1 has also been reported not to alter capillary permeability in isolated perfused lungs of rabbits (Barnard et al., 1991), which may explain why not everyone finds ET-1 to produce edema in lungs from this species (Mann et al., 1991).

Aerosol administration of ET-1 to guinea pig isolated perfused lungs also evokes a dose-dependent increase in PIP but no changes in PPP, edema formation, or TxB_2 release (Pons et al., 1991c). Interestingly, the increase in PIP induced by aerosolized ET-1 is less than that induced by i.a. injection (Pons et al., 1991c).

3. *Other Preparations*

It has been reported that although ET-1 causes concentration-dependent reductions of methacholine- and phenylephrine-induced mucus volume, lysosyme, and albumin output, it has no direct effects on these parameters itself (Yurdakos and Webber, 1991). On the other hand, ET-1, but not ET-2 or ET-3, has been shown to increase the secretion of mucus glycoprotein from isolated submucosal glands of the feline epithelium (Shimura et al., 1992). In agreement with the first report, however, in tracheal explants with epithelium or in isolated submucosal glands in the presence of cultured epithelial cells, ET-1 produces a reduction of mucus secretion, suggesting that ET-1 enhances an epithelial action that inhibits mucus secretion (Shimura et al., 1992). In addition, ET-1 increases the secretion of chloride from epithelial cells isolated from the dog trachea (Plews et al., 1991; Satoh et al., 1992).

TABLE 7

Bronchoconstriction Induced by ET/STX Peptides in Airways

Species	ET/STX	Conc.	Administration	Ref.
Cat	ET-1	0.3 nmol/kg	i.v. bolus	Kadowitz et al., 1991
	ET-1, -2, -3, STX-b	0.1–1 nmol/kg	i.v. bolus	Dyson and Kadowitz, 1991
Guinea pig	Big ET-1	2.5, 5, 10 nmol/kg	i.v. bolus	Noguchi et al., 1991
	Big ET-1	1, 10 nmol/kg	i.v. bolus	Pons et al., 1991a-c
	ET-1	0.25, 0.75, 1.25 μg/kg	i.v. bolus	Macquin-Mavier et al., 1989
	ET-1	0.05-1 nmol/kg	i.v. bolus	Payne and Whittle, 1988
	ET-1	0.5, 1 nmol/kg	i.v. bolus	Touvay et al., 1990
	ET-1	0.5, 1.5 nmol/kg	i.v. bolus	Schumacher et al., 1990
	ET-1	1 nmol/kg	i.v. bolus	Braquet et al., 1989
	ET-1	1 nmol/kg	i.v. bolus	Nagai et al., 1991
	ET-1	1 nmol/kg	i.v. bolus	Pons et al., 1991a-c
	ET-1	1, 3 μg/ml	Aerosol[a]	Boichot et al., 1991b
	ET-1	1, 5, 10 μg/kg	Aerosol	Braquet et al., 1989
	ET-1	1, 5, 10 μg/kg	Aerosol	Lagente et al., 1989; 1990b
	ET-1	4, 40, 400 pmol/kg	i.v. bolus	Franco-Cereceda et al., 1990
	ET-1, -2, -3	1, 2 nmol/kg	i.v. bolus	Pons et al., 1991a-c
Rat	ET-1	1.5 nmol/kg	i.v. bolus	Schumacher et al., 1990
	ET-1	10–1000 pM	i.v. bolus	Raffestin et al., 1991
	ET-1	62–1000 pmol/kg	i.v. bolus	Matsuse et al., 1990
	ET-1, $ET\text{-}1_{16\text{-}21}$	100–40000 nM	Aerosol[a]	DiMaria et al., 1991

[a] 2 min.

B. Effects of ETs *In Vivo*

1. *In the Airways*

The i.v. injection or aerosolized administration of ET/STX peptides to cats, guinea pigs, or rats causes a dose-dependent bronchoconstriction, an increase in lung resistance (e.g., contraction of large airways), and a decrease in dynamic lung compliance (e.g., contraction of small airways and/or decreased elastic recoil of pulmonary tissue; see Table 7). Administration of aerosolized ET-1 also produces a dose-dependent bronchoconstriction in anesthetized guinea pigs (Lagente et al., 1989; Braquet et al., 1989).

ET-1 is very potent at producing these effects. For instance, the increase in pulmonary lung resistance induced by 1 nmol/kg i.v. of ET-1 is comparable to that of 1 μmol/kg i.v. of histamine in anesthetized rats (Matsuse et al., 1990). Similarly, there is no difference in the potency of the effect induced by i.v. ET-1 or NKA in anesthetized guinea pigs (Franco-Cereceda et al., 1990). However, U 46,619, a TxA_2 mimetic, or platelet activating factor (PAF) are more potent than ET-1 at inducing bronchoconstriction in, respectively, anesthetized cats (Dyson and Kadowitz, 1991) and guinea pigs (Payne and Whittle, 1988).

The response to i.v. administration of ET-1 is accompanied by a characteristic biphasic pressor response consisting of a transient hypotension followed by a sustained (>45 min) increase in arterial blood pressure with no significant changes in heart rate (Masaki et al., 1991). On the other hand, aerosolized administration of ET-1 does not cause any systemic or pulmonary vascular responses, suggesting distinct mechanisms of action (Lagente et al., 1989; Touvay et al., 1990). The bronchoconstriction induced by aerosolized or i.v. ET-1 is maximal within 1 to 2 min, and almost completely reversed within 15 min, which is in contrast to the sustained pressor effect induced by the i.v. application of ET-1.

The i.v. administration of ET precursors such as big ET-1 also causes bronchoconstriction in cats (Dyson and Kadowitz, 1991) and guinea pigs (Pons et al., 1991c). However, the response, which is marked and long lasting, develops more slowly, which may be attributed to the time for conversion of the precursor into the mature active peptide (Pons et al., 1991c).

2. *In the Pulmonary Circulation*

In both cats (Lippton et al., 1989; Minkes et al., 1990; Minkes and Kadowitz, 1991) and rats (Raffestin et al., 1991), ET-1, ET-2, ET-3, or STX-b increase pulmonary vascular resistance (PVR). Interestingly, the changes in PVR are biphasic in anesthetized cats. This may be a direct effect or dependent upon the release of other mediators. For instance, in anesthetized ventilated adult sheep, bolus i.v. administration of ET-1 causes an acute rise in mean pulmonary artery pressure (PAP) and PVR that correlates with an increase in plasma TxB_2 levels (Morel et al., 1990).

ET-1 may also cause vasodilatation in the pulmonary circulation of adult pigs (Hemsen et al., 1991), fetal lambs (Chatfield et al., 1990, 1991), and animals where the pulmonary vascular tone is elevated artificially, such as in conscious hypoxic rats (Hasunuma et al., 1990) and in newborn lambs during pulmonary hypertension induced by U 46,619 or alveolar hypoxia (Wong et al., 1993). In cats, with an elevated pulmonary tone induced by U 46,619, the response to intralobar injection of ET-1, ET-2, or ET-3 also causes marked reductions in PVR (Lippton et al., 1991a, b). Some of these responses may be NO-dependent, as pulmonary vasodilations in newborn piglets (Perreault and DeMarte, 1991) and in newborn lambs (Wong et al., 1993) may be inhibited by inhibitors of the NO system. However, in hypoxic rats, pulmonary vasodilatation induced by ET-1 is not dependent on the release of NO, PAF, or eicosanoids but rather involves ATP-sensitive K^+ channels and membrane hyperpolarization (Hasunuma et al., 1990). Activation of potassium channels is also involved in the pulmonary vasodilator response to ET isopeptides in cats (Lippton et al., 1991a, b) and newborn lambs (Wong et al., 1993).

Because of their potent vasoactive properties, it has been suggested that the ETs might be involved in the pathological processes leading to pulmonary hypertension. However, such conclusions should be carefully drawn as the responses induced by ET/STX peptides in the pulmonary circulation are rather complex and species dependent.

3. *Other Effects*

In conscious rats, i.v. administration of ET-1 causes an increase in microvascular permeability in the upper and lower bronchi, but not in the trachea or lung parenchyma (Filep et al., 1991b, 1993c; Sirois et al., 1992). Conversely, i.v. administration of ET-1 to anesthetized guinea pigs does not increase microvascular leakage, produce epithelial damage, or cause the influx of inflammatory cells such as mast cells, eosinophils, macrophages, basophils, and platelets at any sites in the lung (Macquin-Mavier et al., 1989).

V. MECHANISMS OF ACTION

The ET receptor subtypes that might mediate the pulmonary responses induced by ET/STX peptides are now being intensively studied as are the intracellular mechanisms mediating these effects. Currently, it is known, for instance, that ET-1 stimulates the turnover of phosphoinositides in guinea pig (Hay, 1990) and rat (Henry et al., 1992) tracheal smooth muscle, and that both ET-1 and ET-3 increase the formation of inositol phosphates in porcine lung tissue (Hemsen et al., 1991). Similarly, in human cultured bronchial smooth muscle cells, ET-1 also increases the *in vitro* cellular levels of inositol phosphates and diacylglycerol (Mattoli et al., 1991b). Other pathways are also under investigation.

A. Ionic Exchanges

1. *Calcium*

The first paper describing ET-1 proposed that the vasoconstriction it induced involved the activation of L-type voltage-dependent Ca^{2+} channels (Yanagisawa et al., 1988). Further studies supported a role for these channels in mediating the response to ETs in isolated airways (Uchida et al., 1988; Maggi et al., 1989b; Cardell et al., 1990; Sarria et al., 1990; Lee et al., 1990). However, subsequently, it was found that neither the removal of extracellular Ca^{2+} (McKay et al., 1991a) nor the addition of Ca^{2+} channel blockers (Turner et al., 1989b; Secrest and Cohen, 1989; Advenier et al., 1990; Sarria et al., 1990; McKay et al., 1991a) affected contractions induced by ETs in isolated rat, guinea pig, or human airways (see Table 8). Similarly, bronchoconstrictions following aerosol, i.v., or i.a. administration of ET-1 to guinea pig *in vivo* or *in vitro* were unaffected by i.v. administration of Ca^{2+} channel blockers (Lagente et al., 1989; Macquin-Mavier et al., 1989; see Table 9). Possibly these differences are because, as suggested by Advenier and colleagues (1990), low concentrations of ETs induce the opening of voltage-dependent Ca^{2+} channels while high concentrations of ETs produce effects by involving other mechanisms. For instance, it may be that an influx of extracellular Ca^{2+} is necessary to contribute to the sustained elevation of intracellular Ca^{2+} and tone in human airway smooth muscle (Mattoli et al., 1991b), and that receptor-operated Ca^{2+} channels are involved in maintaining responses in the trachea of the rat (Chand et al., 1990). However, it is true to say that contractions induced by low (e.g., ≤ 1 nM) concentrations of ET-1 in airways are primarily mediated by the mobilization of intracellular Ca^{2+} followed by the influx of extracellular Ca^{2+}, and that Ca^{2+} entry is not confined to a particular class of Ca^{2+} channel (Sarria et al., 1990).

2. *Sodium*

Another mechanism that might be involved in mediating contractions induced by ETs in isolated bronchi and trachea of guinea pigs is the sodium-proton exchanger, as selective inhibitors of the entry of Na^+ in exchange for H^+ attenuate ET-1-induced contractions in isolated upper bronchi and tracheal strips of the guinea pig (Battistini et al., 1991).

B. Release of Arachidonic Acid Metabolites

1. *Eicosanoids*

a. *Experiments in Isolated Cells or Tissues*

ET-1 induces the release of eicosanoids, mainly TxA_2, from isolated tracheal and parenchymal strips of the guinea pig (Filep et al., 1990, 1991a), stimulates the formation of TxA_2 and PGD_2 by resident luminal airway cells such as canine alveolar macrophages (Ninomiya et al., 1992), promotes the release of eicosanoids from human isolated nasal mucosal cells in culture (Wu et al., 1992) and from feline tracheal epithelial cells (Wu et al., 1993), and induces the release of PGI_2 and PGE_2,

TABLE 8

Effect of Ca^{2+} Removal or Ca^{2+} Channel Blockers on Contractions Induced by ETs in Isolated Vascular and Nonvascular Pulmonary Preparations

Tissues	Species	ET	Conc. (nM)	Inhibitors	Conc. (μM)	Effects	Ref.
Bronchiolar SM	Human	ET-1	100	Ca^{2+}-free+EGTA	1000	Abolished plateau	1
		ET-1	100	Nicardipine	1	No effect	1
	Ferret	ET-1	100	Verapamil	1	60% Inhibition	2
	Human	ET-1	0.01–300	Ca^{2+}-free+EGTA	1000	No effect	3
		ET-1	0.01–300	Verapamil	10	No effect	3
		ET-1, 2, 3	0.1	Bay K 8644	0.1	Potentiation	4
		ET-1, 2, 3	0.1	Nicardipine	1	Inhibition	4
		ET-1, 2, 3	0.1, 30	Ca^{2+}-free+EDTA	1000	Inhibition	4
		ET-1, 2, 3	1–300	Bay K 8644	0.1	No effect	4
		ET-1, 2, 3	1–300	Nicardipine	1	No effect	4
Pulmonary artery	Guinea pig	ET-1	100	Ca^{2+}-free+EGTA	100	21% Inhibition	5
		ET-1	100	Nifedipine	0.03	15% Inhibition	5
Pulmonary vein	Guinea pig	ET-1	100	Ca^{2+}-free+EGTA	100	9% Inhibition	5
		ET-1	100	Nifedipine	0.03	9% Inhibition	5
	Rabbit	ET-1	0.1–300	$LaCl_3$	3000	Inhibition	6
		ET-1	0.1–300	Nicardipine	0.1	No effect	6
		ET-1	0.1–300	Nifedipine	3	No effect	6
		ET-1	0.1–300	Verapamil	10	No effect	6
		ET-1	30	Ca^{2+}-free+EGTA	1000	Inhibition	6
Trachea	Dog	ET-1	30	Ca^{2+}-free+EGTA	500	No effect	7
		ET-1	30	EGTA	4000	41% Inhibition	7
		ET-1	30	Verapamil	10	15% Inhibition	7
	Ferret	ET-1	0.1–100	Nifedipine	10	Inhibition	8
	Guinea pig	ET-1	0.001–100	Bay K 8644	1	No effect	9
		ET-1	0.001–100	Ca^{2+}-free+EDTA	1000	Inhibition	9
		ET-1	0.001–100	Diltiazem	10	No effect	9
		ET-1	0.001–100	Diltiazem	100	Inhibition	9
		ET-1	0.001–100	$LaCl_3$ or $CdSO_4$	10	Inhib. ≥10 nM	9
		ET-1	0.001–100	Nicardipine	10	Inhib. ≤1 nM	9
		ET-1	0.001–100	Verapamil	10	No effect	9
		ET-1	0.001–100	Verapamil	100	Inhibition	9
		ET-1	0.1–100	Ca^{2+}-free		Inhibition	10
		ET-1	1	Nicardipine	0.01	42% Inhibition	11
		ET-1	1–300	Nifedipine	1	35% inhibition	12
		ET-1	100	Ca^{2+}-free+EGTA	100	24% inhibition	5
		ET-1	100	Nifedipine	0.03	25% inhibition	5
	Rabbit	ET-1	0.1–100	Ca^{2+}-free+EGTA	1000	Abolished	13
		ET-1	0.1–100	Nifedipine	10	Inhibition	13
	Rat	ET-1	0.1–100	Diltiazem	50	No effect	14
		ET-1	0.1–100	Nitrendipine	1	No effect	14
		ET-1	1–100	Ca^{2+}-free		Inhibition	15
		ET-1	20–100	$CdCl_2$	10	Inhibition	15
		ET-1	20–300	Nifedipine	10	Inhibition	15
		ET-1	20–300	Verapamil	10	No effect	15
		ET-1	100–3000	Nicardipine	0.1	No effect	16
		ET-1	1000	Ca^{2+}-free+EGTA	100	57% Inhibition	16

Note: References; 1, Mattoli et al., 1991a; 2, Lee et al., 1990; 3, McKay et al., 1991a; 4, Advenier et al., 1990; 5, Cardell et al., 1990; 6, Steffan and Russell, 1990; 7, Sakata et al., 1989; 8, Yurdakos and Webber, 1991; 9, Sarria et al., 1990; 10, Nagai et al., 1991; 11, Uchida et al., 1988; 12, Maggi et al., 1989b; 13, Grunstein et al., 1991; 14, Secrest and Cohen, 1989; 15, Chand et al., 1990; 16, Turner et al., 1989b.

but not TxA_2, in rabbit isolated trachea (Grunstein et al., 1991). *In vitro* studies examining the roles of the eicosanoids in contractions induced by the ET/STX peptides in isolated airways have yielded contradictory results (see Table 10). For

TABLE 9

Effects of Ca^{2+} Removal and/or Ca^{2+} Channel Blockers on Contractions Induced by ETs in Isolated Perfused Lungs and Anesthetized Animals

Tissues	Species	ET	Conc.	Inhibitors	Conc.	Effects	Ref.
In vivo	Guinea pig	ET-1	0.38 μg/kg i.v.	Nicardipine	30 μg/kg i.v.	No effect	1
		ET-1	10 μg/ml aer.	Nifedipine	50 mg/kg i.p.	No effect	2
		ET-1	10 μg/ml aer.	Nifedipine	50 mg/kg i.p.	No effect	3
		ET-1	10 μg/ml aer.	Verapamil	0.3 mg/kg i.v.	No effect	2
		ET-1	10 μg/ml aer.	Verapamil	0.3 mg/kg i.v.	No effect	3
Perfused lungs	Guinea pig	ET-1	1 nM	Nifedipine	0.1, 10 μM	↓PPP, edema	4
	Rabbit	ET-1	1.2–8 nM	Ca^{2+}-free		↓ PAP	5
		ET-1	1.2–8 nM	Ca^{2+}-free+EGTA	1 mM	↓ PAP	5
		ET-1	1.2–8 nM	Cadmium	20 μM	↓ PAP	5
		ET-1	1.2–8 nM	Verapamil	100–25 μM	↓ PAP, edema	5

Note: References; 1, Macquin-Mavier et al., 1989; 2, Braquet et al., 1989; 3, Lagente et al., 1989; 4, Horgan et al., 1991; 5, Mann et al., 1991.

TABLE 10

Effect of Cyclooxygenase Inhibitors, PAF Antagonists or Tx-Receptor Antagonists on Contractions Induced by ET/STX Peptides in Isolated Vascular and Nonvascular Pulmonary Preparations

Tissues	Species	ET/STX	Conc. (nM)	Antagonist/ Inhibitor	Conc. (μM)	Effects	Ref.
Bronchus	Ferret	ET-1	100	Indomethacin	3	No effect	1
	Guinea pig	ET-1	0.01–300	Indomethacin	5	↑Potency	2
		ET-1	1–100	BM 13,177	10, 50	↓Potency+Em	3
		ET-1	1–100	BM 13,505	1, 5	↓Potency+Em	4,5
		ET-1	1–100	BN 52,021	10	↓Potency+Em	6
		ET-1	1–100	BN 52,021	10	↓Potency+Em	5
		ET-1	1–100	FPL 55,712	19	↓Potency+Em	5
		ET-1	1–100	Indomethacin	10	↓Potency+Em	5
		ET-1	1–100	U 75,302	3	No effect	5
		ET-1	1–100	WEB 2,086	1	↓Potency+Em	5
		ET-1	1–100	YM 16,638	1	↓Potency+Em	5
		ET-3	0.01–300	Indomethacin	5	↑Potency	2
		STX-b	0.01–300	Indomethacin	5	↑Potency	2
	Human	ET-1	0.01–300	Indomethacin	25	No effect	7
		ET-1, 2, 3	0.001–300	Indomethacin	6	No effect	8
Parenchyma	Guinea pig	ET-1	1–100	BM 13,177	10, 50	↓Potency+Em	4
		ET-1	1–100	BM 13,505	1, 5	↓Potency+Em	4,5
		ET-1	1–100	BN 52,021	10	↓Potency+Em	5
		ET-1	1–100	FPL 55,712	19	↓Potency+Em	5
		ET-1	1–100	Indomethacin	10	↓Potency+Em	5
		ET-1	1–100	U 75,302	3	No effect	5
		ET-1	1–100	WEB 2,086	1	↓Potency+Em	5
		ET-1	1–100	YM 16,638	1	↓Potency+Em	5
		ET-1	10, 100	BN 52,021	10	↓TxA_2	5
		ET-1	10, 100	FPL 55,712	19	↓TxA_2	5
		ET-1	10, 100	WEB 2,086	1	↓TxA_2	5
		ET-1	100	Indomethacin	10	↓TxA_2+PGI_2	4,5
Pulm. artery	Guinea pig	ET-1	1–100	Indomethacin	0.1–1	No effect	9
Pulm. vein	Guinea pig	ET-1	1–100	Indomethacin	0.1–1	No effect	9
Trachea	Rabbit	ET-1	0.1–1000	Indomethacin	10	Inhibit relaxation	10
Trachea	Guinea pig	ET-1	0.001–1	Indomethacin	3	Inhibition	11
		ET-1	0.001–100	Indomethacin	1	No effect	11
		ET-1	0.2–20	Indomethacin	10 μg/ml	No effect	12

TABLE 10 (continued)

Effect of Cyclooxygenase Inhibitors, PAF Antagonists or Tx-Receptor Antagonists on Contractions Induced by ET/STX Peptides in Isolated Vascular and Nonvascular Pulmonary Preparations

Tissues	Species	ET/STX	Conc. (nM)	Antagonist/ Inhibitor	Conc. (μM)	Effects	Ref.
		ET-1	1–30	Indo + SQ 30,741	2.8 + 1	No effect	13
		ET-1	1–30	Indomethacin	2.8	No effect	13
		ET-1	1–30	SQ 30,741	1	No effect	13
		ET-1	1–100	BM 13,177	10, 50	↓Potency+Em	3
		ET-1	1–100	BM 13,505	1, 5	↓Potency+Em	4,5
		ET-1	1–100	BN 52,021	10	↓Potency+Em	6
		ET-1	1–100	BN 52,021	10	↓Potency+Em	5
		ET-1	1–100	FPL 55,712	19	↓Potency+Em	5
		ET-1	1–100	Indomethacin	0.1–1	No effect	9
		ET-1	1–100	Indomethacin	10	↓ Potency	5
		ET-1	1–100	Indomethacin	10	↓Potency+Em	5
		ET-1	1–100	U 75,302	3	No effect	5
		ET-1	1–100	WEB 2,086	1	↓Potency+Em	5
		ET-1	1–100	YM 16,638	1	↓Potency+Em	5
		ET-1	1–300	Indomethacin	5	↓Potency	14
		ET-1	1–300	Indomethacin	5	↑Potency+Em	15
		ET-1	10–100	Indomethacin	3	No effect	11
		ET-1	10–300	Indomethacin	5	↑Potency+Em	16
	Mouse	ET-1	0.1–300	Indomethacin	5	↑Potency+Em	16
	Rabbit	ET-1	0.1–1000	Indomethacin	10	↑Em	10
	Rat	ET-1	1–300	Indomethacin	5	No effect	16
		ET-1	20–100	Indomethacin	5	± Potentiation	17

Note: References: 1, Lee et al., 1990; 2, Maggi et al., 1990; 3, Battistini et al., 1990a; 4, Filep et al., 1990; 5, Filep et al., 1991a; 6, Battistini et al., 1990b; 7, McKay et al., 1991a; 8, Advenier et al., 1990; 9, Cardell et al., 1990; 10, Grunstein et al., 1991; 11, Sarria et al., 1990; 12, Nagai et al., 1991; 13, Schumacher et al., 1990; 14, Maggi et al., 1989b; 15, Hay, 1989; 16, Henry et al., 1990; 17, Chand et al., 1990.

instance, cyclooxygenase inhibitors have been reported to inhibit contractions induced by ET-1 in guinea pig isolated airways (Braquet et al., 1988; Battistini et al., 1990a; Filep et al., 1990, 1991a; White et al., 1991). The release of eicosanoids from guinea pig isolated parenchyma strips is also attenuated by indomethacin (Filep et al., 1990, 1991a). In contrast, cyclooxygenase inhibitors produce potentiation of the contractions induced by ET-1 in isolated trachea (Hay, 1989, 1990; Henry et al., 1990) or bronchi (Maggi et al., 1990) of the guinea pig. Furthermore, others have reported no net effect of cyclooxygenase inhibition on contractions of the human isolated bronchus (Advenier et al., 1990; McKay et al., 1991a), ferret bronchus (Lee et al., 1990), rat trachea (Chand et al., 1990; O'Donnell et al., 1990), and rabbit trachea (Grunstein et al., 1991) induced by ET-1 (Table 10). These differences may be explained by wide species variations in the involvement of prostaglandins in mediating contractions and/or relaxations. It is also worth noting that there are marked differences between the release of eicosanoids stimulated by aerosolized or i.v. applied ET-1, suggesting the involvement of multiple mechanisms within the lung.

b. Experiments in Isolated Perfused Lungs

The i.a. administration of ET-1 peptides to isolated lungs of the guinea pig (Antunes et al. 1988; DeNucci et al., 1988; Braquet et al., 1989; Touvay et al., 1990; Pons et al., 1991b, c; Horgan et al., 1991; D'Orléans-Juste et al., 1991), the rat (DeNucci et al. 1988; Barnard et al., 1991), and the dog (Barnard et al., 1991) leads to a large release of eicosanoids (TxA_2, PGI_2, $PGF_{2\alpha}$) into the lung effluent. Big ET-1

TABLE 11

Effect of Cyclooxygenase Inhibitors, PAF Antagonists, or Tx-Receptor Antagonists on Bronchopulmonary Responses Induced by ETs in Isolated Perfused Lungs

Species	ET/STX	Conc.	Inhibitors	Conc. (μM)	Effects	Ref.
Cat	ET-1	0.3 nmol/kg	Meclofenamate	2.5–5 mg/kg	No effect	1
	ET-1	0.3 nmol/kg	SKF 96,148	0.1 mg/kg	No effect	1
Dog	ET-1	10 nM	Ibuprofen	0.08 mg/ml	↑ PVR	2
Guinea pig	ET-1	0.01–10 nM	Indomethacin	10	↓ PPP, edema	3
	ET-1	0.01–10 nM	SQ 29,548	4	↓ PPP, edema	3
	ET-1	3, 6, 10 μg/ml aer.	Indomethacin	5	No effect	4
	ET-1	10 μg/ml aer.	BW 755C	100	No effect	4
	ET-1	10 μg/ml aer.	FPL 55,712	10	No effect	4
	ET-1	10 μg/ml aer.	Indomethacin	100 mg/ml aer.	No effect	4
	ET-1	400 pmol i.a.	BW 755C	10, 100	Inhibition	4
	ET-1	400 pmol i.a.	FPL 55,712	10	No effect	4
	ET-1	400 pmol i.a.	Indomethacin	5	Inhibition	5
	ET-1	400 pmol i.a.	NDGA	50	No effect	4
	ET-1,2,3	400 pmol i.a.	Indomethacin	5	↓ PIP, PPP, edema, TxB_2	4
Rabbit	ET-1	8 nM	Indomethacin	14	↓ PAP	6
	ET-1	8 nM	UK-37	150	↓PAP	6
	ET-1	10 nM	Ibuprofen	0.25 mg/ml	↓PVR	2
Rat	ET-1	0.3 nM	Meclofenamate	3.2	No effect on PPP	7
	ET-1	0.3-3 μg i.a.	BN 52,021	1	No effect	8
	ET-1	0.3-3 μg i.a.	Indomethacin	1	No effect	8
	ET-1	10 nM	Ibuprofen	1.25 mg/ml	↓ PVR, ↑ PG's	2
	ET-1	10 nM	Indomethacin	100	↑ PVR	2

Note: References; 1, Kadowitz et al., 1991; 2, Barnard et al., 1991; 3, Horgan et al., 1991; 4, Pons et al., 1991c; 5, Pons et al., 1991a; 6, Mann et al., 1991; 7, Raffestin et al., 1991; 8, Feng et al., 1990.

also induces the release of eicosanoids in the effluent of guinea pig perfused lungs (D'Orléans-Juste et al., 1991). The ratio of eicosanoids released varies between species. For instance, in rat lungs ET-1 induces a much greater release of prostacyclin than of TxA_2 (DeNucci et al., 1988). Interestingly, aerosol administration of ET-1 to the guinea pig isolated perfused lung does not induce the release of TxB_2, into the perfusate (Pons et al., 1991c) — yet another example of the different responses following administration of ET-1 into the pulmonary vasculature or into airways.

In isolated perfused lungs of guinea pigs, indomethacin, SQ 29548 — a TxA_2-receptor antagonist — or BW 755c — an inhibitor of both cyclooxygenase and lipoxygenase — block the increases in perfusion pressure, PIP, edema, and/or eicosanoid release induced by i.a. administration of ET-1, ET-2, or ET-3 (see Table 11). However, they do not affect the increase in PIP induced by aerosol administration of ET-1 (Pons et al., 1991c).

In rat isolated perfused lungs, cyclooxygenase inhibitors have been reported to both decrease (Barnard et al., 1991) and to leave unaffected the ET-1-induced increase in PPP (Feng et al., 1990). In isolated perfused lungs of cats (Minkes et al., 1990; Kadowitz et al., 1991) or rabbits (Barnard et al., 1991; Nossaman et al., 1992) cyclooxygenase products do not appear to be important in mediating the responses to ETs, while in dog isolated perfused lungs ibuprofen potentiates the increase in total PVR induced by ET-1 by blocking the release of prostacyclin induced by ET-1 (Barnard et al., 1991). Thus, once again, there are clear species differences in the responses to ETs.

TABLE 12

Effect of Cyclooxygenase Inhibitors, PAF Antagonists, or Tx-Receptor Antagonists on Bronchopulmonary Responses Induced by ET/STX Peptides *In Vivo*

Species	ET/STX	Conc.	Inhibitors	Conc. mg/kg	Effect	Ref.
Cat	ET-1	0.3 nmol/kg i.v.	Meclofenamate	5 i.v.	$\downarrow$Airway responses	1
	ET-1	0.3 nmol/kg i.v.	SKF 96,148	5 i.v.	$\downarrow$Airway responses	1
	ET-1, -2, -3, STX-b	0.1–1 nmol/kg i.v.	Meclofenamate	2.5 i.v.	$\downarrow$Airway responses	2
	ET-1, -2, -3, STX-b	0.1–1 nmol/kg i.v.	SKF 96,148	5 i.v.	$\downarrow$Airway responses	2
	ET-1, -3	0.3 nmol/kg i.v.	Meclofenamate	2.5 i.v.	No effect on PAP	3
G. pig	ET-1	0.05–1 nmol/kg i.v.	Indomethacin	10 i.v.	100% inhibition	4
	ET-1	0.38 µg/kg i.v.	Meclofenamate	0.5 i.v.	86% inhibition	5
	ET-1	0.5, 1.5 nmol/kg i.v.	SQ 30,741	1 i.v.	100% inhibition	6
	ET-1	1 nmol/kg i.v.	Indomethacin	1–5 i.p.	Abolished	7
	ET-1	1 nmol/mg i.v.	BN 52,021	20 i.v.	No effect	8
	ET-1	1 nmol/mg i.v.	Esculetin	5 i.v.	No effect	8
	ET-1	1 nmol/mg i.v.	NDGA	12.5 i.v.	No effect	8
	ET-1	1 nmol/mg i.v.	WEB 2,107	1 i.v.	No effect	8
	ET-1	10 µg/ml aerosol	BN 52,021	10 i.v.	52% inhibition	9
	ET-1	10 µg/ml aerosol	BN 52,021	10 i.v.	54% inhibition	10
	ET-1	10 µg/ml aerosol	Indomethacin	10 i.v.	60% inhibition	9
	ET-1	10 µg/ml aerosol	Indomethacin	10 i.v.	62% inhibition	10
	ET-1	10 µg/ml aerosol	Indomethacin	100 aer.	No effect	11
	ET-1, -2, -3	2 nmol/kg i.v.	Indomethacin	1 i.v.	Inhibition	11
Lamb	ET-1	400 pmol/kg i.v.	Indomethacin	1.5 i.v.	Inhibition	12
Rat	ET-1	1.5 nmol/kg i.v.	SQ 30,741	1 i.v.	No effect	6
	ET-1	10–1000 pM	Meclofenamate	6 i.v.	Affect compliance	13
	ET-1	300 pM	Meclofenamate	6 i.v.	No effect on PVR	13

Note: References; 1, Kadowitz et al., 1991; 2, Dyson and Kadowitz, 1991; 3, Minkes et al., 1990; 4, Payne and Whittle, 1988; 5, Macquin-Mavier et al., 1989; 6, Schumacher et al., 1990; 7, Nagai et al., 1991; 8, Lueddeckens et al., 1991; 9, Braquet et al., 1989; 10, Lagente et al., 1989; 11, Pons et al., 1991c; 12, Morel et al., 1990; 13, Raffestin et al., 1991.

c. *Experiments Conducted* In Vivo

As expected from *in vitro* experiments, i.v. administration of ET-1 induces an increase in the content of TxB_2 in the bronchoalveolar lavage fluid (Lueddeckens et al., 1991). The i.v. administration of cyclooxygenase blockers or TxA_2/PGH_2-receptor antagonists also abolishes the bronchoconstrictions induced by i.v. administration of ET-1 to anesthetized guinea pigs (Payne and Whittle, 1988; Macquin-Mavier et al., 1989; Lagente et al., 1989), cats (Dyson and Kadowitz, 1991), and sheep (Morel et al., 1990). Conversely, meclofenamate potentiates the decrease in compliance induced by i.v. ET-1 in anesthetized rats, although it has no effect on the increase in PVR (Raffestin et al., 1991). As a further complicating observation, aerosol administration of indomethacin to guinea pigs has no effect on the increase in PIP induced by aerosolized ET-1 (Pons et al., 1991c) (see Table 12).

The increase in microvascular permeability induced by i.v. administration of ET-1 to conscious rats appears to be mediated, at least in part, by the release of TxA_2, for protein extravasation into upper and lower bronchi is attenuated by indomethacin, BM 13505, a TxA_2-receptor antagonist, or OKY-046, a TxA_2 synthesis inhibitor (Sirois et al., 1992).

2. *Leukotrienes*

Peptidoleukotrienes might be involved in contractions induced by ET-1 as the selective LTC_4-LTD_4-receptor antagonists, YM 16,638 or FPL 55,712, attenuate contractions induced by ET-1 in isolated airways of the guinea pig (Filep et al., 1991a). However, the situation is not so clear *in vivo* as neither FPL 55,712 (Pons et al., 1991a-c) nor esculetin, a lipoxygenase inhibitor (Lueddeckens et al., 1991), has any effect against the bronchoconstrictions induced by i.v. administration of ET-1 in anesthetized guinea pigs (Pons et al., 1991a-c). Similarly, neither FPL 55,712 nor NDGA affect the increase in PIP, PPP, edema, or TxB_2 release induced by i.a. ET-1 *in vivo* or the increase in PIP caused by aerosolized ET-1 in guinea pig isolated perfused lungs (Pons et al., 1991a-c). In addition, i.a. or aerosol ET-1 does not induce the generation of peptido-leukotrienes from isolated perfused lungs of the guinea pig (Pons et al., 1991c) (see Tables 10 to 12). It therefore appears that leukotrienes have only a minor role in mediating the effects of the ETs in guinea pigs.

3. Platelet Activating Factor

PAF may be involved in the mechanical responses to ET-1, as selective PAF-receptor antagonists (BN 52,021 or WEB 2086) significantly inhibit contractions induced by ET-1 in isolated bronchi of guinea pigs (Battistini et al., 1990b; Filep et al., 1991a). In addition, i.v. administration of BN 52,021 inhibits the bronchoconstriction induced by aerosolized administration of ET-1 to anesthetized guinea pigs (Braquet et al., 1989; Lagente et al., 1989). Similarly, cross-desensitization experiments using isolated bronchi of the guinea pig also support a role for PAF in mediating the pulmonary responses to ET-1 (Battistini et al., 1990b). Conversely, i.v. administration of BN 52,021 or WEB 2107 do not affect the increases in PIP induced by i.v. administration of ET-1 to anesthetized and ventilated guinea pigs (Lueddeckens et al., 1991) or the pulmonary vasoconstrictor response to ET-1 in rat isolated perfused lungs (Feng et al., 1990). Interestingly, the increase in microvascular permeability induced by i.v. administration of ET-1 to the lung has also been reported to be mediated in part by the release of PAF, for protein extravasation into lower bronchi is attenuated by BN 52,021 (Filep et al., 1991b) ET-1 and PAF also act in concert to increase vascular permeability in rat airways (Filep et al., 1991b; see Tables 10 to 12).

C. Modulation by the Epithelium

The epithelium of the upper airways (trachea and bronchi) may act in three different ways to modulate contractions induced by the ET/STX peptides: it may (1) act as a diffusion barrier to prevent the agonist reaching the underlying smooth muscle; (2) possess catabolic activity toward peptides; (3) release bronchodilatory and/or contractile mediators. Such effects have been observed with other contractile agonists.

Removal of the epithelium potentiates contractions induced by ET-1, ET-2, and/or STX-b in isolated guinea pig trachea (Hay, 1989; Tschirhart et al., 1991; Filep et al., 1993a) and bronchi (Maggi et al., 1989b, 1990; Filep et al., 1993a), and human bronchi (Hay, 1989; White et al., 1991; Candenas et al., 1992) but not in mouse trachea (Henry et al., 1990) nor for contractions induced by ET-3 in the guinea pig isolated trachea (Tschirhart et al., 1991). This increase in sensitivity may be due to the loss of metabolic enzymes since phosphoramidon, a neutral endopeptidase inhibitor, enhances the contractile response to ET-1, ET-2, or ET-3 in human isolated bronchi (Candenas et al., 1992), although not to ET-3 or STX-b in the isolated trachea of the guinea pig (Tschirhart et al., 1991), or to ET-1 in isolated bronchi of the guinea pig (Noguchi et al., 1991). Conversely, in human and rabbit isolated bronchi, phosphoramidon has no effect on contractions induced by ET-1 (McKay et al., 1992), and captopril, bacitracin, or leupeptin do not affect the contractions induced by ET-1 in the intact guinea pig isolated trachea (Hay, 1989). Thiorphan, captopril, and bestatin do, however, potentiate ET-3-induced contractions of guinea pig isolated bronchi (Maggi et al., 1990). Other protease inhibitors also increase the contractile response induced by ET-1 in the isolated trachea of the guinea pig (DiMaria et al., 1992). Contractions induced by human big ET-1 in isolated bronchi of the guinea pig are potentiated by low concentrations of phosphoramidon but inhibited by higher concentrations of the same inhibitor (Noguchi et al., 1991). In human isolated bronchi, phosphoramidon induces a rightward shift of the concentration-response curve to porcine big ET-1 (Advenier et al., 1992). Thus, metabolizing enzymes appear to have complex and species-variable effects on the airway responses to the ETs.

Metabolizing enzymes may also be relevant *in vivo*, for in anesthetized guinea pigs, pretreatment with aerosolized phosphoramidon markedly enhances the bronchoconstrictions induced by aerosolized ET-1 (Boichot et al., 1991b) while it modifies in a complex fashion the pulmonary response induced by i.v. big ET-1-(1-38) (Noguchi et al., 1991). Similarly, in guinea pig isolated lungs, treatment with aerosolized phosphoramidon also potentiates the response to ET-1 (Boichot et al., 1991b).

As mentioned above, ET-1 can also release cyclooxygenase products such as prostacyclin and PGE_2 from the isolated trachea of the rabbit (Grunstein et al., 1991) and NO from intact tracheal strips of the guinea pig (Filep et al., 1993a). In addition, application of ET-1 to guinea pig isolated trachea results in transient relaxations followed by marked long-lasting contractions, particularly at higher concentrations of ET-1. These transient relaxations are not seen in denuded preparations. In addition, they are inhibited by BQ-123 and they are not produced by ET_B-receptor-selective agonists. This suggests that ET_A-receptors located on epithelial cells mediate this relaxant response, possibly via the release of relaxing prostanoids (Battistini et al., 1994a).

D. Other Mechanisms

In anesthetized guinea pigs, the bronchoconstriction induced by i.v. administration of ET-1 is potentiated by propranolol, a β-adrenoceptor antagonist, or by hexamethonium, but not by propranolol plus atropine, suggesting that the autonomic nervous system modulates bronchopulmonary responses to ET-1 (Macquin-Mavier et al., 1989; Braquet et al., 1989; Touvay et al., 1990). Similarly, increases in PIP induced by ET-1 (40, 120, and 400 pmol i.a.) in isolated perfused lungs of the guinea pig are potentiated by propranolol (Braquet et al., 1989; Touvay et al., 1990). Conversely, in anesthetized cats i.v. administration of hexamethonium or propranolol does not affect the biphasic PVR response to ET-1 or ET-3, suggesting that, at least in this species, the response is not dependent on autonomic reflexes or activation of β-receptors (Minkes et al., 1990). Interestingly, the bronchoconstriction induced by aerosolized administration of ET-1 to guinea pigs is not affected by i.v. administration of propranolol or mepyramine maleate, a histamine H_1-receptor antagonist (Lagente et al., 1989; Braquet et al., 1989). This agrees with much already referred to above, in that it shows that the bronchopulmonary effects induced by ET-1 are at least partly dissociated from the systemic vascular effects.

Unlike many other mediators, histamine release from guinea pig isolated perfused lungs is not stimulated by ET-1 in the vascular effluent (Braquet et al., 1989; Touvay et al., 1990; Pons et al., 1991c). However, ET-1 releases histamine from guinea pig pulmonary mast cells (Uchida et al., 1992). Nevertheless, diphenhydramine, an histamine H_1-receptor antagonist, does not affect contractions induced by ET-1 in guinea pig isolated airways (Filep et al., 1991a). Similarly, neither methysergide, a 5-HT antagonist, propranolol, a β-adrenoceptor antagonist, phentolamine, an α-adrenoceptor antagonist, nor atropine, an anticholinoceptor drug in guinea pig isolated airways (Filep et al., 1991a) nor mepyramine, an H_1-receptor antagonist, cimetidine, an H_2-receptor antagonist, phentolamine, or ketanserin, a 5-HT-receptor antagonist in the guinea pig isolated trachea and pulmonary vessels, affect ET-1-induced contractions (Cardell et al., 1990). Contractions induced by ET-1 in the isolated trachea or bronchi of the guinea pig are also not affected by tetrodotoxin, suggesting that neuronal sodium channels are not involved (Maggi et al., 1989b). Finally, the pulmonary vasoconstrictor response to ET-1 in rat isolated perfused lungs is not inhibited by saralazin, an angiotensin II-receptor antagonist (Feng et al., 1990).

E. ET Receptors Involved in Pulmonary Responses

The initial characterization of the ET receptors involved in pulmonary responses was based on the efficacy of ET agonists (e.g., ET/STX peptides and related analogues) and *in situ* hybridization. The use of the newly available selective ET_A- or ET_B-receptor antagonists and/or nonselective ET_A/ET_B-receptor antagonists has permitted the further characterization of ET receptors involved in various responses in the pulmonary system (see Tables 13 and 14).

TABLE 13

Populations of ET-Receptor Subtypes in the Pulmonary System of Various Species

Tissues	Species	Responses	ET_A	ET_B	ET_X	Ref.
Bronchus	Guinea pig	Contraction	++++	–		1
		Contraction	–	++++	–	2,3
		Contraction	–	++++	–	4
		Contraction	–	++++	?	5
	Human	Contraction	+++	–	–	6
		Contraction	+++	–	–	7
		Contraction	–	++++	–	8,9
		Contraction	–	++++	–	10
	Lamb	Binding	++++	–	–	11
	Rat	Contraction	+++	+	–	12
		In situ hybridization	+++	+	–	13
Parenchyma	Rat	Binding	65	35	–	14
	Guinea pig	Binding	–	++++	–	15
		Binding	15	85	–	14
	Human	Binding	+++	+	–	7
	Lamb	Contraction	++++	–	–	11
	Monkey	Binding	45	55	–	14
	Mouse	Binding	60	40	–	14
	Pig	Binding	++	++	–	16
		Binding	35	65	–	14
	Rat	Binding	55	45	–	14
		Binding	60	40	–	17
		Contraction	+	+++	–	12
		In situ hybridization	++	++	–	13
Pulm. artery	Bovine	Binding; PI hydrolysis	++	++	–	18
	Guinea pig	Contraction	++	–	?	19
		Contraction	+++	–	–	9
		Contraction	+++	–	?	20
		Contraction	++++	–	–	1
		Contraction, cross-des	++	++	–	21
	Human	Contraction	++	++	–	7
		Contraction	+++	–	–	9

TABLE 13 (continued)

Populations of ET-Receptor Subtypes in the Pulmonary System of Various Species

Tissues	Species	Responses	ET_A	ET_B	ET_X	Ref.
		Contraction				22
		SMC proliferation	++	–	–	23
	Pig	Contraction	+++	–	–	24
	Rabbit	Contraction; binding	40	60	–	25
	Rat	Contraction	++	–	–	26
		Contraction	+++	+	–	12
		In situ hybridization	+++	+	–	13
Pulm. vein	Pig	Contraction	–	++	?	24
	Human	Contraction				22
Trachea	Dog	Contraction, PI hydrolysis	++++	–	–	27
	Guinea pig	Binding	++	++	–	28
		Contraction	++	++	–	9
		Contraction	++	+++	–	2,3
		Contraction	–	++++	–	19
		Contraction	–	++++	–	20
		Contraction, binding	++	++	?	29
		Contraction, cross-des	±	+++	–	21
Trachea	Lamb	Contraction, binding	++++	–	–	11
Trachea	Rat	Contraction	++	++	–	30

Note: ET_X: non-ET_A-, non-ET_B-receptors. 1, Franco-Cereceda et al., 1990; 2, Battistini et al., 1994a; 3, Battistini et al., 1994b; 4, Hay et al., 1993c; 5, Maggi et al., 1990; 6, Advenier et al., 1990; 7, Hemsen et al., 1990; 8, Hay et al., 1993a; 9, Hay et al., 1993b; 10, Candenas et al., 1992; 11, Goldie et al., 1994; 12, Nakamichi et al., 1992; 13, Hori et al., 1992; 14, Ihara et al., 1992; 15, Pons et al., 1991a; 16, Hemsen et al., 1991; 17, D'Orléans-Juste et al., 1992; 18, Kent and Keenan, 1994; 19, Cardell et al. 1992; 20, Cardell et al., 1993; 21, Cardell et al., 1991; 22, Brink et al., 1991; 23, Zamora et al., 1993; 24, Sudjarwo et al., 1993; 25, LaDouceur et al., 1993; 26, Bonvallet et al., 1993; 27, Yang et al., 1994; 28, Inui et al., 1994; 29, Tschirhart et al., 1991; 30, Henry, 1993.

TABLE 14

Populations of ET-Receptor Subtypes in the Pulmonary System of Various Species

Tissues	Species	Responses	ET_A	ET_B	ETX	Ref.
In vivo	Guinea pig	↑ PIP	–	++++	–	Noguchi et al., 1993
		↑ PPP, PIP	–	++++	–	Pons et al., 1991c
	Lamb	↑ PIP	++++	–	–	Abraham et al., 1993
	Rat	↑ Microvascular permeability	++	–	–	Filep et al., 1993c
Perf. lungs	Cat	↓ PPP	++	–	++	Lippton et al., 1993
	Guinea pig	↑ PIP	–	++	?	Spinella et al., 1991
	Rat	↑ PGI_2	++	–	–	D'Orléans-Juste et al., 1992
		↑ PPP	++	–	–	Bonvallet et al., 1993

In guinea pig isolated airways, contractions induced by ET/STX peptides in the trachea are mediated by both ET_A- and ET_B-receptors, whereas mostly ET_B-receptors are involved in the bronchus and the lung parenchyma. In rat isolated airways, both receptors are involved. In human isolated bronchus, results are contradictory. In the pulmonary artery of the rabbit, nonselective ET_A/ET_B-receptor agonists (e.g., ET-1, ET-3) and selective ET_B-receptor agonists (e.g., STX-c) are equipotent (LaDouceur et al., 1993). Similarly, both ET-1 and ET-3 contract isolated pulmonary artery and veins (Rodman et al., 1989; Lippton et al., 1991c). These data suggest mediation of these responses by ET_B-receptors. In the pulmonary artery of guinea pigs or rabbits,

TABLE 15

Elements Implicated in the Pathophysiology of Asthma That Are Associated With ET-1

Elements	Response	Species
Airway hyperresponsiveness	No	Guinea pig
Airway secretion		
Mucus	Yes	Cat
Chloride	Yes	Dog
PGs	Yes	Human, cat
Airway smooth muscle hyperplasia and hypertrophy	Yes	Rat, rabbit, ovine
Bronchial microvascular permeability	Yes	Rat
Ciliary beat frequency	Yes	Dog
Decrease airway blood flow	Yes	Dog
Edema	Yes	Guinea pig, rat, dog
Epithelial damage	No	
Increase levels in BAL fluid	Yes	Human
Influx of inflammatory cells	?	
Leukocyte activation		
Leukocyte chemotaxis	Yes	Rat
Neural stimulation	?	
Potent and sustained contractions of airways	Yes	Human
Production of proinflammatory mediators		
TxA_2, PGs	Yes	Guinea pig
Oxygen radicals	Yes	Rat
Superoxide	Yes	Human
15-HETE	Yes	Rat
Pulmonary hypertension	Yes	Human

desensitization experiments also suggest the presence of functional ET_A-receptors (Lippton et al., 1991c; Cardell et al., 1992; LaDouceur et al., 1993). We can conclude, therefore, that both ET_A- and ET_B-receptors mediate contractions induced by ET/STX peptides in the pulmonary artery of rabbits and, most probably, in other species. In perfused lungs or *in vivo*, pulmonary responses in the rat, the cat, or the lamb are mainly mediated by ET_A-receptors, whereas in the guinea pig, ET_B-receptors are involved.

VI. PATHOPHYSIOLOGICAL IMPLICATIONS

ET-1 is well associated with various elements implicated in the pathophysiology of asthma (see Table 15). For instance, ir-ET is present in the bronchoalveolar lavage fluid of asthmatic patients (Nomura et al., 1989), of patients with symptomatic asthma (Mattoli et al., 1991a) or interstitial lung disease (Sofia et al., 1993), and in rats with oleic-induced respiratory distress syndrome (Simmet et al., 1992). Plasma levels of ET-1 are also increased in patients with adult respiratory distress syndrome (Druml et al., 1993) and in patients with acute respiratory failure (Mitaka et al., 1993). ET-1 expression is markedly higher in the airway epithelium and microvascular endothelium in endobronchial biopsies from asthmatic patients (Springall et al., 1991; Vittori et al., 1992), in cryptogenic fibrosing alveolitis patients (Giaid et al., 1993a), and in patients with pulmonary hypertension (Giaid et al., 1993c), compared to control subjects. Interestingly, the concentration of ir-ET measured in the broncho-alveolar lavage fluid of patients after recovery following *status asthmaticus* is much lower than during the acute phase (Nomura et al., 1989) and treatment of asthmatic (chronic bronchitis and airflow obstruction) patients with inhaled β-adrenoceptor agonists and oral corticosteroids causes a threefold reduction in the levels of ir-ET-1

(Mattoli et al., 1991a). Hydrocortisone also decreased the level of ir-ET into the culture medium of asthmatic bronchial epithelial cells (Vittori et al., 1992). Similarly, budesonide abolished the long-lasting inflammation (e.g., edema) produced by intratracheal instillation of Sephadex beads in the rat lung (Andersson et al., 1992). These observations may indicate changes in ET-1 production in the lung that contribute to the pathogenesis of diseases such as asthma and bronchitis. However, it may be that there is actually a decreased activity of neutral endopeptidase which, as mentioned above, modulates ET-1-induced bronchoconstriction (Noguchi et al., 1991).

Despite these data in support of a role for ET-1 in asthma, it is also worthy of note that i.v. ET-1 does not produce airway hyperactivity to histamine or serotonin in anesthetized guinea pigs (Macquin-Mavier et al., 1989). Similarly, aerosolized ET-1 does not alter the responses to aerosolized administered Ach, 3 to 4 or 18 to 24 h later (Lagente et al., 1990a). In addition, aerosolized ET-1 does not alter the responses to aerosolized Ach after even a brief period of 30 min (Lagente et al., 1990a; Boichot et al., 1991a). Therefore, ET-1 cannot be shown to induce bronchial hyperreactivity in these animal models.

ET-1 has also been implicated in other disease states. For instance, the levels of ir- ET and/or ET mRNA in rat lung tissue are increased in acute pulmonary hypoxia (Shirakami et al., 1991). Pulmonary tumors (carcinoma rather than lymphomas or sarcomas) are also associated with high amounts of ir-ET (Giaid et al., 1990, 1991). Plasma levels of ir-ET are also increased in patients with pulmonary hypertension (Cernacek and Stewart, 1989; Yoshibayashi et al., 1991; Stewart et al., 1991) and in rats with idiopathic pulmonary hypertension (Stelzner et al., 1992).

Finally, further roles for ETs in pulmonary pathophysiology are suggested by their growth-promoting properties (see Battistini et al., 1993a). For example, ET-1 increases the proliferation of rabbit tracheal smooth muscle cells (Noveral et al., 1992), human pulmonary artery smooth muscle cells (Zamora et al., 1993), and promotes mitogenesis of lamb airway smooth muscle cells (Glassberg et al., 1994). ET-1 has also been suggested to play a role in the bronchial neovascularization and the pulmonary thickening characteristic of postobstructive pulmonary vasculopathy (Giaid et al., 1993b). Finally, ET-1 and ET-3 have been shown to attract rat pulmonary artery fibroblasts and stimulate their replication (Peacock et al., 1992). Thus, ET-1 may have a role in promoting the airway smooth muscle hyperplasia and hypertrophy seen in patients with bronchial asthma, and the increase in pulmonary blood vessel thickness seen in pulmonary hypertension.

VII. CONCLUSION

Asthma, and other conditions such as adult respiratory distress syndrome, pulmonary edema, and hypertension are complex diseases that involve bronchoconstriction, vascular and nonvascular muscle hyperplasia and hypertrophy, airway hyperresponsiveness, infiltration and activation of inflammatory cells, and the release of inflammatory mediators. ETs are potent bronchoconstrictors and co-mitogens, cause an increase of microvascular permeability in the airways, and modulate the activation of inflammatory cells. In addition, i.v. administration of ET-1 induces the release of oxygen radicals by bronchoalveolar cells (such as neutrophils) and increases the levels of 15-hydroxyeicosatetraenoic acid (15-HETE) in rat bronchoalveolar lavage fluid (Nagase et al., 1990). 15-HETE is a potent agonist at promoting the secretion of mucus and a potent chemotactic agent for inflammatory cells. ET-1 therefore produces many of the effects expected of a mediator of airway disease. Despite all these observations, however, the role(s) of ETs in the airways remains to be firmly established. Further experiments exploiting effective and reproducible animal models of inflamed and hyperresponsive airways, combined with the use of ET receptor antagonists, are required to produce this firm establishment. These may confirm ETs to be multifunctional pulmonary peptides and establish ET receptor antagonists or inhibitors as new prophylactic and/or antiinflammatory drugs for the treatment of pulmonary diseases.

REFERENCES

Abraham, W. M., Ahmed, A., Cortes, A., Spinella, M. J., Malik, A. B. and Andersen, T. T., A specific endothelin-1 antagonist blocks inhaled endothelin-1-induced bronchoconstriction in sheep, *J. Appl. Physiol.*, 74, 2537-2542. 1993.

Advenier, C., Lagente, V., Zhang, Y. and Naline, E., Contractile activity of big endothelin-1 on the human isolated bronchus, *Br. J. Pharmacol.*, 106, 883-887. 1992.

Advenier, C., Sarriá, B., Naline, E., Puybasset, L. and Lagente. V., Contractile activity of three endothelins (ET-1, ET-2 and ET-3) on the human isolated bronchus, *Br. J. Pharmacol.*, 100, 168-172. 1990.

Andersson, S. E., Zackrisson, C., Hemsen, A. and Lundberg, J. M., Regulation of lung endothelin content by the glucocorticosteroid budesonide, *Biochem. Biophys. Res. Commun.*, 188, 1116-1121. 1992.

Anggard, E., Galton, S., Rae, G., Thomas, R., McLoughlin, L., DeNucci, G. and Vane, J. R., The fate of radioiodinated endothelin-1 and endothelin-3 in the rat, *J. Cardiovasc. Pharmacol.*, 13, S46-S49. 1989.

Antunes, E., DeNucci, G. and Vane, J. R., Endothelin releases eicosanoids from and is removed by isolated perfused lungs of the guinea pig, *J. Physiol.*, 407, 40. 1988.

Arai, H., Hori, S., Aramori, I., Ohkubo, H. and Nakanishi, S., Cloning and expression of a cDNA encoding an endothelin receptor, *Nature*, 348, 730-732. 1990

Barnard, J. W., Barman, S. A., Adkins, W. K., Longenecker, G. L. and Taylor, A. E., Sustained effects of endothelin-1 on rabbit, dog, and rat pulmonary circulations, *Am. Physiol. Soc.*, 261, H479-H486. 1991.

Battistini, B., Chailler, P., D'Orléans-Juste, P., Brière, N. and Sirois, P., Growth regulatory properties of endothelins, *Peptides*, 14, 385-399. 1993a.

Battistini, B., Filep, J. G. and Sirois, P., Potent thromboxane-mediated *in vitro* bronchoconstrictor effect of endothelin in the guinea pig, *Eur. J. Pharmacol.*, 178, 141-142. 1990a.

Battistini, B., Filep, J. G., Cragoe, E. J., Fournier, A. and Sirois, P., A role for Na^+/H^+ exchange in contraction of guinea pig airways by endothelin *in vitro*, *Biochem. Biophys. Res. Commun.*, 175, 583-588. 1991.

Battistini, B., Germain, M., Fournier, A. and Sirois, P., Structure-activity relationships of ET-1 and selected analogues in the isolated guinea pig trachea: evidence for the existence of different ET_B receptor subtypes, *J. Cardiovasc. Pharmacol.*, 22, S219-24. 1993b.

Battistini, B., Sirois, P., Braquet, P. and Filep, J. G., Endothelin-induced constriction of guinea pig airways: role of platelet-activating factor, *Eur. J. Pharmacol.*, 186, 307-310. 1990b.

Battistini, B., Warner, T. D., Fournier, A. and Vane, J. R., Characterization of ET_B receptors mediating contractions induced by endothelin-1 or IRL 1620 in guinea pig isolated airways: effects of BQ-123, FR139317 or PD 145065, *Br. J. Pharmacol.*, 111, 1009-1016. 1994a.

Battistini, B., Warner, T. D., Fournier, A. and Vane, J. R., Comparison of PD 145065 and Ro 46-2005 as antagonists of contractions of guinea pig airways induced by endothelin-1 or IRL 1620, *Eur. J. Pharmacol.*, 252, 341-345. 1994b.

Black, P. N., Ghatei, M. A., Takahashi, K., Bretherton-Watt, D., Krausz, T., Dollery, C. T. and Bloom, S. R., Formation of endothelin by cultured airway epithelial cells, *FEBS Lett.*, 255, 129-132. 1989.

Bloch, K. D., Eddy, R. L., Shows, T. B. and Quetermous, T., cDNA cloning and chromosomal assignment of the gene encoding endothelin 3, *J. Biol. Chem.*, 264, 18156-18161. 1989.

Boichot, E., Carre, C., Lagente, V., Pons, F., Mencia-Huerta, J. M. and Braquet, P., Endothelin-1 and bronchial hyperresponsiveness in the guinea pig, *J. Cardiovasc. Pharmacol.*, 17, S329-S331. 1991a.

Boichot, E., Pons, F., Lagente, V., Touvay, C., Mencia-Huerta, J. M. and Braquet, P., Phosphoramidon potentiates the endothelin-1-induced bronchopulmonary response in guinea pigs, *Neurochem. Int.*, 18, 477- 479. 1991b.

Bolger, G. T., Liard, F., Krogsrud, R., Thibeault, D. and Jaramillo, J., Tissue specificity of endothelin binding sites, *J. Cardiovasc. Pharmacol.*, 16, 367-375. 1990.

Bonvallet, S. T., Oka, M., Yano, M., Zamora, M. R., McMurtry, I. F. and Stelzner, T. J., *J. Cardiovasc. Pharmacol.*, 22, 39-43. 1993.

Borges, R., Von Grafenstein, H. and Knight, D. E., Tissue selectivity of endothelin, *Eur. J. Pharmacol.*, 165, 223-230. 1989.

Braquet, P., Touvay, C., Lagente, V., Vilain, B., Pons, F., Hosford, D., Chabrier, P. E. and Mencia-Huerta, J. M., Effect of endothelin-1 on blood pressure and bronchopulmonary system of the guinea pig, *J. Cardiovasc. Pharmacol.*, 13, S143-S146. 1989.

Braquet, P., Touvay, C., Lagente, V., Vilain, B., Pons, F., Lejeune, V. and Mencia-Huerta, J. M., Indomethacin inhibits *in vitro* and *in vivo* endothelin-induced bronchoconstriction in the guinea pig, *Eur. Respir. J.*, 158, 303-304. 1988.

Brink, C., Gillard, V., Roubert, P., Mencia-Huerta, J. M., Chabrier, P. E., Braquet, P. and Verley, J., Effects and specific binding sites of endothelin in human lung preparations, *Pulm. Pharmacol.*, 4, 54-59. 1991.

Candenas, M.-L., Naline, E., Sarria, B. and Advenier, C., Effects of epithelium removal and of enkephalin inhibition on the bronchoconstrictor response to three endothelins of the human isolated bronchus, *Eur. J. Pharmacol.*, 210, 291-297. 1992.

Cardell, L. O., Uddman, R. and Edvinsson, L., A novel ET_A-receptor antagonist, FR 139317, inhibits endothelin-induced contractions of guinea pig pulmonary arteries, but not trachea, *Br. J. Pharmacol.*, 108, 448-452. 1993.

Cardell, L. O., Uddman, R. and Edvinsson, L., Evidence for multiple endothelin receptors in the guinea pig pulmonary artery and trachea, *Br. J. Pharmacol.*, 105, 376-380. 1992.

Cardell, L. O., Uddman, R. and Edvinsson, L., Two functional endothelin receptors in guinea pig pulmonary arteries, *Neurochem. Int.*, 18, 571-574. 1991.

Cardell, L. O., Uddman, R. and Edvinsson, R., Analysis of endothelin-1-induced contractions of guinea pig trachea, pulmonary veins and different types of pulmonary arteries, *Acta Physiol. Scand.*, 139, 103-111. 1990.

Cassin, S., Kristova, V., Davis, T., Kadowitz, P. and Gause, G., Tone-dependent responses to endothelin in the isolated perfused fetal sheep pulmonary circulation *in situ*, *J. Appl. Physiol.*, 70, 1228-1234. 1991.

Cernacek, P. and Stewart, D. J., Immunoreactive endothelin in human plasma: marked elevations in patients in cardiogenic shock, *Biochem. Biophys. Res. Commun.*, 161, 562-567. 1989.

Chand, N., Diamantis, W. and Sofia, R. D., Pharmacologic modulation of endothelin-induced contraction in isolated rat tracheal segments, *Res. Commun. Chem. Pathol. Pharmacol.*, 70, 173-181. 1990.

Chatfield, B. A., McMurtry, I. F., Hall, S. L. and Abman, S. H., Hemodynamic effects of endothelin-1 on ovine fetal pulmonary circulation. *Am. J. Physiol.*, 261, R182-R187. 1991.

Chatfield, B. A., McMurtry, I. F., O'Brien, R. F., Hall, S. L. and Abman, S. H., *In vitro* and *in vivo* effects of endothelin-1 on the fetal pulmonary circulation, *Am. Rev. Respir. Dis.*, 141, A483. 1990.

Crawley, D. E., Liu, S. F., Barnes, P. J. and Evans, T. W., Endothelin-3 is a potent pulmonary vasodilator in the rat, *J. Appl. Physiol.*, 72, 1425-1431. 1992.

D'Orléans-Juste, P., Télémaque, S. and Claing, A., Different pharmacological profiles of big-endothelin-3 and big-endothelin-1 *in vivo* and *in vitro*, *Br. J. Pharmacol.*, 104, 440-444. 1991.

D'Orléans-Juste, P., Télémaque, S., Claing, A., Ihara, M. and Yano, M., Human big-endothelin-1 and endothelin-1 release prostacyclin via the activation of ET1 receptors in the rat perfused lung, *Br. J. Pharmacol.*, 105, 773-775. 1992.

Davenport, A. P., Nunez, D. J., Hall, J. A., Kauman, A. J. and Brown, M. J., Autoradiographical localization of bindig sites for endothelin-1 in human, pigs, and rats: functional relevance in humans, *J. Cardiovasc. Pharmacol.*, 13, S166-S170. 1989.

De Nucci, G., Thomas, R., D'Orléans-Juste, P., Antunes, E., Walder, C., Warner, T. D. and Vane, J. R., Pressor effect of circulating endothelin are limited by its removal in the pulmonary circulation and by the release of prostacyclin and endothelium-derived relaxing factor, *Proc. Natl. Acad. Sci. U.S.A.*, 85, 9797-9800. 1988.

DiMaria, G. U., Bellofiore, S., Malatino, L. S., Maggi, C. A., Torrisi, A. and Mistretta, A., Aerosolized endothelin-1, but not its C-terminal hexapeptide, causes airway narrowing in the rat, *Eur. Respir. J.*, 4, 528- 531. 1991.

DiMaria, G. U., Katayama, M., Borson, D. B. and Nadel, J. A., Neutral endopeptidase modulates endothelin-1-induced airway smooth muscle contraction in guinea pig trachea, *Regul. Pept.*, 39, 137-145. 1992.

Doherty, A.M., Endothelin: a new challenge, *J. Med. Chem.*, 35, 1494-1508. 1992.

Druml, W., Steltzer, H., Waldhausl, W., Lenz, K., Hammerle, A., Vierhapper, H., Gasic, S. and Wagner, O. F., Endothelin-1 in adult respiratory distress syndrome, *Am. Rev. Respir. Dis.*, 148, 1169-1173. 1993.

Dupuis, J., Goresky, C. A. and Stewart, D. J., Pulmonary removal and production of endothelin in the anesthetized dog, *J. Appl. Physiol.*, 76, 694-700. 1994.

Dyson, M. C. and Kadowitz, P. J., Analysis of responses to endothelins 1, 2, 3 and sarafotoxin 6b in airways of the cat, *J. Appl. Physiol.*, 71, 243-251. 1991.

Eglen, R. M., Michel, A. D., Sharif, N. A., Swank, S. R. and Whiting, R. L., The pharmacological properties of the peptide, endothelin, *Br. J. Pharmacol.*, 97, 1297-1307. 1989.

Ehreinreich, H., Anderson, R. W., Fox, C. H., Rieckman, P., Hoffman, G. S., Travis, W. D., Coligan, J. E., Kehrl, J. H. and Fauci, A. S., Endothelins, peptides with potent vasoactive properties, are produced by human macrophages, *J. Exp. Med.*, 172, 1741-1748. 1990.

Feng, C. J., Cai, B. Q., deBoisblanc, B. P., Hyman, A. A., Lippton, H. L. and Summer, W. R., Actions of endothelin (ET-1) and big ET in isolated perfused rat lung, *Am. Rev. Respir. Dis.*, 141, A483. 1990.

Filep, J. G., Battistini, B. and Sirois, P., Endothelin induces thromboxane release and contraction of isolated guinea pig airways, *Life Sci.*, 47, 1845-1850. 1990.

Filep, J. G., Battistini, B. and Sirois, P., Induction by endothelin-1 of epithelium-dependent relaxation of guinea pig trachea *in vitro*: role for nitric oxide, *Br. J. Pharmacol.*, 109, 637-644. 1993a.

Filep, J. G., Battistini, B. and Sirois, P., Pharmacological modulation of endothelin-induced contraction of guinea pig isolated airways and thromboxane release, *Br. J. Pharmacol.*, 103, 1633-1640. 1991a.

Filep, J. G., Sirois, M. G., Rousseau, A., Fournier, A. and Sirois, P., Effects of endothelin-1 on vascular permeability in the conscious rat: interactions with platelet-activating factor, *Br. J. Pharmacol.*, 104, 797-804. 1991b.

Filep, J. G., Battistini, B., Fournier, A. and Sirois, P., Relaxation by dexamethasone of isolated guinea pig airways precontracted with endothelin-1, *Eur. J. Pharmacol.*, 240, 315-318. 1993b.

Filep, J. G., Sirois, M. G., Foldes-Filep, E., Rousseau, A., Plante, G. E., Fournier, A., Yano, M. and Sirois, P., Enhancement by endothelin-1 of microvascular permeability via the activation of ET_A receptors, *Br. J. Pharmacol.*, 109, 880-886. 1993c.

Franco-Cereceda, A., Matran, R., Lou, Y.-P. and Lundberg, J. M., Occurence and effects of endothelin in guinea pig cardiopulmonary tissue, *Acta Physiol. Scand.*, 138, 539-547. 1990.

Germain, M., Battistini, B., Filep, J. G., Sirois, P. and Fournier, A., Endothelin derivatives showing potent effects in the guinea pig trachea, *Peptides*, 14, 613-619. 1993.

Giaid, A., Hamid, Q. A., Springall, D. R., Yanagisawa, M., Shinmi, O., Sawamura, T., Masaki, T., Kimura, S., Corrin, B. and Polak, J. M., Detection of endothelin immunoreactivity and mRNA in pulmonary tumors, *J. Pathol.*, 162, 15-22. 1990.

Giaid, A., Michel, R. P., Stewart, D. J., Sheppard, M., Corrin, B. and Hamid, Q., Expression of endothelin-1 in lung of patients with cryptogenic fibrosing alveolitis, *Lancet*, 341, 1550-1554. 1993a.

Giaid, A., Polak, J. M., Gaitonda, V., Hamid, Q. A., Moscoso, G., Legon, S., Uwanogho, D., Roncalli, M., Shinmi, O. and Sawamura, T., Distribution of endothelin-like immunoreactivity and mRNA in the developing and adult human lung, *Am. J. Respir. Cell Mol. Biol.*, 4, 50-58. 1991.

Giaid, A., Stewart, D. J. and Michel, R. P., Endothelin-1-like immunoreactivity in postobstructive pulmonary vasculopathy, *J. Vasc. Res.*, 30, 333-338. 1993b.

Giaid, A., Yanagisawa, M., Langleben, D., Michel, R. P., Levy, R., Shennib, H., Kimura, S., Masaki, T., Duguid, W. P. and Stewart, D. J., Expression of endothelin-1 in the lungs of patients with pulmonary hypertension, *N. Engl. J. Med.*, 328, 1732-1739. 1993c.

Glassberg, M. K., Ergul, A., Wanner, A. and Puett, D., Endothelin-1 promotes mitogenesis in airway smooth muscle cells, *Am. J. Respir. Cell Mol. Biol.*, 10, 316-321. 1994.

Goldie, R. G., Grayson, P. S., Knott, P. G., Self, G. J. and Henry, P. J., Predominance of $endothelin_A$ (ET_A) receptors in ovine airway smooth muscle and their mediation of ET-1-induced contraction, *Br. J. Pharmacol.*, 112, 749-756. 1994.

Grunstein, M. M., Chuang, S. T., Shramm, C. M. and Pawlowski, N. A., Role of endothelin-1 in regulating rabbit airway contractility, *Am. J. Physiol.*, 260, L75-L82. 1991.

Haendler, B., Hechler, U. and Schleuning, W. D., Molecular cloning of human endothelin (ET) receptors ET_A and ET_B, *J. Cardiovasc. Pharmacol.*, 20, S1-S4. 1992.

Hagiwara, H., Kozuka, M., Eguchi, S., Shibabe, S., Ito, T. and Hirose, S., Solubilization of endothelin receptors from bovine lung plasma membranes in a non-aggregated state and estimation of their minimal functional sizes, *Biochem. Biophys. Res. Commun.*, 172, 576-581. 1990.

Haller, H., Schaberg, T., Lindschau, C., Lode, H. and Distler, A., Endothelin increases $[Ca^{2+}]$, protein phosphorylation, and 0^{2-} production in human alveolar macrophages, *Am. J. Physiol.*, 261, L478-L484. 1991.

Hasunuma, K., Rodman, D. M., O'Brien, R. F. and McMurtry, I. F., Endothelin 1 causes pulmonary vasodilation in rats, *Am. J. Physiol.*, 259, H48-H54. 1990.

Hay, D. W. P., Guinea pig tracheal epithelium and endothelin, *Eur. J. Pharmacol.*, 171, 241-245. 1989.

Hay, D. W. P., Mechanism of ET-induced contraction in guinea pig trachea: comparison with rat aorta, *Br. J. Pharmacol.*, 100, 383-392. 1990.

Hay, D. W. P., Hubbard, W. C. and Undem, B. J., Endothelin-induced contraction and mediator release in human bronchus, *Br. J. Pharmacol.*, 110, 392-398. 1993a.

Hay, D. W. P., Luttmann, M. A., Hubbard, W. C. and Undem, B. J., Endothelin receptor subtypes in human and guinea pig pulmonary tissues, *Br. J. Pharmacol.*, 110, 1175-1183. 1993b.

Hay, D. W. P., Pharmacological evidence for distinct endothelin receptors in guinea pig bronchus and aorta, *Br. J. Pharmacol.*, 106, 759-761. 1992.

Helset, E., Kjaeve, J. and Hauge, A., Endothelin-1-induced increases in microvascular permeability in isolated perfused rat lungs require leukocytes and plasma, *Circ. Shock*, 39, 15-20. 1993.

Hemsen, A., Franco-Cereceda, A., Matran, R., Rudehill, A. and Lundberg, J. M., Occurrence, specific binding sites and functional effects of endothelin in human cardiopulmonary tissue, *Eur. J. Pharmacol.*, 191, 219-328. 1990.

Hemsen, A., Larsen, O. and Lundberg, J. M., Characteristics of endothelin A and B binding sites and their vascular effects in pig peripheral tissues, *Eur. J. Pharmacol.*, 208, 313-322. 1991.

Henry, P. J., Endothelin-1 (ET-1)-induced contraction in rat isolated trachea: involvement of ET_A and ET_B receptors and multiple signal transduction systems, *Br. J. Pharmacol.*, 110, 435-441. 1993.

Henry, P. J., Rigby, P. J., Self, G. J., Preuss, J. M. and Goldie, R. G., Endothelin-1-induced [^{3}H]-inositol phosphate accumulation in rat trachea, *Br. J. Pharmacol.*, 105, 135-141. 1992.

Henry, P. J., Rigby, P. J., Self, G. J., Preuss, J. M. and Goldie, R. G., Relationship between endothelin-1 binding site densities and constrictor activities in human and animal airway smooth muscle, *Br. J. Pharmacol.*, 100, 786-792. 1990.

Hisaki, K., Matsumura, Y., Maekawa, H., Fujita, K., Takaoka, M. and Morimoto, S., Conversion of big ET-1 in the rat lung: role of phosphoramidon-sensitive endothelin-1-converting enzyme, *Am. J. Physiol.*, 266, H422-H428. 1994.

Horgan, M. J., Pinheiro, J. M. B. and Malik, A. B., Mechanism of endothelin-1-induced pulmonary vasoconstriction, *Circ. Res.*, 69, 157-164. 1991.

Hori, S., Komatsu, Y., Shigemoto, R., Mizuno, N. and Nakanishi, S., Distinct tissue distribution and cellular loalization of two messenger ribonucleic acids encoding different subtypes of rat endothelin receptors, *Endocrinology*, 130, 1885-1895. 1992.

Ihara, M., Ishikawa, K., Fukuroda, T., Saeki, T., Funabashi, K., Fukami, T., Suda, H. and Yano, M., *In vitro* biological profile of a highly potent novel endothelin (ET) antagonist BQ-123 selective for the ET_A receptor, *J. Cardiovasc. Pharmacol.*, 20, S11-S14. 1992.

Inui, T., James, A. F., Fujitani, Y., Takimoto, M., Okada, T., Yamamura, T. and Urade, Y., ET_A and ET_B receptors on single smooth muscle cells cooperate in mediating guinea pig tracheal contraction, *Am. J. Physiol.*, 266, L113-L124. 1994.

Ishikawa, S., Tsukada, H., Yuasa, H., Fukue, M., Wei, S., Onizuka, M., Miyauchi, T., Ishikawa, T., Mitsui, K., Goto, K. and Hori, M., Effects of endothelin-1 and conversion of big endothelin-1 in the isolated perfused rabbit lung, *J. Appl. Physiol.*, 72, 2387-2392. 1992.

Kadowitz, P. J., McMahon, T. J., Hood, J. S., Feng, C.-J., Minkes, R. K. and Dyson, M. C., Pulmonary vascular and airway responses to endothelin-1 are mediated by different mechanisms in the cat, *J. Cardiovasc. Pharmacol.*, 17, S374-S377. 1991.

Kanse, S. M., Ghatei, M. A. and Bloom, S. R., Endothelin binding sites in porcine aortic and rat lung membranes, *Eur. J. Biochem.*, 182, 175-179. 1989.

Kent, A. and Keenan, A. K., Evidence for signaling by endothelin ET-A and ET-B receptors in bovine pulmonary artery smooth muscle cells, *Br. J. Pharmacol.*, 112, 549P. 1994.

Kimura, S., Kasuya, Y., Sawamura, T., Shinimi, O., Sugita, Y., Yanagisawa, M., Goto, K. and Masaki, T., Conversion of big endothelin-1 to 21-residue endothelin-1 is essential for expression of full vasoconstrictor activity: structure activity relationships of big endothelin-1, *J. Cardiovasc. Pharmacol.*, 5, S5-S7. 1989.

Kitamura, K., Tanaka, T., Kato, J., Eto, T. and Tanaka K., Chromatographic characterization of immunoreactive endothelin in rat lung, *Life Sci.*, 46, 405-409. 1990a.

Kitamura, K., Tanaka, T., Kato, J., Eto, T. and Tanaka K., Regional distribution of immunoreactive endothelin in porcine tissue: abundance in inner medulla of kidney, *Biochem. Biophys. Res. Commun.*, 161, 348-352. 1989.

Kitamura, K., Yukawa, T., Morita, S., Ichiki, Y., Eto, T. and Tanaka, K., Distribution and molecular form of immunoreactive big endothelin-1 in porcine tissue, *Biochem. Biophys. Res. Commun.*, 170, 497-503. 1990b.

Kozuka, M., Ito, T., Hirose, S., Lodhi, K. M. and Hagiwara, H., Purification and characterization of bovine lung endothelin receptor, *J. Biol. Chem.*, 266, 16892-16896. 1991.

Kundu, G. C. and Misono, K. S., Affinity labeling of endothelin receptors in bovine and rat lung membranes by $N^{\epsilon 9}$-azidobenzoyl-^{125}I-endothelin-1, *Mol. Cell. Endocrinol.*, 79, 85-92. 1991.

Kundu, G. C. and Wilson, I. B., Identification of endothelin in converting enzyme in bovine lung membranes using a new fluorogenic substrate, *Life Sci.*, 50, 965-970. 1992.

Kuwahara, M., Kuwahara, M. and Suzuki, N., Production of endothelin-1 and big endothelin-1 by pleural mesothelial cells, *FEBS Lett.*, 17, 21-24. 1992.

LaDouceur, D. M., Flynn, M. A., Keiser, J. A., Reynolds, E. and Haleen, S. J., ET_A and ET_B receptors coexist on rabbit pulmonary artery vascular smooth muscle mediating contraction, *Biochem. Biophys. Res. Commun.*, 196, 209-215. 1993.

Lagente, V., Boichot, E., Mencia-Huerta, J. and Braquet, P., Failure of aerosolized endothelin (ET-1) to induce bronchial hyperreactivity in the guinea pig, *Fundam. Clin. Pharmacol.*, 4, 275-280. 1990a.

Lagente, V., Chabrier, P. E., Mencia-Huerta, J. M. and Braquet, P., Pharmacological modulation of the bronchopulmonary action of the vasoactive peptide, endothelin, administered by aerosol in the guinea pig, *Biochem. Biophys. Res. Commun.*, 158, 625-632. 1989.

Lagente, V., Touvay, C., Mencia-Huerta, J. M., Chabrier, P. E. and Braquet, P., Bronchopulmonary effects of endothelin, *Clin. Exp. Allergy*, 20, 343-348. 1990b.

Lee, H.-K., Leikauf, G. D. and Sperelakis, N., Electromechanical effects of endothelin on ferret bronchial and tracheal smooth muscle, *J. Appl. Physiol.*, 68, 417-420. 1990.

Liben, A., Stewart, D. J., DeMarte, J. and Perreault, T., Ontogeny of big endothelin-1 effects in newborn piglet pulmonary vasculature, *Am. J. Physiol.*, 34, H139-H145. 1993.

Lippton, H. L., Hauth, T. A., Summer, W. R. and Hyman, A. L., Endothelin produces pulmonary vasoconstriction and systemic vasodilation, *J. Appl. Physiol.*, 66, 1008-1012. 1989.

Lippton, H. L., Cohen, G. A., Knight, M., McMurtry, I. F., Gillot, D., Arena, F., Summer, W. and Hyman, A.L., Evidence for distinct endothelin receptors in the pulmonary vascular bed *in vivo*, *J. Cardiovasc. Pharmacol.*, 17, S370-S373. 1991a.

Lippton, H. L., Cohen, G. A., McMurtry, I. F. and Hyman, A. L., Pulmonary vasodilation to endothelin isopeptides *in vivo* is mediated by potassium channel activation, *J. Appl. Physiol.*, 70, 947-952. 1991b.

Lippton, H. L., Hauth, T. A., Cohen, G. A. and Hyman, A. L., Functional evidence for different receptors in the lung, *J. Appl. Physiol.*, 75, 38-48. 1993.

Lippton, H. L., Ohlstein, E. H., Summer, W. R. and Hyman, A. L., Analysis of responses to endothelins in the rabbit pulmonary and systemic vascular beds, *J. Appl. Physiol.*, 70, 331-341. 1991c.

Lueddeckens, G., Becker, K., Rappold, R. and Förster, W., Influence of aminophylline and ketotifen in comparison to the lipoxygenase inhibitors NDGA and esculetin and the PAF antagonists WEB 2107 and BN 52021 on endothelin-1 induced vaso- and bronchoconstriction, *Prostaglandins Leuko. Essential Fatty Acids*, 44, 155-158. 1991.

MacCumber, M. W., Ross, C. A., Glaser, B. M. and Snyder, S. H., Endothelin: visualization of mRNAs by *in situ* hybridization provides evidence for local action, *Proc. Natl. Acad. Sci. U.S.A.*, 86, 7285-7289. 1989.

Macquin-Mavier, I., Levame, M., Istin, N. and Harf, A., Mechanisms of endothelin-mediated bronchoconstriction in the guinea pig, *J. Pharmacol. Exp. Ther.*, 250, 740-745. 1989.

Maggi, C. A., Giuliani, S., Patachini, R., Rovero, P., Giachetti, A. and Meli, A., The activity of peptides of the endothelin family in various mammalian smooth muscle preparations, *Eur. J. Pharmacol.*, 174, 23-31. 1989a.

Maggi, C. A., Giuliani, S., Patachini, R., Santicioli, P., Giachetti, A. and Meli, A., Further studies on the response of the guinea pig isolated bronchus to endothelins and sarafotoxin S6b, *Eur. J. Pharmacol.*, 176, 1-9. 1990.

Maggi, C. A., Patachini, R., Giuliani, S. and Meli, A., Potent contractile effect of endothelin in isolated guinea pig airways, *Eur. J. Pharmacol.*, 160, 179-182. 1989b.

Mann, J., Farrukh, I. S. and Michael, J. R., Mechanisms by which endothelin 1 induces pulmonary vasoconstriction in the rabbit, *J. Appl. Physiol.*, 71, 410-416. 1991.

Marciniak, S. J., Plumpton, C., Barker, P. J., Huskisson, N. S. and Davenport, A. P., Localization of immunoreactive endothelin and proendothelin in the human lung, *Pulm. Pharmacol.*, 5, 175-182. 1992.

Masaki, T., Kimura, S., Yanagisawa, M. and Goto, K., Molecular and cellular mechanism of endothelin regulation. Implications for vascular function, *Circulation*, 84, 1457-1468. 1991.

Masuda, Y., Miyazaki, H., Kohdoh, M., Watanabe, H., Yanagisawa, M., Masaki, T. and Murakami, K., Two different forms of endothelin receptors in rat lung, *FEBS Lett.*, 257, 208-210. 1989.

Matsuse, T., Fukuchi, Y., Suruda, T., Oouchi, Y. and Orimo, H., Effect of endothelin-1 on pulmonary resistance in rats, *J. Appl. Physiol.*, 68, 2391-2393. 1990.

Mattoli, S., Mezzetti, M., Riva, G., Allegra, L. and Fasoli, A., Specific binding of endothelin on human bronchial smooth muscle cells in culture and secretion of endothelin-like material from bronchial epithelial cells, *Am. J. Respir. Cell. Mol. Biol.*, 3, 145-151. 1990.

Mattoli, S., Soloperto, M., Marini, M. and Fasoli, A., Levels of endothelin in the bronchoalveolar lavage fluid of patients with symptomatic asthma and reversible airflow obstruction, *J. Allergy Clin. Immunol.*, 88, 376-384. 1991a.

Mattoli, S., Soloperto, M., Marini, M. and Fasoli, A., Mechanisms of calcium mobilization and phosphoinositide hydrolysis in human bronchial smooth muscle cells by endothelin 1, *Am. J. Respir. Cell Mol. Biol.*, 5, 424-430. 1991b.

McKay, K. O., Black, J. L. and Armour, C. L., Phosphoramidon potentiates the contractile response to endothelin-3, but not endothelin-1 in isolated airway tissue. *Br. J. Pharmacol.*, 105, 929-932, 1992.

McKay, K. O., Black, J. L. and Armour, C. L., The mechanism of action of endothelin in human lung, *Br. J. Pharamcol.*, 102, 422-428. 1991a.

McKay, K. O., Black, J. L., Diment, L. M. and Armour, C. L., Functional and autoradiographic studies of endothelin-1 and endothelin-2 in human bronchi, pulmonary arteries, and airway parasympathetic ganglia, *J. Cardiovasc. Pharmacol.*, 17, S206-S209. 1991b.

Minkes, R. K. and Kadowitz, P. J., Differential effects of endothelin peptides in the systemic and pulmonary vascular beds of the cat, *Neurochem. Int.*, 18, 547-552. 1991.

Minkes, R. K., Bellan, J. A., Saroyan, R. M., Kerstein, M. D., Coy, D. H., Murphy, W. A., Nossaman, B. D., McNamara, D. B. and Kadowitz, P. J., Analysis of cardiovascular and pulmonary responses to endothelin-1 and endothelin-3 in the anesthetized cat, *J. Pharmacol. Exp. Ther.*, 253, 1118-1125. 1990.

Mitaka, C., Hirata, Y., Nagura, T., Tsunoda, Y. and Amaha, K., Circulating endothelin-1 concentrations in acute respiratory failure, *Chest*, 104, 476-480. 1993.

Moon, D. G., Horgan, M., Andersen, T. T., Krystek, S. R., Fenton, J. W. and Malik, A. B., Endothelin-like pulmonary vasoconstrictor peptide release by α-thrombin, *Proc. Natl. Acad. Sci. U.S.A.*, 86, 9529-9533. 1989.

Morel, D. R., Weber, A., Lacroix, J. S., Jorge-Costa, M. and Pittet, J.-F., Involvement of cyclooxygenase metabolites in the cardiopulmonary vascular and airway response to i.v. bolus of endothelin in sheep, *Am. Rev. Respir. Dis.*, 141, A493. 1990.

Nagai, H., Suda, H., Kitagaki, K. and Koda, A., Effect of tranilast on endothelin-induced bronchoconstriction in guinea pigs, *J. Pharmacobiol-Dyn.*, 14, 309-314. 1991.

Nagase, T., Fukuchi, Y., Jo, C., Teramoto, S., Uejima, Y., Ishida, K., Shimizu, T. and Orimo, H., Endothelin-1 stimulates arachidonate 15-lipoxygenase activity and oxygen radical formation in the rat distal lung, *Biochem. Biophys. Res. Commun.*, 168, 485-489. 1990.

Nakamichi, K., Ihara, M., Kobayashi, M., Saeki, T., Ishikawa, K. and Yano, M., Different distribution of endothelin receptor subtypes in pulmonary tissues revealed by the novel selective ligands BQ-123 and [$Ala^{1,3,11,15}$]ET-1, *Biochem. Biophys. Res. Commun.*, 182, 144-150. 1992

Nakano, J., Takizawa, T., Ohtoshi, T., Shoji, S., Yamaguchi, M., Ishii, A., Yanagisawa, M. and Ito, K., Endotoxin and proinflammatory cytokines stimulate endothelin-1 expression and release by airway epithelial cells, *Clin. Exp. Allergy*, 24, 330-336. 1994.

Namiki, A., Hirata, Y., Ishikawa, M., Moroi, M., Aikawa, J. and Machii, K., Endothelin-1 and endothelin-3-induced vasorelaxation via common generation of endothelium-derived nitric oxide, *Life Sci.*, 50, 677-682. 1992.

Naruse, M., Narase, K., Kurimoto, F., Horiuchi, J., Tsuchiya, K., Kawana, M., Kato, Y., Zeng, Z., Sakurai, H., Demura, H. and Schizume, K., Radioimmunoassay for endothelin and immunoreactive endothelin in culture medium of bovine endothelial cells, *Biochem. Biophys. Res. Commun.*, 160, 662-668. 1989.

Neuser, D., Steinke, W., Dellweg, H., Kazda, S. and Stasch, J. P., ^{125}I-endothelin-1 and ^{125}Ibig endothelin-1 in rat tissues: autoradiographic localization and receptor binding, *Histochemistry*, 95, 621-628. 1991.

Neuser, D., Steinke, W., Theiss, G. and Stasch, J.P., Autoradiographic localization of [^{125}I]endothelin-1 and [^{125}I]atrial natriuretic peptide in rat tissue: a comparative study, *J. Cardiovasc. Pharmacol.*, 13, S67-S73. 1989.

Ninomiya, H., Uchida, Y., Ishii, Y., Nomura, A., Kameyama, M., Saotome, M., Endo, T. and Hasegawa, S., Endotoxin stimulates endothelin release from cultured epithelial cells of guinea pig trachea, *Eur. J. Pharmacol.*, 203, 299-302. 1991.

Ninomiya, H., Yu, X. Y., Hasegawa, S. and Spannhake, E. W., Endothelin-1 induces stimulation of prostaglandin synthesis in cells obtained from canine airways by bronchoalveolar lavage, *Prostaglandins*, 43, 401-411. 1992.

Noguchi, K., Fukuroda, T., Ikeno, Y., Hirose, H., Tsukada, Y., Nishikibe, M., Ikemoto, F., Matsuyama, K. and Yano, M., Local formation and degradation of endothelin-1 in guinea pig airway tissues, *Biochem. Biophys. Res. Commun.*, 179, 830-835. 1991.

Noguchi, K., Noguchi, Y., Hirose, H., Nishikibe, M., Ihara, M., Ishikawa, K. and Yano, M., Role of endothelin ET_B receptors in bronchoconstrictor and vasoconstrictor responses in guinea pigs, *Eur. J. Pharmacol.*, 233, 47-51. 1993.

Nomura, A., Uchida, Y., Kameyama, M., Saotome, M., Oki, K. and Hasegawa, S., Endothelin and bronchial asthma, *Lancet*, ii, 747-748. 1989.

Nossaman, B. D., McMahon, T. J., Ragheb, M. S., Ibrahim, I. N., Babycos, C. R., Hood, J. S. and Kadowitz, P. J., Blockade of thromboxane/endoperoxide receptor-mediated responses in the pulmonary vascular bed of the cat by sulotroban, *Eur. J. Pharmacol.*, 213, 1-7. 1992.

Noveral, J. P., Rosenberg, S. M., Anbar, R. A., Pawlowski, N. A. and Grunstein, M. M., Role of endothelin-1 in regulating proliferation of cultured rabbit airway smooth muscle cells, *Am. J. Physiol.*, 263, L317-L324. 1992.

O'Donnell, S. R., Wanstall, J. C. and Zeng X.-P., Pinacidil antagonism of endothelin-induced contractions of smooth muscle in the lungs: differences between tracheal and pulmonary artery preparations, *J. Pharmacol. Exp. Ther.*, 252, 1318-1323. 1990.

O'Donnell, S. R., Wanstall, J. C. and Zeng X.-P., Tissue selectivity and spasmogen selectivity of relaxant drugs in airway and pulmonary vascular smooth muscle contracted by $PGF_{2\alpha}$ or endothelin, *Br. J. Pharmacol.*, 102, 311-316. 1991.

Ohlstein, E. H., Arleth, A., Ezekiel, M., Horohonich, S., Ator, M. A., Caltabiano, M. M. and Sung, C. P., Biosynthesis and modulation of endothelin from bovine pulmonary arterial endothelial cells, *Life Sci.*, 46, 181-188. 1990.

Opgenorth, T. J., Wu-Wong, J. R. and Shiosaki, K., Endothelin-converting enzymes, *FASEB J.*, 6, 2653-2659. 1992.

Payne, A. N. and Whittle, B. J. R., Potent cyclo-oxygenase-mediated bronchoconstrictor effects of endothelin in the guinea pig *in vivo*, *Eur. J. Pharmacol.*, 158, 303-304. 1988.

Peacock, A. J., Dawes, K. E., Shock, A., Gray, A. J., Reeves, J. T. and Laurent, G. J., Endothelin-1 and endothelin-3 induce chemotaxis and replication of pulmonary artery fibroblasts, *Am. J. Respir. Cell Mol. Biol.*, 7, 492-499. 1992.

Perreault, T. and De Marte, J., Endothelin-1 has a dilator effect on neonatal pig pulmonary vasculature, *J. Cardiovasc. Pharmacol.*, 18, 43-50. 1991.

Pinheiro, J. M. B., Everitt, J. and Malik, A. B., Endothelin is a cyclooxygenase-independent vasodilator in isolated, perfused newborn pig lungs, *Am. Rev. Respir. Dis.*, 141, A483. 1990.

Plews, P. I., Abdel-Malek, Z. A., Doupnik, C. A. and Leikauf, G. D., Endothelin stimulates chloride secretion across canine tracheal epithelium, *Am. J. Physiol.*, 261, L188-L194. 1991.

Pons, F., Loquet, I., Touvay, C., Roubert, P., Chabrier, P.-E., Mencia-Huerta, J. M. and Braquet, P., Comparison of the bronchopulmonary and presssor activities of endothelin isoforms ET-1, ET-2, and ET-3 and characterization of their binding sites in guinea pig lung, *Am. Rev. Respir. Dis.*, 143, 294-300. 1991a.

Pons, F., Touvay, C., Lagente, V., Mencia-Huerta, J. M. and Braquet, P., Bronchopulmonary and pressor effects of big-endothelin-1 in the guinea pig, *Neurochem. Int.*, 18, 481-483. 1991b.

Pons, F., Touvay, C., Lagente, V., Mencia-Huerta, J. M. and Braquet, P., Comparison of the effects of intra-arterial and aerosol administration of endothelin-1 (ET-1) in the guinea pig isolated lung, *Br. J. Pharmacol.*, 102, 791-796. 1991c.

Power, R. F., Wharton, J., Zhao, Y., Bloom, S. R. and Polak, J. M., Autoradiographic localization of endothelin-1 binding sites in the cardiovascular and respiratory systems, *J. Cardiovasc. Pharmacol.*, 13, S50-S56. 1989.

Raffestin, B., Adnot, S., Eddahibi, S., Macquin-Mavier, I., Braquet, P. and Chabrier, P. E., Pulmonary vascular response to endothelin in rats, *J. Appl. Physiol.*, 70, 567-574. 1991.

Ray, S. G., McMurray, J. J., Morton, J. J. and Dargie, H. J., Circulating endothelin is not extracted by the pulmonary circulation in man, *Chest*, 102, 1143-1144. 1992.

Rimar, S. and Gillis, N., Differential uptake of endothelin-1 by the coronary and pulmonary circulations, *J. Appl. Physiol.*, 73, 557-562. 1992.

Rodman, D. M., McMurtry, I. F., Peach, J. L. and O'Brien, R. F., Comparative pharmacology of rat and porcine endothelin in rat aorta and pulmonary artery, *Eur. J. Pharmacol.*, 165, 297-300. 1989.

Rovero, P., Patacchini, R. and Maggi, C. A., Structure-activity studies on endothelin (16-21), the C-terminal hexapeptide of the endothelins, in the guinea pig bronchus, *Br. J. Pharmacol.*, 101, 232-234. 1990.

Rozengurt, N., Springall, D. R. and Polak, J. M., Localization of endothelin-like immunoreactivity in airway epithelium of rats and mice, *J. Pathol.*, 160, 5-8. 1990.

Saito, Y., Mizuno, T., Itakura, M., Suzuki, Y., Ito, T., Hagisawa, H., et al., Primary structure of bovine endothelin ET_B receptor and identification of signal peptidase and metal proteinase cleavage sites, *J. Biol.Chem.*, 266, 23433-23437. 1991.

Sakata, K., Ozaaki, H., Kwon, S. C. and Karaki, H., Effects of ET on the mechanical activity and cytosolic calcium levels of various types of smooth muscle, *Br. J. Pharmacol.*, 98, 483-492. 1989.

Sakurai, T., Yanagisawa, M., Takuwa, Y., Miyazaki, H., Kimura, S., Goto, K. and Masaki, T., Cloning of a cDNA encoding a non-isopeptide-selective subtype of the endothelin receptor, *Nature*, 348, 732-735. 1990.

Saotome, M., Ninomiya, H., Nomura, A., Ohse, H., Endoh, T., Hasegawa, S. and Uchida, Y., Endothelin-1 induced relaxation of guinea pig trachea after an anaphylactic reaction, *Arerugi*, 40, 1377-1383. 1991.

Sarria, B., Naline, E., Morcillo, E., Cortijo, J., Esplugues, J. and Advenier, C., Calcium dependence of the contraction produced by endothelin (ET-1) in isolated guinea pig trachea, *Eur. J. Pharmacol.*, 187, 445-453. 1990.

Satoh, M., Shimura, S., Ishihara, H., Nagaki, M., Sasaki, H. and Takishima, T., Endothelin-1 stimulates chloride secretion across canine tracheal epithelium, *Respiration*, 59, 145-150. 1992.

Sawamura, T., Shinmi, O., Kishi, N., Sugita, Y., Yanagisawa, M., Goto, K., Masaki, T. and Kimura, S., Characterization of phosphoramidon-sensitive metalloproteinases with endothelin-converting enzyme activity in porcine lung membrane, *Biochem. Biophys. Acta*, 1161, 295-302. 1993.

Schumacher, W. A., Steinbacher, T. E., Allen, G. T. and Ogletree, M. L., Role of thromboxane receptor activation in the bronchospastic response to endothelin, *Prostaglandins*, 40, 71-79. 1990.

Secrest, R. J. and Cohen, M. L., Endothelin: differential effects in vascular and nonvascular smooth muscle, *Life Sci.*, 45, 1365-1372. 1989.

Shimura, S., Ishihara, H., Satoh, M., Masuda, T., Nagaki, N., Sasaki, H. and Takishima, T., Endothelin regulation of mucus glycoprotein secretion from feline tracheal submucosal glands, *Am. J. Physiol.*, 262, L208-L213. 1992.

Shiosaki, K., Tasker, A. S., Sullivan, G. M., Sorensen, B. K., von Geldern, T. W., Wu-Wong, J. R., Marselle, C. A. and Opgenorth, T. J., Potent and selective inhibitors of an aspartyl protease-like endothelin converting enzyme identified in rat lung, *J. Med. Chem.*, 36, 468-478. 1993.

Shirakami, G., Nakao, K., Saito, Y., Magaribuchi, T., Jougasaki, M., Mukoyama, M., Arai, H., Hosoda, K., Suga, S., Ogawa, Y., Yamada, T., Mori, K. and Imura, H., Acute pulmonary alveolar hypoxia increases lung and plasma endothelin-1 levels in conscious rats, *Life Sci.*, 48, 969-976. 1991.

Simmet, T., Pritze, S., Thelen, K.I. and Peskar, B.A., Release of endothelin in the oleic acid-induced respiratory distress syndrome in rats, *Eur. J. Pharmacol.*, 211, 319-322. 1992.

Sirois, M. G., Filep, J. G., Rousseau, A., Fournier, A., Plante, G. E. and Sirois, P., Endothelin-1 enhances vascular permeability in conscious rats: role of thromboxane A_2, *Eur. J. Pharmacol.*, 214, 119-125. 1992.

Sofia, M., Mormile, M., Faraone, S., Alifano, M., Zofra, S., Romano, L. and Carratu, L., Increased endothelin-like immunoreactive material on bronchoalveolar lavage fluid from patients with bronchial asthma and patients with interstitial lung disease, *Respiration*, 60, 89-95. 1993.

Spinella, M. J., Makik, A. B., Everitt, J. and Andersen, T. T., Design and synthesis of a specific endothelin 1 antagonist: effects on pulmonary vasoconstriction, *Proc. Natl. Acad. Sci. U.S.A.*, 88, 7443-7446. 1991.

Springall, D. R., Howart, P. H., Counihan, H., Djukanovic, R., Holgate, S. T. and Polak, J. M., Endothelin immunoreactivity of airway epithelium in asthmatic patients, *Lancet*, 337, 697-701. 1991.

Steffan, M. and Russell, J. A., Signal transduction in endothelin-induced contraction of rabbit pulmonary vein, *Pulm. Pharmacol.*, 3, 1-7. 1990.

Stelzner, T. J., O'Brien, R. F., Tanagisawa, M., Sakurai, T., Sato, K., Webb, S., Zamora, M., McMurtry, I. F. and Fisher, J. H., Increased lung endothelin-1 production in rats with idiopathic pulmonary hypertension, *Am. J. Physiol.*, 262, L614-L620. 1992.

Stewart, D. J., Levy, R. D., Cernacek, P. and Langleben, D., Increased plasma ET-1 in pulmonary hypertension. Marker or mediator of disease, *Ann. Int. Med.*, 114, 464-9. 1991.

Sudjarwo, S. A., Hori, M., Takai, M., Urade, Y., Okada, T. and Karaki H., A novel subtype of endothelin B receptor mediating contraction in swine pulmonary vein, *Life Sci.*, 53, 431-437. 1993.

Tod, M. L. and Cassin, S., Endothelin-1-induced pulmonary arterial dilation is reduced by N^w-nitro-L-arginine in fetal lambs, *J. Appl. Physiol.*, 72, 1730-1734. 1992.

Toga, H., Ibe, B. O. and Raj, J. U., *In vitro* responses of ovine intrapulmonary arteries and veins to endothelin-1, *Am. J. Physiol.*, 263, L15-L21. 1992.

Toga, H., Raj, J. U., Hillyard, R., Ku, B. and Anderson, J., Endothelin effects in isolated, perfused lamb lungs: role of cyclooxygenase inhibition and vasomotor tone, *Am. J. Physiol.*, 261, H443-H450. 1991.

Touvay, C., Vilain, B., Pons, F., Chabrier, P. E., Mencia-Huerta, J. M. and Braquet, P., Bronchopulmonary and vascular effect of endothelin in the guinea pig, *Eur. J. Pharmacol.*, 176, 23-33. 1990.

Tschirhart, E. J., Drijfhout, J. W., Pelton, J. T., Miller, R. C., Jones, C. R., Endothelins: functional and autoradiographic studies in guinea pig trachea, *J. Pharmacol. Exp. Ther.*, 258, 381-387. 1991.

Turner, N. C., Dollery, C. T. and Williams, A. J., Endothelin-1-induced contractions of vascular and tracheal smooth muscle: effects of nicardipine and BRL 34915, *J. Cardiovasc. Pharmacol.*, 13, S180-S182. 1989a.

Turner, N. C., Power, R. F., Polak, J. M., Bloom, S. R. and Dollery, C. T., Endothelin-induced contractions of tracheal smooth muscle and identification of specific endothelin binding sites in the trachea of the rat, *Br. J. Pharmacol.*, 98, 361-366. 1989b.

Uchida, Y., Nimomiya, H., Saotome, M., Nomura, A., Ohtsuka, M., Yanagisawa, M., Goto, K., Masaki, T. and Hasegawa, S., Endothelin, a novel vasoconstrictor peptide, as potent bronchoconstrictor, *Eur. J. Pharmacol.*, 154, 227-228. 1988.

Uchida, Y., Ninomiya, H., Sakamoto, T., Lee, J. Y., Endo, T., Nomura, A., Hasegawa, S. and Hirata, F., ET-1 released histamine from guinea pig pulmonary but not peritoneal mast cells, *Biochem. Biophys. Res. Commun.*, 189, 1196-1201. 1992.

Uchida, Y., Saotome, M., Nomura, A., Ninomiya, H., Ohse, H., Hirata, F. and Hasegawa, S., Endothelin-1-induced relaxation of guinea pig trachealis muscles, *J. Cardiovasc. Pharmacol.*, 17, S210-S212. 1991.

Vittori, E., Marini, M., Fasoli, A., De Franchis, R. and Mattoli, S., Increased expression of endothelin in bronchial epithelial cells of asthmatic patients and effect of corticosteroids$_{1-3}$, *Am. Rev. Respir. Dis.*, 146, 1320-1325. 1992.

Wanstall, J. C. and O'Donnell, S. R., Endothelin and 5-hydroxytryptamine on rat pulmonary artery in pulmonary hypertension, *Eur. J. Pharmacol.*, 176, 159-168. 1990.

Wanstall, J. C. and O'Donnell, S. R., Endothelin-induced contractions of rat pulmonary artery are not affected by drugs acting on potassium channels, *J. Pharm. Pharmacol.*, 43, 739-741. 1991.

White, S. R., Hathaway, D. P., Umans, J. G., Tallet, J., Abrahams, C. and Leff, A. R., Epithelial modulation of airway smooth muscle response to endothelin-1, *Am. Rev. Respir. Dis.*, 144, 373-378. 1991.

Wong, J., Vanderford, P. A., Winters, J. W., Chang, R., Soifer, S. J. and Fineman, J. R., Endothelin-1 does not mediate acute hypoxic pulmonary vasoconstriction in the intact newborn lamb, *J. Cardiovasc. Pharmacol.*, 22, S262-S266. 1993.

Wu, T., Mullol, J., Rieves, R. D., Logun, C., Hausfield, J., Kaliner, M. A. and Shelhamer, J. H., Endothelin-1 stimulates eicosanoid production in cultured human nasal mucosa, *Am. J. Respir. Cell Mol. Biol.*, 6, 168-174. 1992.

Wu, T., Rieves, R. D., Larivée, P., Logun, C., Lawrence, M. G. and Shelhamer, J. H., Production of eicosanoids in response to endothelin-1 and identification of specific endothelin-1 binding sites in airway epithelial cells, *Am. J. Respir. Cell. Mol. Biol.*, 8, 282-290. 1993.

Wu-Wong, J. R., Budzik, G. P., Devine, E. M. and Opgenorth, T. J., Characterization of endothelin converting enzyme in rat lung, *Biochem. Biophys. Res. Commun.*, 171, 1291-1296. 1990.

Wu-Wong, J. R., Devine, E. M., Budzik, G. P. and Opgenorth, T. J., Characterization and partial purification of endothelin-converting enzyme in rat lung, *J. Cardiovasc. Pharmacol.*, 17, S20-S25. 1991.

Wypij, D. M., Nichols, J. S., Novak, P. J., Stacy, D. L., Berman, J. and Wiseman, J. S., Role of mast cell chymase in the extracellular processing of big-endothelin-1 to endothelin-1 in the perfused rat lung, *Biochem. Pharmacol.*, 43, 845-853. 1992.

Yanagisawa, M., Kurihara, H., Kimura, S., Tomobe, Y., Kobayashi, Y., Mitsui, Y., Yazaki, Y., Goto, K. and Masaki, T., A novel potent vasoconstrictor peptide produced by vascular endothelial cells, *Nature*, 332, 411-415. 1988.

Yang, C. M., Yo, Y. L., Ong, R. and Hsieh, J. T., Endothelin- and sarafotoxin-induced phosphoinisitide hydrolysis in cultured canine tracheal smooth muscle cells, *J. Neurochem.*, 62, 1440-1448. 1994.

Yoshibayashi, M., Nishioka, K., Nakao, K., Saito, Y., Matsumura, M., Ueda, T., Temma, S., Shirakami, G., Imura, H. and Mikawa, H., Plasma endothelin concentrations in patients with pulmonary hypertension associated with congenital heart defects. Evidence for increased production of endothelin in pulmonary circulation, *Circulation*, 84, 2280-2285. 1991.

Yurdakos, E. and Webber, S. E., Endothelin-1 inhibits pre-stimulated tracheal submucosal gland secretion and epithelial albumin transport, *Br. J. Pharmacol.*, 104, 1050-1056. 1991.

Zamora, M. A., Dempsey, E. C., Walchak, S. J. and Stelzner, T. J., BQ123, an ET_A receptor antagonist, inhibits endothelin-1-mediated proliferation of human pulmonary artery smooth muscle cells, *Am. J. Respir. Cell. Mol. Biol.*, 9, 429-433. 1993.

Chapter **11**

EFFECTS OF ENDOTHELINS ON THE NERVOUS SYSTEM

Maarten van den Buuse

CONTENTS

0-8493-6975-4/97/$0.00+$.50

I. INTRODUCTION

Since the first report on the structure and activity of ET (Yanagisawa et al., 1988), several thousand papers have appeared which describe the effect of these peptides on vascular and nonvascular smooth muscle and other peripheral tissues. Moreover, it has become clear that ETs are present in the brain and may act as neuropeptides to exert a range of effects on nervous tissue. The focus of this chapter will be to give a comprehensive review of the literature on neuronal effects of ETs, highlighting key findings and discussing a number of contradictory results and gaps in our knowledge. Most attention will be paid to ETs and ET receptors in the brain and to effects of ETs *in vivo*.

II. DISTRIBUTION

In order to be able to interpret the effects of ETs and ECE inhibitors in the brain, it is important to first discuss in detail the distribution of ETs and their receptors in the central nervous system (CNS).

A. Distribution of ETs in the CNS

From the combined work on the central distribution of ETs, it becomes clear that these peptides are not only present, but are also synthesized in neurons and glial cells and that this regional distribution does not match that of brain vasculature. The latter is important, as it supports the suggestion that the sole source of ETs in the CNS is not endothelial cells of cerebral vessels.

Differential expression of ET-1, ET-2, and ET-3 has been described, although the relative abundance differs between reports. For example, Firth and Ratcliffe (1992) found that the level of expression of ET-1 and ET-3 in rat whole brain was 3.6 and 53.5%, respectively, of that in pooled rat lung, while the level of ET-2 expression was not detectable. Also MacCumber et al. (1989) observed a predominant expression of ET-3 in brain regions by Northern blot analysis. However, Lee et al. (1990) used the same technique on human brain samples and found significant expression of ET-1 in most of the regions measured. Highest expression was observed in the striatum, area 38 of the cortex, spinal cord, and hypothalamus. In contrast, when these authors used a probe specific for ET-3, expression could only be observed in the hypothalamus.

A more detailed overview of ET gene expression was obtained with *in situ* hybridization by Giaid et al. (1989, 1991). The strongest labeling was observed with a probe for ET-1 in different regions of the spinal cord, cortex, and hippocampus and on Purkinje cells in the cerebellum. A signal was also observed in the paraventricular, periventricular, supraoptic, and lateral hypothalamic nuclei, a finding of interest in view of the proposed role of ETs in endocrine regulation (see Section III.G). In the substantia nigra pars reticulata many cell bodies were found to express ET, but only scattered labeling was found in the substantia nigra pars compacta and caudate nucleus. Some labeled cells were furthermore found in the amygdala, raphe nuclei, and dorsal vagal motor nucleus.

Immunochemical and immunohistochemical localization studies have confirmed and extended these observations. However, as with gene expression studies, there is considerable variance between studies with respect to the relative abundance of different ET isoforms. For example, using dissected pig brain, Hemsén and Lundberg (1991) found the highest levels of ET immunoreactivity in hypothalamus (170 fmol/g) and lower levels in pituitary (110 fmol/g), thoracic spinal cord (99 fmol/g), brain stem (82 fmol/g), cortex (51 fmol/g), and cerebellum (36 fmol/g). Subsequent analysis with HPLC and specific antibodies revealed that in the spinal cord ET-1 and ET-3 were present in comparable amounts, while in the hypothalamus ET-1 was present at higher levels than ET-3. No evidence was obtained for the presence of ET-2 or big-ET-1. Shinmi et al. (1989a,b) found considerably higher levels of ET-1, compared to ET-3, in porcine brain and spinal cord, while according to Matsumoto et al. (1989) concentrations of ET-1 were higher than those of ET-3 in rat cortex (4-fold), cerebellum (1.5-fold), and medulla oblongata (5.5-fold), but lower in pituitary gland (0.3-fold). In contrast to these findings, Samson et al. (1991b,c) suggested that in rat hypothalamus, caudate nucleus, and pituitary ET-3 was much more abundant than ET-1. Similarly, Fuxe et al. (1991) found a 4.5 times higher level of ET-3 immunoreactivity (519 fmol/g) than ET-1 immunoreactivity (116 fmol/g) in rat hypothalamus. In neostriatum the level of ET-3 was substantially higher than that of ET-1 (327 fmol/g vs. 104 fmol/g). Importantly, however, HPLC analysis of the ET immunoreactivity showed one peak coeluting with that of ET-3, but a major peak which coeluted *later* than that of ET-1, leading the authors to suggest that this material might represent an additional form of ET or ET precursor. Whatever the nature of the immunoreactive material, it was present only in cell bodies and in the neostriatum it was colocalized with choline acetyltransferase or somatostatin.

In human brain, Togashi et al. (1991) found the highest levels of ET-1 in cerebral cortex, hippocampus, and thoracic spinal cord, moderate levels in medulla oblongata, cerebellum, and hypothalamus, and low to undetectable levels in thalamus, globus pallidus, and pons. Takahashi et al. (1991b,c), however, found little differentiation between different human brain regions with respect to their ET content: cortex, cerebellum, basal ganglia, brainstem, and hypothalamus had roughly equal concentrations of immunoreactive ET-1 (6 to 10 fmol/g). In contrast, in spinal cord the levels were 2- to 3-fold higher than in these brain regions, while pituitary showed markedly higher concentrations: 15 to 25 times the levels observed in brain. FPLC analysis showed that in hypothalamus ET immunoreactivity consisted mainly of ET-1, while in the pituitary it was mainly ET-3 (Takahashi et al., 1991b). Similar levels and regional distribution were measured by these authors in rat brain (Takahashi et al., 1991c). The latter point is important, as it shows that differences in absolute brain levels of ET immunoreactivity between studies are likely to be caused by the differences between the assay methods used rather than species differences.

The topography of ET immunoreactivity in brain was studied by Giaid et al. (1991) and Yoshizawa et al. (1990). In human brain, the distribution of neurons immunoreactive to ET was similar to that of ET mRNA (Giaid et al., 1991, see above). Thus, immunoreactive cell bodies were found mostly in cortical regions, hippocampus (pyramidal cells), and cerebellum (Purkinje cells). Fewer positive cells were seen in the caudate nucleus, amygdala, hypothalamus, raphe nuclei, substantia nigra, and the dorsal vagal motor nucleus in the medulla oblongata. Immunoreactive fibers were present mainly in the cortex and hypothalamus, and to a lesser extent in brainstem. These authors observed that some cells stained for ET immunoreactivity and showed hybridization for ET mRNA, but there were also several cells that displayed either ET immunoreactivity or mRNA alone. In double-labeling experiments, some cortical cells showed colocalization of ET and neuropeptide Y mRNA, while in the hypothalamus some cells double-stained for ET and neurophysin. The significance of this latter finding becomes clear from the studies of Yoshizawa et al. (1990), who observed strong immunoreactivity for ET-1 in porcine and rat posterior pituitary and paraventricular and supraoptic nuclei of the hypothalamus. These observations suggest a role for ETs in the regulation of vasopressin and oxytocin release (see Section III.G).

The modulation of central levels of ETs has been studied in a variety of experimental models. Primary cultures of rat brain astrocytes released ET-3 into the medium, but regulation of this release was not studied (Ehrenreich et al., 1991). Franco-Cereceda et al. (1991) used dorsal root ganglia as a model and compared ET-1 with calcitonin gene-related peptide (CGRP). They observed that significant levels of both peptides were present in this tissue, but that, unlike CGRP, neither treatment with 6-hydroxydopamine nor capsaicin altered the levels of ET. Furthermore, in contrast to CGRP, ET-1 levels were not altered by ligation or transection of the sciatic or vagal nerves, and no ET-1 release could be demonstrated from dorsal root ganglion cell cultures. These authors concluded that ET showed "remarkable inertness upon various experimental conditions including no evidence for axonal transport." Recently, more positive results were obtained by Hashimoto and colleagues (1994) who studied ET-1 release from neuro-2A cell cultures. These cells, derived from mouse neuroblastoma, released ET-1 spontaneously and time dependently. Addition of glutamate to the medium caused a dose-dependent inhibition of ET-1 release, with up to 80% inhibition at 100 μM glutamate. This effect could be inhibited by pretreatment with pertussis toxin and was thus likely mediated by metabotropic glutamate receptors (Hashimoto et al., 1994). These preliminary results could have important implications as they show that neuronal cells are able to synthesize and release ET and that this release can be modulated by neurotransmitters such as glutamate. ET-like immunoreactivity was increased in the rat hippocampus after local ibotenic acid-induced lesions in this region (Cintra et al., 1989). Similarly, ET levels were increased in rat striatum and cortex (Viossat et al., 1993) or gerbil forebrain (Giuffrida et al., 1992; Willette et al., 1992) after middle cerebral artery occlusion. These increased levels could reflect a growth factor-like action of ETs in damaged neural tissue. A potentially more physiological regulation of ET levels was found in rat posterior pituitary, where ET-1 levels were depleted when the animals were subjected to water deprivation (Ritz et al., 1992; Yoshizawa et al., 1990). Superfusion of the rat spinal cord showed *in vivo* release of ET-3 of 18 ± 2 pg/ml/30 min. This release was significantly enhanced by electrical stimulation of the posterior hypothalamus and by hypotensive hemorrhage, but reduced by infusion of nitroprusside (Knuepfer et al., 1994). Thus, ET-3 was released by a pressor stimulus, which would be compatible with a depressor action of ET-3 in the spinal cord (see Section III.F). Further evidence for *in vivo* release of ETs comes from the observation that human c.s.f. contains significant amounts of ETs (Ando et al., 1991; Hirata et al., 1990; Shirakami et al., 1990; Togashi et al., 1990; Yamaji et al., 1990), the source of which could be neuronal release. Indeed, in rat c.s.f. obtained from the cisterna magna, concentrations of ET-1 were reduced by phenylephrine-induced hypertension, an effect which was not observed after sino-aortic denervation (Mosqueda-Garcia et al., 1992). It should be noted that this observation is not in agreement with findings observed in superfused spinal cord (above).

While ET levels in brain tissue have now been measured in a number of studies, there is relatively little information about other components of the ET synthesis cascade. The presence of preproET-1 and preproET-3 in brain has been demonstrated, although little detail was given regarding regional distribution of these prohormones (Yanagisawa et al., 1990; Imai et al., 1992). The presence of ECE in brain tissue has been suggested by the observation of phosphoramidon-inhibitable ECE in rat and human brain neurons (Warner et al., 1992a,c). The rate of conversion of proET-1 to ET-1 was 0.6 nmol/h/mg protein and the authors suggested that this low conversion rate might be compensated for by the long duration of action of ET (Warner et al., 1992a). The greatest enzyme activity was present in hypothalamus and medulla oblongata, but no conversion of proET-3 to ET-3 could be demonstrated, indicating that the enzyme activity studied had strict substrate requirements (Warner et al., 1992b). It is important to note that, while mature ETs and ECE activity have now been demonstrated in different brain regions, the levels of proETs appear to be very low (e.g., Hemsén and Lundberg, 1991; Takahashi et al., 1991b,c). One explanation for this could be that ETs, once they are produced in their prohormone form, are rapidly converted into the mature peptide, which is then stored or released. However, the low converting activity observed in brain tissue (see above) is not in agreement with this hypothesis. ECE activity appears to be present also in the c.s.f., as the pressor response to central injection of big-ET-1 could be inhibited by pretreatment with phosphoramidon (Hashim and Tadepalli, 1991; Shinyama et al., 1991). In the cisterna magna of anesthetized dogs, injected big-ET-1 was converted to ET-1, an effect which could be inhibited by phosphoramidon (Shinyama et al., 1991).

Taken together, the demonstration of ET mRNA, prohormone forms, and mature forms of ETs and ECE suggests that in the brain the entire pathway required for the production of these peptides is present, although considerable further study appears necessary to fully characterize the synthesis of ETs in brain. Furthermore, there is as yet no detailed map of ET-containing cell groups and projections in the brain.

B. Distribution of ET Receptors in the CNS

Two different ET receptors have been cloned and characterized (see Chapter 2). From binding studies, however, the existence of additional ET receptors or of different forms of the above-mentioned ET receptors has been suggested (reviewed in Gulati, 1992; Huggins et al., 1993). From binding studies it has become clear that ET receptors in brain tissue are mostly, if not exclusively, of the ET_B type (reviewed in Sokolovsky, 1992). Indeed, using *in situ* hybridization with probes specific for ET_A or ET_B receptors, Hori et al. (1992) failed to find expression of ET_A receptors in brain tissue except on endothelial cells of cerebral vessels, while widespread expression of ET_B receptors was found on chorioid plexus and glial cells of the cerebellum, diencephalon, mesencephalon, and lower brain stem. In contrast, in the anterior pituitary mostly ET_A receptors were expressed. These authors failed to observe ET receptor expression on neurons and suggested that ET could play a role in glial proliferation in the brain or modulate brain function through the release of mediators such as nitric oxide from glial cells (Hori et al., 1992). Others have also suggested that glia are important mediators of ET function in the brain (Ehrenreich et al., 1991; MacCumber et al., 1990; Hösli and Hösli, 1991). However, in these and other studies, evidence was obtained for ET receptors on neurons as well (see also Davenport and Morton, 1991).

As early as 1988, Ambar and co-workers showed that rat brain homogenates bound STXs with high affinity. The highest binding density was observed in cerebellum, thalamus, and hypothalamus, with lower densities in striatum and cerebral cortex (Ambar et al., 1988). In contrast, stimulation of phosphoinositide turnover by STX-b was greatest in hypothalamus and caudate putamen and lower in thalamus, cerebellum, and cerebral cortex (Kloog et al., 1989, see Figure 1). Data for the caudate nucleus are of interest as this brain region was found to have an exceptionally high binding affinity value (K_d 0.33 nM vs. 1.5 nM in the other brain regions; Ambar et al., 1988). STXs are now well known to be very similar to ETs and the binding sites measured by these authors are likely to represent brain ET receptors. Indeed, in subsequent papers, several authors showed that STXs such as STX-b could inhibit binding of ET-1 or ET-3 to brain preparations and vice versa (e.g., Ambar et al., 1989; Gulati, 1992). The disparity between binding characteristics and signaling effects in

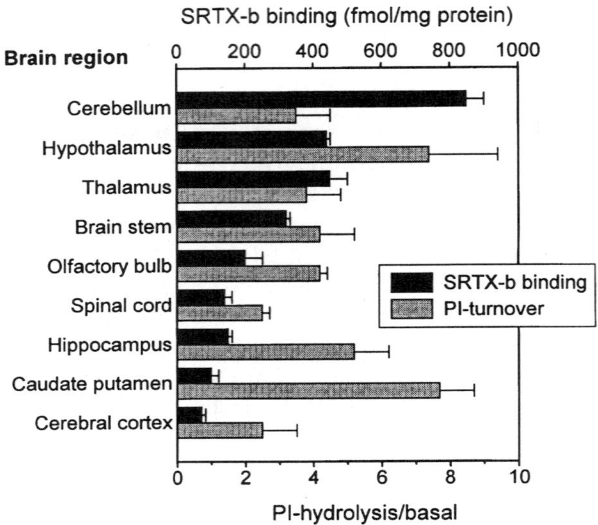

FIGURE 1
Regional distribution of sarafatoxin STX-b binding sites and of sarafatoxin STX-b-induced phosphoinositide hydrolysis in rat brain regions. Data represent mean ± standard deviation and were taken from Kloog, Y. et al., 1989. Note the large disparity between binding density and effects on phosphoinositide turnover in regions such as striatum and hippocampus.

different brain regions has been confirmed with ETs (Crawford et al., 1990). Several papers have furthermore confirmed the relative distribution of ET/STX receptors in brain. However, the absolute values of binding densities vary greatly between reports, most likely due to methodological differences (Bolger et al., 1990, 1992; Nambi et al., 1990). For example, in rat brain, Bolger et al. (1990) observed the highest labeled ET-1 binding density in cerebellum (4378 fmol/mg protein) and brainstem (3104), and lower densities in hippocampus (1143), striatum (1093), and cortex (850). These values are five to ten times higher than those reported by Sokolovsky's group (see above). Interestingly, binding sites for ET-1 and ET-3 showed an identical regional distribution in the rat brain, although binding of ET-3 was always slightly lower than that of ET-1. This finding suggests a single ET receptor type in rat brain, most likely ET_B (Nambi et al., 1990).

Compatible with rat data, Takahashi et al. (1991a) concluded that the human brain contained a single ET binding site (K_d 40 pM). This study found brainstem to contain the highest density of binding sites (B_{max} 5052 fmol/mg protein), and hypothalamus with lower density (B_{max} 963 fmol/mg protein). Cerebral cortex, cerebellum, and basal ganglia gave binding densities similar to those in the hypothalamus. As shown by Scatchard analysis, human pituitary contained two sites, one with an affinity similar to the brain sites, and one with a much higher affinity. These two sites may correspond to ET_A and ET_B receptors, which would agree with *in situ* hybridization data from Hori and colleagues (1992).

Regional distribution of ET binding sites in the brain has been studied with quantitative receptor autoradiography. The most complete overview in rat brain is from Kohzuki et al. (1991) using [^{125}I]ET-1 as a ligand. The highest densities of ET binding sites were found in the Purkinje layer of the cerebellum, chorioid plexus, and median eminence. High densities were also found in the supraoptic and paraventricular nuclei of the hypothalamus, anterior hypothalamic area, ventromedial hypothalamic nucleus, mammillary nuclei, and the glomerular layer of the olfactory bulb. Moderate densities were found in a number of regions including many thalamic

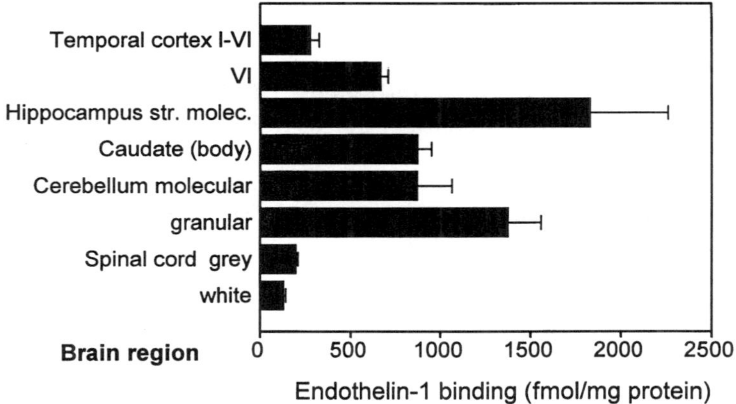

FIGURE 2
Densities of [^{125}I]endothelin binding sites (fmol/mg protein) in human brain. Data represent mean ± SEM and were taken from Jones, C. R., et al., 1989.

nuclei, the pretectal region, interpeduncular nucleus, suprachiasmatic nucleus, raphe nuclei, central gray, reticular nuclei, and the hypoglossal nucleus. Low but significant levels of binding were found in caudate putamen, globus pallidus, substantia nigra, anterior pituitary, median preoptic nucleus, septohypothalamic nucleus, superior colliculus, and area postrema. In the hippocampus only the molecular layer showed significant binding, while in the amygdala only the basal medial nucleus was prominently labeled. Interestingly, in hindbrain regions associated with cardiovascular regulation, including the nucleus tractus solitarius, nucleus ambiguus, dorsal motor nucleus of the vagus, and the C1 group, binding densities are relatively low (Kohzuki et al., 1991). As further described in Chapter 3, these findings have generally been confirmed by several other groups (e.g., Davenport and Morton, 1991; Fuxe et al., 1989a; Hoyer et al., 1989; Jones et al., 1989; Koseki et al., 1989a,b). By contrast, in the cat, Spyer et al. (1991) observed dense binding of labeled ET-1 to dorsomedial regions of the brainstem, including the caudal and intermediate levels of the nucleus tractus solitarius and the dorsal motor nucleus of the vagus.

In human brain, receptor autoradiography showed the highest density of ET binding in cerebellum and hippocampus, high to moderate density in caudate nucleus and hypothalamus, and less prominent binding in amygdala, globus pallidus, and olfactory bulb (Hoyer et al., 1989; Jones et al., 1989, see Figure 2). The marked levels of ET receptors in hippocampus of human brain are unlike those found in rat brain and represent additional examples of species differences, which could be important when central effects of ETs are compared between different species.

In rat spinal cord, the density of ET binding sites was relatively uniform in the gray matter. Values varied from 14 to 16 fmol/mg protein in the sacral region to 9 to 11 fmol/mg protein in the cervical region. The binding density in white matter was 4 to 7 fmol/mg protein (Kar et al., 1991). Similar data were found in pig and human spinal cord, with the highest density in lamina X of the cervical and thoracic region (Niwa et al., 1992).

Regulation of brain ET receptor levels has been little studied. Ibotenic acid-induced lesions of the striatum or hippocampal formation led to a marked reduction of ET binding in the lesioned area (Fuxe et al., 1989a). In rat C6 glioma cells, prolonged exposure to β-adrenergic agonists, such as isoprenaline, caused a time- and dose-dependent decrease in ET receptor density, an effect which was mediated by a cyclic AMP-dependent pathway (Durieu-Trautmann et al., 1991). In cultured rat astrocytes, the addition of dibutyryl cyclic AMP caused a 16-fold increase in B_{max} for ET-1 binding and a marked increase in ET_B gene expression, an effect likely to be indirect through increased differentiation of the cells (Hama et al., 1992). Preliminary results from Ernsberger et al. (1991) indicated that chronic hypoxia increased ET receptor density in rat ventral medulla. Indeed, in ischemia and some pathological states differences in ET receptor levels have been observed when compared to controls (see Section IV).

Taken together, these data suggest that ET receptors are expressed in neuronal or glial elements of the CNS and make it likely that endogenous or exogenously administered ETs will influence neuronal function. Clearly, in the brain, ET receptors (probably (ET_A) are also present on brain microvessels and these vessels furthermore produce ETs (Adner et al., 1993; Durieu-Trautmann et al., 1993; Kadel et al., 1990; Stanimirovic et al., 1994; Yoshimoto et al., 1990). While these vascular receptors could be involved in the regulation of local cerebral blood flow, exogenously injected ETs could act on these receptors and indirectly affect neuronal function via changes in oxygen supply. Vasospasm of cerebral vessels could lead to ischemia of surrounding brain tissue (see Section IV.A). Such effects must be taken into consideration when the central effects of ETs in brain are studied.

III. EFFECTS OF ET

A. Peripheral Nerves

Studies on the effects of ETs on peripheral nerves have mostly dealt with the sympathetic nervous system. Several authors have shown a dual effect of ETs on noradrenergic sympathetic responses: on the one hand, ETs appear to modulate noradrenaline release from sympathetic nerve endings, while, on the other hand, postsynaptic responsiveness is enhanced. An inhibitory effect of ETs on noradrenaline release was observed in guinea-pig femoral (Wiklund et al., 1988) and pulmonary artery (Wiklund et al., 1989), dog renal artery (Suzuki et al., 1992; Takagi et al., 1991), and rat mesenteric artery (Tabuchi et al., 1989, 1990). In contrast, an enhancement of noradrenaline release by ET was found in rabbit ear artery (Wong-Dusting et al., 1989) and rat tail artery (Bucher et al., 1991). In isolated guinea-pig atria, ET-1 did not affect stimulus-evoked noradrenaline release (Reid et al., 1989). As discussed by Bucher et al. (1991), these apparently contradictory results could possibly be explained by the differences in applied stimulus frequencies between the various studies: low stimulus frequencies would result in ET having a facilitatory effect on transmitter release, whereas high stimulus frequencies would result in an inhibitory effect. Enhancement of postsynaptic responses was observed in guinea-pig pulmonary artery (Wiklund et al., 1989), rat mesenteric artery (Tabuchi et al., 1989a,b, 1990) and tail artery (Reid et al., 1991), and rabbit ear artery (Wong-Dusting et al., 1989, 1991). However, La et al. (1990) showed that this effect is not unique to adrenergic responses. ET-1 enhanced the vasoconstrictor responses to vasopressin, serotonin, ATP, and histamine. Thus, it remains uncertain whether ETs alter adrenergic receptor sensitivity or whether they influence intracellular mechanisms shared by adrenergic receptors and receptors for other vasoconstrictor agents.

Few authors have reported effects of ETs on peripheral neuronal responses in nonvascular systems. In anesthetized dogs, ET-1 stimulated adrenal release of noradrenaline and of adrenaline induced by electrical stimulation of the splanchnic nerve (Takeuchi et al., 1992). Basal release was not affected. However, in cultured bovine adrenal chromaffin cells, ET directly stimulated catecholamine release (Boarder et al., 1991). Cholinergic responses may be reduced by ETs, as Kushiki and co-workers (1991) observed that addition of ET-3 dose-dependently reduced acetylcholine release in dog cardiac sympathetic ganglia, while Nishimura et al. (1991) found similar effects in feline colonic parasympathetic ganglia.

There is some evidence that ETs interact with peripheral sensory nerves, although the precise mechanism behind this interaction is still unclear. Intradermal injection of low doses of ET-1 (10 pmol) caused pallor with surrounding axon-reflex flare. Blood flow was markedly increased in the flare area, and because ET-1 failed to directly stimulate release of histamine from isolated skin mast cells the authors suggested that ET-1 stimulated sensory nerves (Brain et al., 1992). However, ET-1 (500 nM or 2 μM) did not affect spontaneous release of substance P or CGRP from isolated sensory nerve endings, although it enhanced the release of these peptides evoked by the addition of capsaicin (Dymshitz and Vasko, 1994). Pretreatment with capsaicin caused only a small, although significant, reduction of the contractile effects of ET-1 on guinea-pig trachea, indicating that an interaction with sensory nerves did not play a major role in these responses (Hay et al., 1993). In all these effects it is difficult to separate a direct interaction of ET with sensory neuropeptide release from an indirect interaction between these peptides through their vasoactive and contractile effects.

These data show that ETs can modulate responses in peripheral nerves via presynaptic and postsynaptic mechanisms. This can involve adrenergic as well as cholinergic and peptidergic responses. While not directly related to the effects of ETs in brain, the peripheral effects of ETs on autonomic or sensory nerves may give important clues about the mechanism of action of these peptides in the central nervous system.

B. Cultured Glial Cells and Neurons

A large number of studies have shown neurochemical effects of ETs on neuronal or glial tissue *in vitro*, either in primary culture or using cell lines, which will be discussed here only briefly (for reviews, see Sokolovsky, 1992; Huggins et al., 1993).

In a rat neuronal NG-108 cell line (neuroblastoma × glioma hybrid), ET-1 caused a rise in intracellular calcium release and a modest increase in phosphoinositide concentrations. This effect appears different from that in C6 glioma cells, where the rise in intracellular calcium was biphasic and a stronger stimulation of phospholipase C was noted (Reiser and Donie, 1990). In C6 glioma cells, the addition of ET-1 caused a 50% reduction in the expression of proenkephalin mRNA, an effect that appeared to be mediated by the rapid induction of *c-fos* and *c-jun* gene expression (Yin et al., 1992). Interestingly, in NG-108 cells ET-1 induced a rise in intracellular cyclic GMP levels, an effect likely to be mediated by increased nitric oxide production (Reiser, 1990). Cultured cerebellar granular cells (Chuang et al., 1991; Morton and Davenport, 1992) and cerebellar slices (Sokolovsky, 1992) also showed stimulated inositol phosphate production and a rise in intracellular calcium levels upon addition of ETs or STXs.

The effects of ETs on astrocytes are generally similar to those on neuronal cells, with the most important signaling system being phosphoinositide hydrolysis and a biphasic increase in intracellular calcium levels (e.g., Supattapone et al., 1989; Marsault et al., 1990; Lin et al., 1991; Reiser and Donnie, 1990; Goldman et al., 1991). Other effects of ETs on astrocytes include stimulation of *c-fos* and *c-jun* gene expression (Ladenheim et al., 1993; Yin et al., 1992), arachidonic acid release (Tencé et al.,

1992), nerve growth factor expression (Ladenheim et al., 1993), and stimulation of cell growth (Supattapone et al., 1989; MacCumber et al., 1990; Zhang et al., 1991). ETs have also been reported to inhibit cyclic AMP production (Couraud et al., 1991; Levin et al., 1992) and glutamine synthetase activity (Hama et al., 1992).

These results indicate that ETs can have direct effects on neuronal and glial cells. This is an important point, as it is often difficult *in vivo* to distinguish direct neuronal effects of ETs from those influenced or mediated by cerebrovascular constriction and ischemia (see Section IV.A) and may help to explain a number of effects of these peptides *in vivo.*

C. Deoxyglucose Studies and Electrophysiological Effects *In Vivo*

The rate of glucose uptake by cells can be assessed by administering radiolabelled deoxyglucose *in vivo* and measuring the distribution of radioactivity by autoradiography. This method gives an indication of metabolic stimulation in different brain regions. The first researchers to use this method to investigate the effect of central injection of ET were Yamashita et al. (1990). They used a dose of 1.25 pmol/rat and found a reduction of glucose uptake in the basolateral amygdala (–12%), caudate putamen (–13%), inferior olive (–24%), and olfactory bulb (–16%).

In more extensive studies, Gross and co-workers detailed the effects of ET on neuronal metabolism in rat brain. In all their studies, a dose of 9 pmol ET-1 was injected i.c.v. This dose induced a pressor response and a number of behavioral manifestations, including barrel-rotation and convulsions; however, electroencephalographic measurements showed no epileptiform response to ET (Gross et al., 1992a). ET-1-induced pronounced hypermetabolism in several periventricular sites and their projection regions without producing neurotoxic effects. Brain regions particularly affected were the caudate nucleus, lateral septal nucleus, corpus callosum and hippocampal fimbria, with increases amounting to 160 to 312% above control levels. More moderate metabolic stimulation (50 to 100% over baseline) was observed in such regions as the medial amygdaloid nucleus, the dorsomedial, lateral and ventromedial hypothalamus, the subfornical organ, periaquaductal gray matter, olfactory tubercle, substantia nigra pars reticulata, and a number of hippocampal and cerebellar regions. ET had no effect on glucose uptake in such regions as the nucleus accumbens, locus coeruleus, or several cortical and hypothalamic regions (Gross et al., 1992a, 1993a,c). Furthermore, these effects of ET-1 were observed despite a significant reduction in cerebral blood flow in the periventricular regions most affected, a finding indicating uncoupling of the normal blood flow/metabolism coupling. Pretreatment with the calcium channel blocker nimodipine inhibited the effects of ET-1 on glucose uptake (Gross et al., 1992a), and nimodipine induced a cerebral vasodilation and inhibited the vasoconstrictor effect of ET. While these studies indicate that several brain regions may be affected by relatively large doses of *exogenously* applied ETs, it is as yet uncertain in which way these results relate to the physiological roles of *endogenous* ETs in the brain.

The first study to examine the electrophysiological effects of ETs was that of Yoshizawa et al. (1989). In an *in vitro* spinal cord preparation from neonatal rat, ET-1 dose-dependently depolarized a ventral root motorneuron potential, an effect which could be blocked by the calcium channel blocker, nicardipine. The effect was also prevented by pretreatment with the substance P antagonist, spantide, indicating that it could have been mediated indirectly via the release of substance P. Removal of ET from the incubation medium caused the ventral root potential to immediately return to baseline, in contrast to vasoconstrictor effects of ET-1, which are long lasting.

Addition of ET-1 or ET-3 caused a concentration-dependent and reversible membrane depolarization in rat spinal cord and brainstem astrocytes in culture (Hösli et al., 1991). Also, Yamashita et al. (1991), using slice preparations of anteroventral third ventricle, observed that the application of 10^{-7} M ET-3 to the bath caused excitation in 18% of the neurons tested, an effect which was reversible and not subject to tachyphylaxis. In this study, the majority of the neurons were unresponsive (76%), while a minority were inhibited (6%). In the supraoptic nucleus, 57% of putative vasopressinergic neurons were inhibited and none excited, while only 19% of putative oxytocinergic neurons were inhibited and none excited (Yamashita et al., 1991). This effect was likely mediated by ET_B receptors as ET-1 and ET-3 were found to be equipotent (Yamamoto et al., 1993). These results are important in the light of neuroendocrine effects of ETs (see Section III.G) and furthermore show that the direct effects of ETs differ markedly depending on cell phenotype.

Ferguson and colleagues studied the effects of systemic administration of low doses (0.1 to 10 pmol) of ET-1. In the area postrema in the medulla oblongata of anesthetized rats, the majority of cells (84% of responsive neurons) were excited (Ferguson and Smith, 1991) and a similar effect was found in the subfornical organ (Wall et al., 1992). In the commissural nucleus tractus solitarius, the majority of responsive cells (84%) were inhibited after this treatment; this might have been mediated by an action of ET-1 via inhibitory area postrema neurons (Ferguson and Smith, 1991). Systemic administration of ET-1 excited putative vasopressinergic and oxytocinergic neurons in the paraventricular and supraoptic nucleus, which was likely mediated through the effects of this treatment on subfornical organ cellular activity (Wall and Ferguson, 1992). These results show that circulating ET may affect central neuroendocrine regulation through an action on circumventricular organs. Thus, neurons in regions such as the paraventricular nucleus or supraoptic nucleus may be influenced by ETs in two ways: direct inhibition (Yamashita et al., 1991; Yamamoto et al., 1993) and indirect excitation (Wall and Ferguson, 1992). A dual action of ET-1 in the brain has also been suggested by Cao et al. (1993), who showed that iontophoretic application of ET-1 to vasomotor neurons in the rostral ventrolateral medulla caused excitation (four out of seven cells tested). However, intracisternal injection of 1 pmol ET-1 or topical application on the ventral medullary surface predominantly caused inhibition of vasomotor neurons, an effect probably mediated by ventral surface elements responsive to ETs in the c.s.f., the identity of which remains to be determined.

D. Neurotransmitter Release *In Vitro* and *In Vivo*

ET-1 dose-dependently stimulated the release of preloaded radiolabelled aspartate from cerebellar granule cells (Lin et al., 1990). At 30 nM ET-1, the release of aspartate was approximately 60% above baseline. This effect could be stimulated by short-term treatment with a phorbol ester or by pertussis toxin, indicating that it was mediated through protein kinase C but inhibited by a pertussis toxin-sensitive mechanism.

The release of dopamine could be modulated by ETs. Local injection of a very high dose of ET-1 (430 pmol) into the striatum caused a marked rise in extracellular dopamine levels, as measured with microdialysis in anesthetized rats (Kurosawa et al., 1991). This effect was associated with a marked reduction in local cerebral blood flow, suggesting that ischemia was a major mediator of this effect. A high dose of locally injected ET-3 (40 pmol) caused a modest stimulation of dopamine release only when the tissue had been depolarized with ouabain or potassium. The outflow of the dopamine metabolites, DOPAC and HVA, and of serotonin and its metabolite 5-HIAA, was not affected by ET-3 (Konya et al., 1992). Also, only a high concentration of ET-3 (10 μM) induced a stimulation of the release of noradrenaline and dopamine from cortical and striatal slices, respectively, *in vitro* (Koizumi et al., 1992). In PC12 pheochromocytoma cells, the addition of 1 to 100 nM ET-3 caused a modest, concentration-dependent stimulation of dopamine release (Koizumi et al., 1994). While it could be argued that high concentrations of ETs are needed because the peptide may not penetrate well into tissue, it is also possible that toxic effects, ischemia, or nonspecific hypermetabolism underlie the effects on neurotransmitter release in these studies (see Section IV.A).

E. Interaction With Catecholaminergic Drugs

An interaction of ETs with central catecholaminergic systems *in vivo* has been postulated (Gulati 1991a; Gulati and Srimal, 1993). The i.v. injection of 20 to 40 pmol/kg of either ET-1, ET-2, or ET-3 completely inhibited the reduction in blood pressure and heart rate induced by clonidine. In addition, i.c.v. injection of ET-1 (10 pmol) or topical application of ET-1 on the ventral surface of the medulla oblongata inhibited the hypotensive effects of peripherally administered clonidine. Importantly, central pretreatment with angiotensin also blocked the effect of clonidine on blood pressure, suggesting that vasoconstriction might have played a role in the effects of the peptides.

The hypertensive response to high doses of clonidine was enhanced after pretreatment with ET-1. Thus, the authors speculated that a peripheral action of ET on adrenoceptor sensitivity might explain the enhanced hypertensive response, which indirectly masked the hypotensive effect of clonidine. In addition, ET could have an inhibitory action on presynaptic adrenoceptor mechanisms, which could explain some of the effects on the clonidine-induced hypotension. These two actions have indeed been demonstrated in peripheral noradrenergic nerves (see Section III.A), but the inconclusive evidence obtained in the brain does not indicate if this could also play a role in the centrally mediated antihypertensive effects of clonidine. While central injection of ETs could have some ischemic effects, this would not provide an explanation for the action of systemically injected ETs on the effects of clonidine.

Recently, we have performed a number of studies in conscious rabbits in order to further investigate the possible interaction of ETs with central catecholaminergic systems. However, neither intracisternal nor i.v. administration of different doses of ET-1 had any significant effect on the hypotensive or bradycardic action of centrally or systemically injected clonidine (Godwin et al., unpublished observations). Moreover, intracisternal injection of ET-1 did not affect the antihypertensive response to central injection of α-methyldopa in conscious rabbits (Van den Buuse, unpublished observations). Thus, we were unable to demonstrate an interaction of ETs with the central action of catecholaminergic drugs on blood pressure, although a species difference cannot be excluded.

F. Blood Pressure and Baroreflex

The first observations that were taken as an indication that ETs could play a direct role in central cardiovascular regulation were the effects of i.c.v. administration of ETs on blood pressure in rats. Several reports have shown a pressor response when the dose of ET-1 was around 10 to 60 pmol/kg, which in conscious animals was associated with severe behavioral effects such as barrel rotation (Siren and

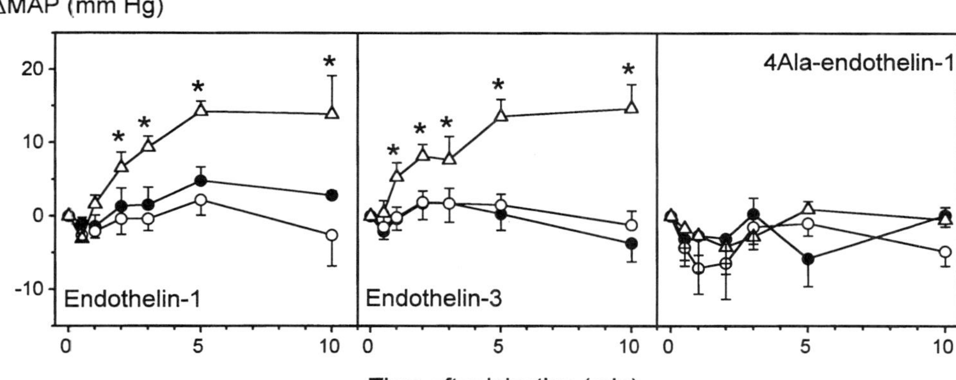

FIGURE 3
The effect of sequential intracerebroventricular injection of saline (closed circles), 10 pmol (open circles) and 50 pmol (open triangles) of endothelin-1, endothelin-3 or [$Ala^{1,3,11,15}$]endothelin-1 (4Ala-endothelin-1) on blood pressure of urethane-anesthetised rats. Data are expressed as mean change in mean arterial pressure (MAP) of 6–8 rats per group. Asterisks indicate significant (p <0.05) changes in blood pressure when compared to baseline values.

Feuerstein, 1989; Ouchi et al., 1989; Kawano et al., 1989; Minamisawa et al., 1989; Nishimura et al., 1990; Lecci et al., 1991; Yoshida et al., 1991; Takahashi et al., 1991a; Yamamoto et al., 1992). The pressor response was mediated by increased sympathetic vasomotor activity (Kawano et al., 1989; Ouchi et al., 1989; Takahashi et al., 1991a) and a rise in plasma vasopressin concentrations (Kawano et al., 1989; Yamamoto et al., 1992). Lower doses of ET injected i.c.v. (1 to 10 pmol/kg) had no effect on blood pressure in acute experiments (Siren and Feuerstein, 1989), but infusion of 10 pmol/h of ET-1 caused a gradual hypertension, maximal at 7 days after the start of the infusion (Nishimura et al., 1991). As further discussed in Section IV.A, it is likely that these pressor responses were caused by negative effects of centrally injected ETs on cerebral blood flow and a resulting ischemia, leading to a reflex rise in blood pressure. Therefore, these results do not indicate whether *endogenous* ETs play a pressor role in the brain. Recently, Webb et al. (1992) injected the selective ET_B receptor agonist IRL 1620 ([Glu^9,$Ala^{11,15}$]-ET-1(8-21]) i.c.v. and produced a moderate and transient pressor response in conscious rats (0.1 to 120 pmol). We have recently compared the effects of i.c.v. injection of ET-1, ET-3, and the ET_B receptor agonist [$Ala^{1,3,11,15}$]ET-1 in anesthetized rats (Van den Buuse, unpublished results). While ET-1 and ET-3 induced a pressor response at 50 pmol, but not 10 pmol, [$Ala^{1,3,11,15}$]ET-1 had little effect on blood pressure (Figure 3). These results could indicate that the pressor response and severe cerebral vasoconstriction caused by centrally injected ET-1 are mediated by ET_A receptors, although this needs to be confirmed using ET_A receptor antagonists such as BQ 123.

Injection of 10 pmol ET-1 or 100 pmol big-ET-1 into the fourth ventricle of anesthetized rats caused a significant decrease in blood pressure. The effect of big-ET-1 was much longer lasting than that of ET-1 (Hashim and Tadepalli, 1991). Intracisternal injection of 100 pmol ET-3 in anesthetized rats caused a rapid but transient increase in blood pressure, followed by a long-lasting hypotension (Kuwaki et al., 1990a). Similar effects were found with ET-1 (Kuwaki et al., 1990b). However, intracisternal injections of ET-1 or ET-3 (2.5 or 25 pmol/kg) in conscious rats did not cause changes in resting blood pressure (Itoh and Van den Buuse, 1991). Intrathecal administration of 1 to 100 pmol ET-1 or ET-3 in urethane-anesthetized rats caused a similar dose-dependent decrease in blood pressure and a hindquarters vasodilation (Han et al., 1991). In another study in urethane-anesthetized rats, however, intrathecal injection of ET-1 (3 to 100 pmol) or ET-3 (3 to 100 pmol) caused pressor responses (Giuliani et al., 1991). In the case of ET-3, the pressor response was shorter lasting and was followed by a reduction in blood pressure. In conscious rats, intrathecal injection of the ET_B receptor selective agonist BQ 3020 produced a weak pressor response, but higher doses of intrathecally injected ET-1 (65 to 650 pmol) or ET-3 (162 to 650 pmol) caused dose-dependent pressor responses, which could be blocked by pretreatment with the ET_A receptor antagonist BQ 123 (Poulat et al., 1994). These data would suggest that injection of low doses of ETs into the hindbrain may lead to an ET_B receptor-mediated decrease in blood pressure in anesthetized rats or no apparent effect in conscious rats, while higher doses would lead to a pressor response which is mediated by ET_A receptors. In the latter response, the involvement of reduced cerebral blood flow and ischemia cannot be excluded.

The local injection of ETs into different regions of the brain has given more detailed information about their site of action than could be derived from i.c.v. injections. However, the results have been difficult to interpret. In agreement with electrophysiological effects in the subfornical organ (see Section III.C), microinjection of 0.5 or 5 pmol, but not lower doses, of ET induced an increase in blood pressure of around 10 mm Hg in urethane-anesthetized rats (Wall et al., 1992).

In the dorsomedial medulla, microinjections of doses of 0.2 and 0.5 pmol ET-1, but not methoxamine, into the area postrema elicited a short-lasting pressor response of 10 to 15 mm Hg, whereas higher doses (2 or 5 pmol) caused a marked fall in blood pressure (Ferguson and Smith, 1990). Microinjections of ET-1 into the nucleus tractus solitarius caused a depressor response as well. However, this response was not dose dependent and, unlike in the area postrema, could be mimicked by local injection of methoxamine. This led the authors to conclude that the effects in the nucleus tractus solitarius were influenced by local vasoconstriction (Ferguson and Smith, 1990). Kuwaki and co-workers injected 2 to 10 pmol ET-3 into the nucleus tractus solitarius and found marked rises in blood pressure and renal nerve activity (Kuwaki et al., 1990a). By contrast, Hashim and Tadepalli (1992) injected lower doses of ET-1 (0.1 and 0.3 pmol) into the nucleus tractus solitarius and observed short-lasting (1 to 5 min) and reproducible depressor effects. Because injection of the same doses of ET-1 into deglutitive sites in this nucleus did not evoke swallowing responses, the authors concluded that the effect of ET on blood pressure was specific. Mosqueda-Garcia et al. (1992) observed depressor responses to microinjections of ET-1 (0.5 to 6 pmol) into the nucleus tractus solitarius or area postrema. Surprisingly, microinjection of ET-3 caused a pressor response in the nucleus tractus solitarius but a depressor response in the area postrema. The effects of ET-1 could be blocked by local pretreatment with BQ 123, but did not appear to be associated with local ischemia, as other vasoconstrictors such as phenylephrine or vasopressin failed to produce similar effects. On the other hand, 2 pmol ET-1 injected into the nucleus tractus solitarius caused a reversible inhibition of hypotensive responses to microinjection of glutamate into the same sites (Mosqueda-Garcia et al., 1992). Thus, ET-1 attenuated neuronal function in this nucleus.

These results agree with those of Lee et al. (1992), who microinjected 1 to 10 pmol ET-1 into the nucleus tractus solitarius and observed, in addition to a rise in blood pressure and heart rate, a marked inhibition of baroreflex responses to i.v. injections of phenylephrine or nitroprusside. This treatment inhibited the hypotensive effects of local injection of L-glutamate, acetylcholine, and *N*-methyl-D-aspartate (NMDA) into the nucleus tractus solitarius, an effect which was similar to that of local lidocaine injections (Lee et al., 1991). Local cerebral blood flow in the nucleus tractus solitarius was markedly reduced (–45%) by microinjection of 1 pmol ET-1 into this region, suggesting that local ischemia might have played a role in the effects of ET-1 on blood pressure and baroreflexes (Lee et al., 1992). In conscious rats, however, we have observed enhanced baroreflex responses to intracisternal injection of ETs (see below). It is likely that the complex effects of ETs in the dorsomedial medulla oblongata (including the area postrema and nucleus tractus solitarius) reflect direct neuronal effects of low doses and indirect effects due to ischemia. In addition, considering the complexity of the nucleus tractus solitarius, it is quite possible that different neuronal elements within this nucleus respond differently to ETs. Finally, it should be noted that the overall density of ET binding sites in the nucleus tractus solitarius is relatively low (Kohzuki et al., 1991). Further studies are needed to more precisely localize and identify ET receptors in this brain region.

Similar considerations apply to the ventrolateral medulla. The ventral surface of the medulla has been shown to be very sensitive to applied ETs (Kuwaki et al., 1991a,b; Hematillake and LaHann, 1992). As little as 1 fmol ET-1 applied to this region caused pressor responses, increases in renal sympathetic nerve activity, and phrenic nerve activity. ET-3 was less effective, whereas higher doses of ET-1 or ET-3 elicited initial excitatory effects followed by long-lasting inhibitory responses (Kuwaki et al., 1991a). Electrophysiological studies (Cao et al., 1993) suggested that neuronal elements on the ventral medullary surface respond differently to ETs when compared to neurons in the ventrolateral medulla itself. Microinjection of ET-1 into the rostral ventrolateral medulla had only moderate effects on blood pressure, but caused respiratory arrest after 12 to 15 min. In artificially ventilated rats, ET-1 (0.5 to 6 pmol) induced a biphasic response on blood pressure: an early pressor response was followed by a depressor response (Mosqueda-Garcia et al., 1992). Thus, it might be that two different sites mediate the effects of ETs in this region. In addition, effects on blood pressure are strongly influenced by effects of ETs on respiration (Fuxe et al., 1989b).

As mentioned above, Lee and co-workers (1992) found that microinjection of ET-1 into the nucleus tractus solitarius caused a marked inhibition of baroreflex responses. Also Kuwaki et al. (1990a,b) found a reduction of baroreflex sensitivity upon intracisternal administration of ET-3 (100 pmol) in anesthetized rats. We have studied the effects of intracisternal injection of ETs on baroreceptor-heart rate reflex of conscious rats as analyzed by the "steady-state" method (Itoh and Van den Buuse, 1991; Van den Buuse and Itoh, 1993). At a dose of 25 pmol/kg, but not 2.5 pmol/kg, both ET-1 and ET-3 induced a significant increase in baroreflex gain. This effect could be attributed to an increase of the curvature coefficient as heart rate range did not change (Figure 4). In experiments in which sympathetic blockade with atenolol or vagal blockade with methylatropine was used, we observed that the action of ET-1 was an enhancement of the sensitivity of the vagal component of the baroreflex only. The effect of ET-1 and of ET-3 amounted to a 38 and 54% increase, respectively, while ET-2 induced only a 17% increase in gain (Itoh and Van den Buuse, 1991). It is unclear why in the anesthetized rats only a reduction of baroreflex sensitivity was found, but it is well known that anaesthesia blunts cardiovascular reflex responses and would therefore make subtle changes more difficult to observe. Unlike some of the studies in anesthetized rats, we failed to find any changes in resting blood pressure or heart rate, which could also have been caused by more effective cardiovascular buffering and compensatory mechanisms.

It remains to be determined through which region of the medulla ET exerts its effect on the baroreflex, but one obvious candidate would be the nucleus tractus solitarius. Studies in spontaneously hypertensive rats (Van den Buuse and Itoh, 1993, see Section IV.B) have suggested that the rostroventrolateral medulla may not be the

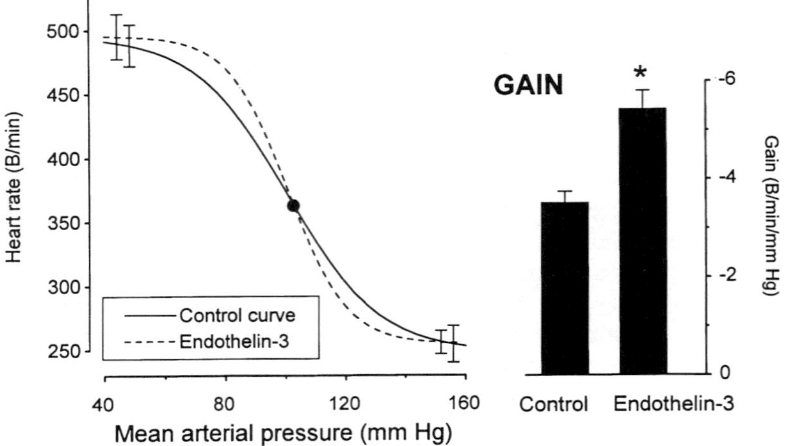

FIGURE 4
The effect of intracisternal injection of 25 pmol/kg of endothelin-3 on baroreceptor-heart rate reflexes in conscious rats. The left panel shows how the mean sigmoidal blood pressure-heart rate curve has become significantly steeper after treatment with endothelin-3 when compared to the control curve. Heart rate range and the position of the two heart rate plateaus were not affected. The right panel shows mean values of average gain (sensitivity) before and after treatment.* $p < 0.05$ for difference with controls. (Data were taken from Itoh, S. and Van den Buuse, M., 1991.)

primary site of action of intracisternally injected ETs on the baroreflex, but further studies are needed to address this point directly.

G. Neuropeptide and Hormone Release

Indirect and direct evidence suggests that central ETs might play an important role in endocrine regulation (Macrae and Bloom, 1992). The indirect evidence includes findings of significant levels of ETs and ET receptors in brain regions classically associated with neuroendocrine control, such as the paraventricular nucleus (Yoshizawa et al., 1990) and the pituitary (see Section II). Interestingly, while ET_B receptors were the abundant subtype in neuronal tissue, in the anterior pituitary the ET_A receptor subtype predominates (Hori et al., 1992; Kanyicska and Freeman, 1993; Samson, 1992). In dispersed anterior pituitary cells, ET induced activation of phospholipase C and a rapid rise in intracellular calcium levels (Stojilkovic et al., 1990; Lewy et al., 1992). Furthermore, i.c.v. injection of ET-1 (9 pmol) increased glucose utilization in the anterior, intermediate, and posterior pituitary by 75, 114 and 92%, respectively (Gross et al., 1991).

More direct evidence for a neuroendocrine role of ETs comes from the observations that these peptides modulate the release of neuropeptides from the hypothalamus and of pituitary hormones. For example, ET-3 induced a concentration-dependent stimulation of the release of atrial natriuretic peptide (ANP) from fetal diencephalic tissue in primary culture (Levin et al., 1991). Similarly, the addition of ET-1 (0.1 or 1 μM) enhanced the release of substance P from hypothalamic blocks and from anterior pituitary tissue *in vitro* by two- to threefold (Calvo et al., 1990). In the latter studies, ET-1 had no effect on the release of vasoactive intestinal peptide, somatostatin, adrenocorticotropin (ACTH), growth hormone (GH), luteinizing hormone (LH), thyroid stimulating hormone (TSH), or prolactin. However, several other studies have shown that the release of these hormones is indeed influenced by ETs.

ET-1 and, to a lesser extent, ET-3 stimulated the release of gonadotropin-releasing hormone (GnRH) from rat hypothalamic neurons in primary culture or from cell lines secreting GnRH in culture (Krsmanovic et al., 1991; Moretto et al., 1993).

At the level of the anterior pituitary, ET-1 was more effective than vasopressin, noradrenaline, or angiotensin, but somewhat less effective than GnRH to stimulate the release of LH. The effect of ET-1 was subject to rapid desensitization caused by endocytosis of ET receptors (Stojilkovic et al., 1990, 1991). Others have also observed transient effects of ETs on LH release, in contrast to long-lasting stimulation of TSH and follicular-stimulating hormone release (Kanyicska et al., 1991; Samson et al., 1991b). The endogenous ET isoform involved in LH release appeared to be ET-3 rather than ET-1. Indeed, gonadotrophs in human anterior pituitary were found to contain immunoreactivity for ET-3, but not ET-1, while cells producing ACTH, TSH, GH, or prolactin did not show immunoreactivity for either isoform (Naruse et al., 1992). The local synthesis of ETs in the anterior pituitary suggests a paracrine role of these peptides in endocrine release (also see Samson et al., 1991b). However, some authors have failed to find an effect of ETs on LH release in static incubations (e.g., Lewy et al., 1992).

Although pituitary prolactin cells do not appear to synthesize ETs (Naruse et al., 1992), several reports have shown that the release of prolactin could be markedly inhibited by the addition of ETs (Domae et al., 1992; Kanyicska et al., 1991; Lewy et al., 1992; Samson and Skala, 1992). This effect was sensitive to treatment with pertussis toxin (Burris et al., 1991). In a number of studies ET-1 was more potent than ET-3 (Dymshitz et al., 1992; Kanyicska and Freeman, 1993) and the selective ET_B receptor agonist STX-c was inactive (Kanyicska and Freeman, 1993), supporting a role of ET_A receptors in these responses (Samson, 1992). In another study, however, ET-3, but not ET-1, inhibited prolactin release but had no effect on the release of LH, GH, or TSH (Samson et al., 1990). It is possible that the ET which is responsible for these effects *in vivo* is ET-3, which is synthesized and released from gonadotrophs (see above). On the other hand, the intermediate lobe could be the source of ET-3 acting on lactotrophs. Neurointermediate lobe extracts were shown to contain a prolactin release-inhibiting factor which was probably ET-3, whereas ET immunoreactivity was predominant in intermediate lobe cells (Samson et al., 1992). A local, paracrine role of ET-3 was suggested by the observation that neither i.c.v. nor i.v. injections of ET-3 affected prolactin release *in vivo* (Samson et al., 1991c). While in static incubations the predominant effect of ETs on prolactin release was an inhibition, in dynamic perfusion experiments a short-lasting stimulation of prolactin release was observed which preceded a longer-lasting inhibition (Samson et al., 1992). The difference in mechanism between these two components of the effect of ET is unclear, but it is tempting to speculate that the rapid stimulation of phosphoinositide turnover induced by ETs in lactotrophs is related to this stimulation of release rather than to the inhibition. Indeed, thyrotropin-releasing hormone induced stimulation of phosphoinositide turnover in lactotrophs and stimulated prolactin release (Lewy et al., 1992). Dymshitz and co-workers (1992) suggested that transient stimulation of prolactin release by ETs could only be observed at relatively high concentrations (>30 nM) and when the pituitary cells were cultured in serum-free medium. Thus, when low concentrations of ETs were used and when cells were cultured in serum-containing medium (regardless of the ET concentration), only an inhibition of prolactin release was found (Dymshitz et al., 1992). The difference in mechanism behind the stimulation c.f. inhibition component remains to be studied in detail.

The injection of ETs into the lateral or third ventricle induced release of vasopressin into the blood (Samson et al., 1991; Matsumura et al., 1991a; Makino et al., 1990, 1992; Antunes-Rodrigues et al., 1993). While it could be argued that ET caused cerebral vasoconstriction and ischemia, there are several indications for a physiological involvement of ETs in vasopressin release. As mentioned above, ET-1 immunoreactivity and ET-1 mRNA could be demonstrated in pig and rat paraventricular nucleus and posterior pituitary (Yoshizawa et al., 1990; Ritz et al., 1992; Nakamura et al., 1993). Vasopressin release from perfused rat hypothalamus *in vitro* was reversibly stimulated by the addition of ET-1 at a dose of 0.1 to 10 nM (Shichiri et al., 1989). Similarly, the release of vasopressin from cultured explants of rat hypothalamo-neurohypophysis complex (Rossi, 1993) and from posterior pituitary nerve endings (Ritz et al., 1992) was enhanced by ET-3. From supraoptic nucleus slices, however, both ET-1 and ET-3 dose-dependently reduced vasopressin release (Yamamoto et al., 1992). It is possible that ETs play a different role in vasopressin release at the level of the cell bodies as compared to the posterior pituitary. In any case, it appears that very low doses of ETs are able to influence endocrine regulation both at the level of releasing hormones from the hypothalamus and at the level of the pituitary. Although further studies are needed to investigate this role in more detail, a physiological involvement of ETs as a local factor in hormone release appears likely.

H. Behavior and Pain Sensitivity

The effects of i.c.v. injections of ETs on blood pressure, hormone release, and other parameters were often associated with changes in behavior such as barrel rotation and convulsions (above). These effects were described in detail for the first time by Moser and Pelton (1989). These authors observed "marked ataxia, flat body posture and tendency to direct movements to one side." This latter postural bias would, at higher doses, lead to longitudinal rolling of the body: barrel rotation. This effect was dose related: at 1.25 pmol/rat none of the animals displayed this rotation, at 2.5 pmol it appeared in 3 out of 8 rats, at 5 pmol it appeared in 10 out of 13 rats (Moser and Pelton, 1989). The motor effects were accompanied by loss of the righting reflex (Lecci et al., 1990), clonus of the eye and facial muscles, as well as nystagmus and exophthalmos (Gross et al., 1993c). In the brain, hypermetabolism occurred in a number of brain nuclei of the visuovestibular and oculomotor systems. It was postulated that ET-1 activated structures such as the periventricular caudate nucleus which, through projections to the substantia nigra pars reticulata, could indirectly cause activation in these systems, which would then be involved in the motor and oculomotor effects of ETs (Gross et al., 1993c). Interestingly, the motor effects of central injection of ET-1 could be inhibited by central pretreatment with the NMDA-receptor antagonist MK-801, the calcium antagonist nimodipine, or the guanylate cyclase inhibitor methylene blue, suggesting the involvement of central glutamatergic pathways, calcium channels, and nitric oxide, respectively (Gross and Weaver, 1993). These authors suggested that the behavioral effects of central injection of ETs could be used as a new experimental model of epilepsy.

Possible actions of ETs injected i.c.v. in the caudate nucleus and effects of ETs on central dopamine release (see Section III.D) suggest that central ETs could modulate dopamine-mediated behaviors. Indeed, we have observed that i.c.v. injection of low doses of ET-1 (2 pmol/kg) caused a significant stimulation of locomotor activity in apomorphine-treated, but not saline-treated rats (Van den Buuse and Corriu, 1991). Similar to the effects of ETs on peripheral adrenergic nerves (see Section III.A), this effect may be explained by a sensitization of postsynaptic dopamine receptors or by reversal of apomorphine-induced presynaptic inhibition of dopamine release. At this dose, ET-1 alone did not cause significant effects on behavior.

The i.c.v. injection of ET-2 or ET-3, but not of the ET_B receptor agonist [$Ala^{1,3,11,15}$]ET-1, inhibited drinking behavior induced by water deprivation, hypertonic saline, or central administration of angiotensin II (Samson et al., 1991a; Samson and Murphy, 1993). The authors used doses of 10 to 40 pmol but did not describe any of the above-mentioned motor effects of ETs. Importantly, the central injection of antiserum against ET *enhanced* the effect of central angiotensin II on water intake, but the antiserum failed to influence water intake in response to water deprivation of hypertonic saline (Samson et al., 1991a). Pretreatment with the ET_A receptor antagonist BQ 123 also enhanced angiotensin II -induced water intake (Samson and Murphy, 1993). These results suggest that, at least in angiotensin II-induced drinking, central ETs play a physiological inhibitory role through an activation of ET_A receptors. This latter finding is remarkable, as in brain the ET_B receptors appear to be the more abundant receptor subtype (Hori et al., 1992). The site of these actions of ETs in the brain is as-yet unclear.

The intrathecal injection of ET-1 (1 or 3 pmol) in mice produced antinociceptive effects, which could be blocked by pretreatment with opiate antagonists such as naloxone (Kamei et al., 1993). Similarly, after i.c.v. injection, ET-1 and ET-3 produced antinociception in a number of test procedures, but in these experiments naloxone was ineffective (Nikolov et al., 1992, 1993a). The analgesic effects of intrathecal or i.c.v. ETs appear to be mediated by different mechanisms and possibly different sites of action in the brain.

These effects of central ETs on behavior expand the spectrum of physiological effects which these peptides appear to have. Much more information is needed, however, before the exact role and relevance of these peptides in behavioral mechanisms will become clear.

IV. PATHOPHYSIOLOGY

The wide variety of physiological and pharmacological effects of ETs in the brain would predict that these peptides play a role in the pathophysiology of cardiovascular, endocrine, and behavioral disorders. Only a limited number of reports address this possibility, either clinically (see Chapter 17) or experimentally. With the potent vasoconstrictor action of ETs on cerebral vessels as background, most of the work on pathophysiology has been focused on models of stroke and subarachnoid hemorrhage.

A. Ischaemia, Stroke, and Subarachnoid Haemorrhage

While for a number of effects of centrally administered ETs an involvement of changes in cerebral blood flow and ischemia cannot be excluded, only a few studies have attempted to directly address this possibility. The intrathecal injection of ET (presumably ET-3) at doses of 10 to 40 pmol produced paresis and lesions in spinal motorneurons (Hökfelt et al., 1989). Local injection of a very high dose of ET-1 (approximately 400 pmol) into the striatum caused ischemic-like lesions (Fuxe et al., 1989c; Agnati et al., 1991). ET-1 had a higher efficacy than ET-3 to produce lesions and the effect could be inhibited by the concomitant local injection of the vasodilator hydralazine. ET-1 produced a local reduction of cerebral blood flow to about 40% of control (Fuxe et al., 1992). Also, the microinjection of ET-1 (1 pmol) into the rat nucleus tractus solitarius (Lee et al., 1992) or cerebral cortex (Willette and Sauermelch, 1990) caused a significant reduction of local cerebral blood flow. Intracisternal injection of 5 to 500 pmol ET-1 into anesthetized dogs or cats induced a marked vasospasm of the basilar artery (Shigeno et al., 1989; Ide et al., 1989; Asano et al., 1990). Furthermore, intracisternal injection of 30 pmol ET-1 caused a pressor response and a marked (up to 85%) reduction of cerebral blood flow in brain regions such as the caudal medulla and cerebellum, but a hyperemia in forebrain structures such as the caudate nucleus (Macrae et al., 1991). These authors concluded that the hypertension caused by ET was a consequence of brainstem ischemia, as lower doses which did not evoke hemodynamic effects also did not alter cerebral blood flow. In further studies, these authors took these findings one step further and developed a stroke model using ET-1, applied on the middle cerebral artery, to induce arterial spasm and central ischemia (Robinson et al., 1991; Macrae et al., 1993). This effect has recently been reproduced in conscious rats (Sharkey et al., 1993). Similarly, intracisternal injection of ET-1 (5 to 10 pmol) enhanced the infarcted area induced by middle cerebral artery occlusion (Nikolov et al., 1993b).

While it is clear that ETs, through their potent vasoconstrictor activity, may cause reduction in cerebral blood flow, it is less clear whether they have direct neurotoxic effects. Kataoka et al. (1989, 1991) showed that in the presence but not in the absence of 10 μM ET, hypoglycemia induced a pathological rise in dopamine release from striatal slices *in vitro*. These authors concluded that ET was "one etiological factor in the development of hypoglycemic/ischemic brain injury" (Kataoka et al., 1989). By contrast, other authors have suggested that ETs did not produce direct neurotoxic effects on cerebral cortical cultures (Lustig et al., 1992) and failed to affect neurotoxic effects of hypoxia or cyanide (Nikolov et al., 1993b). ET-1 did not produce necrosis of brain tissue after i.c.v. injections which caused hypertension and hypermetabolism in the brain (Gross et al., 1992a,b). These results suggest that the ischemic effects of central administration of ETs are the result of vasoconstriction of cerebral vessels rather than direct neurotoxic effects.

The intracisternal injection of blood has been widely used as a model of subarachnoid hemorrhage. This treatment potentiated the cerebral vasoconstrictor effect of ET-1 injected i.a. in goats (Alabadi et al., 1993). It would appear that the injection of blood in some way triggers the synthesis or release of central ET, which then causes cerebral vasoconstriction. Indeed, the reduction of cerebral blood flow caused by the central injection of blood could be markedly inhibited by pretreatment with the ET_A receptor antagonists BQ 123 (Clozel and Watanabe, 1993), BQ 485 (Itoh et al., 1993), or FR 139317 (Nirei et al., 1993), and the nonselective antagonist Ro 46-2005 (Clozel et al., 1993). However, pretreatment with monoclonal antibodies against ET or with the ECE inhibitor phosphoramidon did not affect vasospasm (Shigeno et al., 1991), although in another study phosphoramidon did reduce the deleterious effects of simulated hemorrhage (Matsumura et al., 1991b).

As mentioned above, local lesions and ischemic insults caused an increase in ET concentrations in brain tissue at the site of the lesions (Cintra et al., 1989; Giuffrida et al., 1992; Viossat et al., 1993; Willette et al., 1992). In stroke-prone spontaneously hypertensive rats subjected to bilateral carotid artery occlusion, ET-1 or ET-3 immunoreactivity was markedly increased in hippocampus (Yamashita et al., 1993).

The concentrations of big-ET-1 and of ET-1 in c.s.f. were increased after central injection of blood (Matsumura et al., 1991b), but the increase did not appear to correlate with the vasospasm observed (Shigeno et al., 1991). In patients with subarachnoid hemorrhage, c.s.f. levels of big-ET-1 were not significantly different from those in controls (Hamann et al., 1993), but ET-1 and ET-3 levels in c.s.f. of patients with subarachnoid hemorrhage were significantly increased (Suzuki et al., 1990;

Kraus et al., 1991; Ehrenreich et al., 1992). It is possible that these findings reflect an increased synthesis of ET precursors followed by an increased conversion into mature ETs, thereby leaving the big-ET levels normal but ET levels increased. Measurement of ECE activity in subarachnoid hemorrhage will be needed to confirm this hypothesis. Taken together, these data warrant the conclusion that central ETs may be involved in the pathogenesis of stroke and subarachnoid hemorrhage, although the precise mechanism of this involvement is as yet unclear.

B. Hypertension

The chronic i.c.v. infusion of low doses of ET-1 caused a gradually developing hypertension (Nishimura et al., 1991, see Section III.F). Thus, increased levels or effects of central ETs could be involved in the development of hypertension. This hypothesis has been addressed by a number of studies in spontaneously hypertensive rats. ET-1 levels were measured in cerebral cortex, thalamus, hypothalamus, midbrain, pons, medulla, and cerebellum of 4- to 5-week- and 12- to 14-week-old rats. In the medulla of 4- to 5-week rats, and in all brain regions except the hypothalamus of 12- to 14-week spontaneously hypertensive rats, ET levels were significantly lower than those of normotensive controls (Yoshimi et al., 1991). These changes could be a compensatory response to the developing hypertension, although the early decrease in medullary levels could be involved in the initiation of the rise in blood pressure in these rats.

Binding sites for $[^{125}I]$ET-1 in whole brain of 14-week-old spontaneously hypertensive rats were significantly upregulated when compared to age-matched Wistar-Kyoto controls. B_{max} was increased by 29%, but the K_d was not different between the strains and the Hill coefficient was close to unity in both, suggesting that only a single population of receptors was labeled (Gu et al., 1990). However, in hypothalamus and cerebellum of 6- to 8-month animals, no significant differences were found in B_{max} of $[^{125}I]$ET-1 binding between spontaneously hypertensive rats and Wistar-Kyoto controls (Babasik et al., 1991). In contrast to these studies, in hypothalamus and ventrolateral medulla, but not in cortex, dorsal medulla, or spinal cord of 8-week-old spontaneously hypertensive rats, $[^{125}I]$ET-1 binding was significantly *reduced*. The B_{max} in hypothalamus and ventrolateral medulla was 40 to 50% lower than that in Wistar-Kyoto rats, while K_d values did not differ between the strains (Gulati, 1991b; Gulati and Rebello, 1991, 1992). Subtle differences appeared to exist between spontaneously hypertensive rats and controls in displacement of labeled ET-1 by other ETs (Gulati, 1991), but the binding in these experiments had the characteristics of ET_A receptors and could thus be on vascular elements rather than neurons. Studies with selective agonists or antagonists are needed to further investigate the identity of ET receptors in these rat strains. We have used receptor autoradiography to measure the density of ET binding in hindbrain of spontaneously hypertensive rats. In the ventrolateral medulla, but not other regions including the nucleus tractus solitarius, binding density was significantly lower in spontaneously hypertensive rats when compared to Wistar-Kyoto rats (Van den Buuse and Itoh, 1993, see Figure 5).

Functional studies in spontaneously hypertensive rats suggest little change in responses to centrally administered ETs in these rats. The pressor response to i.c.v. injection of 10 to 40 pmol ET-1 (Yoshida et al., 1991) and the interaction of ET with central voltage-dependent calcium channels (Huguet et al., 1993) were similar in hypertensive and normotensive rats. Similarly, the intracisternal injection of 25 pmol/kg ET-1 or ET-3 caused a significant increase in baroreflex sensitivity which

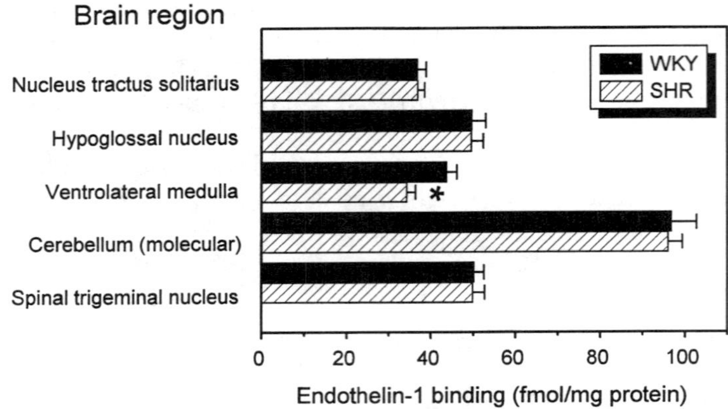

FIGURE 5
Binding density of [^{125}I]endothelin-1 in hindbrain sections of spontaneously hypertensive rats and normotensive wistar-Kyoto controls (n = 6), as measured with quantitative receptor autoradiography. Binding density was similar in the two strains, except in the ventrolateral medulla, where spontaneously hypertensive rats showed lower binding. (Data taken from Van den Buuse, M. and Itoh, S., 1993.)

was identical in spontaneously hypertensive rats and Wistar-Kyoto rats (Van den Buuse and Itoh, 1993). This latter finding is of interest because baseline baroreflex sensitivity is reduced in spontaneously hypertensive rats. However, this baseline difference is associated with a reduction of heart rate range, while the effect of ETs was on curvature, i.e., blood pressure range. Thus, there was a significant difference in heart rate range between spontaneously hypertensive rats and Wistar-Kyoto rats both before *and* after central treatment with ET (Van den Buuse and Itoh, 1993).

In conclusion, while changes in central ET levels and binding were observed in some studies, there are as yet no functional differences in central ET responses between spontaneously hypertensive rats and controls. It would be of particular interest, however, to study the cardiovascular effects of microinjection of ETs into the ventrolateral medulla of these rats.

C. Ageing, Behavioral Disorders, and Miscellaneous

There was no significant difference in ET binding density in cerebral cortex and spinal cord between 4-month, 15-month, or 24-month rats (Bertelsen et al., 1992). Thus, in normal ageing, ET receptors do not appear to be affected. In contrast, in post-mortem cerebral tissue of patients with Alzheimer's disease, markedly higher levels of ET immunoreactivity were observed in reactive astrocytes. A similar increase was observed in tissue from patients with basal ganglia lacunes, frontal lobotomy, or encephalopathy (Jiang et al., 1993). These findings could be related to the observations of increased ET production in brain regions after neurotoxin- or ischemia-induced lesions (see above), suggesting a growth-factor-like role of ET in response to neuronal injury. The findings also emphasize an involvement of glial cells in the central effects of ETs (Supattapone et al., 1989; MacCumber et al., 1990). Indeed, in human astrocytoma and glioblastoma tissue, a high ET binding density was observed (Kurihara et al., 1990). However, c.s.f. levels of ETs in patients with cerebral infarction, subdural haematoma, brain tumors (Suzuki et al., 1990), head injury, intractable epilepsy, or cerebral trauma (Kraus et al., 1991) were not significantly different from those in controls. Similarly, in spinal cerebral ataxia, a genetically determined neurodegenerative disorder of the cerebellum, medulla, and spinal cord, genetic analysis excluded a role of ET in this disease (Cancel et al., 1993).

The c.s.f. levels of ET-1 were around 50% lower in untreated depressive patients (Hoffman et al., 1989), a finding of particular interest in the light of effects of ET on behavior and on neurons in brain structures such as the hippocampus and cerebral cortex. In patients with Alzheimer's disease, c.s.f. levels of ET were significantly reduced as well. These levels were not correlated with age, duration from onset of the disease, blood pressure, or plasma ET-1 levels (Yoshizawa et al., 1992). It remains to be studied, however, which mechanism lies behind the reduced levels of ET in these patients.

V. CONCLUSIONS AND FUTURE PERSPECTIVES

ETs and ET receptors are widely distributed in the central nervous system and there is evidence to suggest a role of these peptides in a variety of central mechanisms. With respect to *distribution* of ETs in the brain (Section II.A), however, a comprehensive mapping of the localization of ET-synthesizing cell bodies, tracing of ET-positive axons, and distribution of ET-releasing terminals is needed. It is possible that ET, unlike other neurotransmitters, is not present as a well-defined system of regions of origin and regions of projections, but this has not yet been studied in any detail. Also electron microscopical studies on the ultrastructural distribution of ET receptors in the brain are needed to determine if these receptors are present presynaptically or postsynaptically on neurons or glial cells. With respect to the *function* of ETs in the brain, several areas are largely incomplete, although, in particular, a role of ETs in neuroendocrine regulation seems justified (Section III.G). The mechanism of action of ETs in the brain appears to be through the modulation of the release of other neurotransmitters or neuropeptides rather than a direct action (Sections III.D, E, and G). Further studies using local injections of ETs instead of intraventricular injections will be needed to better localize brain regions involved in the effects of ETs. Finally, in studies on central ETs there is widespread controversy concerning the identity of ETs present in the brain and of ET receptors involved in these effects. Resolving these issues and that of the regional distribution of ETs in the brain will greatly facilitate studies on the functional role of these neuropeptides. The available evidence so far supports an important role of ETs in brain function and homeostasis.

REFERENCES

Adner, M., You, J. and Edvinsson, L., Characterization of endothelin-A receptors in the cerebral circulation, *Neuroreport*, 4, 441-443, 1993.

Agnati, L. F., Zoli, M., Kurosawa, M., Benfenati, F., Biagini, G., Zini, I., Hallström, A., Ungerstedt, U., Toffano, G. and Fuxe, K., A new model of focal brain ischemia based on the intracerebral injection of endothelin-1, *Ital. J. Neurol. Sci.*, Suppl. 2, 49-53, 1991.

Alabadí, J. A., Salom, J. B., Torregrosa, G., Miranda, F. J., Jover, T. and Alborch, E., Changes in the cerebrovascular effects of endothelin-1 and nicardipine after experimental subarachnoid hemorrhage, *Neurosurgery*, 33, 707-715, 1993.

Ambar, I., Kloog, Y., Kochva, E., Wollberg, Z., Bdolah, A., Oron, U. and Sokolovsky, M., Characterization and localization of a novel neuroreceptor for the peptide sarafatoxin, *Biochem. Biophys. Res. Commun.*, 157, 1104-1110, 1988.

Ambar, I., Kloog, Y., Schvartz, I., Hazum, E. and Sokolovsky, M., A competitive interaction between endothelin and sarafatoxin, binding and phosphoinositides hydrolysis in rat atria and brain, *Biochem. Biophys. Res. Commun.*, 158, 195-201, 1989.

Ando, K., Hirata, Y., Togashi, K., Kawakami, M. and Marumo, F., Endothelin-1- and endothelin-3-like immunoreactivity in human cerebrospinal fluid, *J. Cardiovasc. Pharmacol.*, 17, S434-S436, 1991.

Antunes-Rodrigues, J., Ramalho, M. J., Reis, L. C., Picanco-Diniz, D. W. L., Favaretto, A. L. V., Gutkowska, J. and McCann, S. M., A possible role of endothelin acting within the hypothalamus to induce the release of atrial natriuretic peptide and natriuresis, *Neuroendocrinology*, 58, 701-708, 1993.

Arai, H., Hori, S., Aramori, I., Ohkubo, H. and Nakanishi, S., Cloning and expression of a cDNA encoding an endothelin receptor, *Nature*, 348, 730-732, 1990.

Asano, T., Ikegaki, I., Satoh, S., Suzuki, Y., Shibuya, M., Sugita, K. and Hidaka, H., Endothelin, a potential modulator of cerebral vasospasm, *Eur. J. Pharmacol.*, 190, 365-372, 1990.

Babasik, J. L., Hosick, H., Wright, J. W. and Harding, J. W., Endothelin binding in brain of normotensive and spontaneously hypertensive rats, *J. Pharmacol. Exp. Ther.*, 257, 302-306, 1991.

Bertelsen, G. A., Rebello, S. and Gulati, A., Characteristics of endothelin receptors in the cerebral cortex and spinal cord of aged rats, *Neurobiol. Aging*, 13, 513-519, 1992.

Boarder, M. R. and Marriott, D. B., Endothelin-1 stimulation of noradrenaline and andrenaline release from adrenal chromaffin cells, *Biochem. Pharmacol.*, 41, 521-526, 1991.

Bolger, G. T., Liard, F., Krogsrud, R., Thibeault, D. and Jaramillo, J., Tissue specificity of endothelin binding sites, *J. Cardiovasc. Pharmacol.*, 16, 367-375, 1990.

Bolger, G. T., Berry, R. and Jaramillo, J., Regional and subcellular distribution of [^{125}I]endothelin binding sites in rat brain, *Brain Res. Bull.*, 28, 789-797, 1992.

Brain, S. D., Thomas, G., Crossman, D. C., Fuller, R. and Church, M. K., Endothelin-1 induces a histamine-dependent flare *in vivo*, but does not activate human skin mast cells *in vitro*, *Br. J. Clin. Pharmacol.*, 33, 117-120, 1992.

Bucher, B., Paya, D. and Stoclet, J. C., Modulation of [^{3}H]noradrenaline release by endothelin-1 in the rat tail artery, *Neurochem. Int.*, 18, 455-459, 1991.

Burris, T. P., Kanyicska, B. and Freeman, M. E., Inhibition of prolactin secretion by endothelin-3 is pertussis-toxin sensitive, *Eur. J. Pharmacol.*, 198, 223-225, 1991.

Calvo, J. J., Gonzalez, R., Freire de Carvalho, L., Takahashi, K., Kanse, S. M., Hart, G. R., Ghatei, M. A. and Bloom, S. R., Release of Substance P from rat hypothalamus and pituitary by endothelin, *Endocrinology*, 126, 2288-2295, 1990.

Cancel, G., Khati, C., Stevanin, G., Pages, J. C., Agid, Y., Brice, A. and Cann, H. M., Endothelin 1 is not a candidate gene for spinal cerebellar ataxia 1, *Hum. Mol. Genet.*, 2, 1477-1479, 1993.

Cao, W. H., Kuwaki, T., Unekawa, M., Ling, G. Y., Terui, N. and Kumada, M., Action of endothelin-1 on vasomotor neurons in rat rostral ventrolateral medulla, *J. Cardiovasc. Pharmacol.*, 22, S196-S198, 1993.

Chuang, D. M., Lin, W. W. and Lee, C.Y., Endothelin-induced activation of phosphoinositide turnover, calcium mobilization, and transmitter release in cultured neurons and neurally related cell types, *J. Cardiovasc. Pharmacol.*, 17, S85-S88, 1991.

Cintra, A., Fuxe, K., Änggård, E., Tinner, B., Staines, W. and Agnati, L. F., Increased endothelin-like immunoreactivity in ibotenic acid-lesioned hippocampal formation of the rat brain, *Acta Physiol. Scand.*, 137, 557-558, 1989.

Clozel, M. and Watanebe, H., BQ-123, a peptidic endothelin ET_A receptor antagonist, prevents the early cerebral vasospasm following subarachnoid hemorrhage after intracisternal but not intravenous injection, *Life Sci.*, 52, 825-834, 1993.

Clozel, M., Breu, V., Burri, K., Cassal, J. M., Fischli, W., Gray, G. A., Hirth, G., Löffler, B. M., Müller, M., Neidhart, W. and Ramuz, H., Pathophysiological role of endothelin revealed by the first orally active endothelin receptor antagonist, *Nature*, 365, 759-761, 1993.

Couraud, P. O., Durieu-Trautmann, O., Le Nguyen, D., Marin, P., Glibert, F. and Strosberg, A. D., Functional endothelin-1 receptors in rat astrocytoma C6, *Eur. J. Pharmacol. (Mol. Pharmacol. Sect.)*, 206, 191-198, 1991.

Crawford, M. L. A., Hiley, C. R. and Young, J. M., Characteristics of endothelin-1 and endothelin-3 stimulation of phosphoinositide breakdown differ between regions of guinea pig and rat brain, *Naunyn Schmiedeberg's Arch. Pharmacol.*, 341, 268-271, 1990.

Davenport, A. P. and Morton, A. J., Binding sites for ^{125}I ET-1, ET-2, ET-3 and vasoactive intestinal contractor are present in adult rat brain and neurone-enriched primary cell cultures of embryonic brain cells, *Brain Res.*, 554, 278-285, 1991.

Domae, M., Yamada, K., Hanabusa, Y. and Furukawa, T., Inhibitory effects of endothelin-1 and endothelin-3 on prolactin release, possible involvement of endogenous endothelin isopeptides in the rat anterior pituitary, *Life Sci.*, 50, 715-722, 1992.

Durieu-Trautmann, O., Couraud, P. O., Foignant-Chaverot, N. and Strosberg, A. D., cAMP-dependent downregulation of endothelin-1 receptors on rat astrocytoma C6 cells, *Neurcsci. Lett.*, 131, 175-178, 1991.

Durieu-Trautmann, O., Fédérici, C., Créminon, C., Foignant-Chaverot, N., Roux, F., Claire, M., Strosberg, A. D. and Couraud, P. O., Nitric oxide and endothelin secretion by brain microvessel endothelial cells, regulation by cyclic nucleotides, *J. Cell. Physiol.*, 155, 104-111, 1993.

Dymshitz, J., Laudon, M. and Ben-Jonathan, N., Endothelin-induced biphasic response of lactotrophs cultured under different conditions, *Neuroendocrinology*, 55, 724-729, 1992.

Dymshitz, J. and Vasko, M. R., Endothelin-1 enhances capsaicin-induced peptide release and cGMP accumulation in cultures of rat sensory neurons, *Neurosci. Lett.*, 167, 128-132, 1994.

Ehrenreich, H., Kehrl, J. H., Anderson, R. W., Rieckmann, P., Vitkovic, L., Coligan, J. E. and Fauci, A. S., A vasoactive peptide, endothelin-3, is produced by and specifically binds to primary astrocytes, *Brain Res.*, 538, 54-58, 1991.

Ehrenreich, H., Lange, M., Near, K. A., Anneser, F., Schoeller, L. A. C., Schmid, R., Winkler, P. A., Kehrl, J. H., Schmiedek, P. and Goebel, F. D., Long-term monitoring of immunoreactive endothelin-1 and endothelin-3 in ventricular cerebrospinal fluid, plasma, and 24-h urine of patients with subarachnoid hemorrhage, *Res. Exp. Med.*, 192, 257-268, 1992.

Ernsberger, P., Haxhiu, M. A., and Wendorff, L. A., Endothelin receptors in the ventrolateral medulla, upregulation by hypoxia and role in cardiorespiratory regulation, *Clin. Res.*, 39, 706A, 1991.

Ferguson, A. V. and Smith, P., Cardiovascular responses induced by endothelin microinjection into area postrema, *Regul. Pept.*, 27, 75-85, 1990.

Ferguson, A. V. and Smith, P., Circulating endothelin influences area postrema neurons, *Regul. Pept.*, 32, 11-21, 1991.

Firth, J. D. and Ratcliffe, P. J., Organ distribution of the three rat endothelin messenger RNAs and the effects of ischemia on renal gene expression, *J. Clin. Invest.*, 90, 1023-1031, 1992.

Franco-Cereceda, A., Rydh, M., Lou, Y. P., Dalsgaard, C. J. and Lundberg, J. M., Endothelin as a putative sensory neuropeptide in the guinea pig, different properties in comparison with calcitonin gene-related peptide, *Regul. Pept.*, 32, 253-265, 1991.

Fuxe, K., Änggård, E., Lundgren, K., Cintra, A., Agnati, L. F., Galton, S. and Vane, J., Localization of [^{125}I]endothelin-1 and [^{125}I]endothelin-3 binding sites in the rat brain, *Acta Physiol. Scand.*, 137, 563-564, 1989a.

Fuxe, K., Andbjer, B., Kalia, M. and Agnati, L. F., Centrally administered endothelin-1 produces apnoea in the chloralose-anesthetized male rat, *Acta Physiol. Scand.*, 137, 157-158, 1989b.

Fuxe, K., Cintra, A., Andbjer, B., Änggård, E., Goldstein, M. and Agnati, L. F., Centrally administered endothelin-1 produces lesions in the brain of the male rat, *Acta Physiol. Scand.*, 137, 155-156, 1989c.

Fuxe, K., Tinner, B., Staines, W., Hemsén, A., Hersh, L. and Lundberg, J. M., Demonstration and nature of endothelin-3-like immunoreactivity in somatostatin and choline acetyltransferase-immunoreactive nerve cells of the neostriatum of the rat, *Neurosci. Lett.*, 123, 107-111, 1991.

Fuxe, K., Kurosawa, N., Cintra, A., Hallström, Å., Goiny, M., Rosén, L., Agnati, L. F. and Ungerstedt, U., Involvement of local ischemia in endothelin-1 induced lesions of the neostriatum of the anesthetized rat, *Exp. Brain Res.*, 88, 131-139, 1992.

Giaid, A., Gibson, S. J., Ibrahim, N. B. N., Legon, S., Bloom, S. R., Yanagisawa, M., Masaki, T., Varndell, I. M. and Polak, J. M., Endothelin 1, an endothelium-derived peptide, is expressed in neurons of the human spinal cord and dorsal root ganglia, *Proc. Natl. Acad. Sci. U.S.A.*, 86, 7634-7638, 1989.

Giaid, A., Gibson, S. J., Herrero, M. T., Gentleman, S., Legon, S., Yanagisawa, M., Masaki, T., Ibrahim, N. B. N., Roberts, G. W., Rossi, M. L. and Polak, J. M., Topographical localization of endothelin mRNA and peptide immunoreactivity in neurons of the human brain, *Histochemistry*, 95, 303-314, 1991.

Giuffrida, R., Bellomo, M., Polizzi, G. and Malatino, L. S., Ischemia-induced changes in the immunoreactivity for endothelin and other vasoactive peptides in the brain of the mongolian gerbil, *J. Cardiovasc. Pharmacol.*, 20, S41-S44, 1992.

Giuliani, S., Lecci, A., Maggi, C. A., Rovero, P. and Giachetti, A., Effect of intrathecal administration of ET-1, ET-3 and ET(16-21) on blood pressure and micturition reflex in anesthetized rats, *Neurochem. Int.*, 18, 565-569, 1991.

Goldman, R. S., Finkbeiner, S. M. and Smith, S. J., Endothelin induces a sustained rise in intracellular calcium in hippocampal astrocytes, *Neurosci. Lett.*, 123, 4-8, 1991.

Gross, P. M., Wainman, D. S. and Espinosa, F. J., Differentiated metabolic stimulation of rat pituitary lobes by peripheral and central endothelin-1, *Endocrinology*, 129, 1110-1112, 1991.

Gross, P. M., Wainman, D. S., Espinosa, F. J., Nag, S. and Weaver, D. F., Cerebral hypermetabolism produced by intraventricular endothelin-1 in rats: inhibition by nimodipine, *Neuropeptides*, 21, 211-223, 1992a.

Gross, P. M., Zochodne, D. W., Wainman, D. S., Ho, L. T., Espinosa, F. J. and Weaver, D. F., Intraventricular endothelin-1 uncouples the blood flow, metabolism relationship in periventricular structures of the rat brain, involvement of L-type calcium channels, *Neuropeptides*, 22, 155-165, 1992b.

Gross, P. M. and Weaver, D. F., A new experimental model of epilepsy based on the intraventricular injection of endothelin, *J. Cardiovasc. Pharmacol.*, 22, S282-S287, 1993.

Gross, P. M., Weaver, D. F., Wainman, D. S., Chew, B. H., Espinosa, F. J. and Nag, S., Potent metabolic stimulation of septal gray and cerebral white matter *in vivo* by intraventricular endothelin and nitric oxide, *Biochem. Biophys. Res. Commun.*, 190, 975-981, 1993a.

Gross, P. M., Wainman, D. S., Chew, B. H., Espinosa, F. J. and Weaver, D. F., Calcium-dependent metabolic stimulation of neuroendocrine structures by intraventricular endothelin-1 in conscious rats, *Brain Res.*, 606, 135-142, 1993b.

Gross, P. M., Beninger, R. J., Shaver, S. W., Wainman, D. S., Espinosa, F. J. and Weaver, D. F., Metabolic and neuroanatomical correlates of barrel-rolling and oculoclonic convulsions induced by intraventricular endothelin-1: a novel peptidergic signaling mechanism in visuovestibular and oculomotor regulation?, *Exp. Brain Res.*, 95, 397-408, 1993c.

Gu, X. H., Casley, D. J., Cincotta, M. and Nayler, W. G., ^{125}I-endothelin-1 binding to brain and cardiac membranes from normotensive and spontaneously hypertensive rats, *Eur. J. Pharmacol.*, 177, 205-209, 1990.

Gulati, A., Evidence for antagonistic activity of endothelin for clonidine induced hypotension and bradycardia, *Life Sci.*, 50, 153-160, 1991.

Gulati A., Characteristics of endothelin binding sites in the spinal cord of spontaneously hypertensive rats, *Eur. J. Pharmacol.*, 204, 287-293, 1991.

Gulati, A. and Rebello, S., Down-regulation of endothelin receptors in the ventrolateral medulla of spontaneously hypertensive rats, *Life Sci.* 48, 1207-1215, 1991.

Gulati, A. and Rebello, S., Characteristics of endothelin receptors in the central nervous system of spontaneously hypertensive rats, *Neuropharmacology*, 31, 243-250, 1992.

Gulati, A. and Srimal, R. C., Endothelin mechanisms in the central nervous system: a target for drug development, *Drug Dev. Res.*, 26, 361-387, 1992.

Gulati, A. and Srimal, R. C., Endothelin antagonizes the hypotension and potentiates the hypertension induced by clonidine, *Eur. J. Pharmacol.*, 230, 293-300, 1993.

Hama, H., Sakurai, T., Kasuya, Y., Fujiki, M., Masaki, T. and Goto, K., Action of endothelin-1 on rat astrocytes through the ET_B receptor, *Biochem. Biophys. Res. Commun.*, 186, 55-362, 1992.

Hamann, G., Isenberg, E., Strittmatter, M. and Schimrigk, K., Absence of elevation of big endothelin in subarachnoid hemorrhage, *Stroke*, 24, 383-386, 1993.

Han, S. P., Chen, X., Westfall, T. C. and Knuepfer, M. M., Characterization of the depressor effect of intrathecal endothelin in anesthetized rats, *Am. J. Physiol.*, 260, H1685-1691, 1991.

Hashim, M. A. and Tadepalli, A. S., Functional evidence for the presence of a phosphoramidon-sensitive enzyme in rat brain that converts big endothelin-1 to endothelin-1, *Life Sci.*, 49, PL-207-211, 1991.

Hashim, M. A. and Tadepalli, A. S., Hemodynamic responses evoked by endothelin from central cardiovascular neural substrates, *Am. J. Physiol.*, 262, H1-H9, 1992.

Hashimoto, S., Mizuno, K., Kato, K., Saito, I. and Ono, Y., Evidence for glutamate-regulated release of endothelin from neuronal cells, *J. Hypertens.*, 12 (Suppl. 3), S191, Abstr. 1055, 1994.

Hay, D. W., Hubbard, W. C. and Undem, B. J., Relative contributions of direct and indirect mechanisms mediating endothelin-induced contraction of guinea-pig trachea, *Br. J. Pharmacol.*, 110, 955-962, 1993.

Hematillake, G. and LaHann, T. R., Central modulation of cardiac output by endothelin-1 (ET-1), *FASEB J.*, 6, 1531, 1992.

Hemsén, A. and Lundberg, J. M., Presence of endothelin-1 and endothelin-3 in peripheral tissues and central nervous system of the pig, *Regul. Pept.*, 36, 71-83, 1991.

Hirata, Y., Matsunaga, T., Ando, K., Furukawa, T., Tsukagoshi, H. and Marumo, F., Presence of endothelin-1 like immunoreactivity in human cerebrospinal fluid, *Biochem. Biophys. Res. Commun.*, 166, 1274-1278, 1990.

Hoffman, A., Keiser, H. R., Grossman, E., Goldstein, D. S., Gold, P. W. and Kling, M., Endothelin concentration in cerebrospinal fluid in depressive patients, *Lancet xii.*, 1519, 1989.

Hökfelt, T., Post, C., Freedman, J., Lundberg, J. M. and Terenius, L., Endothelin induces spinal lesions after intrathecal administration, *Acta Physiol. Scand.*, 137, 555-556, 1989.

Hori, S., Komatsu, Y., Shigemoto, R., Mizuno, N. and Nakanishi, S., Distinct tissue distribution and cellular localization of two messenger ribonucleic acids encoding different subtypes of rat endothelin receptors, *Endocrinology*, 130, 1885-1895, 1992.

Hösli, E. and Hösli, L., Autoradiographic evidence for endothelin receptors on astrocytes in cultures of rat cerebellum, brainstem and spinal cord, *Neurosci. Lett.*, 129, 55-58, 1991.

Hösli, L., Hösli, E., Lefkovits, M. and Wagner, S., Electrophysiological evidence for the existence of receptors for endothelin and vasopressin on cultured astrocytes of rat spinal cord and brainstem, *Neurosci. Lett.*, 131, 193-195, 1991.

Hoyer, D., Waeber, C. and Palacios, J. M., [^{125}I]Endothelin-1 binding sites: autoradiographic studies in the brain and periphery of various species including humans, *J. Cardiovasc. Pharmacol.*, 13, S162-S165, 1989.

Huggins, J. P., Pelton, J. T. and Miller, R. C., The structure and specificity of endothelin receptors: their importance in physiology and medicine, *Pharmacol. Ther.*, 59, 55-123, 1993.

Huguet, F., Brisac, A. M., Dubar, M., Ingrand, P. and Piriou, A., Endothelin modulates dihydropyridine receptor decreased binding in hippocampal slices from normotensive and spontaneously hypertensive rats, *Int. J. Dev. Neurosci.*, 11, 295-301, 1993.

Ide, K., Yamakawa, K., Nakagomi, T., Sasaki, T., Saito, I., Kurihara, H., Yosizumi, M., Yazaki, Y. and Takakura, K., The role of endothelin in the pathogenesis of vasospasm following subarachnoid hemorrhage, *Neurol. Res.*, 11, 101-104, 1989.

Imai, T., Hirata, Y., Emori, T., Yanagisawa, M., Masaki, T. and Marumo, F., Induction of endothelin-1 gene by angiotensin and vasopressin in endothelial cells, *Hypertension*, 19, 753-757, 1992.

Itoh, S., Sasaki, T., Ide, K., Ishikawa, K., Nishikibe, M. and Yano, M., A novel endothelin ET_A receptor antagonist, BQ-485, and its preventive effect on experimental cerebral vasospasm in dogs, *Biochem. Biophys. Res. Commun.*, 195, 969-975, 1993.

Itoh, S. and Van den Buuse, M., Sensitization of baroreceptor reflex by central endothelin in conscious rats, *Am. J. Physiol.*, 260, H1106-H1112, 1990.

Jiang, M. H., Höög, A., Ma, K. C., Nie, X. J., Olsson, Y. and Zhang, W. W., Endothelin-1-like immunoreactivity is expressed in human reactive astrocytes, *Neuroreport*, 4, 935-937, 1993.

Jones, C. R., Hiley, C. R., Pelton, J. T. and Mohr, M., Autoradiographic visualization of the binding sites for [^{125}I]endothelin in rat and human brain, *Neurosci. Lett.*, 97, 276-279, 1989.

Kadel, K. A., Heistad, D. D. and Faraci, F. M., Effects of endothelin on blood vessels of the brain and choroid plexus, *Brain Res.*, 518, 78-82, 1990.

Kamei, J., Hitosugi, H., Kawashima, N., Misawa, M. and Kasuya, Y,. Antinociceptive effects of intrathecally administered endothelin-1 in mice, *Neurosci. Lett.*, 153, 69-72, 1993.

Kanyicska, B., Burris, T. P. and Freeman, M. E., Endothelin-3 inhibits prolactin and stimulates LH, FSH and TSH secretion from pituitary cell culture, *Biochem. Biophys. Res. Commun.*, 174, 338-343, 1991.

Kanyicska, B. and Freeman, M. E., Characterization of endothelin receptors in the anterior pituitary gland, *Am. J. Physiol.*, 265, E601-E608, 1993.

Kar, S., Chabot, J. G. and Quiron, R., Quantitative autoradiographic localization of [^{125}I]endothelin-1 binding sites in spinal cord and dorsal root ganglia of the rat, *Neurosci. Lett.*, 133, 117-120, 1991.

Kataoka, Y., Koizumi, S., Kumakura, K., Kurihara, M., Niwa, M. and Ueki, S., Endothelin-triggered brain damage under hypoglycemia evidenced by real-time monitoring of dopamine release from striatal slices, *Neurosci. Lett.*, 107, 75-80, 1989.

Kataoka, Y., Koizumi, S., Ueki, S. and Niwa, M., Is an endothelin a neurotoxic factor in the brain, *Neurochem. Int.*, 18, 503-506, 1991.

Kawano, Y., Yoshida, K., Yoshimi, H., Kuramochi, M. and Omae, T., The cardiovascular effect of intracerebroventricular endothelin in rats, *J. Hypertens.*, 7, S22-S23, 1989.

Kloog, Y., Ambar, I., Kochva, E., Wollberg, Z., Bdolah, A. and Sokolovsky, M., Sarafatoxin receptors mediate phosphoinostied hydrolysis in various rat brain regions, *FEBS Lett.*, 242, 387-390, 1989.

Knuepfer, M. M., O'Brien, D., Hoang, D., Gan, Q. and Song, C., Central sympathetic control of spinal endothelin release in the rat, *Eur. J. Pharmacol.*, 259, 305-308, 1994.

Kohzuki, M., Chai, S. Y., Paxinos, G., Karavas, A., Casley, D. J., Johnston, C. I. and Mendelsohn, F. A. O., Localization and characterization of endothelin receptor binding sites in the rat brain visualized by *in vitro* autoradiography, *Neuroscience*, 42, 245-260, 1991.

Koizumi, S., Kataoka, Y., Niwa, M. and Kumakura, K., Endothelin-3 stimulates the release of catecholamine from cortical and striatal slices of the rat, *Neurosci. Lett.*, 134, 219-222, 1992.

Koizumi, S., Inoue, K., Kataoka, Y., Niwa, M. and Takanaka, A., Endothelin-3 activates a voltage-gated Ca channel via a pertussis toxin sensitive mechanism leading to dopamine release from PC12 cells, *Neurosci. Lett.*, 166, 191-194, 1994.

Konya, H., Nagai, K., Masuda, H. and Kakishita, E., Endothelin-3 modification of dopamine release in anesthetized rat striatum, an *in vivo* microdialysis study, *Life Sci.*, 51, 499-506, 1992.

Koseki, C., Imai, M., Hirata, Y., Yanagisawa, M. and Masaki, T., Autoradiographic localization of [^{125}I]-endothelin-1 binding sites in rat brain, *Neurosci. Res.*, 6, 581-585, 1989.

Koseki, C., Imai, M., Hirata, Y., Yanagisawa, M. and Masaki, T., Autoradiographic distribution in rat tissues of binding sites for endothelin: a neuropeptide, *Am. J. Physiol.*, 256, R858-R866, 1989.

Kraus, G. E., Bucholz, R. D., Yoon, K. W., Knuepfer, M. M. and Smith, K. R. Jr., Cerebrospinal fluid endothelin-1 and endothelin-3 levels in normal and neurosurgical patients: a clinical study and literature review, *Surg. Neurol.*, 35, 20-29, 1991.

Krsmanovic, L. Z., Stojilkovic, S. S., Balla, T., Al-Damluji, S., Weiner, R. I. and Catt, K. J., Receptors and neurosecretory actions of endothelin in hypothalamic neurons, *Proc. Natl. Acad. Sci. U.S.A.*, 88, 11124-11128, 1991.

Kurihara, M., Ochi, A., Kawaguchi, T., Niwa, M., Kataoka, Y. and Mori, K., Localization and characterization of endothelin receptors in human gliomas: a growth factor, *Neurosurgery*, 27, 275-281, 1990.

Kurosawa, M., Fuxe, K., Hallström, Å., Goiny, M., Cintra, A. and Ungerstedt, U., Responses of blood flow, extracellular lactate, and dopamine in the striatum to intrastriatal injection of endothelin-1 in anesthetized rats, *J. Cardiovasc. Pharmacol.*, 17, S340-S342, 1991.

Kushiku, K., Ohjimi, H., Yamada, H., Tokunaga, R., Furukawa, T., Endothelin-3 inhibits ganglionic transmission at preganglionic sites through activation of endogenous thromboxane A_2 production in dog cardiac sympathetic ganglia, *J. Cardiovasc. Pharmacol.*, 17, S197-S199, 1991.

Kuwaki, T., Koshiya, N., Takahashi, H., Terui, N. and Kumada, M., Modulatory effects of rat endothelin on central cardiovascular control in rats, *Jpn. J. Physiol.*, 40, 97-116, 1990a.

Kuwaki, T., Koshiya, N., Cao, W. H., Takahashi, H., Terui, N. and Kumada, M., Modulatory effects of endothelin-1 on central cardiovascular control in rats, *Jpn. J. Physiol.*, 40, 827-841, 1990b.

Kuwaki, T., Cao, W. H., Unekawa, M., Terui, N. and Kumada, M., Endothelin-sensitive areas in the ventral surface of the rat medulla, *J. Auton. Nerv. Syst.*, 36, 149-158, 1991a.

Kuwaki, T., Koshiya, N., Terui, N. and Kumada, M., Endothelin-1 modulates cardiorespiratory control by the central nervous system, *Neurochem. Int.*, 18, 519-524, 1991b.

La, M., Wong-Dusting, H. K. and Rand, M. J., Endothelin enhances responses to sympathetic nerve stimulation and vasoconstrictor agonists in the rabbit ear artery, *Eur. J. Pharmacol.*, 183, 1801-1802, 1990.

Ladenheim, R. G., Lacroix, I., Foignant-Chaverot, N., Strosberg, A. D. and Couraud, P.O., Endothelins stimulate *c-fos* and nerve growth factor expression in astrocytes and astrocytoma, *J. Neurochem.*, 60, 260-266, 1993.

Lecci, A., Maggi, C. A., Rovero, P., Giachetti, A. and Meli, A., Effect of endothelin-1, endothelin-3 and c-terminal hexapeptide, endothelin(16-21) on motor activity in rats, *Neuropeptides*, 16, 21-24, 1990.

Lecci, A., Giuliani, S., Santicioli, P., Rovero, P., Maggi, C. A. and Giachetti, A., Intracerebroventricular administration of endothelins: effects on the supraspinal micturition reflex and blood pressure of the anesthetized rat, *Eur. J. Pharmacol.*, 199, 201-207, 1991.

Lee, C. C., Ohta, H. and Talman, W. T., Endothelin-1 micro-injected into nucleus tractus solitarii of rat blocks arterial baroreflexes, in *Genetic Hypertension*, Sassard, J., Ed., Colloque INSERM/John Libbey Eurotext, Paris, 1992, 109-111.

Lee, M. E., De la Monte, S. M., Ng, S. C., Bloch, K. D. and Quertermous, T., Expression of the potent vasoconstrictor endothelin in the human central nervous system, *J. Clin. Invest.*, 86, 141-147, 1990.

Levin, E. R., Isackson, P. J. and Hu, R. M., Endothelin increases atrial natriuretic peptide production in cultured rat diencephalic neurons, *Endocrinology*, 128, 2925-2930, 1991.

Levin, E. R., Frank, H. J. L. and Pedram, A., Endothelin receptors on cultured fetal rat diencephalic glia, *J. Neurochem.*, 58, 659-666, 1992.

Lewy, H., Galron, R., Bdolah, A., Sokolovsky, M. and Naor, Z., Paradoxical signal transduction mechanism of endothelins and sarafatoxins in cultured pituitary cells: stimulation of phosphoinositide turnover and inhibition of prolactin release, *Mol. Cell. Endocrinol.*, 89, 1-9, 1992.

Lin, W. W., Lee, C. Y. and Chuang, D. M., Endothelin-1 stimulates the release of preloaded [^{3}H]D-aspartate from cultured cerebellar granule cells, *Biochem. Biophys. Res. Commun.*, 167, 593-599, 1990.

Lin, W. W., Lee, C. Y. and Chuang, D. M., Endothelin- and sarafatoxin-induced phosphoinositide hydrolysis in cultured cerebellar granule cells: biochemical and pharmacological characterization, *J. Pharmacol. Exp. Ther.*, 257, 1053-1061, 1991.

Lustig, H. S., Chan, J. and Greenberg, D. A., Comparative neurotoxic potential of glutamate, endothelins, and platelet-activating factor in cerebral cortical cultures, *Neurosci. Lett.*, 139, 15-18, 1992.

MacCumber, M. W., Ross, C. A., Glaser, B. M. and Snyder, S. H., Endothelin: visualization of mRNAs by *in situ* hybridization provides evidence for local action, *Proc. Natl. Acad. Sci. U.S.A.*, 86, 7285-7289, 1989.

MacCumber, M. W., Ross, C. A. and Snyder, S. H., Endothelin in brain: Receptors, mitogenesis, and biosynthesis in glial cells, *Proc. Natl. Acad. Sci. U.S.A.*, 87, 2359-2363, 1990.

Macrae, A. D. and Bloom, S. R., Endothelin. An endocrine role, *Trends Endocrinol. Metab.*, 3, 153-157, 1992.

Macrae, I. M., Robinson, M. J., McAuley, M. A., Reid, J. L. and McCulloch, J., Effects of intracisternal endothelin-1 injection on blood flow to the lower brain stem, *Eur. J. Pharmacol.*, 203, 85-91, 1991.

Macrae, I. M., Robinson, M. J., Graham, D. I., Reid, J. L. and McCulloch, J., Endothelin-1-induced reductions in cerebral blood flow: dose dependency, time course, and neuropathological consequences, *J. Cereb. Blood Flow Metab.*, 13, 276-284, 1993.

Makino, S., Hashimoto, K., Hirasawa, R., Hattori, T., Kageyama, J. and Ota, Z., Central interaction between endothelin and brain natriuretic peptide on pressor and hormonal responses, *Brain Res.*, 534, 117-121, 1990.

Makino, S., Hashimoto, K., Hirasawa, R., Hattori, T. and Ota, Z., Central interaction between endothelin and brain natriuretic peptide on vasopressin secretion, *J. Hypertens.*, 10, 25-28, 1992.

Marsault, R., Vigne, P., Breittmayer, J. P. and Frelin, C., Astrocytes are targets for endothelins and sarafatoxin, *J. Neurochem.*, 54, 2142-2144, 1990.

Matsumoto, H., Suzuki, N., Onda, H. and Fujino, M., Abundance of endothelin-3 in rat intestine, pituitary gland and brain, *Biochem. Biophys. Res. Commun.*, 164, 74-80, 1989.

Matsumura, K., Abe, I., Tsuchihashi, T., Tominaga, M., Kobayashi, K. and Fujishima, M., Central effect of endothelin on neurohormonal responses in conscious rabbits, *Hypertension*, 17, 1192-1196, 1991a.

Matsumura, Y., Ikegawa, R., Suzuki, Y., Takaoka, M., Uchida, T., Kido, H., Shinyama, H., Hayashi, K., Wataneba, M. and Morimoto, S., Phosphoramidon prevents cerebral vasospasm following subarachnoid hemorrhage in dogs: the relationship to endothelin-1 levels in the cerebrospinal fluid, *Life Sci.*, 49, 841-848, 1991b.

Minamisawa, K., Hashimoto, R., Ishii, M. and Kimura, F., Complicated central effects of endothelin on blood pressure in rats, *Jpn. J. Physiol.*, 39, 825-832, 1989.

Moretto, M., Lopez, F. J. and Negro-Villar, A., Endothelin-3 stimulates luteinizing hormone-releasing hormone (LHRH) secretion from LHRH neurons by a prostaglandin-dependent mechanism, *Endocrinology*, 132, 789-794, 1993.

Morton, A. J. and Davenport, A. P., Cerebellar neurons and glia respond differentially to endothelins and sarafatoxin S6b, *Brain Res.*, 581, 299-306, 1992.

Moser, P. C. and Pelton, J. T., Behavioural effects of centrally administered endothelin in the rat, *Br. J. Pharmacol.*, (Suppl. 96), 347P, 1989.

Mosqueda-Garcia, R., Inagami, T., Appalsamy, M., Sugiura, M. and Robertson, R. M., Endothelin as a neuropeptide. Cardiovascular effects in the brainstem of normotensive rats, *Circ. Res.*, 72, 20-35, 1992.

Nakamura, S., Naruse, M., Maruse, K., Shioda, S., Nakai, Y. and Uemura, H., Colocalization of immunoreactive endothelin-1 and neurohypophyseal hormones in the axons of the neural lobe of the rat pituitary, *Endocrinology*, 132, 530-533, 1993.

Nambi, P., Pullen, M. and Feuerstein, G., Identification of endothelin receptors in various regions of rat brain, *Neuropeptides*, 16, 195-199, 1990.

Naruse, M., Naruse, K., Nishikawa, T., Yoshihara, I., Ohsumi, K., Suzuki, N., Demura, R. and Demura, H., Endothelin-3 immunoreactivity in gonadotrophs of the human anterior pituitary, *J. Clin. Endocrinol. Metab.*, 74, 968-972, 1992.

Nikolov, R., Maslarova, J., Semkova, I. and Moyanova, S., Intracerebroventricular endothelin-1 (ET-1) produces Ca^{2+}-mediated antinociception in mice, *Meth. Find. Exp. Clin. Pharmacol.*, 14, 229-233, 1992.

Nikolov, R., Semkova, I., Maslarova, J. and Moyanova, S., Antinociceptive effect of centrally administered endothelin-1 and endothelin-3 in the mouse, *Meth. Find. Exp. Clin. Pharmacol.*, 15, 447-453, 1993a.

Nikolov, R., Rami, A. and Krieglstein, J., Endothelin-1 exacerbates focal cerebral ischemia without exerting neurotoxic action *in vitro*, *Eur. J. Pharmacol. (Environ. Toxicol. Pharmacol.)*, 248, 205-208, 1993b.

Nirei, H., Hamada, K., Shoubo, M., Sogabe, K., Notsu, Y. and Ono, T., An endothelin ET_A receptor antagonist, FR 139317, ameliorates cerebral vasospasm in dogs, *Life Sci.*, 52, 1869-1874, 1993.

Nishimura, M., Takahashi, H., Matsusawa, M., Ikegaki, I., Nakanishi, T., Hirabayashi, M. and Yoshimura, M., Intracerebroventricular injections of endothelin increase arterial pressure in conscious rats, *Jpn. Circ. J.*, 54, 662-670, 1990.

Nishimura, M., Takahashi, H., Matsusawa, M., Ikegaki, I., Sakamoto, M., Nakanishi, T., Hirabayashi, M. and Yoshimura, M., Chronic intracerebroventricular infusions of endothelin elevate arterial pressure in rats, *J. Hypertens.*, 9, 71-76, 1991.

Niwa, M., Kawaguchi, T., Himeno, A., Fujimoto, M., Kurihara, M., Yamashita, K., Kataoka, Y., Shigematsu, K. and Taniyama, K., Specific binding sites for ^{125}I-endothelin-1 in the porcine and human spinal cord, *Eur. J. Pharmacol. (Mol. Pharmacol. Sect.)*, 225, 281-289, 1992.

Ouchi, Y., Kim, S., Souza, A. C., Ijima, S., Hattori, A., Orimo, H., Yoshizumi, M., Kurihara, H. and Yazaki, Y., Central effect of endothelin on blood pressure of conscious rats, *Am. J. Physiol.*, 256, H1747-1751, 1989.

Poulat, P., D'Orleans-Juste, P., De Chaplain, J., Yano, M. and Couture, R., Cardiovascular effects of intrathecally administered endothelins and big endothelin-1 in conscious rats: receptor characterization and mechanism of action, *Brain Res.*, 648, 239-248, 1994.

Reid, J. J., Wong-Dusting, H. K. and Rand, M. J., The effect of endothelin on noradrenergic transmission in rat and guinea-pig atria, *Eur. J. Pharmacol.*, 168, 93-96, 1989.

Reid, J. J., Vo, P. A., Lieu, A. T., Wong-Dusting, H. K. and Rand, M. J., Modulation of norepinephrine-induced vasoconstriction by endothelin-1 and nitric oxide in rat tail artery, *J. Cardiovasc. Pharmacol.*, 17, S272-S275, 1991.

Reiser, G. and Donié, F., Endothelin induces a rise of inositol 1,4,5-triphosphate, inositol 1,3,4,5-tetrakisphosphate levels and of cytosolic Ca^{2+} activity in neural cell lines, *Eur. J. Neurosci.*, 2, 769-775, 1990.

Reiser, G., Endothelin and a Ca^{2+} ionophore raise cyclic GMP levels in a neuronal cell line via formation of nitric oxide, *Br. J. Pharmacol.*, 101, 722-726, 1990.

Ritz, M. F., Stuenkel, E. L., Dayanithi, G., Jones, R. and Nordmann, J. J., Endothelin regulation of neuropeptide release from nerve endings of the posterior pituitary, *Proc. Natl. Acad. Sci. U.S.A.*, 89, 8371-8375, 1992.

Robinson, M. J., Macrae, I. M., Todd, M., Reid, J. L. and McCulloch, J., Reduction in local cerebral blood flow induced by endothelin-1 applied topically to the middle cerebral artery in the rat, *J. Cardiovasc. Pharmacol.*, 17, S354-S357, 1991.

Rossi, N., Effect of endothelin-3 on vasopressin release *in vitro* and water excretion *in vivo* in Long-Evans rats, *J. Physiol.*, 461, 501-511, 1993.

Sakurai, T., Yanagisawa, M., Takuwa, Y., Miyazaki, H., Kimura, S., Goto, K. and Masaki, T., Cloning of a cDNA encoding a non-isopeptide-selective subtype of the endothelin receptor, *Nature*, 348, 732-735, 1990.

Samson, W. K., Skala, K. D., Alexander, B. D. and Huang, F. L. S., Pituitary site of action of endothelin: selective inhibition of prolactin release *in vitro*, *Biochem. Biophys. Res. Commun.*, 169, 737-743, 1990.

Samson, W. K., Skala, K., Huang, F. L. S., Gluntz, S., Alexander, B. and Gómez-Sánchez, C. E., Central nervous system action of endothelin-3 to inhibit water drinking in the rat, *Brain Res.*, 539, 347-351, 1991a.

Samson, W. K., Skala, K. D., Alexander, B. D. and Huang, F. L. S., Possible neuroendocrine actions of endothelin-3, *Endocrinology*, 128, 1465-1473, 1991b.

Samson, W. K., Skala, K. D., Alexander, B. D. and Huang, F. L. S., Hypothalamic endothelin: presence and effects related to fluid and electrolyte homeostasis, *J. Cardiovasc. Pharmacol.*, 17, S346-S349, 1991c.

Samson, W. K., The endothelin-A receptor subtype transduces the effects of the endothelins in the anterior pituitary gland, *Biochem. Biophys. Res. Commun.*, 187, 590-595, 1992.

Samson, W. K. and Skala, K. D., Comparison of the pituitary effects of the mammalian endothelins: vasoactive intestinal contractor (endothelin-, rat endothelin-2) is a potent inhibitor of prolactin secretion, *Endocrinology*, 130, 2964-2970, 1992.

Samson, W. K., Skala, K. D., Alexander, B. D., Huang, F. L. S. and Gomez-Sanchez, C., A prolactin release inhibiting activity isolated from neurointermediate lobe extracts is an endothelin-like peptide, *Regul. Pept.*, 39, 103-112, 1992.

Samson, W. K. and Murphy, T. C., Antidipsogenic actions of endothelins are exerted via the endothelin-A receptor in the brain, *Am. J. Physiol.*, 265, R1212-R1215, 1993.

Sharkey, J., Ritchie, I. M. and Kelly, P. A. T., Perivascular microapplication of endothelin-1: a new model of focal cerebral ischemia in the rat, *J. Cereb. Blood Flow Metab.*, 13, 865-871, 1993.

Shichiri, M., Hirata, Y., Kanno, K., Ohta, K., Emori, T. and Marumo, F., Effect of endothelin-1 on release of arginine-vasopressin from perfused rat hypothalamus, *Biochem. Biophys. Res. Commun.*, 163, 1332-1337, 1989.

Shigeno, T., Mima, T., Takakura, K., Yanagisawa, M., Saito, A., Goto, K. and Masaki, T., Endothelin-1 acts in cerebral arteries from the adventitial but not from the luminal side, *J. Cardiovasc. Pharmacol.*, 13, S174-S176, 1989.

Shigeno, T., Mima, T., Yanagisawa, M., Saito, A., Fujimori, A., Shiba, R., Goto, K., Kimura, S., Yamashita, K., Yamasaki, Y., Masaki, T. and Takakura, K., Possible role of endothelin in the pathogenesis of cerebral vasospasm, *J. Cardiovasc. Pharmacol.*, 17, S480-S483, 1991.

Shinmi, O., Kimura, S., Yoshizawa, T., Sawamura, T., Uchiyama, Y., Sugita, Y., Kanazawa, I., Yanagisawa, M., Goto, K. and Masaki, T., Presence of endothelin-1 in porcine spinal cord: isolation and sequence determination, *Biochem. Biophys. Res. Commun.*, 162, 340-346, 1989a.

Shinmi, O., Kimura, S., Sawamura, T., Sugita, Y., Yoshizawa, T., Uchiyama, Y., Yanagisawa, M., Goto, K., Masaki, T. and Kanazawa, I., Endothelin-3 is a novel neuropeptide: isolation and sequence determination of endothelin-1 and endothelin-3 in porcine brain, *Biochem. Biophys. Res. Commun.*, 164, 587-593, 1989b.

Shinyama, H., Uchida, T., Kido, H., Hayashi, K., Watanebe, M., Matsumura, Y., Ikegawa, R., Takaoka, M. and Morimoto, S., Phosphoramidon inhibits the conversion of intracisternally administered big endothelin-1 to endothelin-1, *Biochem. Biophys. Res. Commun.*, 178, 24-30, 1991.

Shirakami, G., Nakao, K., Saito, Y., Magaribuchi, T., Nagata, H., Jougasaki, M., Mukoyama, M., Arai, H., Hosoda, K., Suga, S., Ogawa, Y., Mori, K. and Imura, H., Endothelin in human cerebrospinal fluid, *Ther. Res.*, 11, 85-90, 1990.

Sirén, A. L. and Feuerstein, G., Hemodynamic effects of endothelin after systemic and central nervous system administration in the conscious rat, *Neuropeptides*, 14, 231-236, 1989.

Sokolovsky, M., Endothelins and sarafatoxins: physiological regulation, receptor subtypes and transmembrane signaling, *Pharmacol. Ther.*, 54, 129-149, 1992.

Spyer, K. M., McQueen, D. S., Dashwood, M. R., Sykes, R. M., Daly, M., de B. and Muddle, J. R., Localization of [^{125}I]endothelin binding sites in the region of the carotid bifurcation and brainstem of the cat: possible baro- and chemoreceptor involvement, *J. Cardiovasc. Pharmacol.*, 17, S385-S389, 1991.

Stanimirovic, D. B., Yamamoto, T., Uematsu, S. and Spatz, M., Endothelin-1 receptor binding and cellular signal transduction in cultured human brain endothelial cells, *J. Neurochem.*, 62, 592-601, 1994.

Stojilkovic, S. S., Merelli, F., IIda, T., Krsmanovic, L. Z. and Catt, K. J., Endothelin stimulation of cytosolic calcium and gonadotropin secretion in anterior pituitary cells, *Science*, 248, 1663-1666, 1990.

Stojilkovic, S. S., Iida, T., Merelli, F. and Catt, K. J., Calcium signaling and secretory responses in endothelin-stimulated anterior pituitary cells, *Mol. Pharmacol.*, 39, 762-770, 1991.

Supattapone, S., Simpson, A. W. M. and Ashley, C. C., Free calcium rise and mitogenesis in glial cells caused by endothelin, *Biochem. Biophys. Res. Commun.*, 165, 1115-1122, 1989.

Suzuki, H., Sato, S., Suzuki, Y., Oka, M., Tsuchiya, T., Iino, I., Yamanaka, T., Ishihara, N. and Shimoda, S., Endothelin immunoreactivity in cerebrospinal fluid of patients with subarachnoid hemorrhage, *Ann. Med.*, 22, 233-236, 1990.

Suzuki, Y., Matsumura, Y., Umekawa, T., Hayashi, K., Takaoka, M. and Morimoto, S., Effects of endothelin-1 on antidiuresis and norepinephrine overflow induced by stimulation of renal nerves in anesthetized dogs, *J. Cardiovasc. Pharmacol.*, 19, 905-910, 1992.

Tabuchi, Y., Nakamura, M., Rakugi, H., Nagano, M. and Ogihara, T., Endothelin-1 enhances adrenergic vasoconstriction in perfused rat mesenteric arteries, *Biochem. Biophys. Res. Commun.*, 159, 1304-1308, 1989a.

Tabuchi, Y., Nakamura, M., Rakugi, H., Nagano, M., Mikami, H. and Ogihara, T., Endothelin inhibits presynaptic adrenergic neurotransmission in rat mesenteric artery, *Biochem. Biophys. Res. Commun.*, 161, 803-808, 1989b.

Tabuchi, Y., Nakamura, M., Rakugi, H., Nagano, M., Higashimori, K., Mikami, H. and Ogihara, T., Effects of endothelin on neuroeffector junction in mesenteric arteries of hypertensive rats, *Hypertension*, 15, 739-743, 1990.

Takagi, H., Hisa, H. and Satoh, S., Effects of endothelin on adrenergic neurotransmission in the dog kidney, *Eur. J. Pharmacol.*, 203, 291-294, 1991.

Takahashi, H., Nishimura, M., Nakanishi, T., Habuchi, Y., Tanaka, H., Ikegaki, I. and Yoshimura, M., Effects of intracerebroventricular and intravenous injections of endothelin-1 on blood pressure and sympathetic activity in urethane-anesthetized rats, *J. Cardiovasc. Pharmacol.*, 17, S287-S289, 1991a.

Takahashi, K., Ghatei, M. A., Jones, P. M., Murphy, J. K., Lam, H. C., O'Halloran, D. J. and Bloom, S. R., Endothelin in human brain and pituitary gland: presence of immunoreactive endothelin, endothelin messenger ribonucleic acid, and endothelin receptors, *J. Clin. Endocrinol. Metab.*, 72, 693-699, 1991b.

Takahashi, K., Ghatei, M. A., Jones, P. M., Murphy, J. K., Lam, H. C., O'Halloran, D. J. and Bloom, S. R., Endothelin in human brain and pituitary: comparison with rat, *J. Cardiovasc. Pharmacol.*, 17, S101-S103, 1991c.

Takeuchi, A., Kimura, T. and Satoh, S., Enhancement by endothelin-1 of the release of catecholamines from the canine adrenal gland in response to splanchnic nerve stimulation, *Clin. Exp. Pharmacol. Physiol.*, 19, 663-666, 1992.

Tencé, M., Cordier, J., Glowinski, J. and Prémont, J., Endothelin-evoked release of arachidonic acid from mouse astrocytes in primary culture, *Eur. J. Neurosci.*, 4, 993-999, 1992.

Togashi, K., Hirata, Y., Ando, K., Matsunaga, T., Kawakami, M. and Marumo, F., Abundance of endothelin-3 in human cerebrospinal fluid, *Biomed. Res.*, 11, 243-246, 1990.

Togashi, K., Ando, K., Kameya, T. and Kawakami, M., Regional distribution of immunoreactive endothelin-1 in the human central nervous system, *Biomed. Res.*, 12, 161-163, 1991.

Van den Buuse, M. and Corriu, C., The effect of endothelin-1 on locomotor activity in rats: interaction with central dopamine systems, *Br. J. Pharmacol.*, (Suppl. 102), 228P, 1991.

Van den Buuse, M. and Itoh, S., Central effects of endothelin on baroreflex of spontaneously hypertensive rats, *J. Hypertens.*, 11, 379-387, 1993.

Viossat, I., Duverger, D., Chapelat, M., Pirotzky, E., Chabrier, P. E. and Braquet, P., Elevated tissue endothelin content during focal cerebral ischemia in the rat, *J. Cardiovasc. Pharmacol.*, 22, S306-S309, 1993.

Wall, K. M., Nasr, M. and Ferguson, A. V., Actions of endothelin at the subfornical organ, *Brain Res.*, 570, 180-187, 1992.

Wall, K. M. and Ferguson, A. V., Endothelin acts at the subfornical organ to influence the activity of putative vasopressin and oxytocin-secreting neurons, *Brain Res.*, 586, 111-116, 1992.

Warner, T. D., Mitchell, J. A., D'Orleans-Juste, P., Ishii, K., Forstermann, U. and Murad, F., Characterization of endothelin-converting enzyme from endothelial cells and rat brain, detection of the formation of biologically active endothelin-1 by rapid bioassay, *Mol. Pharmacol.*, 41, 399-403, 1992a.

Warner, T. D., Budzik, G. P., Matsumoto, T., Mitchell, J. A., Förstermann, U. and Murad, F., Regional differences in endothelin converting enzyme activity in rat brain, inhibition by phosphoramidon and EDTA, *Br. J. Pharmacol.*, 106, 948-952, 1992b.

Warner, T. D., Schmidt, H. H. H. W., Kuk, J., Mitchell, J. A. and Murad, F., Human brain contains a metalloprotease that converts big endothelin-1 to endothelin-1 and is inhibited by phosphoramidon and EDTA, *Br. J. Pharmacol.*, 106, 505-506, 1992c.

Webb, R. L., Okada, T., Barclay, B. W. and Lappe, R. W., Central and peripheral actions of IRL 1620, an ET_B selective endothelin (ET) receptor agonist in conscious rats, *Hypertension*, 20, 425, 1992.

Wiklund, N. P., Öhlén, A. and Cederqvist, B., Inhibition of adrenergic neuroeffector transmission by endothelin in the guinea-pig femoral artery, *Acta Physiol. Scand.*, 134, 311-312, 1988.

Wiklund, N. P., Öhlén, A. and Cederqvist, B., Adrenergic neuromodulation by endothelin in guinea-pig pulmonary artery, *Neurosci. Lett.*, 101, 269-273, 1989.

Willette, R. N. and Sauermelch, C. F., Abluminal effects of endothelin in cerebral microvasculature assessed by laser-Doppler flowmetry, *Am. J. Physiol.*, 259, H1688-1693, 1990.

Willette, R. N., Ohlstein, E. H., Pullen, M., Sauermelch, C. F., Cohen, A. and Nambi, P., Transient forebrain ischemia alters acutely endothelin receptor density and immunoreactivity in gerbil brain, *Life Sci.*, 52, 35-40, 1992.

Wong-Dusting, H. K., Reid, J. J. and Rand, M. J., Paradoxical effects of endothelin on cardiovascular noradrenergic neurotransmission, *Clin. Exp. Pharmacol. Physiol.*, 16, 229-233, 1989.

Wong-Dusting, H. K., La, M. and Rand, M. J., Endothelin-1 enhances vasoconstrictor responses to sympathetic nerve stimulation and noradrenaline in the rabbit ear artery, *Clin. Exp. Pharmacol. Physiol.*, 18, 131-136, 1991.

Wong-Dusting, M. L. H. K. and Rand, M. J., Endothelin-1 enhances responses to sympathetic nerve stimulation in endothelium-denuded and endothelium-intact rabbit ear arteries, *Neurochem. Int.*, 18, 465-469, 1991.

Yamaji, T., Joshita, H., Ishibashi, M., Takaku, F., Ohno, H., Suzuki, N. and Matsumoto, H., Endothelin family in human plasma and cerebrospinal fluid, *J. Clin. Endocrinol. Metab.*, 71, 1611-1615, 1990.

Yamamoto, S., Morimoto, I., Yamashita, H. and Eto, S., Inhibitory effects of endothelin-3 on vasopressin release from rat supraoptic nucleus *in vitro*, *Neurosci. Lett.*, 141, 147-150, 1992.

Yamamoto, S., Inenaga, K., Kannan, H., Eto, S. and Yamashita, H., The actions of endothelin on single cells in the anteroventral third ventricular region and supraoptic nucleus in rat hypothalamic slices, *J. Neuroendocrinol.*, 5, 427-434, 1993.

Yamamoto, T., Kimura, T., Ota, K., Shoji, M., Inoue, M., Sato, K., Ohta, M. and Yoshinaga, K., Central effects of endothelin-1 on vasopressin release, blood pressure, and renal solute excretion, *Am. J. Physiol.*, 262, E856-E862, 1992.

Yamashita, H., Yamamoto, S., Inenaga, K. and Kannan, H., Endothelin-3 directly affects neurons in the anteroventral third ventricle region and supraoptic nucleus of the rat hypothalamus *in vitro*, *J. Cardiovasc. Pharmacol.*, 17, S200-S202, 1991.

Yamashita, K., Moser, P. C. and Jones, C. R., Cerebral 2-deoxyglucose uptake in rats following intravenous and intracerebroventricular administration of endothelin-1, *Eur. J. Pharmacol.*, 183, 493-494, 1990.

Yamashita, K., Kataoka, Y., Niwa, M., Shigematsu, K., Himeno, A., Koizumi, S. and Taniyama, K., Increased production of endothelins in the hippocampus of stroke-prone spontaneously hypertensive rats following transient forebrain ischemia: histochemical evidence, *Cell. Mol. Neurobiol.*, 13, 15-23, 1993.

Yanagisawa, M., Kurihara, H., Kimura, S., Tomobe, Y., Kobayashi, M., Mitsui, Y., Yazaki, Y., Goto, K. and Masaki, T., A novel potent vasoconstrictor peptide produced by vascular endothelial cells, *Nature*, 322, 411-415, 1988.

Yanagisawa, M., Inoue, M., Imai, T., Matsushita, Y. and Masaki, T., The human endothelin family: cDNA cloning, complete primary structure and differential tissue expression of the three precursors, *Eur. J. Pharmacol.*, 183, 1628-1629, 1990.

Yin, J., Lee, J. A. and Howells, R. D., Stimulation of *c-fos* and *c-jun* gene expression and downregulation of proenkephalin gene expression in C_6 glioma cells by endothelin-1, *Mol. Brain Res.*, 14, 213-220, 1992.

Yoshida, K., Kawano, Y., Yoshimi, H., Kuramochi, M. and Omae, T., Blood pressure responses to intravenous or intracerebroventricular endothelin-1 in spontaneously hypertensive rats, *J. Cardiovasc. Pharmacol.*, 17, S297-S299, 1991.

Yoshimi, H., Kawano, Y., Akabane, S., Ashida, T., Yoshida, K., Kinoshita, O., Kuramochi, M. and Omae, T., Immunoreactive endothelin-1 contents in brain regions from spontaneously hypertensive rats, *J. Cardiovasc. Pharmacol.*, 17, S417-S419, 1991.

Yoshimoto, S., Ishizaki, Y., Kurihara, H., Sasaki, T., Yoshizumi, M., Yanagisawa, M., Yazaki, Y., Masaki, T., Takakura, K. and Murota, S., Cerebral microvessel endothelium is producing endothelin, *Brain Res.*, 508, 283-285, 1990.

Yoshizawa, T., Kimura, S., Kanazawa, I., Uchiyama, Y., Yanagisawa, M. and Masaki, T., Endothelin localizes in the dorsal horn on the spinal neurons, possible involvement of dihydropyridine-sensitive calcium channels and substance P release, *Neurosci. Lett.*, 102, 179-184, 1989.

Yoshizawa, T., Shinmi, O., Giaid, A., Yanagisawa, M., Gibson, S. J., Kimura, S., Uchiyama, Y., Polak, J. M., Masaki, T. and Kanazawa, I., Endothelin: a novel peptide in the posterior pituitary system, *Science*, 247, 462-464, 1990.

Yoshizawa, T., Iwamoto, H., Mizusawa, H., Suzuki, N., Matsumoto, H. and Kanazawa, I., Cerebrospinal fluid endothelin-1 in Alzheimer's disease and senile dementia of Alzheimer's type, *Neuropeptides*, 22, 85-88, 1992.

Zhang, W., Sakai, N., Fu, T., Okano, Y., Hirayama, H., Takenaka, K., Yamada, H. and Nozawa, Y., Diacylglycerol formation and DNA synthesis in endothelin-stimulated C_6 glioma cells: the possible role of phosphatidylcholine breakdown, *Neurosci. Lett.*, 123, 164-166, 1991.

Chapter 12

Urogenital Actions of Endothelins

Kathleen S. Sang and John P. Huggins

CONTENTS

0-8493-6975-4/97/$0.00+$.50

I. INTRODUCTION

ETs are expressed and have potent actions in the urogenital system, some of which may have relevance to clinical disease situations. This chapter concentrates on the pharmacology of ETs in urinary bladder, uterus, and vas deferens with mention also of urethra, prostate, corpus cavernosum, and seminal vesicle.

II. BLADDER

A. Responses to ETs

ETs produce concentration-dependent contractions of rat (Wicklund et al.,1989; Bolger et al., 1990; Secrest and Cohen, 1989; Lecci et al., 1991; Donoso et al., 1994), rabbit (Secrest and Cohen, 1989; Garcia-Pascual et al., 1990; Saenz de Tejada et al., 1992; Traish et al., 1992; Garcia-Pascual et al., 1993), human (Le Monnier de Gouville et al., 1989; Maggi et al., 1989a,b,c, 1990; Kondo et al., 1992; Kondo and Morita, 1993; Morita et al., 1993), and pig (Persson and Andersson, 1992), but not guinea pig (Wicklund et al., 1989; Eglen et al., 1989) urinary bladder.

The rabbit bladder body was found to be more sensitive than that from rat (Secrest and Cohen, 1989). All contractions elicited by ETs were slow to develop, sustained, and difficult to wash out. The major excitatory innervation of the bladder is cholinergic. However, participation of cholinergic nerves in responses to ETs in human bladder can be excluded as these responses are resistant to atropine (Le Monnier de Gouville et al., 1989; Maggi et al., 1989b; Morita et al., 1993). However, ET may regulate cholinergic neurotransmission by a paracrine mechanism in this tissue.

Transmural electrical stimulation of rabbit bladder nerves causes frequency-dependent contractions of the bladder that are greatly attenuated by atropine (Saenz de Tejada et al., 1992). However, there is also an atropine-resistant component that can be completely blocked by the combination of atropine and α,β-methylene ATP (which desensitizes purinergic receptors). ET-1 caused a 20% reduction of the contraction elicited by transmural electrical stimulation of strips, with this effect being abolished after muscarinic receptor blockade by atropine, but still present after desensitization of purinergic receptors with α,β-methylene ATP. ET-1 failed to inhibit contractions elicited by either acetylcholine or ATP. Therefore, in this tissue, ET affects cholinergic neurotransmission via a prejunctional mechanism. Results from Donoso et al. (1994) suggest that, in rat bladder, the potentiation of tone by ET-1 involves the participation of endogenous ATP as a neurotransmitter, and they hypothesize that ET-1 may play a role as a local modulator of the purinergic component of neurotransmission. ATP receptors in the bladder are of the P_{2x} type. ET-1 was found to selectively increase motor activity evoked by purinergic agonists, but not to alter the potency of nonpurinoceptor agonists. This effect was similar to the facilitation of nonadrenergic noncholinergic (NANC) nerve-evoked responses by ET-1, suggesting that the effect of ET-1 involves participation of endogenous ATP as a bladder neurotransmitter. Tetrodotoxin was without effect on this facilitation, suggesting that ET-1 acts at a postjunctional site. The increase in bladder tone by ET-1 is not obligatory for potentiation of the ATP responses, and the mechanisms involved in the increase in tone elicited by NANC transmission or exogenous ATP are different.

B. Sources of Calcium for ET Responses

Multiple Ca^{2+} channels exist in smooth muscle (see Karaki and Weiss, 1988). To determine the role of Ca^{2+} in bladder contractions elicited by ETs, the effects of Ca^{2+} channel blockers have been investigated.

Secrest and Cohen (1989) studied the role of extracellular Ca^{2+} in ET-induced contractions of rabbit bladders by using the Ca^{2+} channel blockers diltiazem (a benzothiazepine) and nitrendipine (a dihydropyridine), and found both to decrease the maximum response to ET-1 without a change in EC_{50}, indicating involvement of extracellular Ca^{2+} in this species. In agreement, Garcia-Pascual et al. (1990) found that the ET-1-induced contraction was inhibited by the Ca^{2+} entry blocker nifedipine in this tissue, with Ca^{2+}-free medium also abolishing this contraction. Also, in the presence of ET-1, Ca^{2+}-induced contractions were blocked by nifedipine while ET-1 had no effect on responses induced by electrical stimulation. Hence, in rabbit bladder, contractions elicited by ET-1 are dependent on extracellular Ca^{2+}, with probable involvement of voltage-operated Ca^{2+} channels. This conclusion is supported by results from Saenz de Tejada et al. (1992), who found that in rabbit bladder strips the Ca^{2+} channel blockers nifedipine or verapamil, or removal of extracellular Ca^{2+}, greatly reduced or eliminated spontaneous tone and attenuated (but did not block) ET-1-induced contractions. Rat bladder tone was also abolished in the absence of external Ca^{2+} (Donoso et al., 1994).

In contrast, nifedipine had no effect on responses elicited by ET-1 in human urinary bladder (Le Monnier de Gouville et al., 1989; Maggi et al., 1989b,c, 1990) whereas it abolished the contractions produced by potassium chloride (KCl), carbachol (acting on muscarinic receptors), or neurokinin A (acting on NK-2 tachykinin receptors; Maggi et al., 1989b,c). Contractile responses elicited by ET-1 do not involve the opening of voltage-sensitive Ca^{2+} channels as a primary signal transduction event (Maggi et al., 1989c; Van Renterghem et al., 1988).

In mucosa-free strips of human urinary bladder dome, Maggi et al. (1989c) tested a range of Ca^{2+} channel blockers on contractions elicited by KCl, carbachol, neurokinin A, and ET-1. Nifedipine fully antagonized responses induced by KCl, but only partially inhibited responses produced by carbachol or neurokinin A, while having no effect on those produced by ET-1, indicating a range of dependencies on extracellular Ca^{2+} for these agents. The responses to ET-1 were also unaffected by the dihydropyridine L-type Ca^{2+} channel activator, Bay k 8644, reinforcing the view that these channels are not involved in ET-1-induced contractions. These channels may be indirectly activated as a result of membrane depolarization subsequent to receptor activation. Nifedipine-resistant components of contractions elicited by carbachol and neurokinin A, and also the response to ET-1, are probably due to influx of Ca^{2+} from the extracellular space or from a pool located at the plasma membrane that can refill from extracellular Ca^{2+} via a nifedipine-resistant $LaCl_3$-sensitive pathway. ET-1 responses were ~20% $LaCl_3$ resistant, suggesting Ca^{2+} mobilization from intracellular stores. Indeed, in Ca^{2+}-free medium containing EDTA, high concentrations of carbachol, neurokinin A, and ET-1 were still able to produce contraction, indicating the use of such a store. Procaine failed to inhibit ET-1-induced contractions in high K^+, Ca^{2+}-free medium, also supporting the involvement of Ca^{2+} arising from an intracellular pool.

The effect of ET-1 on InsP turnover in rabbit detrusor muscle has been studied by Garcia-Pascual et al. (1993) who found that the responses to ET-1 were similar to those observed by this group in rabbit urethral smooth muscle (see Section IV).

C. ET-Receptor Subtypes

Binding studies using [^{125}I]ET-1 have shown that binding sites in human bladder are mainly in the outer longitudinal muscle layer, blood vessels, and submucosa, with the highest density in vessels and submucosa (Garcia-Pascual et al., 1990). A larger amount of binding was seen in human bladder dome compared with base. K_D values for the two areas were not significantly different (4 ± 3 pM for base, 7 ± 3 pM for dome) but B_{max} values were 2.74 ± 2.81 fmol/mg protein for base and 47.3 ± 9.52 for dome (Kondo et al., 1992). Saturation experiments with [^{125}I]ET-1 revealed a single class of saturable, high-affinity binding sites in homogenates of human bladder (base and dome) (Kondo et al., 1993). Traish et al. (1992) described an ET binding site with a pharmacological profile characteristic of an ET_A-receptor. This group showed that the radiolabelled ET isoforms ET-1, ET-2, and ET-3, bound reversibly to rabbit bladder membrane sites specifically, with high affinity (K_D between 0.4 to 0.6 nM) and limited capacity (60 to 240 fmol/mg protein). Competition binding experiments revealed two populations of receptors; one with high affinity for ET-1 and ET-2 and low affinity for ET-3 (characteristic of ET_A-receptors), and another with high affinity for all three peptides (ET_B-receptors). The number of binding sites detected with [^{125}I]ET-1 and [^{125}I]ET-2 were approximately four- to fivefold greater than those detected with [^{125}I]ET-3. Threshold concentrations of ET-3 required to elicit smooth muscle contractions were found to be higher than those of ET-1 or ET-2. In membrane preparations from rabbit bladder base and dome, unlabeled ET-1 and ET-2 competed effectively for binding of [^{125}I]ET-1 and [^{125}I]ET-2. Unlabeled ET-1, ET-2, and ET-3 competed effectively for binding of [^{125}I]ET-3. ET-1 and ET-2 were probably acting on the same receptor as they had similar affinity for ET receptors, with similar binding densities and distributions. Chemical affinity labeling showed that [^{125}I]ET-1 and [^{125}I]ET-2 cross-link to three proteins (75, 52, and 34 kDa) while [^{125}I]ET-3 mainly cross-links to a 34-kDa protein, supporting the concept that ET-3 acts on different receptors to those that bind ET-1 and ET-2. These results indicate that at least two receptor subtypes exist in bladder tissue, and that activation of these receptors by ETs plays a role in regulating bladder smooth muscle tone.

Maggi et al. (1989b) found ET-1, ET-2, ET-3, and STX-b to be agonists in contracting mucosa-free strips from human urinary bladder dome, with ET-2 being the most potent peptide, a result that is incompatible with the current ET receptor classification (see Chapter 2). The C-terminal fragment ET-(16-21) (Maggi et al., 1989d) however, was only weakly active. By assuming that the ET-(16-21) is only active on ET_B-receptors, this group suggested that this tissue contains mainly ET_A-receptors. In further agonist studies (Maggi et al., 1990), ET-1 and STX-b were found to be more potent than ET-3, also supporting involvement of ET_A-receptors.

In rat bladder, Donoso et al. (1994) have also demonstrated involvement of ET_A-receptors in the responses elicited by ETs of increasing basal tension and facilitating purinoceptor mechanisms. They obtained an order of potency of ET-2 > ET-1 ≫ ET-3, and also found ET-(16-21), STX-c, AGETB-9, and AGETB-89 (ET_B-receptor agonists) all to be inactive in this tissue. ET-1 analogues missing one or both disulfide bonds (that have intrinsic activity for ET_B-receptors) were also inactive. The actions of ET-1 could be blocked by the ET_A-receptor antagonists BE-18257B and FR139,317. However, again it can be observed that ET-3 was more potent than ET-1, a response that cannot be elicited from cloned ET receptors.

It is not yet known if the same ET_A-receptor is involved in increasing rat bladder tension and potentiating contractions elicited by ATP, but it could be either that different ET_A-receptor subtypes are involved, or that the same ET_A-receptors act via different signal transduction pathways.

D. Physiological Roles

In human and rabbit bladder, ET-like immunoreactivity (IR) has been localized in transitional epithelium, in serosal mesothelium, in vascular endothelium of blood vessels, and in vascular and nonvascular (detrusor) smooth muscle. Northern blot analysis of RNA extracted from whole bladder demonstrates the expression of a 2.1- to 2.2-kb band corresponding to the transcript size of preproET, confirming the expression of preproET-1 mRNA (Saenz de Tejada et al., 1992). mRNA encoding preproET-1 had a similar localization, with the same cellular distribution as that of ET-like immunoreactivity. Imai et al. (1992) cloned and sequenced full-length bovine preproET-1 cDNA and used this as a probe in Northern blot analysis to demonstrate a single 2.3-kb preproET-1 mRNA expressed in bovine urinary bladder. Together, these results suggest that ETs may act as autocrine hormones to regulate bladder wall structure and smooth muscle tone.

ETs may also act as neuropeptides in the CNS to regulate bladder function. Studies by Lecci et al. (1991) showed that ET-1 and ET-3 administered i.c.v. inhibit the supraspinal micturation reflex (SMR) in urethane-anesthetized rats, and this was not affected by guanethidine pretreatment, vagotomy, or forced ventilation. ET-1 blocked SMR after i.v. administration with an 8-min delay. Guanethidine pretreatment revealed a stimulatory effect of i.v. ET-1 on the frequency of SMR, and also reduced late inhibition, indicating that ET-1 activated the sympathetic inhibitory drive to the bladder. Since the effect on isolated bladder is stimulatory, it is likely that ETs do not enhance presynaptic activity of sympathetic nerves in the bladder. Sympathetic inhibition of SMR produced by i.v. ET-1 could be due to both a reflex response triggered by a sustained increase in bladder muscle tone and catecholamine release from the adrenals. The delayed inhibition of the SMR by i.v. ET-1 occurred at the same time as the maximal increase in blood pressure, suggesting that ET-1-induced vasoconstriction of bladder vessels could be involved.

E. Pathological Roles

The pathophysiological condition of bladder outlet obstruction is characterized by dense infiltration of the subserosal layer with fibroblasts and collagen deposition. Saenz de Tejada et al. (1992) measured the amount of immunoreactive ET in bladder tissue from a patient with bladder cancer, as well as bladder outlet obstruction, and found that in tissue slices the fibroblasts that had infiltrated the thickened subserosal layer stained intensely for ET. Guidry and Hook (1991) investigated which exogenous factors stimulate collagen gel contraction by fibroblasts and identified the ETs as the major endothelial cell-secreted peptide agent. ETs have also been shown to induce fibroblast mitogenesis (Muldoon et al., 1989, 1990; Takuwa et al., 1989). Taken together, these results may indicate an involvement of ETs in the pathophysiology of bladder outlet obstruction.

III. RENAL PELVIS

In longitudinal muscle strips of human renal pelvis, ET-1, ET-3, and STX-c potently elicited concentration-dependent contractions. All agonists had similar effects; however, at higher agonist concentration (30 nM and above) the order of response magnitude became STX-c > ET-1 > ET-3, pointing towards ET_B-receptor involvement (Maggi et al., 1990). Nifedipine did not inhibit contractions induced by the agonists. Therefore, these contractile responses do not involve the direct opening of voltage-sensitive Ca^{2+} channels, although these channels may become activated by depolorization of the cell membrane subsequent to ETs interacting with their receptors.

IV. URETHRA

ET-1 causes concentration-related contractions of rabbit urethral smooth muscle (Garcia-Pascual et al., 1990, 1993; Andersson et al., 1992) and also a concentration-dependent increase in InsP turnover, both of these effects being slow in onset. Electrically stimulated contractions of urethral smooth muscle are tetrodotoxin sensitive, and can be blocked by α-adrenoceptor antagonists (Garcia-Pascual et al., 1990). ET-1 pretreatment increased these nerve-induced contractions but had no effect when exogenous noradrenaline was applied nor affected spontaneous or stimulation-induced [^{3}H]noradrenaline release. Electrical stimulation relaxed ET-1-precontracted preparations in a frequency-dependent manner.

ET-1-induced contractions of rabbit urethral muscle were not affected by phentolamine, indomethacin, or nifedipine and, in the presence of ET-1, Ca^{2+}-induced contractions were not blocked by nifedipine. Although the contractions were dependent on extracellular Ca^{2+}, voltage-operated Ca^{2+} channels were therefore probably not involved. Nifedipine did not reduce InsP turnover, but no InsP turnover was obtained when tissues were incubated in Ca^{2+}-free medium. Therefore, both InsP formation and contraction depended on extracellular Ca^{2+} (Garcia-Pascual et al., 1993). Pretreatment with H-7 (a protein kinase inhibitor with specificity for protein kinase C *in vitro*) produced a nonparallel shift of the ET- 1 concentration-response curve, with a reduction in maximum response. ET-1-induced contractions in urethral preparations were inhibited by Ni^{2+} at 0.3 mM (Garcia-Pascual et al., 1993). These results suggest that ET-1 stimulation leads to Ca^{2+} influx through T-type rather than L-type channels.

V. UTERUS

A. Effects of ETs

ET-1 strongly contracts uterine smooth muscle from rats (Bousso-Mittler et al., 1989; Kozuka et al., 1989; Sakata and Karaki, 1992; Do Khac et al., 1994), guinea pigs

(Sakata et al., 1989; Eglen et al., 1989; Borges et al., 1989), and humans (Maher et al., 1991; Breuiller-Fouche et al., 1994; Heluy et al., 1995), and is one of the most potent oxytocic agents known in rat uterus.

In rat uterus, contractions in response to ETs consisted of increased muscle tonic tension, increased frequency of spontaneous rhythmic contractions, and developing tonic contractions seen at higher agonist concentrations (Bousso-Mittler et al., 1989; Kozuka et al., 1989; Rae et al., 1993). The rhythmic contractions were found to be dependent on extracellular Ca^{2+} and sensitive to voltage-dependent Ca^{2+} channel blockers. However, the developing contractions were nifedipine resistant but absolutely dependent on extracellular Ca^{2+} (Kozuka et al., 1989). These results suggest that in the contraction of rat uterine smooth muscle there are multiple ET-sensitive pathways that regulate Ca^{2+} influx. However, not all groups found rat uterine contractions to be nifedipine resistant; Borges et al. (1989) observed full antagonism with nifedipine of ET-1-induced contractions.

In guinea-pig uterus, the Ca^{2+} channel blocker verapamil only partially inhibited the ET-1-stimulated rise in intracellular free Ca^{2+} (Ca_i^{2+}) and there was dissociation between Ca_i^{2+} and muscle tension in the presence of EGTA, suggesting that part of the ET-1-induced contraction was due to a Ca^{2+}-independent mechanism (Sakata et al., 1989). Borges et al. (1989) also found nifedipine to be only a partial antagonist of ET-1-induced contractions in guinea-pig uterus.

Maher et al. (1991) observed an increase in Ca_i^{2+} after ET-1 or oxytocin stimulation in cultured human myometrial cells. This response was partly due to Ca^{2+} influx through Ca^{2+} channels, as Ca^{2+} channel inhibitors reduced both transient and post-transient Ca_i^{2+}. ET-1 and oxytocin were shown to act through distinct receptors. The ET-1 increase in Ca_i^{2+} involved Ca^{2+} mobilization from intracellular stores as well as Ca^{2+} influx through Ca^{2+} channels.

B. Effects of Proendothelins

Big-ET-1 has similar effects to that of ET-1 in that, in rat uterus, it increases the rate of spontaneous contractions and causes a tonic contraction. However, big-ET-1 was sevenfold less potent than ET-1 in causing contraction, and its action could be blocked by the endopeptidase inhibitor phosphoramidon, indicating the presence of a phosphoramidon-sensitive ET-converting enzyme in this tissue (Rae et al., 1993). This group demonstrated that ET-1 and big-ET-1 contract rat uterus via activation of a homogeneous population of ET_A-receptors.

C. ET Receptors

Bousso-Mittler et al. (1989) have demonstrated specific receptors for ET/STX peptides in rat uterus that specifically bind [^{125}I]STX-b, ET-1, ET-3, and STX-c, and are coupled to InsP hydrolysis. The receptors showed lower affinity for ET-3 and the least potent peptide was STX-c, indicating the presence of ET_A-receptors in this tissue. This is supported by Do Khac et al. (1994) who showed that in estrogen-treated rat myometrium, ET-1 interacts with specific ET_A-receptors to activate phospholipase C via a pertussis toxin-insensitive G protein, leading to degradation of $PtdInsP_2$, and to inhibit the adenylylcyclase pathway via a pertussis toxin-sensitive G_i protein and thus leading to uterine contraction. BQ 123 caused a decrease in InsP turnover in myometrium prestimulated with ET-1, and the rank order of agonist potency for contractile activity was ET-1 > STX-b >> ET-3. These results suggest ET_A-receptor subtype involvement. This group has also demonstrated a stimulatory effect of ET-1 on phospholipase D activity that could also be inhibited by BQ 123, indicating ET_A-receptors are also involved here, as is a pertussis toxin-insensitive G protein (Do Khac et al., 1995).

Results from Sakata and Karaki (1992) also lend support to ET_A-receptor involvement in ET-1-induced contraction of rat uterus. They found that the sensitivity of rat uterus to contractile effects of ET- 1, ET-2, or STX-b were nearly identical, but the sensitivity to ET-3 was 100 times less. In nonpregnant rat uterus, they observed an increase in Ca_i^{2+} elicited by ET-1 that was dependent on influx of extracellular Ca^{2+} via L-type Ca^{2+} channels, as verapamil abolished the increase in Ca_i^{2+} and ensuing contraction. This increase in Ca_i^{2+} is not thought to be achieved by release of Ca^{2+} from intracellular storage sites as ET-1 failed to increase Ca_i^{2+} or muscle tension in Ca^{2+}-free medium.

In pregnant uterus, verapamil only partially reduced the ET-1-induced Ca_i^{2+} increase and muscle contraction, suggesting that non-L-type Ca^{2+} channels may be involved here and that hormonal changes may alter expression of Ca^{2+} channel types in rat uterus (Sakata and Karaki, 1992). Calixto and Rae (1991) studied the responses of nonpregnant rat uteri to ET agonists. They found that preparations from pregnant rats were 2- to 3-fold more sensitive to ET-1 compared with preparations from nonpregnant rats, and also in pregnant preparations Bay k 8644 was 40-fold more potent. Bay k 8644 and ET-1 showed different patterns of contraction; Bay k 8644 produced phasic-rhythmic contractions whereas ET-1 produced typically tonic contractions that were only partially blocked by nicardipine — in contrast to those induced by Bay k 8644, that were fully blocked by nicardipine. These results confirm that ETs do not act by directly activating voltage-dependent L-type-Ca^{2+} channels. In Ca^{2+}-free medium, both agonists failed to cause contraction. Therefore, both agonist responses are dependent on extracellular Ca^{2+}. In the case of ET-1, Ca^{2+} enters via nicardipine-insensitive nonselective cation channels, the resulting depolarization leading to the activation of voltage-dependent L-type channels, thus accounting for the nicardipine-sensitive component of the ET-1 response.

D. Coexistence of ET Receptors

Competition of binding studies using [^{125}I]ET-1 in cultured human myometrial cells indicates that these cells exclusively express ET receptors of the ET_A subtype. ET-1 and ET-3 activated these receptors, which were coupled by phospholipase C to InsP turnover. FR 139,317 totally suppressed ET-1-induced InsP turnover, whereas BQ 123 was only a partial inhibitor, suggesting that several classes of ET_A-receptors coexist in these cells (Heluy et al., 1995). The ET receptors mediating contractions of guinea pig uterus are also not thought to be exclusively of the ET_A-receptor subtype (Kousides et al., 1995) due to the relative potencies of ET agonists and the absence of blockade of contraction by BQ 123.

Indeed, in human uterine tissue Breuiller-Fouche et al. (1994) have demonstrated the coexistence of ET_A- and ET_B-receptors, the stimulation of both leading to the activation of phospholipase C and InsP generation. Binding experiments using [^{125}I]ET-1 and [^{125}I]ET-3 have revealed that human myometrium contains 75% ET_A-receptors and 25% ET_B-receptors but, although ET-1 and ET-3 induce uterine contraction, it is only achieved via stimulation of ET_A- and not ET_B-receptors (Breuiller-Fouche et al., 1995). Maggi et al. (1991) found that the concentration of ET_A-receptors was ninefold higher than the concentration of ET_B-receptors in rabbit myometrium.

In human myometrium, mRNA encoding ET-1 is predominantly expressed, with ET-3 mRNA being detected only at low levels (MacMahon et al., 1993). In rat uterus

ET_A-receptor mRNA is predominantly expressed (Hori et al., 1992). ET_A-receptor RNA has also been detected in rat fallopian tube smooth muscle (Iwai et al., 1993).

E. Location of ET Production

ET-1 is synthesized by rat corpus lutea (Usuki et al., 1991), rabbit and human endometrium (Orlando et al., 1990; Economos et al., 1992; Maggi et al., 1991, 1993), and by porcine and human placenta (Onda et al., 1990). Economos et al. (1992) demonstrated gene expression and biosynthesis of ET-1 in human myometrium. Using autoradiography, Davenport et al. (1991) revealed a higher density of binding sites for $[^{125}I]$ET-1 and $[^{125}I]$ET-3 in the endometrium compared with myometrium. The presence of ET-1 binding sites in human myometrium has been confirmed by Svane et al. (1993). Schiff et al. (1993) demonstrated a homogeneous pattern of distribution of ET-1 binding sites in myometrium throughout the menstrual cycle. Maggi et al. (1991, 1993) reported that in rabbit uterus ET was produced by endometrium, but specific ET receptors were located in myometrium, suggesting that ETs have a paracrine role.

F. Pathophysiological Roles

ET-1, at concentrations insufficient to cause contraction, potentiates $PGF_{2\alpha}$-induced contraction. This is important because, in humans, $PGF_{2\alpha}$ generated in large amounts by decaying endometrial cells during menstruation causes dysmenorrhea (Chan, 1983). ETs, in addition to being involved in myometrial activity, may also play a role in triggering and/or mediating labor. The occurrence of functional ET receptors in uterine smooth muscle and specific high-affinity binding sites in placental membrane, as well as the fact that human umbilical vein endothelial cells can synthesize ET-1 (Fischli et al., 1989), and the placenta can synthesize ET- 3 (Onda et al., 1990), all lend support for this role.

VI. VAS DEFERENS

In addition to acting directly on specific receptors located on most smooth muscle to cause constriction, the ET/sarafotoxin peptides also affect the autonomic neurotransmission in various tissues, including the vas deferens, where they potentiate the contractions evoked by electrical field stimulation in a concentration-dependent manner (Table 1), and are therefore thought to be potent modulators of neuroeffector transmission.

A. Neurotransmission

The vas deferens is a useful model to choose to study peripheral autonomic nerve transmission, as its smooth muscle layers are densely sympathetically innervated. Noradrenaline and adenosine 5′-triphosphate (ATP), stored as a complex in granules in the sympathetic nerve terminal, are released upon nerve stimulation (Sneddon et al. 1982; Meldrum and Burnstock, 1983; Sneddon and Burnstock, 1984; Sneddon and Westfall, 1984; Stjärne and Astrand, 1985). Noradrenaline activates nonselective prejunctional α_2-adrenoceptors to regulate its own release by negative feedback inhibition (a process generally thought to involve Ca^{2+}). Noradrenaline also acts on postjunctional α_1-adrenoceptors, and along with the action of ATP on P_{2x} purinoceptors (Hogaboom et al., 1980; Burnstock and Kennedy, 1985; Fedan and Lamport, 1990) evokes smooth muscle contraction. The type of response obtained depends on the type of stimulation administered. Short pulse trains activate mainly NANC nerves to produce twitch contraction. Field stimulation at higher frequencies elicits a biphasic response consisting of a phasic contraction, probably due to NANC transmission, with ATP thought to be the neurotransmitter, followed by a more sustained contraction being partly NANC and partly noradrenaline mediated (Stjärne and Astrand, 1985). (This pattern was described by Wicklund et al. [1990] as a twitch and hump.)

As well as potentiating the neuronally induced muscle twitching and ATP-induced contractions, ET-1 also increases basal muscle tone (Table 1). This increase in tension is not a prerequisite for the twitch response, and the two types of response to ET-1 are thought to occur via separate mechanisms, perhaps involving more than one ET receptor subtype. Evidence for this comes from the time taken for ET-1 to exert its effects; it is slow to affect the potentiation of neuronally induced muscle twitching and ATP-induced contraction and has a long duration of action that is difficult to wash out, and is not dependent on extracellular Ca^{2+}. By contrast, in increasing basal tone ET-1 acts rapidly and reversibly and is antagonized by Ca^{2-} channel blockers. Also, the slower effects occur in both epididymal and prostatic portions of rat vas deferens, whereas the rapid effects have only been observed in the epididymal portion (Donoso et al., 1992).

B. Pre- and Postjunctional Mechanisms

In the vas deferens, ET-1 may be acting pre- and postjunctionally. Modulation of neurotransmission by ET has been studied at both these sites.

In the study of prejunctional mechanisms, Wicklund et al. (1990) measured the effect of ET-1 on the fractional release of [^{3}H]noradrenaline on the electrical stimulation of vas deferens, and found ET-1 to have an inhibitory effect while augmenting contractile responses and basal tone. Felodipine, a Ca^{2+} channel blocker, inhibited the potentiation by ET-1 of twitch and hump induced by electrical stimulation, and of basal tone, but not the inhibition of [^{3}H]noradrenaline release, suggesting that postjunctional stimulation is not critically involved in the prejunctional effects of ET-1.

The mechanism of inhibition of [^{3}H]noradrenaline release by ET-1, although not fully understood, may be due to a prejunctional influence on the cotransmitter, antagonism of neuronal uptake of transmitter, and/or modification of transmitter or signal transduction at the postjunctional site. Cyclic AMP is not thought to be involved, as forskolin did not antagonize [^{3}H]noradrenaline release or contractile responses. However, cyclic AMP may be involved in the regulation of basal noradrenaline release from autonomic nerves, as forskolin alone did stimulate [^{3}H]noradrenaline release from presynaptic nerves and attenuated contractile responses elicited by nerve stimulation.

To study postjunctional mechanisms involved in the augmentation by ET-1 of contraction, its effects on contractions elicited by exogenously applied noradrenaline or ATP have been studied. ET-1 enhances in a concentration-dependent manner the contractions evoked by ATP, but not noradrenaline (Wicklund et al., 1990, 1991; Donoso et al., 1992), supporting the suggestion that ET's actions are via purinergic motor activity at a postjunctional site. That ET-1 did not facilitate the release of [^{3}H]noradrenaline (Wicklund et al., 1990, 1991; Donoso et al., 1992; Warner et al.,

TABLE I

Effect of ETs on Vas Deferens

Species	Peptide	Potentiation of Contraction				Ref.
		Electrically Induced	Basal	ATP-induced	NA-induced	
Rat	ET-1	Yes	No			1[b,c],2[b],3[b]
		Yes	No	Yes	No	4[b]
		Yes	Yes			5[b],6[b],7[d], 8[e] 9[c],10[c]
		Yes	Yes	Yes	Yes	11[b]
		Yes	Yes	Yes	No	4[c], 12[f]
	ET-2	Yes	Yes			5[b], 6[b]
	ET-3	Yes				13[b]
		Yes	No			2[b], 3[b]
		Yes	Yes			5[b], 6[b]
	STX-b	Yes	No			3[b]
		Yes	Yes			5[b]
	STX-c	No	No			1[b]
		No				13[b]
	Big ET-1	Yes	No			2[b]
		Yes	Yes[a]			6[b]
	Big ET-2	Yes	Yes[a]			6[b]
	Big ET-3	No	No			2[b]
		Yes	Yes[a]			6[b]
	ET(16-21)	Yes	Yes			5[b]
	ET(16-21)NH_2	No	Yes			5[b]
	[$Ala^{3,11}$]ET-1	Yes	Yes			14[f]
	[$Ala^{1,15}$]ET-1	No	Yes			14[f]
Guinea pig	ET-1	Yes	Yes	Yes	No	8e
Mouse	ET-1	Yes	No	Yes	Yes	15[f]
	ET-2	Yes	No			15[f]
	ET-3	Yes	No			15[f]

Note: References: 1, Borges et al. (1989); 2, Télémaque and D'Orléans-Juste (1991); 3, Warner et al. (1993); 4, Donoso et al. (1992); 5, Maggi et al. (1989a); 6, Mattera et al. (1993); 7, Shoji and Goto (1990); 8, Wicklund et al. (1990); 9, Mutafova-Yambolieva et al. (1992); 10, Radomirov and Mutafova-Yambolieva (1993); 11, Mutafova-Yambolieva and Radomirov (1993); 12, Wicklund et al. (1991); 13, Eglezos et al. (1993); 14, Hiley et al. (1989); 15, Rae and Calixto (1990).

[a] At high concentrations.
[b] Prostatic.
[c] Epididymal.
[d] Urethral.
[e] Mid-portion.
[f] Portion unspecified.

NA, noradrenaline.

1993), and that the potentiation of ATP-induced contraction and basal tone by ET-1 was unaltered by denervation (Donoso et al., 1992), indicates that these actions of ATP occur postjunctionally and that additional neuronal mechanisms are not involved in ET-evoked potentiation of ATP-induced contraction. However, Mutafova-Yambolieva and Radomirov (1993) observed contractions of vas deferens by noradrenaline concentrations higher than 3 μM, suggesting some adrenergic involvement in the contractions induced by ET-1.

The postjunctional mechanisms for ET-1-induced potentiation may be due to its action on the P_{2x}-purinoceptor; namely, increasing its number and/or affinity or facilitating its transduction mechanism. ET-1 may also act to alter muscle metabolism and/or electrical activity to potentiate ATP-induced contractions. Indeed, electrophysiological studies by Shoji and Goto (1990) suggest that the ET-1-induced increase in twitch contraction and basal tone, evoked by field stimulation, can be partly attributed to concomitant depolarization of the postsynaptic membrane.

ET-1 decreases the inhibition by clonidine of electrically stimulated contractions (Wicklund et al., 1990, 1991; Donoso et al., 1992; Mutafova-Yambolieva et al., 1992), this being due to functional antagonism with clonidine rather than a direct action at the α_2-adrenoceptor. In agreement with this, yohimbine (an α_2-adrenoceptor antagonist), albeit at high concentrations, blocked the inhibitory effect of clonidine on electrically stimulated contractions but had the opposite effect on the response to ET-1. The mechanism of action of ET-1 may be inhibition of neuronal uptake causing increased noradrenaline release, thus decreasing the prejunctional inhibitory effect of clonidine. It is unlikely that ETs act by blocking ATP degradation, as ET-1 also potentiates the action of a nondegradable ATP analogue, AMP-PNP.

Mutafova-Yambolieva and Radomirov (1993) also studied purinergic and adrenergic components of contractile responses elicited by electrical field stimulation of prostatic rat deferens and found the α-adrenoceptor antagonist, prazosin, to have no effect on this response or its modulation by ET-1. After desensitizing purinergic receptors with α,β-methylene ATP, ET-1 was found to enhance only the α,β-methylene ATP-resistant component of electrically evoked contractions. Guanethidine or tetrodotoxin had no effect on enhancement by ET-1 of ATP-induced contractions, but inhibited the enhancement by ET-1 of noradrenaline-induced contractions at high (>3 μM) but not low (0.01 to 1 μM) concentrations of the latter. Hence, ET-1 produces a potentiating effect via postjunctional mechanisms, mainly through purinergic and partially through adrenergic contractile responses. This is in agreement with Wicklund et al. (1990), who found that after α,β-methylene ATP a few of their guinea-pig vas deferens preparations still showed purely adrenergic (phentolamine-sensitive) contractile responses.

Differences were found in epididymal preparations (Radomirov and Mutafova-Yambolieva, 1992). Guanethidine or tetrodotoxin inhibited the potentiation by ET-1 of electrically evoked contraction, with no effect on increased basal tone. Clonidine did not alter the effects of ET-1, whereas yohimbine decreased the EC_{50} value of ET-1. It was concluded that the effects of ET-1 do not depend on the state of sympathetic innervation, and are mainly via a postsynaptic mechanism.

In summary, ET-1 potentiates both components of the biphasic response by acting on pre- and postjunctional sites, as well as potentiating ATP- and noradrenaline-elicited contractions.

C. ET Receptors

Maggi et al. (1989a) found that ET-1, ET-2, ET-3, and STX-b produced concentration-dependent potentiation of responses in the electrically stimulated rat vas deferens. The effects of these peptides developed slowly (between 6 to 10 min) and were difficult to remove upon washout, whereas the C-terminal fragment ET-(16-21) produced a potentiating effect rapidly (1 to 2 min) that faded rapidly upon washout. ET-1-(16-21) was used in several tissues from different species to discriminate between ET_A- and ET_B-receptors. By assuming that the peptide exerts full agonist activity on ET_B-receptors but is inactive or weakly active at ET_A-receptors, this group concluded that the vas deferens contained ET_B-receptors. Télémaque and D'Orléans-Juste (1991) obtained similar potencies for ET-1 and ET-3 in the field-stimulated prostatic portion of rat vas deferens, supporting the suggestion of the involvement of ET_B-receptors in this response. However, this is contested by Eglezos et al. (1993). Although ET-1 was three times more potent than ET-3 (also observed by Wicklund et al., 1991), with both peptides inducing similar maximal facilitatory effects on the twitch response to field stimulation, the ET_B-receptor-selective agonist STX-c was found to possess very weak activity (inconsistent with an ET_B-receptor type [Williams et al., 1991]). Furthermore, BQ 123 (an ET_A-selective antagonist [Ihara et al., 1992]) antagonized responses elicited by ET-3 but not ET-1, suggesting that the responses to these two peptides are mediated by receptors different to the ET_A- and ET_B-subtypes.

Warner et al. (1993) obtained a relative order of potency of STX-b > ET-1 > ET-3 in the rat vas deferens, producing evidence that these peptides act via an ET_A-receptor. In agreement with Eglezos et al. (1993), they also reported a lack of effect of STX-c. In the presence of 10 μM BQ 123 or PD 142,893 (a nonselective ET receptor antagonist, although found by this group to have little activity against ET_B-receptors), STX-b and ET-3 no longer potentiated the response to nerve stimulation. However, BQ 123, but not PD 142,893, failed to inhibit the response to ET-1, leading them to suggest that the response may not be mediated by ET_A-receptors either. They also found that concentration-response curves for ET-1 and ET-3 were very close, but that at the highest concentration used, ET-3 produced a greater potentiation of the twitch response than ET-1 (also observed by Télémaque and D'Orléans-Juste, 1991). This lack of parallelism between the ET-1 and ET-3/STX-b concentration-response curves was explained by suggesting that ET receptors are present both pre- and postsynaptically, the activation of presynaptic ET receptors causing inhibition of noradrenaline release, and the activation of postsynaptic ET receptors causing an increased responsiveness for nerve stimulation. At lower concentrations, ET-1 was suggested to potentiate transmission by a postsynaptic effect that at higher concentrations becomes limiting due to presynaptic inhibition, leading to a flattened ET-1 concentration-response curve. The presynaptic ET receptor may be of the ET_A-type, explaining why ET-1 at higher concentrations reduces transmitter release, whereas ET-3 did not (supported by Wicklund et al. [1991] who measured presynaptic [^{3}H]noradrenaline release, and found that ET-1 but not ET-3 inhibited its release). PD 142,893, but not BQ 123, antagonized the effects of ET-1 on this receptor, but potentiated the response of ET-1 at higher concentrations. The authors conclude that different ET receptors may be mediating the responses elicited by ET-1 to those elicited by ET-3/STX-b in the rat vas deferens.

Donoso et al. (1992) also propose multiple ET receptor types and/or different signal transduction pathways coupled to the same receptor. Depolarization studies also support this idea (Shoji and Goto, 1990), as well as the differences observed in the time taken to wash out the effects of ET-1 on the contraction of rat deferens in comparison with the short time taken to wash out and reverse the effects on basal tone. Hiley et al. (1989) tested the ET analogues [$Ala^{3,11}$]ET-1 and [$Ala^{1,15}$]ET-1 (in which the cysteine residues forming disulfide bridges have been replaced by Ala residues) and found that [$Ala^{3,11}$]ET-1, like ET-1, dose-dependently increased the twitch response to electrical stimulation in rat vas deferens, whereas [$Ala^{1,15}$]ET-1 had no effect. However, both analogues had an additive effect with ET-1, suggesting that either they do not have a common mechanism of action or ET-1 is a partial agonist relative to [$Ala^{3,11}$]ET-1.

D. Calcium Channels and Signal Transduction

There is conflicting evidence whether Ca^{2+} channels are involved in contractions elicited by ET-1 in vas deferens, as some groups have found that postjunctional effects are blocked by Ca^{2+} channel antagonists (Borges et al., 1989; Wicklund et al.,

1990; Rae and Calixto, 1990) while others have found Ca^{2+} channel blockers to be ineffective in potentiating contractions (Shoji and Goto, 1990) and inhibiting prejunctional noradrenaline release (Wicklund et al., 1990).

Donoso et al. (1992) reported nifedipine and/or removal of external Ca^{2+} using Ca^{2+}-free buffer containing EGTA (that causes a reduction in ATP motor activity) to have no effect on the ET-1-induced potentiation of ATP-induced contractions, and suggested that ET-1 may mobilize intracellular (perhaps mitochondrial) Ca^{2+} stores to facilitate the ATP motor response.

Bay k 8644, in addition to producing a potentiation in contraction evoked by field stimulation in rat vas deferens, also augments excitatory junction potentials (EJPs) without causing any postsynaptic membrane depolarization (unlike ET-1), perhaps by acting presynaptically on L-type Ca^{2+} channels on the axolemma to allow more Ca^{2+} to enter and thus enhancing the exocytotic release of neurotransmitter from granules (Shoji and Goto, 1990). Rae and Calixto (1990) found that ET-1, ET-2, and Bay k 8644 all potentiated contractions in the electrically stimulated mouse vas deferens. ETs and dihydropyridines both activate L-type Ca^{2+} channels but do not compete for a common binding site — ETs acting indirectly via stimulation of their receptors (Gu et al., 1989). Prior potentiation of twitches with ET-1 or Bay k 8644 resulted in a decrease in the ability of nicardipine to inhibit nerve-mediated responses. Therefore, activation of these L-type Ca^{2+} channels either directly using Bay k 8644, or indirectly with ET-1, decreases their susceptibility to blockage by nicardipine.

Both noradrenaline and ETs are linked to InsP metabolism. However, the Ca^{2+} released from intracellular pools by $InsP_3$ is not thought to be involved in ET-1-induced potentiation of contractions elicited by ATP, as noradrenaline, known to activate $InsP_3$ synthesis, is inactive in Ca^{2+}-free medium. Also, contractions elicited by ETs and noradrenaline are not synergistic. However, $InsP_3$ may be involved in ET-1-induced increases in basal tension (Donoso et al., 1992).

E. Effect of Proendothelins

The effects of the ETs have been compared with those of their respective proETs (Télémaque and D'Orléans-Juste, 1991) on nerve-induced contractile responses of the prostatic portion of rat vas deferens. Big ET-1 was only twofold less potent than ET-1 (also observed by Mattera et al. [1993]; see below). Phosphoramidon, an ECE inhibitor, significantly reduced these effects of big ET-1 but not ET-1, indicating the presence of conversion activity by ET converting enzyme (ECE) in this tissue. ET-1 and ET-3 were equipotent in these responses, but big ET-3 was inactive. It was therefore concluded that the ECE present in rat vas deferens does not convert big ET-3 to ET-3.

However, Mattera et al. (1993) reported that while ET-1 and ET-2 were equipotent and 3 to 4 times more potent than ET-3, and big ET-2 was 3 times less potent than ET-2, big ET-3 was also active, albeit with a potency 20 times less than that of ET-3. Phosphoramidon and thiorphan were used to determine whether the enzyme involved was a phosphoramidon-sensitive neutral endopeptidase, or NEP 24.11, a thiorphan/phosphoramidon-sensitive neutral endopeptidase. Neither inhibitor affected the electrically stimulated vas deferens or its responses to ET-1 or ET-3, but the E_{max} of the response elicited by ET-2 was potentiated by both drugs without an increase in potency, implying that NEP-like activity may be involved in the degradation of ET-2 at high peptide concentrations. Phosphoramidon, but not thiorphan, inhibited the increase in contraction of the electrically stimulated vas deferens produced by big ET-1 and big ET-2. Therefore, the enzyme for this conversion is probably not NEP 24.11, but a phosphoramidon-sensitive ECE. Thiorphan reduced the effects of big ET-3, indicating that conversion of this proET to ET-3 is probably by a different, thiorphan-sensitive, ECE.

VII. PROSTATE

A. Introduction

The prostate gland is composed primarily of smooth muscle (~40%), connective tissue, and secretory epithelium. ET-like immunoreactivity is localized to the epithelium, being prominent in the glandular epithelium of the human prostate, with minimal ET-like immunoreactivity in prostatic stroma (Langenstroer et al., 1993). ETs elicit potent, concentration-dependent contractions in strips of human prostatic smooth muscle (Morita et al., 1993; Langenstroer et al., 1993; Kobayashi et al., 1994b). These are not inhibited by indomethacin (Morita et al., 1993; Langenstroer et al., 1993), tetrodotoxin, atropine, phentolamine, propranalol (Morita et al., 1993), terazosin, or nifedipine (Langenstroer et al., 1993), suggesting that ET-1 mediated contractions are not mediated by α_1-adrenoceptors, prostaglandins, or dihydropyridine-sensitive Ca^{2+} channels. A Ca^{2+}-free buffer abolished contractile responses elicited by ET-1 (Langenstroer et al., 1993) indicating the influx of Ca^{2+} via non-L-type channels.

ETs affect the pathology of benign prostatic hypertrophy through changes in receptor density in both bladder and prostate. Kondo and Morita (1993) measured [^{125}I]ET-1 binding to membranes prepared from bladder and prostate in patients with and without benign prostatic hypertrophy and found the K_D values between normal and diseased groups to be the same, but B_{max} values in bladder (base and dome) were lower in the group with prostatic hypertrophy, while B_{max} values in the prostate were higher in this group.

B. ET-Receptor Subtypes

Using quantitative autoradiography, ET receptors have been localized in tissue sections of human prostate (Kobayashi et al., 1994a), using STX-c and BQ 123 to distinguish between ET_A- and ET_B-subtypes. In the stroma, the predominant ET receptor was the ET_A-type, as the specific radioactive densities of ET_A- and ET_B-receptor binding sites were 7.57 ± 0.65 and 2.98 ± 0.81 nCi/mg (mean ± SEM), respectively. In contrast, in glandular epithelium, the ET_B-receptor subtype predominated, with densities of ET_A- and ET_B-receptor binding sites being 1.59 ± 0.15 and 7.87 ± 1.35 nCi/mg, respectively.

Further binding studies (Kobayashi et al., 1994b) using [^{125}I]ET-1 demonstrated saturable and high-affinity binding to human prostate tissue sections obtained from patients with benign prostatic hyperplasia with a K_D value of 0.72 ± 0.13 nM and a B_{max} value of 40.4 ± 6.9 fmol/mg of wet weight. The Hill coefficient was 0.99, indicating binding to a single population of binding sites. With competition binding experiments using unlabeled ET-1, STX-c, and BQ 123, two ET-1 binding sites could be demonstrated. K_i values for STX-c at its higher (ET_B) and lower (ET_A) affinity binding sites were 0.5 ± 0.09 nM and 0.84 ± 0.28 μM, respectively. Approximately two thirds of the ET-1 binding sites were estimated to be of the ET_A-type using BQ 123.

Functional studies show both receptor subtypes to be involved in mediating prostatic smooth muscle contraction. Both ET-1 and STX-c potently contracted human prostate smooth muscle. However, BQ 123 only affected responses elicited by ET-1, indicating that ET_A-receptors mediate ET-1-induced contractions and that STX-c-induced contractions are mediated via ET_B-receptors (Kobashi et al., 1994b). This group suggests that endogenous ETs may be involved in the pathophysiology of benign prostatic hyperplasia in which the bladder outlet is obstructed, so concluding that ET antagonists could be used in its treatment.

VIII. CORPUS CAVERNOSUM

ET-1 causes slowly developing contractions of strips of rabbit and human corpus cavernosum (Saenz de Tejada et al., 1989; Holmquist et al., 1990, 1992a,b) that can only be partially inhibited by nimodipine or using a Ca^{2+}-free medium (Holmquist et al., 1990), with the remaining contraction failing to be abolished by the removal of noradrenaline- or caffeine-sensitive Ca^{2+} stores. However, the residual contraction could be abolished by H-7, which had no effect on the ET-1-induced contraction in normal medium (Holmquist et al., 1990). ET-1 stimulates InsP turnover in strips of rabbit corpus cavernosum (Holmquist et al., 1992a). These results suggest that ET receptors in corpus cavernosum are functionally coupled to phospholipase C, forming $InsP_3$ and diacylglycerol. The receptor subtype present in this tissue still remains to be classified.

IX. SEMINAL VESICLE

ET-1 causes long-lasting tonic contractions of rat seminal vesicle via stimulation of ET_A-receptors, and it selectively potentiates motor responses to ATP, but not noradrenaline or acetylcholine (Rae et al., 1995).

X. CONCLUSION

It has become clear that the physiological effects of ET are important in nonvascular as well as vascular smooth muscle. ETs are produced in many tissues of the male and female urogenital tracts, and by acting on specific ET receptor subtypes they produce various effects, the main one being smooth muscle contraction via several signal transduction mechanisms. ETs act in an autocrine or paracrine fashion and are also involved in the modulation of neurotransmitters. As well as having important physiological actions, they may be important in many pathological conditions associated with the reproductive tract.

REFERENCES

Andersson, K.E., Garcia-Pascual, A., Persson, K., Forman, A. and Tottrup, A., Electrically induced, nerve-mediated relaxation of rabbit urethra involves nitric oxide, *J. Urol.*, 147, 253-9. 1992.

Bolger, G.T., Liard, F. and Jaramillo, J., Tissue selectivity and calcium dependence of contractile responses to ET, *J. Cardiovasc Pharmacol.*, 15, 946-58. 1990.

Borges, R., Von Grafenstein, H. and Knight, D.E., Tissue selectivity of endothelin, *Eur. J. Pharmacol.*, 165, 223-230. 1989.

Bousso-Mittler, D., Kloog, Y., Wollberg, Z., Bdolah, A., Kochva, E. and Sokolovsky, M., Functional endothelin/sarafotoxin receptors in rat uterus, *Biochem. Biophys. Res. Commun.*, 162, 952-957. 1989.

Breuiller-Fouche, M., Heluy, V., Fournier, T. and Ferre, F., Endothelin receptors: binding and phosphoinositide breakdown in human myometrium, *J. Pharmacol. Exp. Ther.*, 270, 973-978. 1994.

Breuiller-Fouche, M., Heluy, V., Fournier, T. and Ferre, F., Only activation of ET_A receptors induces contraction in human myometrium, in Proc. 4th Int. Conf. Endothelin, London, U.K., P109. 1995.

Burnstock, G. and Kennedy, C., Is there a basis for distinguishing two types of P_2-purinoceptors?, *Gen. Pharmacol.*, 16, 433-440. 1985.

Calixto, J.B. and Rae, G.A., Effects of endothelins, Bay k 8644 and other oxytocics in nonpregnant and late pregnant rat isolated uterus, *Eur. J. Pharmacol.*, 192, 109-116. 1991.

Chan, W.Y., Prostaglandins and nonsteroidal antiinflammatory drugs in dysmenorrhoea, *Annu. Rev. Pharmacol. Toxicol.*, 29, 71. 1983

Davenport, A.P., Cameron, I.T., Smith, S.K. and Brown, M.J., Binding sites for iodinated endothelin-1, endothelin-2 and endothelin-3 demonstrated on human uterine glandular epithelial cells by quantitative high resolution autoradiography, *J. Endocrinol.*, 129, 149-154. 1991.

Do Khac, L., Naze, S. and Harbon, S., Endothelin receptor type A signals both the accumulation of inositol phosphates and the inhibition of cyclic AMP generation in rat myometrium. Stimulation and desensitization, *Mol. Pharmacol.*, 46, 485-494. 1994.

Do Khac, L., Naze, S., Le Stunff, H., Thomas, G. and Harbon, S., Endothelin A receptors mediate activation of both phospholipases C and D in rat myometrium, in Proc. 4th Int. Conf. Endothelin, London, U.K., P90. 1995.

Donoso, M.V., Montes, C.G., Lewin, J., Fournier, A., Calixto, J.B. and Huidobro-Toro, J.P., ET-1 (ET)-induced mobilization of intracellular Ca^{2+} stores from the smooth muscle facilitates sympathetic cotransmission by potentiation of adenosine 5'-triphosphate (ATP) motor activity:studies in the rat vas deferens, *Peptides*, 13, 831-840. 1992.

Donoso, M.V., Salas, C., Sepulveda, G., Lewin, J., Fournier, A. and Huifobro-Toro, J.P., Involvement of ET_A-receptors in the facilitation by ET-1 of non-adrenergic non-cholinergic transmission in the rat urinary bladder, *Br. J. Pharmacol.*, 111, 473-482. 1994.

Economos, K., MacDonald, P.C. and Casey, M.L., Endothelin-1 gene expression and protein biosynthesis in human endometrium: potential modulator of endometrial blood flow, *J. Clin. Endocrinol. Metab.*, 74, 14-19. 1992.

Eglen, R.M., Michel, A.D., Sharif, N.A., Swank, S.R. and Whiting, R.L., The pharmacological properties of the peptide, ET, *Br. J. Pharmacol.*, 97, 1297-1307. 1989.

Eglezos, A., Cucchi, P., Patacchini, R., Quartara, L., Maggi, C.A. and Mizrahi, J., Differential effects of BQ 123 against ET-1 and ET-3 on the rat vas deferens: evidence for an atypical ET-receptor, *Br. J. Pharmacol.*, 109, 736-738. 1993.

Fedan, J.S. and Lamport, S.J., Two dissociate phases in the contractile response of the guinea-pig vas deferens to adenosine triphosphate, *J. Pharmacol. Exp. Ther.*, 253, 1-9. 1990.

Fischli, W., Clozel, M. and Guilly, C., Specific receptors for endothelin on membranes from human placenta. Characterization and use in a binding assay, *Life Sci.*, 44, 1429. 1989.

Garcia-Pascual, A., Larsson, B. and Andersson, K.E., Contractile effects of ET-1 and localization of ET binding sites on rabbit lower urinary tract smooth muscle, *Acta Physiol. Scand.*, 140, 545-55. 1990.

Garcia-Pascual, A., Persson, K., Holmquist, F. and Andersson, K.E., ET-1-induced phosphoinositide hydrolysis and contraction in isolated rabbit detrusor and urethral smooth muscle, *Gen. Pharmacol.*, 24, 131-8. 1993.

Guidry, C. and Hook, M., ETs produced by endothelial cells promote collagen gel contraction by fibroblasts, *J. Cell Biol.*, 115, 873-80. 1991.

Gu, X.H., Liu, J.J., Dilon, J.S. and Naylor, W.G., The failure of endothelin to displace bound, radioactively labeled, calcium antagonists (PN 200/110, D888 and diltiazem), *Br. J. Pharmacol.*, 96, 262-264. 1989.

Heluy, V., Breuiller-Fouche, M., Cavaille, F., Fournier, T. and Ferre, F., Characterization of type A endothelin receptors in cultured human myometrial cells, *Am. J. Physiol.*, 268, E825-E831, 1995.

Hiley, C.R., Pelton, J.T. and Miller, R.C., Effects of endothelin on field-stimulated rat vas deferens and guinea-pig ileum, *Br. J. Pharmacol.*, 96, 104P. 1989.

Hogaboom, G.K., O'Donnell, J.P. and Fedan, J.S., Purinergic receptors: photoaffinity analogue of adenosine triphosphate in a specific adenosine triphosphate agonist, *Science*, 208, 1273-1276. 1980.

Holmquist, F., Andersson, K.E., and Hedlund, H., Actions of endothelin on isolated corpus cavernosum from rabbit and man, *Acta Physiol. Scand.*, 139, 113-122. 1990.

Holmquist, F., Persson, K., Garcia-Pascual, A. and Andersson, K.E., Phospholipase C activation of endothelin-1 and noradrenaline in isolated penile erectile tissue from rabbit, *J. Urol.*, 147, 1632-1635. 1992a.

Holmquist, F., Hedlund, H. and Andersson, K.E., Characterization of inhibitory neurotransmission in the isolated corpus cavernosum from rabbit and man, *J. Physiol.*, 449, 295-311. 1992b.

Hori, S., Komatsu, Y., Shigemoto, R., Mizuno, N. and Nakanishi, S., Distinct tissue distribution and cellular localization of two messenger ribonucleic acids encoding different subtypes of rat endothelin receptors, *Endocrinology*, 130, 1885-1895. 1992.

Ihara, M., Noguchi, K., Saeki, T., Fukuroda, T., Tsuchida, S., Kimura, S., Fukami, T., Ishikawa, K., Nishikibe, M. and Yano, M., Biological profiles of highly potent endothelin antagonists selective for the ET_A-receptor, *Life Sci.*, 50, 247-255. 1992.

Imai, T., Hirata, Y., Emori, T., Yanagisawa, M., Masaki, T. and Marumo, F., Induction of endothelin-1 gene by angiotensin and vasopressin in endothelial cells, *Hypertension*, 19, 753-7. 1992.

Iwai, M., Hori, S., Shigemoto, R., Kanzaki, H., Mori, T. and Nakanishi, S., Localization of endothelin receptor messenger ribonucleic acid in rat ovary and fallopian tube by *in situ* hybridization, *Biol. Reprod.*, 49, 675-680. 1993.

Karaki, H. and Weiss, G.B., Minireview: calcium release in smooth muscle, *Life Sci.*, 42, 111-122. 1988.

Kobayashi, S., Tang, R., Wang, B., Opgenorth, T., Stein, E., Shapiro, E. and Lepor, H., Localization of ET-receptors in the human prostate, *J. Urol.*, 151, 763-6. 1994a.

Kobayashi, S., Tang, R., Wang, B., Opgenorth, T., Langenstroer, P., Shapiro, E. and Lepor, H., Binding and functional properties of ET-receptor subtypes in the human prostate, *Mol. Pharmacol.*, 45, 306-311. 1994b.

Kondo, S., Fushimi, E., Morita, T. and Tashima, Y., Direct measurement of ET-receptor in human bladder base and dome using ^{125}I-ET, *Tohoku J. Exp. Med.*, 167, 159-61. 1992.

Kondo, S. and Morita, T., Effect of benign prostatic hypertrophy on the ET-1-receptor density in human urinary bladder and prostate, *Nippon Hinyokika Gakkai Zasshi*, 84, 1821-7. 1993.

Kousides, M., Pennefather, J.N. and Storey, M.E., Endothelin receptors mediating contraction of guinea-pig uterus, in Proc. 4th Int. Conf. Endothelin, London, U.K., P108. 1995.

Kozuka, M., Ito, T., Hirose, S., Takahashi, K. and Hagiwara, H., Endothelin induces two types of contractions of rat uterus: phasic contractions by way of voltage-dependent calcium channels and developing contractions through a second type of calcium channel, *Biochem. Biophys. Res. Commun.*, 159, 317-323. 1989.

Langenstroer, P., Tang, R., Shapiro, E., Divish, B., Openorth, T. and Lepor, H., ET-1 in the human prostate; tissue levels, source of production and isometric tension studies, *J. Urol.*, 150, 495-9. 1993.

Le Monnier de Gouville, A.-C., Lippton, H.L., Cavero, I., Summer, W.R. and Hyman, A.L., ET — A new family of endothelium-derived peptides with widespread biological properties, *Life Sci.*, 45, 1499-14513. 1989.

Lecci, A., Giuliani, S., Santicioli, P., Rovero, P., Maggi, C.A. and Giachetti, A., Intracerebroventricular administration of ETs; effects on the supraspinal micturition reflex and blood pressure in the anesthetized rat, *Eur. J. Pharmacol.*, 199, 201-7. 1991.

MacMahon, L.P., Redman, C.W.G. and Firth, J.D., Expression of the three endothelin genes and plasma levels of endothelin in pre-eclamptic and normal gestations, *Clin. Sci.*, 85, 417-424. 1993.

Maggi, C.A., Giuliani, S., Patacchini, R., Rovero, P., Giachetti, A. and Meli, A., The activity of peptides of the ET family in various mammalian smooth muscle preparations, *Eur. J. Pharmacol.*, 174, 23-31. 1989a.

Maggi, C.A., Guiliani, S., Patacchini, R., Santicioli, P., Turini, D., Barbanti, G. and Meli, A., Potent contractile activity of ET on the human isolated urinary bladder, *Br. J. Pharmacol.*, 96, 755-757. 1989b.

Maggi, C.A., Guiliani, S., Patacchini, R., Turini, D. Barbanti, G., Giachetti, A. and Meli, A., Multiple sources of calcium for contraction of the human urinary bladder muscle, *Br. J. Pharmacol.*, 98, 1021-1031. 1989c.

Maggi, C.A., Guiliani, S., Patacchini, R., Santicioli, P., Rovero, P., Giachetti, A. and Meli, A., The C-terminal hexapeptide, ET-(16-21), discriminates between different ET-receptors, *Eur. J. Pharmacol.*, 166, 121-122. 1989d.

Maggi, C.A., Guiliani, S., Patacchini, R., Barbanti, G., Turini, D. and Meli, A., Contractile responses of the human urinary bladder, renal pelvis and renal artery to ETs and sarafotoxin STX-b, *Gen. Pharmacol.*, 21, 247-249. 1990.

Maggi, M., Vannelli, G.G., Peri, A., Brandi, M.L., Fantoni, G., Giannini, S., Torrisi, C., Guardabasso, V., Barni, T., Toscano, V., Massi, G. and Serio, M., Immulocalization, binding, and biological activity of endothelin in rat uterus: effects of ovarian steroids, *Am. J. Physiol.*, 260, E292-E305. 1991.

Maggi, M., Fantoni, G., Peri, A., Rossi, S., Baldi, E., Magini, A., Massi, G. and Serio, M., Oxytocin-endothelin interactions in the uterus, *Regul. Peptides*, 45, 97-101. 1993.

Mattera, G.G., Eglezos, A., Renzetti, A.R. and Mizahi, J., Comparison of the cardiovascular and neural activity of ET-1, -2, -3 and respective proETs: effects of phosphoramidon and thiorphan, *Br. J. Pharmacol.*, 110, 331-337. 1993.

Maher, E., Bardequez, A., Gardner, J.P., Goldsmith, L., Weiss, G., Mascarina, M. and Aviv, A., Endothelin- and oxytocin-induced calcium signaling in cultured human myometrial cells, *J. Clin. Invest.*, 87, 1251-1258. 1991.

Meldrum, L.A. and Burnstock, G., Evidence that ATP acts as a cotransmitter with noradrenaline in sympathetic nerves supplying the guinea-pig vas deferens, *Eur. J. Pharmacol.*, 92, 161-163. 1983.

Morita, T., Ando, M., Kihara, K., Matsumura, T., Kamai, T., Oshima, H., Tsuchiya, N. and Kondo, S., Effects of ET-1 on the smooth muscle contractility of human urinary bladder, spermatic cord and prostatic adenoma, *Nippon Hinyokika Gakkai Zasshi*, 84, 1649-54. 1993.

Muldoon, L.L., Rodland, K.D., Forsythe, M.L. and Magun, B.E., Stimulation of phosphatidylinositol hydrolysis, diacylglcerol release, and gene expression in response to endothelin, a potent new agonist for fibroblast and smooth muscle cells, *J. Biol. Chem.*, 264, 8529-8536. 1989.

Muldoon, L.L., Pribnow, D., Rodland, K.D. and Magun, B.E., Endothelin-1 stimulates DNA synthesis and anchorage-independent growth of Rat-1 fibroblasts through a protein kinase C-dependent mechanism, *Cell Regul.*, 1, 379-390. 1990.

Mutafova-Yambolieva, V., Petkov, O., Staneva-Stoytcheva, D. and Lasova, L., Interactions between the effects of ET-1, clonidine and yohimbine on electrically induced contractions in rat vas deferens, *Gen. Pharmacol.*, 23, 529-534. 1992.

Mutafova-Yambolieva, V. and Radomirov, R., Effects of ET-1 on postjunctionally mediated purinergic and adrenergic components of rat vas deferens contractile responses, *Neuropeptides*, 24, 35-42. 1993.

Onda, H., Okubo, S., Ogi, K., Kosawa, T., Kimura, C., Matsumoto, H., Suzuki, N. and Fujino, M., One of the endothelin gene family, endothelin-3 gene, is expressed in the placenta, *FEBS Lett.*, 261, 327. 1990.

Orlando, C., Brandi, M.L., Peri, A., Giannini, S., Fantoni, G., Calabresi, E., Serio, M. and Maggi, M., Neurohypophyseal hormone regulation of endothelin secretion from rabbit endometrial cells in primary culture, *Endocrinology*, 126, 1780. 1990.

Persson, K. and Andersson, K.E., Nitric oxide and relaxation of pig lower urinary tract, *Br. J. Pharmacol.*, 106, 416-22. 1992.

Radomirov, R. and Mutafova-Yambolieva, V., Drug-induced modulation of ET-1 effects on the tone of electrically stimulated rat vas deferens, *Arch. Int. Pharmacodyn. Ther.*, 318, 86-96. 1992.

Rae, G.A. and Calixto, J.B., Effects of ETs on nerve-mediated contractions of the mouse vas deferens, *Life Sci.*, 47, PL83-PL89. 1990.

Rae, G.A., Calixto, J.B. and D'Orleans-Juste, P., Big-endothelin-1 contracts rat isolated uterus via a phosphoramidon-sensitive endothelin ET_A receptor mediated mechanism, *Eur. J. Pharmacol.*, 240, 113-119. 1993.

Rae, G.A., Luciano, L.G., D'Orleans-Juste, P. and Calixto, J.B., Endothelin-1 contracts and increases motor responses to ATP in rat seminal vesicle, in Proc. 4th Int. Conf. Endothelin, London, U.K., P108. 1995.

Saenz de Tejada, I., Carson, M.P., Traish, A., Eastman, E.H. and Golstein, L., Role of endothelin, a novel vasoconstrictor peptide, in the local control of penile smooth muscle, *J. Urol.*, 141, 22A. 1989.

Saenz de Tejada, I., Mueller, J.D., De las Morenas, A., Machado, M., Moreland, R.B., Krane, R.J., Wolfe, H.J. and Traish, A.M., ET in the urinary bladder. I. Synthesis of ET-1 by epithelia, smooth muscle and fibroblasts suggests autocrine and paracrine cellular regulation, *J. Urol.*, 148, 1290-1298. 1992.

Sakata, K., Ozaki, H., Kwon, S.-C. and Karaki, H., Effects of endothelin on the mechanical activity and cytosolic calcium levels of various types of smooth muscle, *Br. J. Pharmacol.*, 98, 483-492. 1989.

Sakata, K. and Karaki, H., Effects of endothelin on cytosolic Ca^{2+} level and mechanical activity in rat uterine smooth muscle, *Eur. J. Pharmacol.*, 221, 9-15. 1992.

Schiff, E., Ben-Baruch, G., Galron, R., Mashiach, S. and Sokolovsky, M., Endothelin-1 receptors in human myometrium. Evidence for different binding properties in postmenopausal as compared to premenopausal and pregnant women, *Clin. Endocrinol.*, 38, 321-324. 1993.

Secrest R.J. and Cohen M.L., ET: differential effects in vascular and nonvascular smooth muscle, *Life Sci.*, 45, 1365-1372. 1989.

Shoji, T. and Goto, K., Comparison of the effects of ET-1 and Bay k 8644 on twitch contractions of the field-stimulated rat vas deferens, *Eur. J. Pharmacol.*, 193, 371-374. 1990.

Sneddon, P. and Burnstock, G., Inhibition of excitatory junction potentials in guinea-pig vas deferens by α,β methylene-ATP: further evidence for ATP and noradrenaline as cotransmitters, *Eur. J. Pharmacol.*, 100, 85-90. 1984.

Sneddon, P., Westfall, D.P. and Fedan, J.S., Cotransmitters in the motor nerves of the guinea-pig vas deferens: electrophysiological evidence, *Science*, 218, 693-695. 1982.

Sneddon, P. and Westfall, D.P., Pharmacological evidence that adenosine triphosphate and noradrenaline are co-transmitters in the guinea-pig vas deferens, *J. Physiol.*, 347, 561-580. 1984.

Stjärne, L. and Astrand, P., Relative pre- and postjunctional roles of noradrenaline and adenosine 5'-triphosphate as neurotransmitters of sympathetic nerves of guinea-pig and mouse vas deferens, *Neuroscience*, 14, 929-946. 1985.

Svane, D., Larsson, B., Alm, P., Andersson, K.E. and Foreman, A., Endothelin-1: immunohistochemistry, localization of binding sites and contractile effects in human uteroplacental smooth muscle, *Am. J. Obstet. Gynecol.*, 168, 233-241. 1993.

Takuwa, N., Takuwa, Y., Yanagisawa, M., Yamashita, K. and Masaki, T., A novel vasoactive peptide endothelin stimulates mitogenesis through inositol lipid turnover in Swiss 3T3 fibroblasts, *J. Biol. Chem.*, 264, 7856-7861. 1989.

Télémaque, S. and D'Orléans-Juste, P., Presence of a phosphoramidon-sensitive ET-converting enzyme which converts big ET-1, but not big ET-3, in the rat vas deferens, *Naunyn-Schmiedeberg's Arch. Pharmacol.*, 344, 505-507. 1991.

Traish, A., Moran, E., Krane, R.J. and Saenz de Tejada, I., ET in the urinary bladder. II. Characterization of ET- receptor subtypes, *J. Urol.*, 148, 1299-1306. 1992.

Usuki, S., Suzuki, H., Matsumoto, M., Yanagisawa, M. and Masaki, T., Endothelin-1 in luteal tissue, *Mol. Cell. Endocrinol.*, 80, 147. 1991.

Van Renterghem, C., Vigne, P., Barhanin, J., Schmid-Alliana, A., Frelin, C. and Lazdunski, M., Molecular mechanisms of action of the vasoconstrictor peptide endothelin, *Biochem. Biophys. Res. Commun.*, 157, 977-985. 1988.

Warner, T.D., Allcock, G.H. Mickley, E.J. and Vane, J.R., Characterization of ET-receptors mediating the effects of the ET/sarafotoxin peptides on autonomic neurotransmission in the rat vas deferens and guinea-pig ileum, *Br. J. Pharmacol.*, 110, 783-789. 1993.

Wicklund, N.P., Ohlen, A., Wicklund, C.U., Cederqvist, B., Hedqvist, P. and Gustafsson, L.E., Neuromuscular actions of ET on smooth, cardiac and skeletal muscle from guinea pig, rat and rabbit, *Acta Physiol. Scand.*, 137, 399-407. 1989.

Wicklund, N.P., Ohlen, A., Wicklund, C.U., Hedqvist, P. and Gustafsson, L.E., ET modulation of neuroeffector transmission in rat and guinea-pig vas deferens, *Eur. J. Pharmacol.*, 185, 25-33. 1990.

Wicklund, N.P., Wicklund, C.U., Cederqvist, B., Ohlen, A., Hedqvist, P. and Gustafsson, L.E., ET modulation of neuroeffector transmission in smooth muscle, *J. Cardiovasc. Pharmacol.*, 17 (Suppl. 7), S335-S339. 1991.

Williams, D.L. Jr., Jones, K.L., Pettibone, D.J., Lis, E.V. and Clineschmidt, B.V., Sarafotoxin STX-c: an agonist which distinguishes between ET-receptor subtypes, *Biochem. Biophys. Res. Commun.*, 175, 556-561. 1991.

Chapter 13

ENDOTHELIN IN THE EYE

Neville N. Osborne

CONTENTS

I. INTRODUCTION

The first significant study related to ET in the eye was reported in 1989 (MacCumber et al., 1989). Northern blot analysis revealed a 3.7-kb form of ET mRNA in the eye. *In situ* hybridization was used to detect the message and this was correlated with the autoradiographical localization of ET binding sites in various ocular tissues (MacCumber et al., 1989). ET binding sites and mRNA were highly concentrated in the iris. High densities of ET binding sites were also observed in the endothelial layer of the cornea, the choroid, and the retina. However, no message was apparent in the cornea and only low levels were present in the choroid. Only low levels of ET mRNA and binding sites were localized in the ciliary body (MacCumber et al., 1989). These studies clearly showed that ET may potentially elicit important functions in a variety of ocular tissues and that the peptide can be released a distance away from the eventual site of action. Some of the work on these finds is discussed below.

0-8493-6975-4/97/$0.00+$.50

II. OCULAR EFFECTS OF ET

A. Ocular Effects in Various Species

The ocular effects of an injection of ETs into the eyes of cats (Granstam et al., 1992), monkeys (Erickson-Lamy et al., 1991), and rabbits (MacCumber et al., 1991; Granstam et al., 1991) have been described. An intracameral (10 µl into the aqueous fluid) injection of ET-1 (range 0.1 to 100 nM) into the monkey eye increased the outflow facility (range 22 to 77%), induced a modest accommodation (maximally 2.43 diopters) and did not alter pupil diameter (Erickson-Lamy et al., 1991). Intravitreal injection of ET-1 (0.4 nmol) into the cat eye caused a reduction of retinal blood flow without influencing blood flow in the ciliary body, iris, and choroid. However, when the peptide was administered intracamerally (1 pmol) a reduction in the cat's pupil size, an increase in the aqueous humor protein, and an increase in the concentration of prostaglandin E_2 in the aqueous humor were recorded (Granstam et al., 1992). As the effect of ET-1 on pupil size in the cat (which does not occur in monkey) was abolished by indomethacin pretreatment, it would appear that the ET-1 effect is mediated by an arachidonic acid metabolite. In the rabbit, ETs injected into either the vitreous humor (MacCumber et al., 1991) or intracamerally (Granstam et al., 1991) caused an increase in intraocular pressure (IOP), a breakdown of the blood-aqueous barrier (Granstam et al., 1991), and a vasodilation in the anterior uvea (Granstam et al., 1991). It was concluded from a variety of evidence that the ocular effects of the ETs when injected intracamerally into the rabbit eye are to a large extent mediated by arachidonic acid metabolites (Granstam et al., 1991).

B. Localization and Quantification of ET in Ocular Tissue

ET-1-like and ET-3-like immunoreactivity is highest in the iris and ciliary body, lower in the retina and choroid, and lowest in the cornea of the rabbit eye (MacCumber et al., 1991). In all of the areas measured in the rabbit eye, ET-3 exceeded that of ET-1, by twofold in the iris and ciliary body, 5-fold in the retina and choroid, and 20-fold in the cornea. ET-like immunoreactivity has also been found in the aqueous humor from human and bovine eyes, the concentration being two to three times greater than the corresponding plasma levels (Lepple-Wienhues et al., 1992).

C. ET-Induced Contraction of Ocular Muscle Tissue

ET-1 induces a concentration-dependent contraction of isolated retinal arteries from human (Haefliger et al., 1992) and bovine (Nyborg et al., 1991) eyes. In the case of the bovine retinal arteries, the ET-1 contractions were dependent on extracellular calcium (Nyborg et al., 1991), whereas this was not the case for the human ophthalmic artery (Haefliger et al., 1992).

ETs have been shown to induce contractions in isolated ciliary muscle (Lepple-Wienhues et al., 1992; Wiederholt et al., 1993), isolated trabecular meshwork (Lepple-Wienhues et al., 1992; Wiederholt et al., 1993), and isolated iris sphincter muscle (Osborne and Barnett, 1992; Osborne et al., 1993; Abdel-Latif and Zhang, 1991; El-Mowafy and Abdel-Latif, 1994). However, as shown in the studies by Abdel-Latif and Zhang (1991), the degree of contraction of the isolated sphincter iris muscle elicited by applied ETs depends on the animal species — human, bovine, and monkey muscles are unresponsive to a concentration of ET-1 which drastically contracts tissues from dog and rabbit. Such studies confirm the need for caution when comparing data derived from ocular tissues from different animals.

III. ET RECEPTORS AND SECOND MESSENGERS

A. ET-Induced Actions on Second Messengers

ET-1, ET-2, and ET-3 stimulate phosphoinositide metabolism in rabbit iris/ciliary tissues in a concentration-dependent manner, with the order of potency being ET-1 > ET-2 > ET-3 (Osborne et al., 1993). Similar findings for rabbit tissues have been reported by Abdel-Latif and co-workers (Abdel-Latif and Zhang, 1991; El-Mowafy and Abdel-Latif, 1994). However, it is again clear that the effectiveness of the ETs in stimulating phosphoinositide metabolism in iris/ciliary tissues is species dependent, with bovine, monkey, and human tissues unresponsive to a concentration of ET-1 which drastically stimulates accumulation of inositol phosphate in rabbit, dog, cat, and pig tissues (Abdel-Latif and Zhang, 1991).

ET-1 has also been reported to activate phospholipase A_2 (Abdel-Latif et al., 1991), phospholipase D (Zhang and Abdel-Latif, 1992), and adenylate cyclase (Abdel-Latif and Zhang, 1991; El-Mowafy and Abdel-Latif, 1994) activities in rabbit iris sphincter tissues. No evidence has been found for ET-1 stimulating guanylate cyclase activity in the rabbit iris/ciliary processes (Figure 1). In addition, Osborne et al. (1993) provided evidence that showed that ET-1 reduces, rather than increases, adenylate cyclase activity in the rabbit iris sphincter tissue. Why the adenylate cyclase studies on rabbit tissues by Osborne et al. (1993) and Abdel-Latif and Zhang (1991) have yielded such opposite data is difficult to rationalize. ETs have been reported to be even more potent stimulators of adenylate cyclase activity in iris sphincter muscle tissues of animals like cow, monkey, and dog than in rabbit (Abdel-Latif and Zhang, 1991; El-Mowafy and Abdel-Latif, 1994).

B. ET Receptors in Ocular Tissues

Two distinct types of ET receptors have been cloned and named ET_A and ET_B; ET_A is an ET-1-specific receptor subtype and ET_B is a nonisopeptide-selective subtype (Kondoh et al., 1991; Sakurai et al., 1992). Moya et al. (1993) recently concluded from a study on rat retinal tissue that ET_B receptors were present in that tissue. This was based on the characterization of [^{125}I]-ET-1 binding to retinal tissues and the nature of cross-linking of the radioactive peptide to retinal particulate fractions.

Osborne et al. (1993) also concluded from their study that ET_B receptors were present in the iris/ciliary processes in the rabbit retina. This was based primarily on studies utilizing the polymerase chain reaction (PCR) (see Figure 2). The data were supported by autoradiography which showed that [^{125}I]-ET-1 binding sites are associated with the iris/ciliary processes (Figure 3). The biochemical characterization of these binding sites has since been described (El-Mowafy and Abdel-Latif, 1994) although it was concluded in this study that both ET_A and ET_B receptors are present. The PCR studies carried out by Osborne et al. (1993) failed to demonstrate the existence of ET_A receptors in rabbit iris/ciliary processes (see Figure 2). Interestingly, it was concluded from a series of indirect experiments that ET_A-type receptors are present in the rabbit cornea (Takagi et al., 1994) (see below).

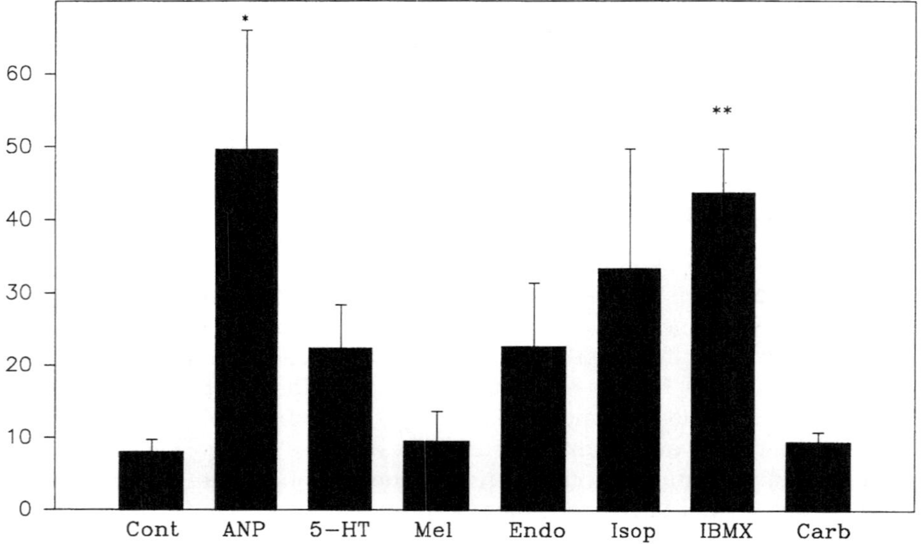

FIGURE 1
Comparative *in vitro* effects of atrial natriuretic factor (ANF, 10 nM), serotonin (5-HT, 100 μM), carbachol (Carb, 100 μM), melatonin (Mel, 100 μM), isoproterenol (Isop, 100 μM), isobutylmethylxanthine (IBMX, 1 nM), and endothelin-1 (Endo, 1 μM) on the stimulation of cyclic GMP production in rabbit iris/ciliary processes. Of the substances tested, only ANF and IBMX caused a significant increase in cyclic GMP compared with controls (Cont).

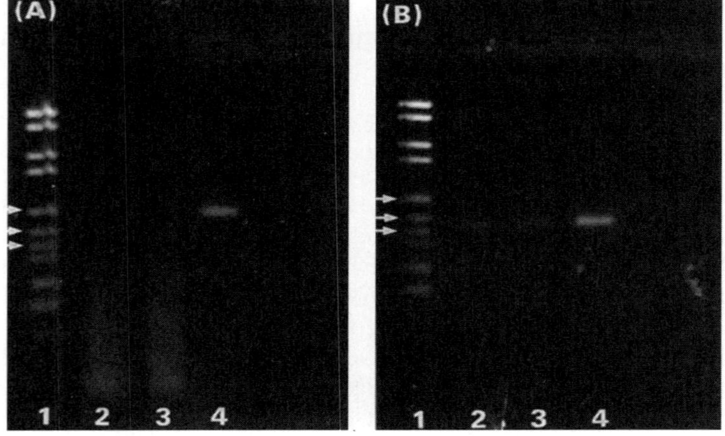

FIGURE 2
PCR data showing the presence of ET_B (B) and absence of ET_A (A) receptors in rat (lane 2) and rabbit (lane 3) iris ciliary processes. Lane 4 shows human ET_A and ET_B (B) receptor plasmids amplified with primer designed for human ET_A and (A) ET_B receptor RNAs, respectively. The arrows show the positions of specific marker base pairs (lane 1) from top to bottom being 653, 517 and 453. (From Osborne, N.N., Barnett, N. L., and Luttmann, W., *Exp. Eye Res.*, 56, 721, 1993. With permission.)

C. ET and the Cornea

Autoradiography has shown that [^{125}I]-ET-1 binding sites are clearly associated with rabbit corneal endothelial cells, with little or no binding to epithelial cells (Figure 3). However, ETs have been reported to cause transient increases in intracellular calcium,

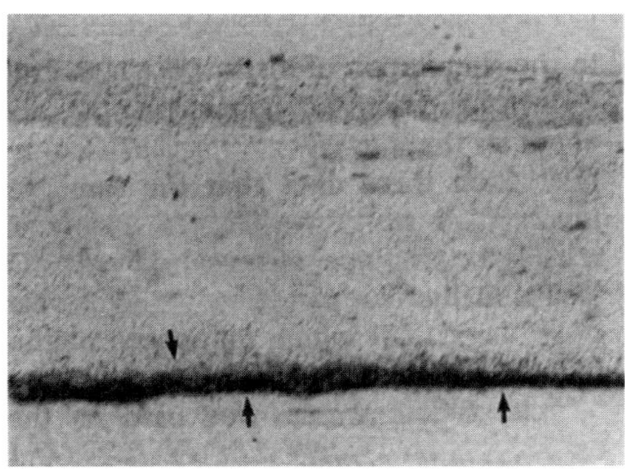

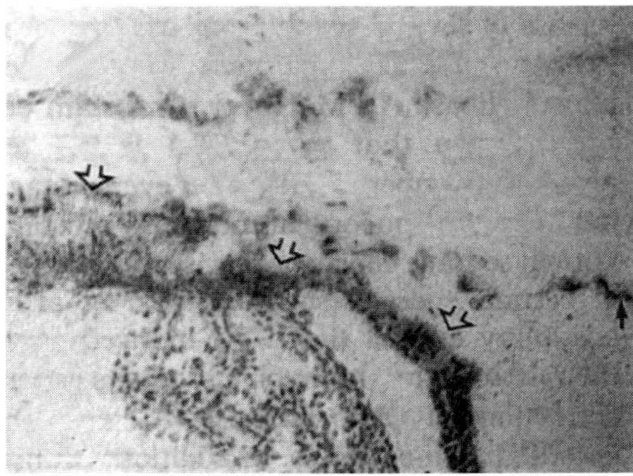

FIGURE 3
Autoradiographical localization of specific [^{125}I]-ET-1 binding sites to rabbit cornea and iris ciliary processes. It can be seen that radioactive binding of ET-1 was confined to the endothelial cells of the cornea (closed arrows), ciliary epithelium (open arrows), and parts of the musculature (open arrows). (From Osborne, N. N., Barnett, N. L., and Luttmann, W., *Exp. Eye Res.*, 56, 721, 1993. With permission.)

stimulate inositol phosphates, and also stimulate cAMP formation in corneal cultured epithelial cells, causing the authors to suggest the presence of ET_A receptors (Takagi et al., 1994). These data and those shown in Figure 3 appear to be incompatible and one can only surmise that the cultured corneal epithelial cells and the same cells in the intact cornea possess different amounts of ET receptors.

Further evidence that the corneal endothelium has an abundance of ET-1 binding sites has been provided by studies on cultured bovine corneal endothelial cells (Chollet et al., 1993). Moreover, in this study it was shown that ET receptors are involved in proliferation and migration, with ETs thus acting as a growth factor.

D. Influence on Retinal Capillary Pericytes

The pericytes of the retinal capillaries have many features that indicate that they are contractile cells, regulating microvascular calibre and tone, similar to the smooth muscle cells of larger vessels. Studies by Takahashi et al. (1989) have shown that cultured retinal capillary endothelial cells secrete ET-1 and corresponding pericytes bear receptors for this peptide. These studies suggest that a specific, interactive system may exist between capillary endothelial cells and pericytes for the regulation of retinal capillary blood flow, independent of possible extrinsic innervation. Bovine retinal capillary pericytes do respond *in vitro* to ET-1 stimulation by prolonged contraction which can be maintained for up to 2 h after removal of the agonist

(Chakravarthy et al., 1992). ETs have also been shown to be very effective in inducing a release of intracellular calcium from cultured bovine microvascular pericytes (Ramachandran et al., 1993). Thus, there is evidence that ET plays a role in the interaction between endothelial cells and pericytes, with consequent effects on the retinal microcirculation. This could have implications in the treatment of diseases such as diabetes and hypotension, where there may be a derangement in the interaction between endothelial cells and pericytes.

REFERENCES

Abdel-Latif, A.A. and Zhang, Y., Species differences in the effects of endothelin-1 on myo-inositol trisphosphate accumulation, cyclic AMP formation and concentration of isolated iris sphincter of rabbit and other species, *Invest. Ophthalmol. Vis. Sci.*, 32, 2432-2438, 1991.

Abdel-Latif, A.A., Zhang, Y. and Yousufzai, S., Endothelin-1 stimulates the release of arachidonic acid and prostaglandins in rabbit iris sphincter smooth muscle: activation of phospholipase A_2, *Curr. Eye Res.*, 259-265, 1991.

Chakravarthy, U., Gardiner, T.A., Anderson, P., Archer, D.B. and Trimble, E.R., The effect of endothelin-1 on the retinal microvascular pericyte, *Microvasc. Res.*, 43, 242-254, 1992.

Chollet, P., Malecaze, F., Gouzi, L., Arne, J.L. and Plouet, J., Endothelin 1 is a growth factor for corneal endothelium, *Exp. Eye Res.*, 57, 595-600, 1993.

De, J.J., Moya, F.J., De, L.M., Fernandez-Cruz, A. and Ferandex-Durango, R., Identification and characterization of endothelin receptor subtype B in rat retina, *J. Neurochem.*, 61, 1113-1119, 1993.

El-Mowafy, A.M. and Abdel-Latif, A.A., Characterization of iris smooth muscle endothelin receptor subtypes which are coupled to cyclic AMP formation and polyphosphoinositide hydrolysis, *J. Pharmacol. Exp. Ther.*, 268, 1343-1351, 1994.

Erickson-Lamy, K., Korbmacher, C., Schuman, J.S. and Nathanson, J.A., Effect of endothelin on outflow facility and accommodation in the monkey eye *in vivo*, *Invest. Ophthalmol. Vis. Sci.*, 32, 492-495, 1991.

Granstam, E., Wang, L. and Bill, A., Ocular effects of endothelin-1 in the cat, *Curr. Eye Res.*, 11, 325-332, 1992.

Granstam, E., Wang, L. and Bill, A., Effects of endothelins (ET-1, ET-2 and ET-3) in the rabbit eye: role of prostaglandins, *Eur. J. Pharmacol.*, 194, 217-223, 1991.

Haefliger, I.O., Flammer, J. and Luscher, T.F., Nitric oxide and endothelin-1 are imporant regulators of human ophthalmic artery, *Invest. Ophthalmol. Vis. Sci.*, 33, 2340-2343, 1992.

Inoue, A., Yanigisawa, M., Kimura, S., Kasauya, Y., Miyauchi, T., Goto, K. and Maxaki, T., The human endothelin family: three structurally and pharmacologically distinct isopeptides predicted by three separate genes, *Proc. Natl. Acad. Sci. U.S.A.*, 86, 2863-2867, 1989.

Kondoh, M., Miyazaki, H., Uchiyama, Y., Yanagisawa, Y., Maskai, T. and Murakami, K., Solubilization of two types of endothelin receptors, ET_A and ET_B from rat lung with retention of binding activity, *Biomed. Res.*, 12, 417-423, 1991.

Lepple-Wienhues, A., Becker, M., Stahl, F., Berweck, S., Hensen, J., Noske, W., Erichhorn, M. and Wiederholt, M., Endothelin-like immunoreactivity in the aqueous humor and in conditioned medium from cultured ciliary epthelial cells, *Curr. Eye Res.*, 11, 1041-1046, 1992.

MacCumber, M.W., Rose, C.A., Glaser, B.M. and Snyder, S.H., Endothelin: visualisation of mRNAs by *in situ* hybridization provides evidence for local action, *Proc. Natl. Acad. Sci. U.S.A.*, 86, 7285-7289, 1989.

MacCumber, M.W., Jampel, H.D. and Snyder, S.H., Ocular effects of endothelin, *Arch. Ophthalmol.*, 109, 705-709, 1991.

Nyborg, N.C.B., Prieto, D., Benedito, S. and Nielsen, P.J., Endothelin-1 induced contraction of bovine retinal small arteries is reversible and abolished by nitrendipine, *Invest. Ophthalmol. Vis. Sci.*, 32, 27-31, 1991.

Osborne, N.N. and Barnett, N.L., Endothelin-1 stimulates phosphotidylinositol hydrolysis in the iris-ciliary complex and is a potent constrictor of the sphincter muscle, *Exp. Eye Res.*, 54, 285-290, 1992.

Osborne, N.N., Barnett, N.L. and Luttmann, W., Endothelin receptors in the cornea, iris and ciliary processes. Evidence from binding, secondary messenger and PCR studies, *Exp. Eye Res.*, 56, 721-728, 1993.

Ramachandran, E., Frank, R.N and Kennedy, A., Effects of endothelin on cultured bovine retinal microvascular pericytes, *Invest. Ophthalmol. Vis. Sci.*, 34, 586-595, 1993.

Sakurai, T., Yanagisawa, M and Masaki, T., Molecular characterization of endothelin receptors, *TIPS*, 13, 103-108, 1992.

Takagi, H., Reinach, P.S., Tachado, S.D. and Yosimura, N., Endothelin-mediated cell signaling and proliferation in cultured rabbit corneal epithelial cells, *Invest. Ophthalmol. Vis. Sci.*, 35, 134-142, 1994.

Takahashi, K., Brooks, K.R.A., Kanse, S.M., Ghatei, M.A., Kohner, E.M., and Bloom, S., Production of endothelin-1 by cultured bovine retinal endothelial cells and presence of endothelin receptors on associated pericytes, *Diabetes*, 38, 1200-1202, 1989.

Wiederholt, M., Lepple-Wienhues, A. and Stahl, F., Contractile properties of trabecular meshwork and ciliary muscle, in: *Basic Aspects of Glaucoma Research III*, Lütjen-Drecoll, E., Ed., Schattauer, Stuttgart, pp 287-306, 1993.

Zhang, Y. and Abdel-Latif, A.A., Activation of phospholipase D by endothelin-1 and other pharmacological agents in rabbit iris sphincter smooth muscle, *Cell. Signal.*, 4, 777-786, 1992.

Part IV

Ligand Structure and Activity Relationships

Chapter 14

Conformational Studies of Endothelins and Analogs

John T. Pelton

CONTENTS

0-8493-6975-4/97/$0.00+$.50

I. INTRODUCTION

The endothelins, sarafotoxins, mouse vasoactive intestinal contractor (VIC), and bibrotoxin all belong to a family of highly conserved bicyclic 21-amino-acid peptides (Table 1). All members of this family are characterized by a free amino terminus, a hydrophobic C-terminal "tail" with a free carboxylic acid function, and disulfide bonds between Cys^1-Cys^{15} and Cys^3-Cys^{11} forming two rings of 28 and 29 atoms (Figure 1). Nearly half the amino acids in this peptide family are always conserved, namely residues 1, 3, 8, 10, 11, 15, 16, 18, 20, and 21. In addition, there are limited or conservative substitutions at residues 2 (Ser or Thr), 9 (Lys or Glu), 12 (Val or Leu), 13 (Tyr or Asn), 14 (Phe or Tyr), 17 (Leu or Gln), and 19 (Ile or Val). Yet despite this conservation in primary structure, the individual members of this family display differences in their distribution, receptor selectivity, and biological activities. ET-1, for example, has high affinity for both ET_A and ET_B receptors. ET-3, in contrast, has much lower affinity for ET_A receptors than for ET_B receptors, despite the high conservation in sequence and structure. As a consequence, a wide variety of spectroscopic methods have been applied to better understand those structural and conformational features that are important for receptor recognition and activation.

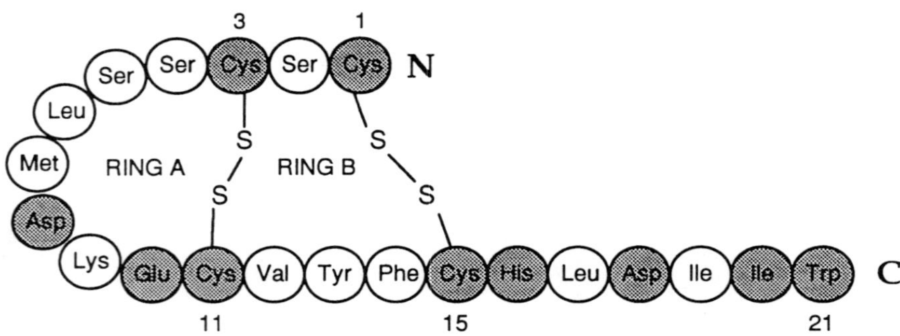

FIGURE 1
Endothelin-1 is characterized by two disulfide bonds involving Cys^1-Cys^{15} and Cys^3-Cys^{11}, forming two rings of (A) 29 and (B) 28 atoms. Nearly half the residues are always conserved within the endothelin/sarafotoxin peptide family (shaded circles) and all peptides have free amino and carboxy termini.

II. UV SPECTROSCOPY AND CIRCULAR DICHROISM

A. UV Spectroscopy

The near-UV spectrum of the endothelins and sarafotoxins is dominated by the absorption of the tyrosine and tryptophan side chains and by the disulfide bonds (Figure 2). ET-1 has an absorption maximum at $\lambda = 279$ nm and a molar extinction

TABLE 1

Amino Acid Sequence of Endothelin/Sarafotoxin Family of Peptides

	1		3		5		7		9		11		13		15		17		19		21
ET-1	Cys	Ser	Cys	Ser	Ser	Leu	Met	Asp	Lys	Glu	Cys	Val	Tyr	Phe	Cys	His	Leu	Asp	Ile	Ile	Trp
ET-2	Cys	Ser	Cys	Ser	Ser	Trp	Leu	Asp	Lys	Glu	Cys	Val	Tyr	Phe	Cys	His	Leu	Asp	Ile	Ile	Trp
ET-3	Cys	Thr	Cys	Phe	Thr	Tyr	Lys	Asp	Lys	Glu	Cys	Val	Tyr	Tyr	Cys	His	Leu	Asp	Ile	Ile	Trp
S6a	Cys	Ser	Cys	Lys	Asp	Met	Thr	Asp	Lys	Glu	Cys	Leu	Asn	Phe	Cys	His	Gln	Asp	Val	Ile	Trp
S6b	Cys	Ser	Cys	Lys	Asp	Met	Thr	Asp	Lys	Glu	Cys	Leu	Tyr	Phe	Cys	His	Gln	Asp	Val	Ile	Trp
S6c	Cys	Thr	Cys	Asn	Asp	Met	Thr	Asp	Glu	Glu	Cys	Leu	Asn	Phe	Cys	His	Gln	Asp	Val	Ile	Trp
S6d	Cys	Thr	Cys	Lys	Asp	Met	Thr	Asp	Lys	Glu	Cys	Leu	Tyr	Phe	Cys	His	Gln	Gly	Ile	Ile	Trp
VIC[a]	Cys	Ser	Cys	Asn	Ser	Trp	Leu	Asp	Lys	Glu	Cys	Val	Tyr	Phe	Cys	His	Leu	Asp	Ile	Ile	Trp
BI[a]	Cys	Ser	Cys	Ala	Asp	Met	Thr	Asp	Lys	Glu	Cys	Leu	Tyr	Phe	Cys	His	Gln	Asp	Val	Ile	Trp

[a] Vic = vasoactive intestinal contractor peptide; BI = bibrotoxin.

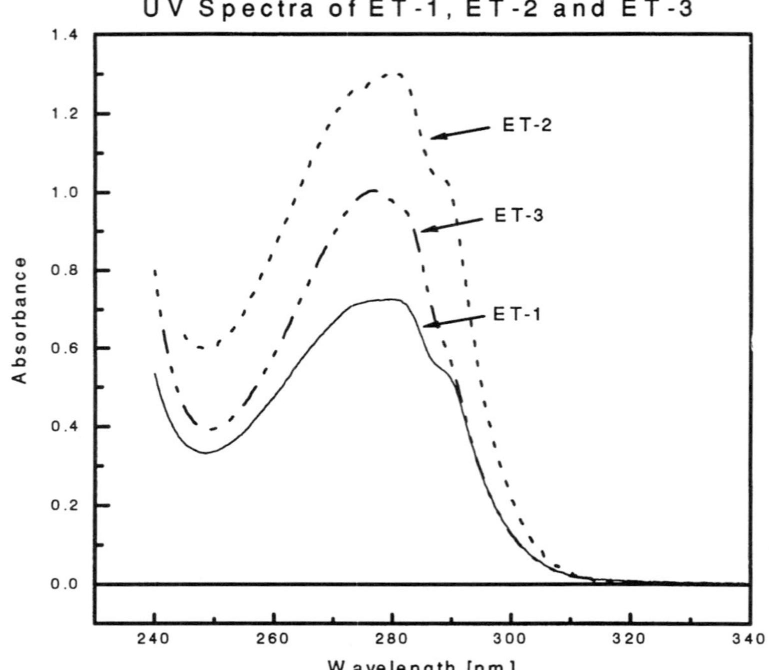

FIGURE 2
UV spectrum of 100 μM endothelin-1, endothelin-2, and endothelin-3 in 20 mM phosphate buffer, pH 7.4 at 22°C. Peptides have measured molar extinction coefficients of ε_{279} = 7,250 $M^{-1}\cdot cm^{-1}$, ε_{280} = 13,080 $M^{-1}\cdot cm^{-1}$, and ε_{277} = 10,090 $M^{-1}\cdot cm^{-1}$ for ET-1, ET-2, and ET-3, respectively.

coefficient of ε = 7250 $M^{-1}\cdot cm^{-1}$ at this wavelength while ET-2, with an additional tryptophan residue at position 6, has the largest extinction coefficient within this peptide family (ε_{280} = 13,100 $M^{-1}\cdot cm^{-1}$).

B. Circular Dichroism of Endothelins and Sarafotoxins

Circular dichroism studies of the endothelins and sarafotoxins have generally supported the presence of a helix-like structure, based on the presence of a negative band (shoulder) at about 223 nm (the helical trough) and a positive band around 192 nm (Figure 3). The shoulder at 223 nm was assigned to the $n\rightarrow\pi^*$ transition of the nonbonded electrons of the amide carbonyl oxygen in a helical conformation. From the intensity of this band, Harris (1993) estimated 29% helical content for ET-1 in water. The weakness of the negative 222 to 225 nm band in ET-1 could be explained by the presence of two disulfide bonds and their positive band around 228 nm, thus decreasing the minimum at 223 nm (Hider et al., 1988). Based on spectra of disulfide-linked peptides with short α-helical segments such as apamine and mast-cell degranulating peptide, Perkins et al., (1990) estimated a somewhat higher helical content in ET-1 of about 35%. Using secondary structure prediction methods, they proposed a single α-helix involving residues 9-15, in good agreement with the NMR studies (see below). Alternatively, the band around 223 nm might arise from $\pi\rightarrow\pi^*$ transitions associated with the indole ring of tryptophan and, indeed, the fluorescence excitation spectrum (Figure 4, panel A) of endothelin-1 shows such a band in this region. However, a similar shoulder has also been observed for [Phe^{21}]ET-1, which does not

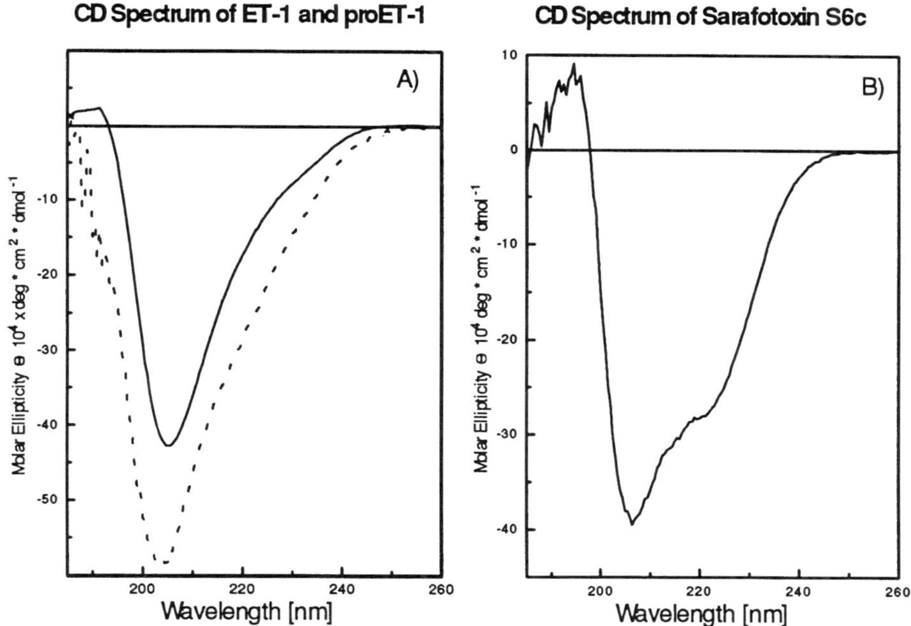

FIGURE 3
CD spectra of (A) endothelin-1 (solid line) and human big-endothelin-1 (dotted line), and (B) sarafotoxin S6c. Peptide concentrations were below 20 μM in 20 mM phosphate buffer, pH 7.5; reported in molar ellipticity. The shoulder at 223 nm associated with the presence of an α-helix is more pronounced in the spectrum of sarafotoxin S6c.

contain an indole ring (Pelton, 1991). Moreover, the band at 223 nm becomes more pronounced (Saudek et al., 1991; Andersen et al., 1992a; Tamaoki, et al., 1992; Harris, 1993) in aqueous trifluoroethanol (TFE) solutions, as expected since TFE is known to induce/stabilize helical structure.

Based on the CD pattern of the endothelins and sarafotoxins in the far UV, Tamaoki et al. (1992) observed that these peptides fell into three groups: (1) ET-1, ET-2, VIC; (2) the sarafotoxins; and (3) ET-3. ET-3 appeared to be more flexible with less tendency of the backbone to form a helical structure in aqueous TFE solutions than the other peptides, and was more sensitive to changes in pH than were the other endothelins or VIC. The bands characteristic of helical structure were most prevalent in the sarafotoxins, with sarafotoxin S6c showing a very pronounced shoulder at 223 nm (Figure 3, panel B). These results indicate that individual members within the endothelin and sarafotoxin family do adopt somewhat different conformations, which may account to some extent for the variety of helical structures observed by other techniques.

C. Circular Dichroism of Big-ET-1

Considerable interest has also been shown in the conformation of big-ET-1, particularly in the region of the Trp^{21}-Val^{22} cleavage site. Such information could provide insight into the structural and conformational features required for recognition by the putative endothelin cleavage enzyme (ECE). The CD spectrum of human big-ET-1 is shown in Figure 3, panel A. The spectrum is nearly identical to that of ET-1 itself, indicating that the C-terminal portion of this peptide does not

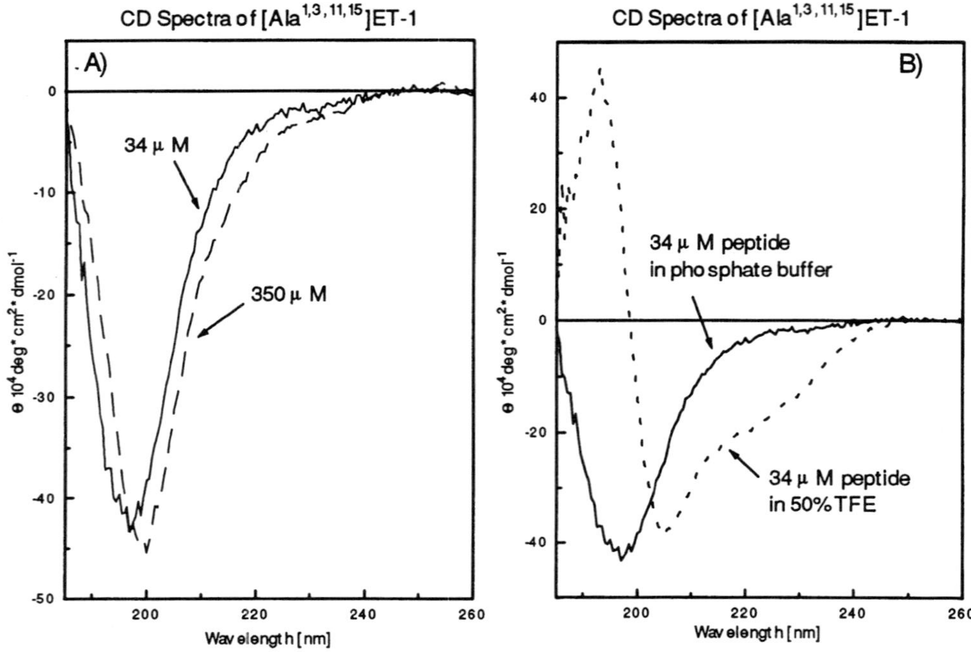

FIGURE 4
$[Ala^{1,3,11,15}]$ET-1 shows (A) some minor changes in the CD spectrum as a function of concentration, while (B) the addition of 50% trifluoroethanol results in a major shift in the bands with evidence for increased helical character.

adopt a well-defined conformation in solution. Nevertheless, Fabbrini et al. (1993) recently showed that in *Xenopus* oocytes the cleavage at the Trp^{21}-Val^{22} site was unaffected by several mutations at these positions and that the processing efficiency of the mutants was comparable to that of wild type. They interpreted these results to suggest that only a conformational requirement exists for proper cleavage at this site.

D. Circular Dichroism of BQ 123

CD studies (Atkinson and Pelton, 1992) of BQ 123 (cyclo[D-Asp-Pro-D-Val-Leu-D-Trp]) indicate that this small cyclic ET_A receptor antagonist adopts a well-defined conformation in both water and in aqueous-organic solvents. The far UV CD spectrum is characterized by a minimum at 185 nm, a rather broad and intense maximum at 204 nm that contains a shoulder around 212 nm, and another minimum at 234 nm. By comparison with the UV absorption spectrum of this peptide, the ^1B_b band for the indole ring of tryptophan appears to contribute to the broad positive ellipticity observed in the region between 225 nm and 200 nm. The negative band centered at 234 nm was tentatively assigned to the indole ^1C transition.

E. Solvent and Concentration Effects Studied by CD Spectroscopy

Brown et al. (1990) reported an increase in helical content for ET-1 in 100% TFE and for the peptide bound to dimyristoylphosphatidyl glycerol (DMPG) vesicles. Similar behavior has been noted by a number of other groups (Tamaoki et al., 1992;

Pelton, 1991; Saudek et al., 1991; Harris, 1993) for ET-1 and related peptides. Figure 4, panel B shows the effect of 50% TFE in phosphate buffer on the acyclic, ET_B-receptor-selective ligand [$Ala^{1,3,11,15}$]ET-1. In this analog, the alanine residues remove any possible effects from the disulfide bonds and the increase in helical content is clearly visible. This is in contrast to several other reports (Bennes et al., 1990; Calas et al., 1992) who concluded that there was no significant amount of α-structure in ET-1, and to Perkins et al. (1990) who reported that TFE did not cause a significant change in the CD spectrum of ET-1. However, in most solvents employed for NMR studies (aqueous acetonitrile, aqueous ethylene glycol, and DMSO), only minor changes in the CD spectrum are reported.

In aqueous solutions, Tamaoki et al. (1992) and Atkins et al. (1994) reported that ET-1 and ET-2 showed only a slight pH dependence in the range pH 3.0 to 7.5, while ET-3 was more sensitive. Andersen et al. (1992a) also remarked on the similarity of the CD spectra of ET-1 throughout the pH range 3 to 7 in their aqueous ethylene glycol mixtures.

Concentration effects have been reported by some (Bennes et al., 1990) but not others (Brown et al., 1990; Andersen et al., 1992a; Harris, 1993). Bennes et al. (1990) reported that ET-1 self-associates in a wide range of solvents, but not DMSO, with a critical micelle concentration of about 22 μM. This result was supported by conductivity and surface tension measurements. However, Andersen et al. (1992a) observed no changes in the CD spectrum of ET-1 in 4 and 54% aqueous ethylene glycol solutions over the range 13.4 μM to 1.4 mM. Similarly, Brown et al. (1990) saw no concentration-dependent changes in the CD signal for ET-1 or big-endothelin in various solvents in the concentration range 5 to 500 μM. In 20 mM phosphate buffer, pH 7.5, we have seen minor concentration-dependent changes in the CD spectrum of the endothelins and analogs (Figure 4, panel A) and some aggregation could be detected by light scattering. However, the aggregation was highly variable and appeared to depend on sample history (supplier, sample age, previous exposure to other solvents, etc.). A sedimentation equilibrium analysis of the endothelins by Atkins et al. (1994) also detected peptide aggregation. The aggregation was most pronounced for ET-3, even at low concentrations (24 μM), but aggregation could be reversed by the addition of acetonitrile. These results suggest some caution in sample preparation, particularily in pure aqueous conditions.

III. NMR STUDIES

A. ^{1}H NMR Studies

The conformations of the endothelins/sarafotoxins have been extensively studied by NMR spectroscopy. Table 2 summarizes many of the different peptides and the conditions under which they have been investigated. In addition, a number of reports have appeared detailing the solution conformation of endothelin receptor antagonists.

B. NMR Studies of Endothelins and Sarafotoxins

All NMR studies of the endothelins and sarafotoxins, whether in predominantly aqueous or organic solvents, report the presence of a helix-like structure within the peptide. In DMSO, Saudek et al. (1989, 1991) reported the presence of a right-handed helix-like structure extending from Lys^9 to Cys^{15}, as did the studies by Endo et al.

TABLE 2

NMR Studies of Endothelins, Sarafotoxins, and Analogs

Peptide	Solvent	Ref.
Endothelin-1	DMSO	Saudek et al.,1989, 1991
Endothelin-1	DMSO	Endo et al., 1989
Endothelin-1	DMSO	Munro et al., 1991
Endothelin-1	DMSO	Brown et al., 1990
Endothelin-1	CD_3CN/H_2O	Reily and Dunbar, 1991
Endothelin-1	Aq. ethylene glycol	Krystek et al., 1991
Endothelin-1	Aq. ethylene glycol	Andersen et al., 1992a
Endothelin-1	Aq. acetic acid	Tamaoki et al., 1991
Endothelin-1	Aq. acetic acid	Donlan et al., 1991
Endothelin-1	Aq. acetic acid	Dalgarno et al., 1992
Endothelin-1	Aqueous	Ragg et al., 1994
$[Nle^7]$ET-1	DMSO and CD_3CN/H_2O	Aumelas et al., 1991b
$[Aba^{1,15}]$ET-1[a]	CD_3CN/H_2O	Coles et al., 1993
$[Pen^{3,15}, Nle^7]$ET-1[a]	Aq. ethylene glycol	Andersen et al., 1992b
Endothelin-3	Aq. acetic acid	Bortman et al., 1991
Endothelin-3	H_2O	Mills et al., 1992
Sarafotoxin S6b	CD_3CN/H_2O	Aumelas et al., 1991a
Sarafotoxin S6b	H_2O	Mills et al., 1991
Sarafotoxin S6c	H_2O/100 mM NaCl	Mills et al., 1994
VIC[a]	CD_3CN/H_2O	Aumelas et al., 1992
Big-endothelin	Undefined	Inooka et al., 1991
Big-endothelin	Sodium acetate, pH 3.0	Donlan et al., 1992

[a] Aba = aminobutyric acid; VIC = vasoactive intestinal contractor.

(1989) and Brown et al. (1990). Munro et al. (1991), however, found a helical structure involving the residues between Leu^6 and Cys^{11}, and a well-defined but nevertheless nonhelical structure in the region Cys^{11} to Cys^{15}. In predominantly aqueous media (aqueous acetonitrile, aqueous ethylene glycol, or aqueous acetic acid solutions), a helical region is also observed from Glu^{10} to Cys^{15} (Baumer et al., 1991), Lys^9 to Cys^{15} (Reily and Dunbar, 1991; Tamaoki et al., 1991; Andersen et al., 1992a; Aumelas et al., 1992), or Lys^9 to His^{16} (Krystek et al., 1991; Ragg et al., 1994), with several groups suggesting some "fraying" at both the C- and N-terminal ends. Finally, a somewhat longer helical region was observed in the more conformationally constrained $[Pen^{3,15},Nle^7]$ET-1 in aqueous ethylene glycol (Andersen et al., 1992b) extending from Lys^9 to about Asp^{18}.

Interestingly, in apamin — a small, neurotoxic peptide that is a component of honey bee venom — the cysteine residues are located at the same positions as in the endothelins and sarafotoxins ($Cys^{1,3,11,15}$), but the disulfide bond pairing is $Cys^{1\text{-}11}$, $Cys^{3\text{-}15}$. Nevertheless, NMR studies (Freeman et al., 1986; Pease and Wemmer, 1988) of this peptide in water indicate the presence of a helix-like region involving residues 9-17. From these and other data, it was proposed (Pease et al., 1990; Kobayashi et al., 1991) that the arrangement of cysteine residues in a Cys-X-X-X-Cys crosslinked to a Cys-X-Cys pattern, represents a structural motif which may correlate with the bioactivity of neurotoxins. These results are in agreement with their CD spectra (see above) and structure-activity studies of ET-1, in which many of the residues most sensitive to substitution such as Asp^8, Glu^{10}, Val^{12}, Tyr^{13}, and Phe^{14} are located within or adjacent to this helix-like region. In addition, Harris (1993) found that the receptor affinity of some monocyclic ET-1 analogs substituted with helix-breaking residues at position 11 correlated well with the observed helix content of the peptides, as detected by CD. However, Panek et al. (1993) reported that $[Pro^{12}]$ET-1 was nearly as potent as the native peptide in receptor binding assays, although somewhat less potent as an agonist in functional assays. As proline should disrupt the helical structure between Lys^{9} and Cys^{15}, these authors concluded that this feature was not critical for receptor binding or agonist activity.

In addition to a helix-like structure, several studies have identified other regions within the peptide that adopt a well-defined conformation, although many of these results are more controversial. A reverse turn involving the Ser^{5} to Asp^{8} residues has been reported by several groups for endothelin-1 (Krystek et al., 1991; Tamaoki et al., 1991; Andersen et al., 1992a) but not observed by others (Reily and Dunbar, 1991; Munro et al., 1991; Saudek et al., 1991). A β-turn structure involving residues 5-8 was also observed in big-ET-1 (Inooka et al., 1991; Donlan et al., 1992) but not in ET-3 (Bortmann et al., 1991; Mills et al., 1992). In sarafotoxin S6b, the β-turn involved residues 3-6 (Mills et al., 1991; Aumelas, 1991a), indicating a conformation in this region of the peptide that is different to that of ET-1. In VIC, Aumelas et al. (1992) observed a turn structure involving Asn^{4}-Leu^{7}. This last result suggests that despite major replacements within the outer loop of VIC relative to ET-1, the peptide adopts a very similar conformation.

More controversial still has been the conformation of the C-terminal hexapeptide. Saudek et al. (1989, 1991), Baumer et al. (1991), and Ragg et al., 1994 reported structures in which the C-terminal hexapeptide of ET-1 was folded back towards the bicyclic portion of the molecule, while others observed no long-range interactions between the C-terminal "tail" and the main body of endothelin. These groups argued for a more random conformation in this portion of the molecule (Munro et al., 1991; Endo et al., 1989; Reily and Dunbar, 1991). Andersen et al. (1992a) observed that within the C-terminus certain unusual chemical shifts are almost always observed, in particular an upfield shift of the Ile^{19}-βMe in the endothelins or Val^{19}-γMe groups in sarafotoxin S6b. In addition, some unusual nOe connectivities were also noted, suggesting the presence of some preserved structural feature in the C-terminal tail region. They proposed the existence of at least two discreet conformers that interconvert with lifetimes in the microsecond domain. This was supported by data indicating that it was the N-terminal loop involving the Ser^{2} to Ser^{5} residues, and not the C-terminus, that was the most mobile portion of the ET-1 molecule.

C. NMR Studies of BQ 123 and Analogs

The recent development of small, cyclic peptide antagonists, such as BQ 123 (cyclo[D-Trp-D-Asp-Pro-D-Val-Leu]), has generated considerable excitement as conformational studies of these peptides would likely help to define those features necessary for receptor recognition and inhibition. While the endothelins and sarafotoxins are large, flexible molecules that apparently adopt several conformations in solution (see above), the conformationally constrained BQ 123 and similar peptides are likely to have a much more defined structure. ^{1}H NMR studies of BQ 123 (Atkinson and Pelton 1992; Krystek et al., 1992; Reily et al., 1992; Gonnella et al., 1994; Bradley et al., 1994; Verheyden et al., 1994; Bean et al., 1994), BQ 153 (Bogusky et al., 1993), and BE 18257B (Coles et al., 1993) have found that these peptides adopt a stable structure. Their backbone conformation is characterized by an inverse γ-turn involving the D-Asp-Pro-D-Val residues (or D-Cys($SO_3^-Na^+$)-Pro-D-Val in BQ 153) with a hydrogen bond between the D-Val amide proton and the D-Asp (D-Cys[$SO_3^-Na^+$]) carbonyl oxygen. Most groups (Atkinson and Pelton, 1992; Krystek et al., 1992; Gonnella et al., 1994; Coles et al., 1993; Verheyden et al., 1994) also reported the presence of a type II beta turn involving the D-Val-Leu-D-Trp-D-Asp residues. A type II beta turn was also reported by Bogusky et al. (1993) for the analog BQ 153, cyclo[D-Trp-D-Cys(SO_3-Na^+)-Pro-D-Val-Leu], and by Coles et al. (1993) for BE 18257B, cyclo[D-Glu-Ala-D-allo-Ile-Leu-D-Trp]. These studies are in agreement with previous observations (Pease and Watson, 1978; Karle, 1978) that cyclic pentapeptides with alternating D and L chirality amino acids adopt a backbone structure consisting of a type II β-turn and an inverse γ-turn. Nevertheless, the NMR study of BQ 123 by Reily et al. (1992) did not report a β-turn. These authors, basing their structure only on ^{1}H-^{1}H and ^{1}H-^{13}C coupling constants and not nOe data, argued for a somewhat different conformation. However, their data did report a low $\Delta\delta/\Delta T$ coefficient for the Asp amide proton, consistent with the presence of a β-turn.

In all of these peptides, χ_1 of D-Trp is well defined and in the *gauche*$^+$ orientation, although Gonnella et al. (1994) recently reported that the indole ring of BQ 123 can undergo a 180° flip, depending on the solvent employed. Interestingly, Verheyden et al. (1994) noted a high field shift for the Leu CγH in BQ 123 at low temperature (193 K). This results from a change in χ_1 of Leu from *gauche*$^+$ to *gauche*$^-$ and brings the leucine side chain near to the indole ring.

D. NMR Studies of Big-Endothelin

^{1}H NMR studies of pro-ET-1 (Inooka et al., 1991; Donlan et al., 1992) have examined the solution conformation of the pro-hormone to determine the effect of the extended C-terminal tail on the conformation of the endothelin part of the peptide. In addition, any structural features that may be present in and about the Trp^{21}-Val^{22} cleavage site could provide insight as to specific conformational requirements for the endothelin cleaving enzyme. Chou-Fasman secondary structural predictions of big-ET-1 suggest a high probability for a beta turn structure near this processing site, in agreement with several theories that propose such a conformational feature as part of the signaling system for endoproteases. While these NMR studies observe a well-defined structure for the endothelin part of the prohormone, residues 17-34 appear to adopt a series of conformers that are in dynamic equilibrium with no indication of tertiary structure. These results are in agreement with the CD studies noted above. Of course, in association with the processing enzyme, big-ET-1 may still adopt some preferred conformation as suggested by the results of Fabbrini et al. (1993) above, but additional studies will evidently be needed before such a structure can be assigned.

IV. FLUORESCENCE SPECTROSCOPY

A. Fluorescence Studies of ET-1

The fluorescence emission spectrum of endothelin-1 (Figure 5, panel A), is dominated by the C-terminal tryptophan ring system which has an emission maximum (λ = 351 nm) and quantum yield (Φ_{trp} = 0.099), consistent with a residue that is fully accessible to solvent. No evidence of tryptophan fluorescence quenching by either the disulfide bonds or the histidinium moiety is observed, indicating that the indole ring is not located near to these groups. The fluorescence excitation spectrum in the near and far UV is shown in Figure 5 (panel B).

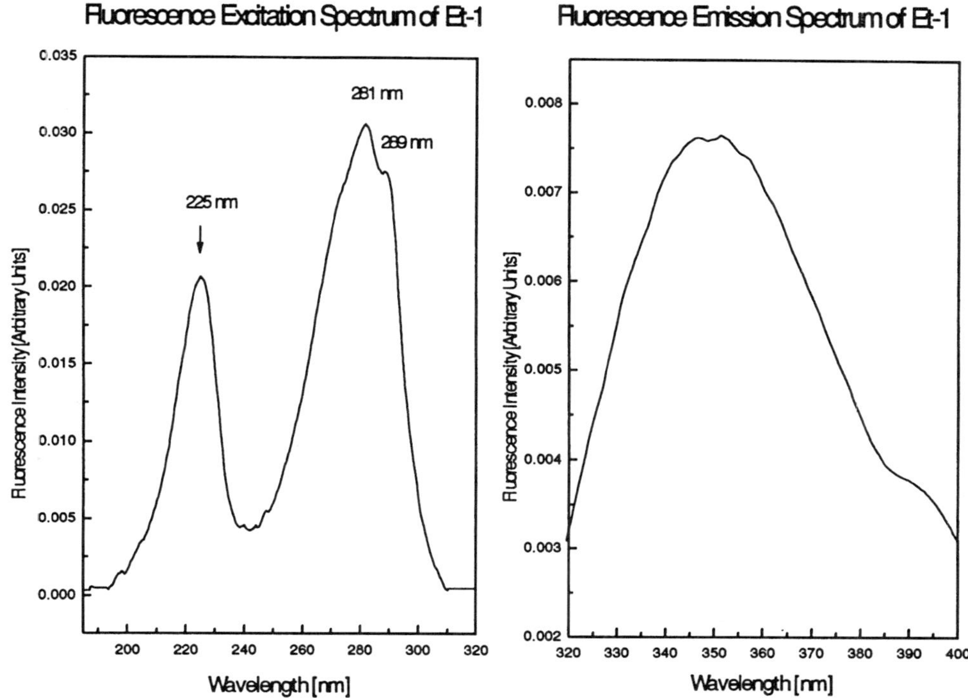

FIGURE 5
(A) Fluorescence excitation spectrum of ET-1 in the near and far UV (emission monitored at 350 nm). The far UV region is characterized by a strong band centered at 225 nm. (B) The fluorescence emission spectrum of ET-1 with excitation at 278 nm. The fluorescence emission maximum is centered at about 351 nm, indicative of an indole ring that is fully accessible to the solvent.

B. Fluorescence Energy Transfer in ET-1

A fluorescence energy transfer study of endothelin-1 in water (Pelton, 1991) indicated that there was some association of the C-terminus with the bicyclic portion of the peptide. A Förster critical distance in ET-1 of *Ro* = 11.9 ± 0.5 Å was calculated from the quantum yield of tyrosine in the absence of energy transfer using the analog [Phe^{21}]ET-1. From the ET-1 donor (Tyr^{13}) fluorescence quenching, the efficiency of energy transfer was 0.43, giving a Tyr^{13}-Trp^{21} distance of 12.5 ± 0.5 Å. The efficiency of energy transfer can also be obtained from measuring the relative enhancement of the acceptor fluorescence. A comparison of the fluorescence excitation spectra of the acceptor in the presence and absence of the donor with the UV absorption spectrum of the donor-acceptor pair gave an efficiency of transfer of 0.36 and a Tyr^{13}-Trp^{21} distance of 13.1 ± 0.6 Å. This distance is considerably shorter than the 25 Å obtained if ET-1 is modeled with a fully extended hexapeptide tail. By either method, the fluorescence energy transfer measurements indicated that the C-terminal indole ring of tryptophan is not completely free and that the side chain must bend back and associate, at least part of the time, with the bicyclic core containing the tyrosine side chain.

These results are consistent with the ^{1}H NMR studies of Bortmann et al. (1991), who also noted a close association of the C-terminus of ET-3 to the bicyclic ring of the peptide, and those of Aumelas et al. (1992) who reported a β-turn conformation for Leu^{17}-Ile^{20} in VIC, with C-terminal folding. Finally, cleavage of ET-1 by proteolytic enzymes appears to occur in two steps: (1) opening of the outer ring by cleavage of the Ser^5-Leu^6 bond, followed by (2) cleavage in the C-terminus between the Asp^{18}-Ile^{19} bond (Sokolovsky et al., 1990). This result was interpreted as an indication that ET-1 assumes a compact structure in which the C-terminal tail is not accessible to proteolytic enzymes until after the outer ring has been opened.

C. Fluorescence Anisotropy Studies

Fluorescence anisotropy measurements might be expected to provide some insight into the conformational and rotational freedom available to the C-terminal indole ring of endothelin. A recent study (Cowley and Pelton, 1995) found two rotational relaxation times for the C-terminal residue. The first component was a subnanosecond relaxation time that corresponded to a rapid, local segment conformational flux or to internal bond rotation(s). This rotation occurs at a rate comparable to that observed for tryptophan in the C-terminal hexapeptide, $ET_{16\text{-}21}$. The second rotational lifetime component was in the nanosecond range and could relate to the overall Brownian rotation rate of the whole peptide, increasing in an expected manner from about 2.0 ns for ET-1 to 3.3 ns for the 38-residue big-endothelin. An analysis of the fractional contributions of these two components in the fluorescence anisotropy decay suggested that the indole ring must be appreciably involved in conformations for which its freedom of movement was restricted. This could occur if the C-terminal tail interacted or associated, at least transiently but with a persistence of several nanoseconds, with the bicyclic core of the peptide.

In big-endothelin, the fluorescing tryptophan residue 21 in the middle of the C-terminal chain region experiences considerable conformational constraint, such that its motion was linked frequently with the slower tumbling of this larger molecule. It was suggested that the chain, involving residues 16 to 38, is loosely folded and associated with the rigid bicyclic core to form a contracted pseudo-globular peptide for appreciable periods of time. These results are in agreement with ^1H NMR studies of Donlan et al. (1992), which indicated nonrandomly distributed chain conformations for residues 17-34.

V. FT-IR AND LASER RAMAN STUDIES

A. FT-IR

Brockel and Pelton (1995) have analyzed the amide I band of ET-1 and the acyclic $[Ala^{1,3,11,15}]$ET-1 by Fourier transform infrared spectroscopy. Analysis of the spectrally deconvoluted amide I′ band of ET-1 in D_2O indicated about $16 \pm 2\%$ helical structure for this peptide in solution and a similar value ($17 \pm 5\%$) was obtained for the peptide as a lyophilized powder. While this value is somewhat lower than the 29% estimated by Harris (1993) from the CD spectrum (see Section II.B above) or that observed by NMR (about 30%), it is significantly different from that reported for the X-ray crystal structure (about 60%). The low helical content observed by FT-IR is consistent with the observation that the ends of helical segments do not contribute to the spectroscopically observed band at about 1654 cm^{-1}. This may be particularly true for ET-1, where the helical segment is not only short (~6 to 7 amino acids) but usually described in most NMR studies as somewhat irregular with considerable fraying at each end.

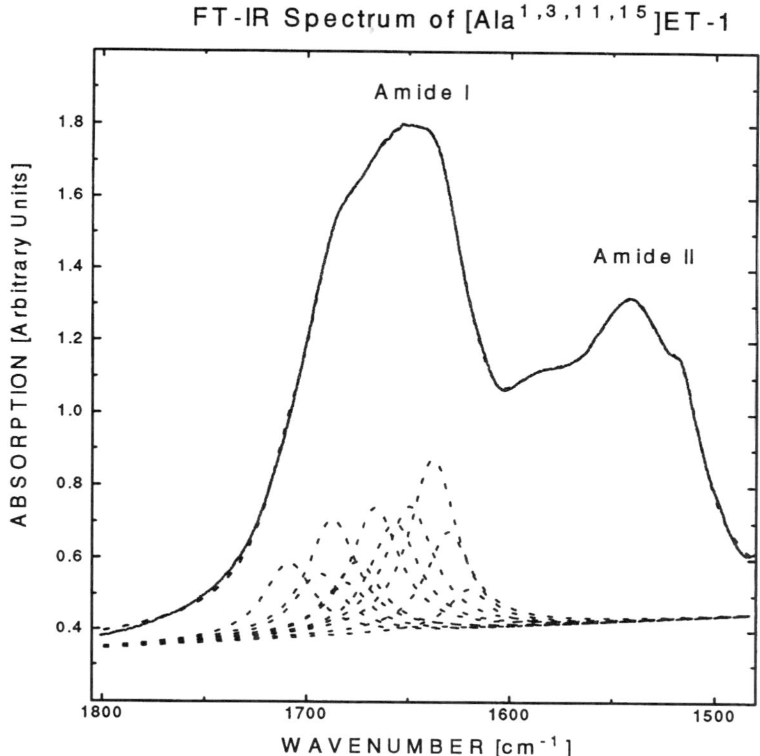

FIGURE 6
Amide I band from the FT-IR spectrum of lyophilized [Ala^{1,3,11,15}]ET-1. Curve fitting of the amide I band indicates 12 ± 2% helix, 35 ± 9% β-sheet structure, and 52 ± 5% turn and unordered structure.

The acyclic peptide [Ala^{1,3,11,15}]ET-1 behaved similarily, although the conformation was less well defined with reduced helical content in solution (13 ± 5%) and as a lyophilized powder (12 ± 2%). In contrast, no helical structure was observed in the CD spectrum of this peptide in aqueous solutions, although a clear propensity for the peptide to adopt some helical character in certain solvents (Figure 4) was noted. These results suggest that [Ala^{1,3,11,15}]ET-1 has the propensity to form a helix and that some residues may be transiently involved in such a structure. The helical content detected by FT-IR for the linear peptide may be important for the peptide's ability to recognize and bind to ET_B receptors.

B. Laser Raman

Raman spectroscopic (Brockel and Pelton, 1993) studies of the endothelins and sarafotoxins have been limited due to the poor solubility of these peptides in aqueous solutions. However, the Raman spectrum of crystalline ET-1 and lyophilized endothelins and sarafotoxins have been presented (Figure 6). Spectral deconvolution of the amide I band in the crystal structure of ET-1 resulted in major peaks at 1654 cm^{-1}, 1665 cm^{-1}, and 1672 cm^{-1}, indicative of 48 ± 8% unordered, 35 ± 6% sheet/turn, and 17 ± 5% helical structure. Nearly identical values (53 ± 5, 28 ± 8, and 19 ± 7%, respectively) were obtained for the ET-1 as a lyophilized powder (Brockel and Pelton, 1995). These results are quite similar to those obtained by FT-IR. Particularly interesting

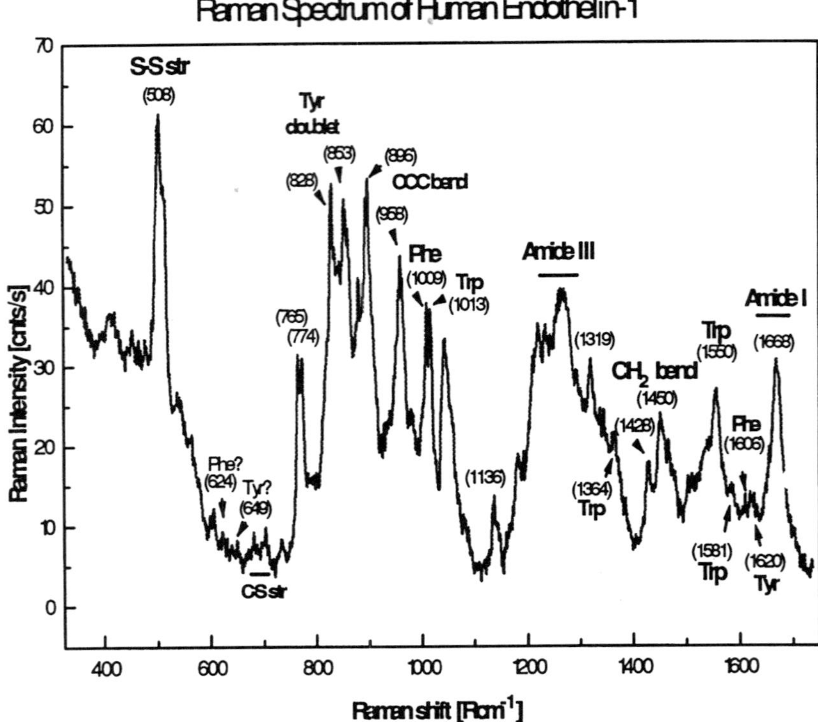

FIGURE 7
Laser Raman spectrum of crystaline ET-1. Estimation of the secondary structure is obtained from an analysis of the amide I and III bands, while other bands provide information regarding the environment of some of the side chains.

is the helical content (17 ± 5%) for the crystal structure. This value is substantially different from the approximately 60% observed in the X-ray crystal structure (see Section VII.A), even if one assumes that the ends of the helices are not contributing to the observed amide I vibrational band. These results suggest that there may be some crystal forms in which the ET-1 adopts a conformation that is closer to that deduced from the NMR and CD studies.

The tyrosine "doublet" bands at 853 and 828 cm^{-1} arise from Fermi resonance and are considered to be useful for determining the environment of the ring system. The greater intensity observed for the 828 cm^{-1} band is suggestive of a somewhat buried residue. Finally, the conformation of the disulfide bonds in endothelin may be determined from the position of the S-S stretch observed between 500 and 550 cm^{-1}. ET-1 shows a single strong band at 508 cm^{-1}, consistent with an all *gauche* conformation for both disulfide bonds.

VI. MOLECULAR MODELING

A. Molecular Modeling of ET-1

Early approaches to modeling the structure of endothelin-1 employed homology studies, secondary structure prediction programs, hydropathic considerations, and CD spectroscopic data. Spinella et al. (1989) first proposed a model of endothelin based on a combination of these methods. Their proposed model consisted of a rigid disulfide-stabilized core region containing two turns involving residues 1-4 and 8-11. No helical structure was predicted, although an extended sheet structure was suggested for the hydrophobic C-terminal "tail", possibly stabilized by intermolecular hydrogen bonding. Partition studies using a water/ether system supported an amphipathic structure for endothelin-1, as did the observation that microcrystals of ET-1 grow at the organic-aqueous interface in a two-phase system. Using atomic coordinates for apamin, Perkins et al. (1990) predicted an α-helix involving residues 9-15. Their model also appears to have a loose "turn" structure that comprises residues 5-8 and an unstructured, hydrophobic C-terminal tail.

Hempel et al. (1993) generated an *in vacuo* conformational search for low energy conformers of ET-1 using both molecular dynamics simulations and a Monte Carlo/minimization approach involving torsion angle space. Low-energy structures were then subjected to a search using an electrostatics model. The most favored conformer contained a helical stretch involving residues 9-13 with the C-terminal carboxylate and the negatively charged side chains of Asp^{8}, Glu^{10}, and Asp^{18} adjacent in space. The positively charged ε-amino group of Lys^{9} and the N-terminus are positioned well away from these groups, leading to enhanced solvation of the peptide. They concluded that the electrostatic contributions to the solvation energy of ET-1 are significant and lead to conformational preferences which are not predicted by searches using the neutral, unionized peptide.

Recently, Janes and Wallace (1994) modeled the structures of ET-2 and ET-3 based on their X-ray crystal structure of human ET-1. Only the amino acid side chains of those residues in ET-2 and ET-3 that are different from ET-1 were allowed to adopt new orientations during the energy minimization procedure. In the case of ET-2, they observed that neither substitution of the larger hydrophobic residue Trp for Leu at position 6, nor Leu for Met at position 7, in ET-2 altered the χ_1 values or the percent surface accessibility for these amino acids. In contrast, nearly all the residues in ET-3 that are different from ET-1 adopt new χ_1 angles after energy minimization. They suggested that in addition to electrostatic and steric differences in ET-3 vs. ET-1 noted by others, the difference in receptor selectivity of ET-3 may also be due to the different χ_1 rotamers employed by this peptide.

B. Molecular Modeling of BQ 123

A molecular modeling study of BQ 123 by Satoh and Barlow (1992) proposed that the conformation of this cyclic peptide antagonist was similar to that reported for residues Leu^{6}-Met^{7}-Asp^{8} of ET-1. They suggested that ET_A receptor recognition required two hydrogen bond donor sites, two nonspecific hydrophobic binding sites, and a third site bearing a positive charge. Structure-activity studies showing the importance of the tertiary structure of ET-1 for ET_A receptor binding (especially the loops arising from the two disulfide bridges) and of the Asp^{8} carboxylic acid function and its stereochemistry (L configuration) were noted as supporting this model. However, others (Miller et al., 1993) have suggested that BQ 123 may be mimicking the C-terminus of ET-1. Structure-activity studies have clearly shown the importance of Trp for receptor recognition and the need for a C-terminal carboxylic acid function. Although the cyclic BQ 123 peptide has no "C-terminus", the side chain carboxylic acid function of the adjacent D-Asp in BQ 123 could fulfill this requirement. In addition, the linear peptide antagonist PD 142893 (Ac-D-Dip-Leu-Asp-Ile-Ile-Trp), which contains the C-terminal Leu^{17}-Trp^{21} sequence of ET-1, is reported to have high affinity (15 nM) for ET_A receptors. These results clearly indicate that ET_A receptor recognition does not require the tertiary structure formed by the two disulfide bridges.

C. Modeling Antagonist Pharmacophores

Cordova et al. (1994) examined the conformational preferences of three different ET_A receptor antagonists as a means of identifying shared pharmacophores. BQ-123 (cyclo[D-Asp-Pro-D-Val-Leu-D-Trp]), the linear hexapeptide Ac-D-Dip-Leu-Asp-Ile-Ile-Trp, and a natural product steroid isolated from the bayberry *Myrica cerifera*, were overlaid and their common, shared, and excluded volumes were determined. Putative bioactive conformations were those where the common and shared volumes were maximal and the excluded volume minimal. One such conformation placed the side chain of D-Asp in BQ123, Asp3 in the linear hexapeptide, and a carboxylic acid group present in the steroid in a similar region of the pharmacophore. This orientation also matched, among others, the hydrophobic isopropyl side chain of Val^3 in BQ123 to a similar region in ring A of the steroid. This model was refined by Hempel et al. (1995) using Apex-3D to identify compact overlays followed by a search of the MDDR-3D compound database to identify molecules consistent with the defined 3D pharmacophore. This approach should prove useful as more selective and conformationally constrained analogs are developed.

VII. X-RAY CRYSTALLOGRAPHY

Crystallization conditions for ET-1 have been reported (Waller et al., 1992) and a recent X-ray crystal structure for this peptide has appeared (Janes et al., 1994). The structure is characterized by an N-terminal extended beta-strand, a turn involving residues 7-11, followed by a long and somewhat irregular helix that extends to the C-terminal tryptophan residue. While this structure contains some conformational elements reported by other spectroscopic techniques, a number of entirely new features are observed. A "bulge" between residues 5 and 7 is seen in the X-ray structure, while the reverse turn involving Ser^5-Asp^8, reported in a number of NMR studies (see Section III.B above), is absent. These differences may, however, be due to crystal contacts involving the N-terminus of one molecule with residues Ser^5-Asp^8 of another. A turn involving the carbonyl oxygen of Met^7 and the amide proton of Cys^{11} is also observed in the crystal structure. Interestingly, this turn is one of two predicted in an early model by Spinella et al. (1989). In addition, a number of hydrogen bonds involving both side chain residues and main chain amide protons in the "core" region are inferred from the X-ray structure, but are not observed in any of the NMR studies.

The most striking feature of the X-ray structure, however, is the long helix involving residues Cys^{11} to Trp^{21}. While a helix is observed by a number of spectroscopic methods, none of them indicates that the helix involves more than 50% of the residues in ET-1. Harris (1993) estimated 29% helical content for ET-1 in water based on CD studies (see Section II.C above), increasing to about 40% in the helix-forming solvent TFE. NMR studies in a variety of solvents place the helical content between these values. While less accurate, laser Raman studies of ET-1 crystals (crystallized according to the method reported by Waller et al., 1992) also indicate a modest amount (~15%) of helix present in the peptide. The X-ray structure is consistent with (1) the fluorescence energy transfer measurements involving Tyr^{13} and Trp^{21}, and (2) the fluorescence anistropy studies which indicate a partially constrained Trp^{21} indole ring. A comparison of the X-ray and NMR structures for ET-1 has recently appeared (Wallace et al., 1995).

VIII. CONCLUSION

A wide array of spectroscopic techniques have been applied towards understanding the structure of the endothelin and sarafotoxin peptides. These studies reveal the presence of some highly conserved conformational features within this family, as well as some differences that help explain the observed biological activities. In spite of this success, there remains considerable debate regarding the conformation of the C-terminal tail and its role in receptor recognition and activation/inhibition. The structural and conformational features present in ET-1 and ET-3 that are responsible for selective receptor recognition are also not well understood, nor do we completely understand the critical features present in the antagonists now emerging in the literature that permit receptor recognition without activation. Clearly, despite considerable progress in recent years, much remains to be discovered.

REFERENCES

Andersen, N.H., Chen, C., Marschner, T.M., Krstek, S.R., Jr. and Bassolino, D.A., Conformational isomerism of endothelin in acidic aqueous media: a conformational NOESY analysis, *Biochemistry*, 31, 1280-1295, 1992a.

Andersen, N.H., Cao, B. and Chen, C., Peptide/protein structure analysis using the chemical shift index method: upfield α-CH values reveal dynamic helices and α_L sites, *Biochem. Biophys. Res. Commun.*, 184, 1008-1014, 1992b.

Atkins, A.R., Ralston, G.B. and Smith, R., Conformational stability of the endothelin/sarafotoxin family of peptides, *Int. J. Peptide Protein Res.*, 44, 372-377, 1994.

Atkinson, R.A. and Pelton, J.T., Conformational study of cyclo[D-Trp-D-Asp-Pro-D-Val-Leu], an endothelin-A receptor-selective antagonist, *FEBS Lett.*, 296, 1-6, 1992.

Aumelas, A., Chiche, L., Brun, E., Mahé, E., Le-Nguyen, D., Sizun, P., Berthault, P. and Perly, B., Vasoactive intestinal contractor (VIC) peptide structure in solution as determined by proton NMR, *J. Chem. Phys.*, 89, 175-181, 1992.

Aumelas, A., Chiche, L., Mahé, E., Le-Nguyen, D., Sizun, P., Berthault, P. and Perly, B., ^{1}H NMR study of the solution structure of sarafotoxin-S6b, *Neurochem. Int.*, 18, 471-475, 1991a.

Aumelas, A., Chiche, L., Mahé, E., Le-Nguyen, D., Sizun, P., Berthault, P. and Perly, B., Determination of the structure of [Nle^7]-endothelin by ^{1}H NMR, *Int. J. Peptide Protein Res.*, 37, 315-324, 1991b.

Baumer, L., Bolis, G., Gauaragna, A., Mondelli, R., Penco, S. and Ragg, E., Determination of the conformation of endothelin in water by nuclear magnetic resonance and molecular modeling. In: *Peptides, Chemistry and Biology, Proc. 12th Am. Peptide Symp.*, Smith, J.A. and Rivier, J.E., Eds., ESCOM, Leiden, 1991.

Bean, J. W., Peishoff, C.E. and Kopple, K.D., Conformations of cyclic pentapeptide endothelin receptor antagonists, *Int. J. Peptide Protein Res.*, 44, 223-232, 1994.

Bennes, R., Calas, B., Chabrier, P.-E., Demaille, J. and Heitz, R., Evidence for aggregation of endothelin 1 in water, *FEBS Lett.*, 276, 21-24, 1990.

Bogusky, M.J., Brady, S.F., Sisko, J.T., Nutt, R.F. and Smith, G.M., Synthesis and solution conformation of cyclic[D-Trp-D-Cys(SO_3-Na^+)-Pro-D-Val-Leu), a potent endothelin-A receptor antagonist, *Int. J. Peptide Protein Res.*, 42, 194-203, 1993.

Bortmann, P., Hoflack, J., Pelton, J.T. and Saudek, V., Solution conformation of endothelin-3 by ^{1}H NMR and distance geometry calculations, *Neurochem. Int.*, 18, 491-496, 1991.

Bradley, E.K., Ng, S.C., Simon, R.J. and Spellmeyer, D.C., Synthesis, molecular modeling and NMR structure determination of four cyclic peptide antagonists of endothelin, *Bioorg. Med. Chem.*, 2, 279-296, 1994.

Brown, S.C., Donlan, M.E. and Jeffs, P.W., Structural studies of endothelin by CD and NMR. In: *Peptides, Chemistry, Structure and Biology, Proc. 11th Am. Peptide Symp.*, Rivier, J.E. and Marshall, G.R., Eds., ESCOM, Leiden, 1990, pp 595-597.

Brockel, C. and Pelton, J.T., Structural investigation of endothelin-1 and [$Ala^{1,3,11,15}$]endothelin-1 by laser Raman spectroscopy, Int. Symp. Structure and Function in Biological Systems, Würzburg, Germany, Sept. 20-23, 1993.

Brockel, C. and Pelton, J.T., Structural studies of endothelin-1 and [$Ala^{1,3,11,15}$]endothelin-1 by FT-IR and laser Raman spectroscopy, submitted for publication.

Calas, B., Harricane, M.-C., Gulmard, L., Heitz, F., Mendre, C., Chabrier, P.E. and Bennes, R., Endothelin-1: conformation and aggregation, *Peptide Res.*, 5, 97-101, 1992.

Coles, M., Munro, S.L.A. and Craik, D.J., The solution structure of a monocyclic analogue of endothelin [1,15 Aba]-ET-1, determined by ^{1}H NMR spectroscopy, *J. Med. Chem.*, 37, 656-664, 1994.

Coles, M., Sowemimo, V., Scanlon, D., Munro, S.L.A. and Craik, D.J., A conformational study by ^{1}H NMR of a cyclic pentapeptide antagonist of endothelin, *J. Med. Chem.*, 36, 2658-2665, 1993.

Cordova, T., Hassan, M., Hempel, J.C., Koerber, S.C., Vorpagel, E.R. and Hagler, A.T., Endothelin ET_A receptor antagonist pharmacophore hypotheses, in: *Peptides, Chemistry and Biology, Proc. 13th Am. Peptide Symp.*, Hodges, R.S. and Smith, J.A., Eds., ESCOM, Leiden, 1994, pp. 660-663.

Cowley, D.J. and Pelton, J.T., Solution conformational dynamics of the C-terminal residues in endothelin-1 and some analogues: a time-resolved fluorescence study, *Int. J. Peptide Protein Res.*, 46, 56-64, 1995.

Dalgarno, D.C., Slater, L., Chackalamannil, S. and Senior, M.M., Solution conformation of endothelin and point mutants by nuclear magnetic resonance spectroscopy, *Int. J. Peptide Protein Res.*, 40, 515-523, 1992.

Donlan, M.L., Brown, S.C. and Jeffs, P.W., Solution conformation of human big endothelin-1, *J. Cell. Biochem.*, S15G, 85, 1991.

Donlan, M.L., Brown, F.K. and Jeffs, P.W., Solution conformation of human big endothelin-1, *J. Biomol. NMR*, 2, 407-420, 1992.

Endo, S., Inooka, H., Ishibashi, Y., Kitada, C., Mizuta, E. and Fujino, M., Solution conformation of endothelin determined by nuclear magnetic resonance and distance geometry, *FEBS Lett.*, 257, 149-154, 1989.

Fabbrini, M.S., Vitale, A., Pedrazzini, E., Nitti, G., Zamai, M.,Tamburin, M., Caiolfa, V.R., Patrono, C., and Benatti, L., *In vivo* expression of mutant preproendothelins. Hierarchy of processing events but no strict requirement of Trp-Val at the processing site, *Proc. Natl. Acad. Sci. U.S.A.*, 90, 3923-3927, 1993.

Freeman, C.M., Catlow, C.R.A., Hemmings, A.M. and Hider, R.C., The conformation of apamin, *FEBS Lett.*, 197, 289-296, 1986.

Gonnella, N.C., Zhang, X., Jin, Y., Prakash, O., Paris, C.G., Kolossvary, I., Guida, W.C., Bohacek, R.S., Vlattas, I. and Sytwu, T., Solvent effects on the conformation of cyclo(D-Trp-D-Asp-Pro-D-Val-Leu), an NMR spectroscopy and molecular modeling study, *Int. J. Peptide Protein Res.*, 43, 454-462, 1994.

Harris, S.C., Solution Conformations of Medium Size Peptides: A Comparison of the Information Content of CD and NMR Applied to C-Peptide Analogs and Endothelins, Ph.D. thesis, University of Washington, Seattle, 1993.

Hempel, J.C., Fine, R.M., Guaragna, A., Hassan, M., Koerber, S.C. and Hagler, A.T., Effects of solvation on the conformational preferences of endothelin-1, in: *Peptides, Chemistry and Biology, Proc. 13th Am. Peptide Symp.*, Hodges, R.S. and Smith, J.A., Eds., ESCOM, Leiden, 1994, pp. 890-892.

Hempel, J.C., Cordova de Sintjago, T., Osman, G., Hassan, M., Koerber, S.C., Vorpagel, E.R. and Hagler, A.T., Pharmacophore hypotheses and peptidomimetic design strategies: endothelin ET_A antagonists, Proc. XIIIth Int. Symp. Medicinal Chemistry, *Eur. J. Med. Chem.*, S30, 391s-395s, 1995.

Hider, R.C., Kupreyszweski, G., Rekowski, P. and Lammek, B., *Biophys. Chem.*, 31, 45-51, 1988.

Inooka, H., Endo, S., Kikuchi, T., Wakimasu, M., Mizuta, E. and Fujino, M., Solution conformation of human endothelin-1 (big ET-1), *Peptide Chemistry*, 28, 409-414, 1991.

Janes, R.W., Peapus, D.H. and Wallace, B.A., The crystal structure of human endothelin, *Nat. Struct. Biol.*, 1, 311-319, 1994.

Janes, R.W. and Wallace, B.A., Modeling the structures of the isoforms of human endothelins based on the crystal structure of human endothelin-1, *Biochem. Soc. Trans.*, 22, 1037-1043, 1994

Karle, I.L., Crystal structure and conformation of cyclo[Gly-Pro-Gly-D-Ala-Pro] containing 4→1 and 3→1 intramolecular hydrogen bonds, *J. Am. Chem. Soc.*, 100, 1286-1289, 1978.

Kobayashi, Y., Sato, A., Takashima, H., Tamaoki, H., Nishimura, S., Kyogoku, Y., Ikenaka, K., Kondo, T., Mikoshiba, K., Hojo, H., Aimoto, S. and Moroder, L., A new α-helical motif in membrane active peptides, *Neurochem. Int.*, 18, 525-534, 1991.

Krystek, S.R., Jr., Bassolino, D.A., Novotny, J., Chen, C., Marschner, T.M. and Andersen, N.H., Conformation of endothelin in aqueous ethylene glycol determined by ^{1}H NMR and molecular dynamics simulations, *FEBS Lett.*, 281, 212-218, 1991.

Krystek, S.R., Jr., Bassolino, D.A., Bruccoleri, R.E., Hunt, J.T., Porubcan, M.A., Wandler, C.F. and Andersen, N.H., Solution conformation of a cyclic pentapeptide endothelin antagonist. Comparison of structures obtained from constrained dynamics and conformational search, *FEBS Lett.*, 299, 255-261, 1992.

Miller, R.C., Pelton, J.T. and Huggins, J.P., Endothelins — from receptors to medicine, *TiPS*, 141, 54-60, 1993.

Mills, R.G., Atkins, A.R., Harvey, T., Junius, F.K., Smith, R. and King, G.F., Conformation of sarafotoxin-6b in aqueous solution determined by NMR spectroscopy and distance geometry, *FEBS Lett.*, 282, 247-252, 1991.

Mills, R.G., O'Donoghue, S.I., Smith, R. and King, G.F., Solution structure of endothelin-3 determined using NMR spectroscopy, *Biochemistry*, 31, 5640-5645, 1992.

Mills, R.G., Ralston, G.B. and King, G.F., The solution structure of sarafotoxin-c, *J. Biol. Chem.*, 269, 23415-23419, 1994.

Munro, S., Craik, D., McConville, C., Hall, J., Searle, M., Bicknell, W., Scanlon, D. and Chandler, C., Solution conformation of endothelin, a potent vasoconstricting bicyclic peptide. A combined use of ^{1}H NMR spectroscopy and distance geometry calculations, *FEBS Lett.*, 278, 9-13, 1991.

Panek, R.L., Major, T.C., Taylor, D.G., Hingorani, G.P., Dunbar, J.B., Doherty, A.M. and Rapundalo, S.T., Importance of secondary structure for endothelin binding and functional activity, *Biochem. Biophys. Res. Commun.*, 183, 572-576, 1992.

Pease, J.H.B. and Wemmer, D.E., Solution structure of apamin determined by nuclear magnetic resonance and distance geometry, *Biochemistry*, 27, 8491-8498, 1988.

Pease, J.H.B., Storrs, R.W. and Wemmer, D.E., Floding and activity of hybrid sequence, disulfide-stabilized peptides, *Proc. Natl. Acad. Sci. U.S.A.*, 87, 5643-5647, 1990.

Pease, L.G. and Watson, C., Conformational and ion binding studies of a cyclic pentapeptide. Evidence for β- and γ-turns in solution, *J. Am. Chem. Soc.*, 100, 1279-1286, 1978.

Pelton, J.T., Fluorescence studies of endothelin-1, *Neurochem. Int.*, 18, 485-489, 1991.

Perkins, T.D.T., Hider, R.C. and Barlow, D.J., Proposed solution structure of endothelin, *Int. J. Peptide Protein Res.*, 36, 128-133, 1990.

Ragg, E., Mondelli, R., Penco, S., Bolis, G., Baumer, L. and Guaragna, A., Tertiary structure of Endothelin-1 in water by ^{1}H NMR and molecular dynamics studies, *J. Chem. Soc. Perkin Trans.*, 2, 1317-1326, 1994.

Reily, M.D., Thanabal, V., Omecinsky, D.O., Dunbar, J.B., Jr., Doherty, A.M. and DePue, P.L., The solution structure of a cyclic endothelin antagonist, BQ-123, based on ^{1}H-^{1}H and ^{13}C-^{1}H bond coupling constants, *FEBS Lett.*, 300, 136-140, 1992.

Reily, M.D. and Dunbar, J.B., The conformation of endothelin-1 in aqueous solution: NMR-derived constraints combined with distance geometry and molecular dynamics calculations, *Biochem. Biophys. Res. Commun.*, 178, 570-577, 1991.

Satoh, T. and Barlow, D., Molecular modeling of the structures of endothelin antagonists. Identification of a possible determinant for ET-A receptor binding, *FEBS Lett.*, 310, 83-87, 1992.

Saudek, V., Hoflack, J. and Pelton, J.T., Solution conformation of endothelin-1 by ^{1}H NMR, CD and molecular modeling, *Int. J. Peptide Protein Res.*, 37, 174-179, 1991.

Saudek, V., Hoflack, J. and Pelton, J.T., ^{1}H NMR study of endothelin, sequence-specific assignment of the spectrum and a solution structure, *FEBS Lett.*, 257, 145-148, 1989.

Sokolovsky, M., Galron, R., Kloog, Y., Bdolah, A., Indig, F.E., Blumberg, S. and Fleminger, G., Endothelins are more sensitive than sarafotoxins to neutral endopeptidase: possible physiological significance, *Proc. Natl. Acad. Sci. U.S.A.*, 87, 4702-4706, 1990.

Spinella, M.J., Krystek, S.R., Jr., Peapus, D.H., Wallace, B.A., Bruner, C., and Andersen, T.T., A proposed structural model of endothelin, *Peptide Res.*, 2, 191-286, 1989.

Tamaoki, H., Kyogoku, Y., Nakajima, K., Sakakaibara, S., Hayashi, M. and Kobayashi, Y., Conformational study of endothelins and sarafotoxins with the cystine-stabilized helical motif by means of CD spectra, *Biopolymers*, 32, 353-357, 1992.

Tamaoki, H., Kobayashi, Y., Nishimura, S., Ohkubo, T., Kyogoku, Y., Nakajima, K., Kumagaye, S.-I., Kimura, T. and Sakakibara, S., Solution conformation of endothelin determined by means of ^{1}H NMR spectroscopy and distance geometry calculations, *Protein Eng.*, 4, 509-518, 1991.

Verheyden, P., Van Assche, I., Brichard, M.-H., Demaude, T., Paye, I., Scarso, A., and Van Binst, G., Conformational study of three endothelin antagonists with ^{1}H NMR at low temperature and molecular dynamics, *FEBS Lett.*, 344, 55-60, 1994.

Wallace, B.A., Janes, R.W., Bassolino, D.A. and Krystek, S.R., A comparison of X-ray and NMR structures for human endothelin-1, *Prot. Sci.*, 4, 75-83, 1995.

Waller, D., Cudney, R., Wolff, M., Day, J., Greenwood, A., Larson, S. and McPherson, A., Crystallization and preliminary X-ray analysis of human endothelin, *Acta Crystallogr. Sect. B Struct. Sci.*, 48, 239-240, 1992.

Chapter 15

Endothelin Agonists

John T. Pelton

CONTENTS

I. INTRODUCTION

Early evidence that the presence of endothelium enhances the contraction of pig pulmonary artery was reported by Holden and McCall (1983). Subsequently, serum-free media, conditioned by endothelial cell growth, was found to contain a proteinaceous agent that induced vasocontraction (Agricola et al., 1984; Hickey et al., 1985).

0-8493-6975-4/97/$0.00+$.50

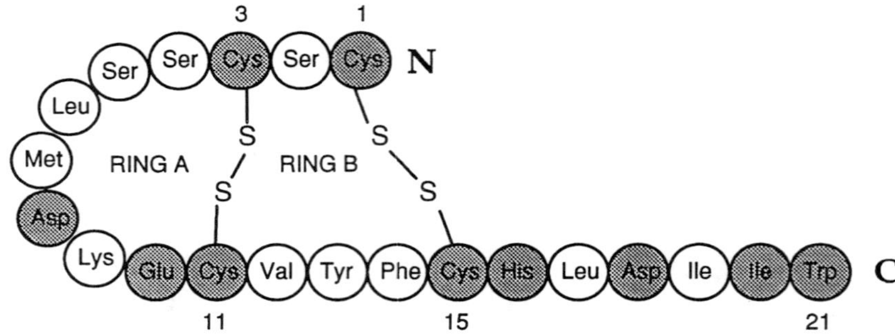

FIGURE 1
Amino acid sequence of ET-1. Shaded circles identify conserved residues within the endothelin and sarafotoxin peptide family.

Finally, in 1988, Yanagisawa et al. reported the isolation and characterization of a peptide, which they termed endothelin (ET), from cultured porcine aortic endothelial cells that produced a potent and long-lasting vasoconstriction.

ET is a 21-amino-acid peptide (Figure 1) characterized by the presence of two disulfide bridges involving cysteines at positions 1-15 and 3-11. The two rings that are formed are nearly identical in size, having 28 and 29 atoms each (the former being defined by the two disulfide bonds). As with many peptides, ET is the product of proteolytic cleavage of an inactive propeptide termed big endothelin, which is in turn generated from a much larger precursor protein (see Chapter 6). From a search of the mammalian genome (Inoue et al., 1989), encoding sequences of two additional endothelin-like peptides were obtained. These peptides, termed ET-2 and ET-3 (the original ET was renamed ET-1), were found to have similar overall structure, but slightly different amino acid sequences and biological activities.

The endothelins are also related in structure to a family of five peptides isolated from the venom of two burrowing asps, *Atractaspis engaddensis* and *Atractaspis bibroni*. These 21-amino-acid peptides, termed sarafotoxins (from the Israeli *A. engaddensis*) or bibrotoxin (from the South African *A. bibroni*), have disulfide bonds at positions identical to those of the endothelins but are distinguished from them at positions 5, 6, 7, and 17 (see Table 4). Sarafotoxin 6a and 6b (STXa and STXb) are only marginally different in biological activity from ET-1, whereas STXc and STXd are much weaker vasoconstrictors (see Section III below). Bibrotoxin is most similar in sequence to sarafotoxin STXb and has comparable receptor affinity and *in vitro* activity.

A novel peptide homologous to ET-1, vasoactive intestinal contractor (VIC), was identified by Southern blot analysis of mouse genomic DNA. This peptide differs from ET-1 at positions 4, 6, and 7, but its disulfide bond pairing (Saida et al., 1989) and overall conformation (Aumelas et al., 1992) is the same as the endothelins. VIC appears to be expressed only in the intestine and was not found in other tissues or endothelial cells. However, the peptide can evoke a strong contractile response in guinea-pig ileum similar to that of endothelin-1 (Ishida et al., 1989).

II. ENDOTHELIN PEPTIDES

A. Early Observations

Shortly after the first report of the isolation and characterization of endothelin, Nakajima et al. (1989a) reported a structure-activity relationship (SAR) study of this peptide. Their work revealed that three regions of ET-1 were sensitive to modification:

(1) the N-terminus; (2) the C-terminus; and (3) the bicyclic system with the 1-15, 3-11 disulfide bond pairing.

1. *N-Terminus*

N-terminal extension, as in Lys-Arg-ET-1, is about 500-fold less potent than ET-1 itself in rat pulmonary artery ring preparations, indicating that correct processing of the prohormone was important for endothelin activity. N-terminal acetylation and alkylations are also poorly tolerated and results in a several-hundredfold decrease in vasoconstrictor activity (Table 1). N^{α}-Ac-ET-1 retained only 0.5% the activity of the parent peptide in a rat pulmonary artery ring preparation (Nakajima et al., 1989a), 1% of ET-1 activity in the isolated rabbit renal artery (Johansen et al., 1990), and had low micromolar affinity for ET receptors on A10 vascular smooth muscle cell membranes (Hunt et al. 1991). Magazine et al. (1991) reported that the contractile activity of acetylated ET-1 analogs in carotid artery strips showed the following order: ET-1 $\gg$ [Ac-N^{ε}-Lys^{9}]ET-1 > [Ac-N^{α}-Ser^{1}]ET-1 > [Ac-N^{α}-Ser^{1},Ac-N^{ε}-Lys^{9}]ET-1. Similarly, a dicarba analog of ET-1, in which the terminal amino group was completely missing, was reported as inactive in the same assay (Nakajima et al., 1989a).

2. *C-Terminus*

In general, modifications of the C-terminal carboxylic acid function are poorly tolerated. Endothelin amide is about 20-fold less active than ET-1 in both functional and receptor binding assays while esters have greatly reduced activity or are inactive. C-terminally extended peptides also display poor activity. Porcine big-ET-1 (ET 1-39) has only 1 to 2% the constrictor activity of the mature peptide and porcine big-$ET\text{-}1_{1\text{-}25}$ is also weakly active. These results again clearly demonstrate the importance of correct processing of the prohormone with cleavage at the Trp^{21}-Val^{22} bond.

3. *Disulfide Bonds*

Studies of ET-1 in which the cysteines were protected with the acetamidomethyl (Acm) group suggested that the bicyclic system (and presumably the constrained conformation of the peptide) was important for biological activity. $[Cys(Acm)^{3,11}]$ET-1, $[Cys(Acm)^{1,15}]$-ET-1, and $[Cys(Acm)^{1,3,11,15}]$ET-1 were all found to be inactive in a rat pulmonary artery ring preparation. Similarly, reduction of the two disulfide bonds followed by carboxamidomethylation (CAM) results in peptides with very weak partial agonist activity relative to ET-1 in a porcine coronary artery assay (Kimura et al., 1988).

It was noted early on that the location of the cysteine residues at positions 1, 3, 11, and 15 in endothelins is identical to that found in apamin, a small neurotoxic peptide that is a component of honey bee venom. In apamin, however, the pairing of the disulfide bonds is 1-11 and 3-15 compared to 1-15 and 3-11 found in the endothelin/sarafotoxin family. In spite of the differences in disulfide linkages, ^{1}H NMR studies indicate that apamin and the endothelins both form a helix-like structure that involves residues 9-16. However, the pharmacology of apamin is substantially different from that of the endothelins indicating that a compact bicyclic core containing a helical segment is not of itself sufficient to elicit endothelin activity.

B. Structure-Activity Relationships

1. *Native Sequences*

Early biological studies of the endothelins quickly revealed that ET-3 had the weakest vasoconstrictor activity within the family of naturally occurring peptides, prompting a comparison of amino acid sequences for clues regarding the residues responsible. Bdolah et al. (1989) initially ascribed the decrease in constrictor activity of ET-3 to the substitution of threonine for serine at position 2, by analogy with the equally weak constrictor peptides STXc and STXd. However, from a study of chimera peptides, Takasaki et al., (1991) demonstrated that this substitution was not responsible for the weak vasoconstrictor activity. Lamthanh et al. (1994) came to substantially similar conclusions, although they noted that substitution of threonine for serine at position 2 could decrease the vasoconstrictor activity in some systems.

Most of the remaining variable residues were observed to cluster within the 29-atom ring system at positions 4, 5, 6, and 7. While serine occurs at position 4 in both ET-1 and ET-2, ET-3 has the aromatic amino acid residue phenylalanine. Serine is again present at position 5 in ET-1 and ET-2, while ET-3 contains the somewhat more hydrophobic residue threonine. Several variations are present at position 6, leucine in ET-1, tryptophan in ET-2, and tyrosine in ET-3. Finally, at position 7, a hydrophobic residue occurs in both ET-1 (methionine) and ET-2 (leucine), but the strongly basic amino acid lysine appears at this position in ET-3. From these observations, it was suggested that the net charge within the Cys^3-Cys^{11} loop may be important. However, STXd also possesses extremely low activity and has the same charge distribution within the Cys 3-11 ring as the more potent STXb. This indicates that the net charge in this region cannot be the only cause for decreased activity.

2. *ET Analogs*

Alanine scans of ET-1 (a series of endothelin analogs containing sequential single-point alanine mutations) have provided useful information regarding the side chain requirements at each position within the peptide (de Castiglione et al., 1992; Tam et al., 1994). In a rabbit vena cava assay, L-alanine could substitute for residues 2, 4, 5, 6, 7, 9, 16, and 19 with little or no loss in contractile activity. Similar results were reported by Nakajima et al. (1989b) in a rat pulmonary artery ring preparation: residues 4, 5, 6, 7, and 9 could be substituted or modified with little loss in smooth muscle constricting activity. For example, the methionine at position 7 in ET-1 can be replaced by Gly, Ala, βAla, Nle, N^{ε}-biotinylaminocaproyl-Lys or oxidized to the more hydrophilic sulfoxide, all without loss of potency or activity (Table 1). Even multiple substitutions at these "tolerant" positions could be made. Hunt et al. (1991), reported that $[Ala^{3,4,5,6,7,11}]$ET-1 had an EC_{50} of 36 nM (Table 2) in a rabbit carotid artery ring assay.

In contrast, positions 8 and 10 were found to be sensitive to modification. $[Ala^8]$ET-1 was reported to be 125-fold less active that ET-1 itself, and $[Ala^{10}]$ET-1 was found to be a partial agonist, both in the rabbit vena cava assay. Even the $[Asn^8]$ET-1 and $[Gln^{10}]$ET-1 analogs, where the only change is a transformation of the side-chain acid to an amide function, were found to be nearly inactive (Nakajima et al., 1989a; Watanabe et al., 1991). As these results would predict, introduction of a positively charged residue, as in $[Lys^{10}]$ET-1, also results in an analog with greatly reduced activity. Taken together, these studies indicate that the negatively charged carboxylic acid functions at positions 8 and 10 are critical. In contrast, the positively charged N^{ε}-amino group of the lysine residue at position 9 is not important. Both $[Ala^9]$ET-1 and $[Leu^9]$ET-1 were reported to be full agonists. Even substitution of lysine with a negatively charged residue as in $[Glu^9]$ET-1 resulted in fully active analogs.

Valine is conserved in both VIC and the endothelin family at position 12 and a neutral, hydrophobic residue appears to be important. Thus, the less hydrophobic $[Ala^{12}]$ET-1 analog is also less potent than ET-1 whereas $[Leu^{12}]$ET-1 is quite active.

TABLE 1

Endothelin and Monosubstituted ET-1 Analogs

	EC_{50} [nM]			
Peptide	Rat Pulmonary Artery Rings	Rabbit Vena Cava	ET_A Receptor Binding^{+} [nM]	ET_B Receptor Binding^{+} [nM]
ET-1	0.68/0.65	0.289/0.94/0.5*	0.008/0.16/0.47	0.83/0.008/0.015
ET-2	1.5/1.1		0.024/0.22	1.1/0.11
ET-3	40.6/39.6		8.0/5.7	1.1/0.11/0.09
VIC	48$^{\pm}$			
D-ET-1$^{\#}$	>1000		>1000	
ET-1 amide	11.7/12	1.9/3.8	1.4/2.9/13	
N^{α}-Acetyl-ET-1	130	36/30	21/13	
Lys-Arg-ET-1	365			
Gly-ET-1		30	25	
[Ala^{2}]ET-1	2.6	0.41/38	0.47	
[D-Ser^{2}]ET-1		290	470	
[Thr^{2}]ET-1		0.85	0.75	
[D-Thr^{2}]ET-1		481	470	
[Ala^{4}]ET-1	1.5/1.3	0.56	0.25	
[Ala^{5}]ET-1	2.68	1.17	0.3	
[D-Ser^{5}]ET-1		0.63	0.82	
[Gly^{6}]ET-1	0.83			
[Ala^{6}]ET-1	0.9	0.4	0.47	
[Pro^{6}]ET-1	90			
[D-Leu^{6}]ET-1		0.5	0.37	
[Met(O)^{7}]ET-1	0.94			
[D-Met^{7}]ET-1		1.4	0.31	
[Ala^{7}]ET-1		0.32*	0.13	
[Nle^{7}]ET-1	0.8/1.3	0.44/0.3	0.1/0.13	
[Ser^{7}ET-1		1.4	0.65	
[Hse^{7}]ET-1		0.94	0.09	
[Leu^{7}]ET-1		0.3	0.29	
[Asn^{8}]ET-1	81	6	3.5	
[Ala^{8}]ET-1	170	36	0.47	
[Gly^{8}]ET-1	230			
[Lys^{8}]ET-1	Inactive			
[D-Asp^{8}]ET-1	120	48	17	
[Ala^{9}]ET-1	0.8	0.51	0.09	
[Glu^{9}]ET-1	1.6			
[Leu^{9}]ET-1	1.2			
[D-Lys^{9}]ET-1		0.069	0.22	
[Ala^{10}]ET-1		0.23 (p.a.)	0.29	
[Lys^{10}]ET-1	Inactive			
[Gln^{10}]ET-1	Inactive			
[D-Glu^{10}]ET-1		0.54	0.31	
[Ala^{12}]ET-1	Inactive	1.0	3.1	
[D-Val^{12}]ET-1	0.81	1.3	1.7	
[Pro^{12}]ET-1		27.8*	2.1/3.6	
[Ala^{13}]ET-1	6.5	0.3 (p.a.)	58	
[D-Tyr^{13}]ET-1	3	3	5.9	
[Phe^{13}]ET-1	1.02			
[Ala^{14}]ET-1	Inactive	2	28/100	
[D-Phe^{14}]ET-1	Inactive	20 (p.a.)	21	
[Ala^{16}]ET-1		0.076	0.14	
[Phe^{16}]ET-1	0.21	1.5*	0.3/0.16	
[Pro^{16}]ET-1		>900	>1000	
[D-His^{16}]ET-1	Inactive	290	157	
[D-Phe^{16}]ET-1		>300	>1000	
[des-His^{16}]ET-1		>900	>1000	

TABLE 1 (continued)

Endothelin and Monosubstituted ET-1 Analogs

Peptide	EC_{50} [nM] Rat Pulmonary Artery Rings	EC_{50} [nM] Rabbit Vena Cava	ET_A Receptor Binding$^+$ [nM]	ET_B Receptor Binding$^+$ [nM]
[Ala^{17}]ET-1	80	32	1.9	
[Gly^{17}]ET-1	Inactive			
[D-Leu^{17}]ET-1	65	290	4	
[Ala^{18}]ET-1		2.1	0.9	
[Asn^{18}]ET-1	7		148	
[D-Asp^{18}]ET-1		18	29	
[Ala^{19}]ET-1	3	0.23	0.12	
[D-Ile^{19}]ET-1		≈100	940	
[des-Ile^{19}]ET-1		>900	>1000	
[Trp^{20}]ET-1			19.5	0.17
[D-Trp^{21}]ET-1	72	74	20	
[Ala^{20}]ET-1		1.3	2.1/30	0.4
[D-Ile^{20}]ET-1		≈6 (p.a.)	26	0.40
[D-Val^{20}]ET-1			1	0.89
[Asp^{21}]ET-1	>1000		>1000/	
[Ala^{21}]ET-1	>500/>1000	>300/298	89/530	
[Tyr^{21}]ET-1	2.1/28		0.56	2.65
[Phe^{21}]ET-1	3.5/41	2.9	3.5/4.7	16.5
[His^{21}]ET-1	400		2.8	13.3
[Gly^{21}]ET-1	>1000		5	23.6
[Ser^{21}]ET-1	>1000		2	9.4
[Lys^{21}]ET-1	>1000		280	425
[Ile^{21}]ET-1				16.6
[Glu^{21}]ET-1				5500
[Pro^{21}]ET-1				Inactive
[Nal^{21}]ET-1			1.7	
[des-Trp^{21}]ET-1	>1000		>300	

Note: (+) inhibition of [^{125}I]ET-1 binding; data restricted to those studies where responses could be assigned to one of the two endothelin receptors with some degree of confidence; (*) rabbit pulmonary artery (ET-1 EC_{50} = 5.5 nM); (±) porcine coronary artery strips (ET-1 EC_{50} = 35 nM); (#) endothelin peptide composed of all D-amino acids; p.a. - partial agonist.

Nearly all conformational studies of endothelin have featured a helical region involving residues 9 to 15 (see Chapter 14). To probe the importance of the helix for ET activity, Panek et al. (1992b) substituted proline for valine at position 12, which would normally introduce a "kink" in the middle of the helix and disrupt the secondary structure. Surprisingly, [Pro^{12}]ET-1 exhibited high-affinity binding to rabbit pulmonary artery, rat aorta, and rat left atria, similar to that of the native hormone, although it had somewhat reduced activity in functional assays. This result would suggest that a helix itself is not a feature critical for either receptor recognition or activation. Subsequently, however, Andersen et al. (1995) found receptor binding affinity of some monocyclic ET-1 analogs did correlate with inducible helix length. They concluded that the helix-stabilizing effect of the two disulfide bonds present in ET-1 may have been sufficient to maintain the helix despite the Pro^{12} substitution.

Substitution of alanine for tyrosine in [Ala^{13}]ET-1 gave an analog with ET-3 like activity in the rabbit vena cava: weak constriction at very low concentrations (10 pM to 1 nM) but full agonist effect at high concentrations (>10 nM). In the rat pulmonary artery ring preparation, [Ala^{14}]ET-1 was inactive and only weakly active in the isolated rabbit renal artery (Johansen et al., 1990) and rabbit vena cava (Tam et al., 1994). These results stress the importance of aromatic residues at positions 13 and

14. Although tyrosine is conserved within VIC and the endothelin family at position 13, the phenolic hydroxyl group of this residue is not essential.

Within the tail region of endothelin, the substitution of histidine at position 16 with its optical isomer substantially decreased its vasoconstrictor activity in rat pulmonary artery preparations, whereas the removal of the basic imidazole ring altogether, as in [Ala^{16}]ET-1, increased both the binding affinity to human vascular smooth muscle cells and contractile activity in the rabbit vena cava assay threefold. Similarly, Panek et al. (1992b) reported that [Phe^{16}]ET-1 was more potent than ET-1 in both binding affinity and contractile activity and was about tenfold more selective for the ET_A receptor. Leu^{17}, and to a lesser extent Ile^{19}, along with the previously reported Trp^{21} position, were sensitive to alanine substitution with 0.9, 22.5, and <0.1% ET-1 potency in the rabbit vena cava, respectively (EC_{50} for ET-1 = 0.289 nM). These results are in agreement with those found by Watanabe et al. (1991) where substitution of glycine for leucine at position 17 resulted in a significant reduction in potency in the rat pulmonary arterial rings. At position 18, substitution of asparagine for aspartic acid results in a substantial loss in binding affinity to ET_A receptors and prompted Koshi et al. (1992) to suggest that this residue may be involved in hydrogen bonding to the ET_A receptor.

The nature of the side chain at the C-terminus of endothelin also plays an important role in the vasoconstricting activity of this peptide. Substitution of alanine, which maintains a chiral amino acid at the C-terminus but eliminates the indole ring, results in an inactive analog. Charged amino acids are not tolerated at the C-terminus, as [Asp^{21}]ET-1 and [Lys^{21}]ET-1 are more than 10,000-fold less active than the parent peptide. Introduction of other aromatic groups such as phenylalanine, tyrosine, or naphthylalanine leads to analogs with approximately the same vasoconstricting activity as the native hormone. These last results indicate that the indole nitrogen of tryptophan is not a critical element for biological activity, and indeed, [N-formyl-Trp^{21}]ET-1 is also active.

While an alanine scan can provide useful information regarding the importance of the side chain group at each position, D-amino acid scans have yielded information regarding the stereochemical requirements at each position within the peptide. Galantino et al. (1992) observed that inversion of configuration within the Cys$^{3\text{-}11}$ loop was generally well tolerated, as judged by activity in the rabbit vena cava assay, with the exception of the [D-Asp^{8}]ET-1 analog. This peptide had only 0.6% the activity of ET-1. Interestingly, [D-Lys^{9}]ET-1 was reported to be four times as active at ET-1. Within the Cys$^{1\text{-}15}$ loop, Ser^{2} and the two aromatic residues D-Tyr^{13} and D-Phe^{14} were less potent, with 0.1, 9.5, and 1.5% the activity of ET-1, respectively. All the residues within the C-terminal "tail" region were quite sensitive to inversion of the side chain configuration, the most active being only a partial agonist (22%) [D-Ile^{20}]ET-1, with but 5% the potency of ET-1 itself. Koshi et al. (1992) reported that [D-Ile^{20}]ET-1 was a weak antagonist at the ET_A receptor. Inversion of configuration at the C-terminus, as in [D-Trp^{21}]ET-1, reduces activity in the rabbit vena cava assay (Kimura et al., 1988; Galantino et al., 1992).

3. *Disulfide Bond Modifications*

Galantino et al. (1992) also examined the effect of D-amino acid substitution on disulfide bond formation. Oxidation of the cysteine residues by air oxidation gave exclusively the natural Cys$^{1\text{-}15}$, Cys$^{3\text{-}11}$ isomer for [D-Ser^{2}]ET-1, [D-Lys^{9}]ET-1, and [D-Val^{12}]ET-1, while [D-Ser^{4}]ET-1 gave only the inverted [Cys$^{1\text{-}11}$, Cys$^{3\text{-}15}$] isomer. This last result clearly shows the dependence of disulfide isomer formation on the linear endothelin sequence and that oxidation of tetrathiol analogs may not necessarily give the natural disulfide pairing.

TABLE 2

Disulfide Bond Modified Endothelin Analogs

Peptide	Carotid Artery Rings ED_{50} [nM]	Guinea-Pig Trachea ED_{50} [nM]	Receptor Binding[a] K_d [nM]
ET-1	0.68/0.94	1.36	0.15/0.2
[Cys(Acm)^{3,11}]ET-1	>1000		
[Cys(Acm)^{1,15}]ET-1	>1000		
[Cys(Acm)^{1,3,11,15}]ET-1	>1000		
[Cys^{1,11}, Cys^{3,15}]ET-1	88		
[Cys^{1,3}, Cys^{11,15}]ET-1	180		
[Ala^{3}, Nle^{7}, Aib^{11}]ET-1	1.5		2.0
[Pen^{1,11}, Nle^{7}]ET-1	7.5 pa		4.5
[Pen^{3,11}, Nle^{7}]ET-1	0.9		
[Pen^{3,15}, Nle^{7}]ET-1	0.9		0.7
[Pen^{1,15}, Nle^{7}]ET-1	>1000		400
[Pen^{1,15}, Nle^{7}]ET-1*	880*		87*
[Ala^{1,15}]ET-1	170	45	
[Ala^{3,11}]ET-1	2.3	10	11
[Ala^{3,11}, Nle^{7}]ET-1	3.3		2.1
[Ala^{2,3,11}, Nle^{7}]ET-1	38		9.6
[Ala^{3,4,5,6,7,11}]ET-1	36		130
[Ala^{3,8,11}, Nle^{7}]ET-1	~1000		3074
[Ala^{3,10,11}, Nle^{7}]ET-1	>1000		188
[Ala^{3,11,12}, Nle^{7}]ET-1	130		31
[Ala^{3,11,13}, Nle^{7}]ET-1	>1000		3920
[Ala^{3,11,14}, Nle^{7}]ET-1	>1000		646
[Ala^{3,11,16}, Nle^{7}]ET-1	31		52
[Ala^{3,11,17}, Nle^{7}]ET-1	⩾1000		
[Ala^{3,11,18}, Nle^{7}]ET-1	⩾1000		
[Ala^{3,11,19}, Nle^{7}]ET-1	182		96
[Ala^{3,11,20}, Nle^{7}]ET-1	>1000		2109
[Ala^{3,11,21}, Nle^{7}]ET-1	⩾1000		>20,000
[Ala^{3}, Nle^{7}, D-Ala^{11}]ET-1	6.0		76
[Ala^{3}, Nle^{7}, Pro^{11}]ET-1	170		730
[Ala^{3,11,17}, Nle^{7}]ET-1	>1000		930
[Ala^{1,3,11,15}]ET-1	Inactive	270	~2000

Note: [a] ET_A receptor; pa = partial agonist; *disulfide isomer is 1-11, 3-15. Data from Nakajima et al. (1989a,b); Pelton and Miller, (1991); Hunt et al., (1991, 1992, 1993); Huggins et al., (1993); Andersen et al. (1995).

Recently, Hunt, et al. (1992) used the known tendency of penicillamine (β,β-dimethyl cysteine) to form mixed disulfides, to produce disulfide isomers of ET-1. Random air oxidation of the mixed tetrathiols [Pen^{1,11}, Nle^{7}]ET-1 and [Pen^{3,15}, Nle^{7}]ET-1 produced the 1-15, 3-11 isomer in >20:1 ratio. [Pen^{1,11}, Nle^{7}]ET-1 was found to be a partial agonist in rabbit carotid artery ring preparations with an EC_{50} of 7.5 nM, while the [Pen^{3,15}, Nle^{7}]ET-1 analog was a full agonist (EC_{50} = 0.9 nM). This work was subsequently extended (Andersen et al., 1995) to some monocyclic [Ala^{3}, Nle^{7}, Xaa^{11}]ET-1 analogs. Using (1) the known helix-promoting α-aminoisobutyric acid (Aib) residue, and (2) the helix-breaking amino acid proline at position 11, they observed that receptor binding affinity correlated with inducible helix length, as detected by circular dichroism. This is in contrast to the report by Panek et al. (1992b) who also employed proline as a helix disrupter, but at position 12. In this latter case, however, the helical structure may well have been retained due to the stabilizing Cys-X-X-X-Cys/Cys-X-Cys motif (Tamaoki et al., 1992) despite the presence of the proline residue.

Nakajima et al. (1989a) reported that incorrect crosslinking of the disulfide bonds in ET-1 resulted in only a reduction in vasoconstricting potency, not inactivation. However, reduction of the disulfide bonds followed by alkylation or chemical synthesis of ET-1 analogs employing the HF-stable acetamidomethyl protection on the cysteine side chain leads to compounds with markedly reduced biological activity, as noted above (Section II.A). However, substitution of the cysteines with pseudo-isosteric alanine residues (Topouzis et al., 1989; Pelton et al., 1990; de Castiglione et al., 1991; Galantino et al., 1992) gave ET analogs that still retain biological activity. For example, $[Ala^{3,11}]$ET-1 and $[Ala^{1,15}]$ET-1 both display high affinity for rat cerebellum membranes (Pelton et al., 1990) and are full agonists in rat isolated aorta preparations (Topouzis et al., 1989). The linear endothelin peptide $[Ala^{1,3,11,15}]$ET-1 was reported to bind with high affinity to rat cerebellar homogenates (Hiley et al., 1990), but was neither an agonist nor an antagonist in isolated rat aortic ring preparations (Topouzis et al., 1991). Of the two disulfide bonds, results from both the alanine analogs and the Cys(Acm)-substituted endothelin peptides indicate that the $Cys^{1,15}$ bridge is the more critical linkage. Sokolovsky et al. (1990) reported that enzymatic cleavage of the ET-1 on the carboxy side of Ser^5 with bovine kidney neutral endopeptidase (NEP) resulted in a monocyclic peptide with biological activity comparable to the intact peptide.

Based on previous structure-activity studies, Spinella et al. (1991) synthesized $[Dpr^1, Asp^{15}]$ET-1. In this peptide, the outer disulfide bond is replaced with an amide linkage (Dpr = diaminopropionic acid). The outer loop was chosen for replacement as this bridge is more important in maintaining endothelin activity than is the Cys^3-Cys^{11} bridge and may play a more direct role in ET activity than just structural support. Dpr was employed at position 1 to retain the N-terminal amino function, which previous studies had shown was important for biological activity. In this analog, an additional methylene group was inserted between the N-terminus and the alpha carbon and the number of bridging atoms was reduced from 4 to 3. In a competition binding assay to rat pulmonary artery smooth muscle cells in culture, $[Dpr^1, Asp^{15}]$ET-1 had an EC_{50} of 2 nM (EC_{50} for ET-1 was 0.05 nM in the same binding assay). This result appears reasonable in view of the solution conformation of $[Aba^{1,15}]$ET-1, also a moncyclic analog of ET-1 (Aba = α-amino butyric acid). A ^1H NMR study of this peptide indicated that the removal of the disulfide bridge between residues 1 and 15 did not cause a major structural change in either the helical or turn regions (Coles et al., 1994). This is in contrast to the analog reported by Nakajima et al. (1989a), in which the Cys^1-Cys^{15} bridge was also replaced by an amide link, but found to be inactive. However, in this peptide the critical N-terminal amino group is missing, the second disulfide bond was eliminated, and the cysteines were replaced by alanine.

C. Endothelin Fragment Peptides

As noted above, early SAR studies of ET-1 identified the C-terminal region as important due to (1) the highly conserved nature of these residues within the ET/STX family, and (2) the critical nature of the terminal tryptophan residue to biological activity.

Maggi et al. (1989) reported that a peptide comprising the final six residues of ET-1, ET-1[16-21], was a full agonist in the guinea-pig bronchus assay with an EC_{50} of 228 nM, but devoid of any activity in the rat thoracic artery preparation. This

TABLE 3

Endothelin Fragment Peptides

Peptide	Receptor Binding IC_{50} [nM]*	Guinea-Pig Bronchus [nM]	Rat Pulmonary Artery Ring [nM]
ET-1	0.008/0.11	1.6(3.1+)	0.68
ET-3	0.008/0.07	11.5 (13+)	40
N-Terminal Deletions			
[Asu^{1,15},Ala^{3,11}]ET-1			Inactive
[Ala^{11,15}]ET-1[6-21]	0.32		
[Ala^{11,15}]ET-1[6-20]	22,000		
[Ala^{7,11,15}]ET-1[7-21]	0.10		
[Ala^{11,15}]ET-1[8-21]	0.083/1.2		
Suc-[Ala^{11,15}]ET-1[8-21]	0.083		
[Glu^{9},Ala^{11,15}]ET-1[8-21]	0.083		
Suc[Glu^{9},Ala^{11,15}]ET-1[8-21]	0.016	(28+)	
Suc-[Ala^{9,11,15}]ET-1[8-21]	0.054		
[Ala^{8,11,15}]ET-1[8-21]	5.0		
Ac-[Ala^{11,15}]ET-1[10-21]	12		
[Cys^{11,15}]ET-1[11-21]			
[Ala^{11,15}]ET-1[11-21]	140/1200		
Suc-[Ala^{11,15}]ET-1[11-21]	2		
[Ala^{15}]ET-1[12-21]	>500		
Suc-[Ala^{15}]ET-1[12-21]	4		
[Ala^{15}]ET-1[13-21]	>500		
[Ala^{15}]ET-1[14-21]	>500		
ET-1[16-21]	> 50,000/>10,000	228	>2,000
ET-1[16-21]-amide		>100,000	
Ac-ET-1[16-21]	43,000		
Ac-[D-His^{16}]ET-1[16-21]	9,100		
Ac-[Phe^{16}]ET-1[16-21]	>50,000		
Ac-[D-Phe^{16}]ET-1[16-21]	2,000		
[Lys^{16}]ET-1[16-21]		>100,000	
[Gln^{17}]ET-1[16-21]		>100,000	
[Asn^{18}]ET-1[16-21]		>100,000	
[Pro^{19}]ET-1[16-21]			
[Val^{19}]ET-1[16-21]		1,230	
[D-Trp^{21}]ET-1[16-21]		>100,000	
ET-1[17-21]		25,000	
ET-1[18-21]		>100,000	
C-Terminal Deletions			
ET-1[1-20]	>30,000±	>10,000	3,000/inactive
ET-1[1-19]			10,000
ET-1[1-16]	Inactive		>30,000
ET-1[1-15]	Inactive		>2,000
ET-1[1-15]-amide	>30,000±	Inactive	>2,000

Note: *Inhibition of [^{125}I]ET-3 binding to either rat cerebellum or porcine lung membranes (principally ET_B receptors); ± inhibition of [^{125}I]ET-1 binding to rabbit cardiac tissue; + ET_{30} guinea-pig trachea.

finding suggested the presence of at least two endothelin receptor subtypes, which were provisionally termed ET_A (aorta) and ET_B (bronchus) and that ET-1[16-21] was a selective agonist for this latter receptor [NOTE: this definition of ET_A and ET_B receptors should not be confused with the current definition based on the presence or absence of selectivity in ET-1 and ET-3 binding]. However, Tschirhart et al. (1991) noted that this peptide failed to compete with [^{125}I]ET-1 at endothelin binding sites on guinea-pig tracheal epithelium, submucosa, or smooth muscle, even at concentrations as high as 10,000 nM, casting some doubt that the hexapeptide was binding to endothelin receptors. Subsequently, Panek et al. (1992a) reported that in the rabbit pulmonary artery, a tissue rich in ET_B receptors (current definition of ET receptors), the IC_{50} for ET-1[16-21] binding was 71,000 nM. This finding was followed by reports that N^{α}-acetyl-endothelin[16-21] had an IC_{50} of >50,000 nM for binding to ET_B receptors in the rat cerebellum (Cody et al. 1992a), while in an isolated rabbit renal artery preparation, Johansen et al. (1990) reported that the C-terminal hexapeptide ET-1[16-21] was inactive but had binding affinities to rabbit cardiac tissue in excess of 30,000 nM. The poor affinity of ET-1[16-21] for endothelin receptors, as opposed to the nanomolar activity reported by Maggi et al. (1989), continued to fuel speculation that the C-terminal hexapeptide was not acting via an endothelin receptor in the guinea-pig bronchus preparation.

Nevertheless, these early reports stimulated considerable interest in endothelin fragment peptides, particularly as structure-activity studies of endothelin's C-terminal hexapeptide in the guinea-pig isolated bronchus assay revealed aspects in common with the SAR of ET-1 (Rovero et al., 1990). For example, while ET-1[16-21] had an EC_{50} of 228 nM in this tissue, ET-1[16-21]-amide was much less potent, as was the [D-Trp^{21}] analog. These results mirror those found for the same modifications made to ET-1 itself. Of several analogs reported by Rovero et al. (1990) [Val^{19}]ET-1[16-21] and ET-1[17-21] were found to be full agonists when compared with ET-1[16-21], while shorter peptides lost activity very rapidly. Interestingly, [Gln^{17}]ET-1[16-21], which corresponds to the hexapeptide tail of STX-c, was found to be inactive in their guinea-pig isolated bronchus assay.

In 1991, Doherty et al. reported IC_{50} values of 74,000, 71,000, and 44,000 nM for ET-1[16-21] binding to rabbit aorta, rabbit pulmonary artery, and rat heart ventricles, respectively. They noted that in these assays the hexapeptide Ac-D-His-Leu-Asp-Ile-Ile-Trp was the most active with affinities of about 4,000 nM to each tissue. However, none of these short hexapeptide analogs demonstrated either agonist or antagonist effects at concentrations to 30,000 nM in the rabbit isolated pulmonary artery or on ET-1-induced inotropic responses in the rat isolated left atria, again feeding speculation that these peptides may not be acting at ET receptors.

Finally, in a second-messenger assay measuring IP_3 accumulation, the hexapeptides with D-amino acids at position 16 were reported as ET receptor antagonists (Doherty et al., 1992). This work was extended to include a number of unusual aromatic amino acids at position 16 leading to more potent ET receptor antagonists. Spellmeyer et al. (1993a) reported an extensive SAR study of the hexapeptide Ac-D-Phe-Orn-Asp-Ile-Ile-Trp-OH (binding affinity to ET_A receptors: IC_{50} <800 nM) in which each amino acid was systematically replaced with one of 50 alternative residues. As expected, only peptides containing L-Trp derivatives at the C-terminus were active. Their work is similar to that of Cody et al. (1992b) and is more fully discussed in Chapter 16.

III. SARAFOTOXINS/BIBROTOXIN

The sarafotoxins and bibrotoxin are members of a family of bicyclic peptides isolated from the venom of the burrowing asp. The sarafotoxin peptides were first isolated by Takasaki et al. (1988) from *Atractaspis engaddensis* while bibrotoxin was isolated from the related asp, *Atractaspis bibroni*. Both venoms are extremely toxic and lead to a strong vasoconstriction of the coronary vessels that is accompanied by changes in the electrocardiogram associated with atrioventricular block and cardiac arrest (Weiser et al., 1984). Gel filtration of the crude sarafotoxin venom on Sephadex

TABLE 4

Amino Acid Sequences of Sarafotoxins and Bibrotoxin

	1				5					10					15					20	
ET-1	C	S	C	S	S	L	M	D	K	E	C	V	Y	F	C	H	L	D	I	I	W
STXa	C	S	C	K	D	M	T	D	K	E	C	L	N	F	C	H	Q	D	V	I	W
STXb	C	S	C	K	D	M	T	D	K	E	C	L	Y	F	C	H	Q	D	V	I	W
STXc	C	T	C	N	D	M	T	D	E	E	C	L	N	F	C	H	Q	D	I	V	W
STXd[a]	C	T	C	K	D	M	T	D	K	E	C	L	Y	F	C	H	Q	G	I	I	W
BTX	C	S	C	A	D	M	T	D	K	E	C	L	Y	F	C	H	Q	D	V	I	W

[a] The sequence of STXd originally published by Bdolah et al. (1989) was subsequently revised (Lamthanh et al., 1994). Given here is the revised sequence, which is the same as "STXe" reported by Ducancel et al. (1993).

G-50 yielded six fractions, with the major cardiotoxic activity located in fraction 6 (STX). DEAE-cellulose chromatography of this fraction, followed by reverse phase HPLC, ultimately gave four peptides of this family denoted as sarafotoxin: STXa, STXb, STXc, and STXd. The sequences for these peptides are given in Table 4 (Takasaki et al., 1988).

A. Structure-Activity Relationships

1. *Native Sequences*

The sarafotoxins are most distinguished from the endothelins at residues 5, 6, 7, and 17, where the residues in the sarafotoxins are invariably aspartic acid, methionine, threonine, and glutamine, respectively. Most of these distinguishing substitutions occur within the inner loop formed by the disulfide bond between Cys^{3} and Cys^{11}. Within the sarafotoxin family, STXc and STXd show low potency in vasoconstriction, are weakly toxic in animals, but have high binding affinity to receptors in rabbit aorta and brain (Bdolah et al., 1989; Sokolovsky, 1992). The sequence of STXd, originally described by Bdolah et al. (1989), was reanalyzed and found to be identical to STXe reported by Ducancel et al. (1993). In this report, STXe/d are identified as STXd.

2. *STX Analogs*

Like ET-3, whose pharmacology is different from the other endothelins, STXc and STXd contain a threonine for serine substitution at position 2. Bdolah et al. (1989) suggested that the low toxicity and potency in vasoconstriction assays and weak receptor binding affinity of sarafotoxin STXc might be due to the Ser^{2} to Thr^{2} substitution. This was substantiated, in part, by the synthesis of [Thr^{2}]sarafotoxin STXb, whose potency in eliciting contractions in both endothelium-denuded rat aortic rings and isolated perfused mesentery was indeed less than the parent STXb peptide (Takasaki et al. 1991). However, the major cause of the low activities of sarafotoxin STXc was ascribed to the Lys^{9} to Glu^{9} substitution. [Glu^{9}] STXb was completely inactive in both the rat aortic ring and perfused mesentery at 10^{-7} M, although it was only eightfold less lethal than STXb when injected into mice. The authors also examined the asparagine for lysine substitution at position four of STXc. STXb and [Asn^{4}] STXb elicited similar vasoconstrictor activity in both the rat aortic ring and perfused mesenteric artery and had similar binding affinities to rat aortic membranes. These results are consistent with those found for the endothelins, where substitution of Ser^{4} by alanine or other residues has little effect on vasoconstrictor activity (Watanabe et al., 1991).

TABLE 5

Sarafotoxin Structure-Activity Relationships

Peptide	Rat Aorta [nM]	Rat Mesentery [nM]	Rabbit Aorta [nM]	[^{125}I]ET-1 Binding [nM]† ET$_A$	ET$_B$
ET-1	12	0.27		0.16/0.19$^{\pm}$	0.83
				0.47	0.015
				0.11$^{\approx}$	0.015$^{\approx}$
Sarafotoxins					
STXa	30	3.5	3.5	9.4	5.5
				9.7$^{\pm}$	
STXb	5	0.3	6.4	0.95/0.81$^{\pm}$	2
			7.9	18	3.8
				0.24$^{\approx}$	0.013$^{\approx}$
STXc	>100	>100	18	>200	25
				>5000$^{\approx}$	0.016$^{\approx}$
				1220$^{\pm}$	
STXd				80*	
[Thr^2]STXb	12	1.2	39	3.0$^{\pm}$	
[Asn^4]STXb	5	0.3		0.75$^{\pm}$	
[Glu^9]STXb	>100	>100		115$^{\pm}$	3.5
[Lys^9]STXc					22
[Arg^{17}]STXd					1.7
[Glu^{17}]STXd					13
STXb [5-9]		Inactive			
STXb [1-19]		Inactive			
STXb [1-18]		Inactive			
STXb [8-21]		Inactive			
BTX					3.2#

Note: * Inhibition of [^{125}I]STXb, ED$_{50}$ STXb = 80 nM; ± Inhibition of [^{125}I]STXb, IC$_{50}$ STXb = 0.81 nM; # K$_i$ value, in the same assay K$_i$ for STXb was 4.2 nM; †K$_i$ values, ET$_A$ (rat aorta), ET$_B$ (rat cerebellum); ≈ IC$_{50}$ value. Data taken from: Takasaki et al., 1991; Huggins et al., 1993; Mahé et al., 1992; Kitazumi et al., 1990, Panet et al., 1992, Douglas and Hiley, 1990; Saeki et al., 1991; Williams et al., 1991, Nakamichi et al., 1992; Topouzis et al., 1989; Bdolah et al., 1989; Lamthanh et al., 1994; Becker et al., 1992.

The glutamine for leucine substitution in the otherwise invariant C-terminal tail is also noteworthy. However, one would not expect this substitution to greatly alter the steric bulk of the C-terminal region and, although glutamine is capable of forming a hydrogen bond via the side-chain amide functionality, it is also quite hydrophobic. In addition, STXb, which contains a glutamine residue at position 19, is only marginally different in activity from ET-1 at both ET$_A$ and ET$_B$ receptors.

The report (Maggi et al., 1989) that the C-terminal hexapeptide of endothelin discriminates between different endothelin receptors has led to a study of the C-terminal region of the sarafotoxins (Doherty et al., 1991). [^{125}I]ET-1 binding to receptors in rabbit aorta, rabbit pulmonary artery, and rat heart ventricle could be inhibited (IC$_{50}$ ≈ 160 μM) by the sarafotoxin C-terminal hexapeptide His-Gln-Asp-Val-Ile-Trp. This peptide was twofold less potent, however, than the corresponding ET[16-21] fragment, which was largely attributed to the Gln^{17} residue. Interestingly, at concentrations up to 3×10^{-5} M, none of the C-terminal hexapeptide fragments showed either agonist or antagonist effects in the rabbit isolated pulmonary artery or on ET-1-induced ionotropic responses in the rat isolated left atria.

Among the toxins, STXc is unique in its selectivity for ET$_B$ receptors. STXc has micromolar binding affinity to endothelin receptors in both the rat aorta and rat atria

(ET_A receptors), but low nanomolar affinity to ET_B receptors in rat cerebellum and hippocampus (Williams et al., 1991). In contrast, STXb and ET-1 displayed high affinity to all receptors. In second-messenger assays, STXc was a potent activator of turnover in rat hippocampus but not in rat atria, while sarafotoxin STXb was equipotent in both tissues. From the structure-activity studies noted above, it would appear that the substitution of glutamic acid for lysine at position 9 in STXc is largely responsible for the observed selectivity.

IV. RECEPTOR-SELECTIVE PEPTIDES

Since the classification of endothelin receptors as either ET_A or ET_B, considerable effort has been made to determine those residues and structural features that are important for selectively binding to and activating/inhibiting individual endothelin receptor subtypes.

The linear endothelin peptide [$Ala^{1,3,11,15}$]ET-1 was reported to bind with high affinity to rat cerebellar homogenates (Hiley et al., 1990) but was neither an agonist nor an antagonist in isolated rat aortic ring preparations (Topouzis et al., 1991). This led to the suggestion that [$Ala^{1,3,11,15}$]ET-1 may be an ET_B receptor-selective agonist (Saeki et al., 1991) — subsequently supported by studies of the peptide in the anesthetized rat (Bigaud and Pelton, 1992). Takai et al. (1992) reported the development of linear ET-1 fragment analogs that are potent and specific agonists at the ET_B receptor. [$Ala^{11,15}$]ET-1[8-21] was reported to have good selectivity for ET_B receptors in porcine lung membranes with K_i [ET_A] of 1,100 nM, K_i [ET_B] of 0.083 nM, and a selectivity ratio of about 13,000. When the aspartic acid residue at position 8 was replaced with alanine ([$Ala^{8,11,15}$]ET-1[8-21]) affinity for the ET_B receptor decreased 60-fold, indicating that the carboxylic acid group at position 8 may be important for binding to the ET_B receptor. Selectivity was also increased when lysine at position 9 was replaced by alanine or glutamic acid. Within this series, IRL 1620, (Suc-[Glu^9, $Ala^{11,15}$]ET-1[8-21]) was the most potent and selective ET_B agonist peptide, with K_i for ET_A = 1,900 nM; K_i for ET_B = 0.016 nM, and a selectivity of 120,000 in binding to porcine lung membrane receptors. In the isolated guinea-pig trachea preparation, IRL 1620 induced contractions in a concentration-dependent manner with an EC_{30} of 28 nM (EC_{30} is the concentration of peptide that gives 30% of the effect of 60 mM KCl, which for ET-1 = 3.1 nM and ET-3 = 13 nM in this same preparation). Recently, [^{125}I]IRL 1620 has been suggested for use as an ET_B-selective radioligand.

Williams et al. (1991) reported that STXc, like ET-3, was a more potent inhibitor of [^{125}I]ET-1 binding in the rat hippocampus and cerebellum than in the rat atria and aorta, and suggested that STXc is also an ET_B-selective agonist. Although both ET-3 and STXc contain a threonine residue at position 2, previous SAR studies have shown that this residue alone does not result in ET_B receptor selectivity and that other residues within the bicyclic portion of the molecule are likely to be responsible for receptor differentiation. Watanabe et al. (1991) noted that ET-3 and [Ala^{13}]ET-1 were much more potent in the rat tracheal ring preparation than in the pulmonary artery, while de Castiglione et al. (1992) found that [Ala^{13}]ET-1 displayed a biphasic dose-response curve in the rabbit vena cava assay similar to that of ET-3. Hunt et al. (1991) observed that substituting alanine for valine at position 12 in the monocyclic peptide [$Ala^{3,11}$]ET-1 resulted in a peptide that could discriminate well between two populations of receptors, while Panek et al. (1992a) reported that substitution of proline for valine at position 12 in ET-1 resulted in a peptide with greater ET_B receptor selectivity. This last result indicates that ET_B receptor recognition and activation may not require a helical conformation in this region (see, however, Section II.B above).

This is supported by circular dichroism studies (Pelton and Miller, 1990) of [$Ala^{1,3,11,15}$]ET-1, which is also a potent and selective ET_B receptor ligand (Hiley et al., 1990; Saeki et al., 1991; Panek et al., 1992a). Unlike ET-1, which contains a shoulder at 222 nm, consistent with some helical character present within the peptide, no such structural element is present in the linear analog.

Saeki et al. (1991) also studied the requirements for ET_B receptor selectivity through the preparation of a series of linear peptides. Unlike ET-1, where N-terminal extension or alkylation resulted in a substantial decrease in vasoconstrictor activity, N^{α}-acetyl-[$Ala^{1,3,11,15}$]ET-1 was as potent and selective as [$Ala^{1,3,11,15}$]ET-1 in binding to porcine cerebelum membranes, indicating that a free amino terminus is not a requirement for ET_B receptor binding. In addition, deletion of the five N-terminal residues still gave peptides that retained good ET_B receptor affinity and selectivity. Further deletions, however, resulted in a prompt decrease in receptor binding, indicating the importance of Asp^8, Lys^9, and especially Glu^{10} for ET_B receptor recognition. As expected, a terminal tryptophan (or similar) residue was required for ET_B receptor affinity.

Similar results were obtained by Takai et al. (1992) in a series of N^{α}-succinyl peptide analogs of [$Ala^{1,3,11,15}$]ET-1. They also found that deletion of amino acids from the N-terminus caused a gradual reduction in binding affinity to ET_B receptors. Within this series, Suc[$Ala^{11,15}$]ET-1[10-21] was the shortest peptide that still retained high binding and selectivity for ET_B receptors, supporting the previous report of the importance of the Glu^{10} residue. ET_B receptor selectivity could also be increased when Lys^9 was replaced by alanine or glutamic acid. Similarly, Jarvis et al. (1994) reported that BQ3020 (N-Ac-[$Ala^{11,15}$]ET-1[6-21]) was a potent and selective ET_B receptor agonist.

Watanabe et al. (1991) found that substitution of histidine at position 16 in ET-1 with its optical isomer substantially decreased its vasoconstrictor activity in rat pulmonary artery, but tracheal contracting activity was three times that of ET-1. Galantino et al. (1992) reported that this same analog had markedly reduced binding affinity to human vascular smooth muscle cells and decreased contractile activity on the rabbit vena cava. Removal of the basic imidazole ring of histidine as in [Ala^{16}]ET-1, however, resulted in a threefold increase in binding affinity and contractile activity in these same assays. Subsequently, Panek et al. (1992b) reported that [Phe^{16}]ET-1 displayed increased ET_A receptor selectivity and potency in both functional and binding assays.

Finally, nonpeptidic ET agonists have recently been reported by Rivero et al. (1995). From a screening of their sample collection, they identified a bicyclic diacid system (8,9-dicarboxyldibenzo[2,3:5,6] bicyclo[5.2.0]nonan-4-one) with moderate (micromolar) affinity for both ET_A and ET_B receptors. Structure-activity studies further improved the activity and provided analogs having nanomolar affinity for ET receptors with a range of receptor selectivity. Conformational analysis of these and similarly constrained molecules should provide new insight into the features required for receptor recognition and activation.

V. CONCLUSION

The studies outlined above clearly indicate that ET_A and ET_B receptors recognize and/or bind to different regions of the endothelin molecule. In addition, there are probably conformational differences in the requirements for ET_A and ET_B receptor recognition. These results supported early efforts in the design of receptor-selective agonists and antagonists that have greatly aided our understanding of endothelin pharmacology.

REFERENCES

Agricola, K.M., Rubanyi, G., Paul, R.J. and Highsmith, R.F., Characterization of a potent coronary artery vasoconstrictor produced by endothelial cells in culture, *Fed. Proc.*, 43, 899, 1984.

Andersen, N.H., Harris, S.M., Lee, V.G., Liu, E.C.-K., Moreland, S. and Hunt, J.T., The receptor binding affinity of monocyclic [Ala3, Xaa[11]]endothelin-1 analogs correlates with inducible helix length, *Bioorg. Med. Chem.*, 3, 113-124, 1995.

Aumelas, A., Chiche, L., Brun, E., Mahé, E., Le Nguyen, D., Sizun, P., Berthault, P., and Perly, B., Vasoactive intestinal contractor (VIC) peptide structure in solution as determined by proton NMR, *J. Chem. Phys.*, 89, 175-181, 1992.

Bdolah, A., Wollberg, Z., Fleminger, G. and Kochva, E., SRTX-d, a new native peptide of the endothelin/sarafotoxin family, *FEBS Lett.*, 256, 1-3, 1989.

Becker, A., Dowdle, E.B., Hechler, U., Kauser, K., Donner, P. and Schleuning, W.-D., Bibrotoxin, a novel member of the endothelin/sarafotoxin peptide family, from the venom of the burrowing asp *Atractaspis bibroni*, FEBS Lett., 315, 100-103, 1993.

Bigaud, M. and Pelton, J.T., Discrimination between ET_A- and ET_B-receptor-mediated effects of endothelin-1 and [Ala[1,3,11,15]]endothelin-1 by BQ-123 in the anesthetized rat, *Br. J. Pharmacol.*, 107, 912-918, 1992.

Cody, W.L. and Doherty, A.M., The development of potent peptide agonists and antagonists for the endothelin receptors, *Biopolymers*, 37, 89-104, 1995a.

Cody, W.L., Doherty, A.M., He, J.X., Topliss, J.G., Haleen, S.J., LaDouceur, D., Flynn, M.A., Hill, K.E. and Reynolds, E.E., Structure-activity relationships in the C-terminus of endothelin-1 (ET-1). The discovery of potent antagonists, in: *Peptides 1992*, Proc. Eur. Peptide Symp., Schneider, C.H. and Eberle, A.N. Eds., ESCOM, Leiden, 1992, 687-688.

Cody, W.L., Doherty, A.M., He, J.X., DePue, P.L., Rapundalo, S.T., Hingorani, G.A., Major, T.C., Panek, R.L., Dudley, D.T., Haleen, S.J., LaDouceur, D., Hill, K.E., Flynn, M.A. and Reynolds, E.E., Design of a functional hexapeptide antagonist of endothelin, *J. Med. Chem.*, 35, 3301-3303, 1992b.

Cody, W.L., He, J.X., DePue, P., Rapundalo, S.T., Hingorani, G.P., Dudley, D.T., Hill, K.E., Reynolds, E.E. and Doherty, A.M., Structure-activity relationships in a series of monocyclic endothelin analogues, *Bioorg. Med. Chem. Lett.*, 4, 567-472, 1994.

Cody, W.L., He, J.X., DePue, P.L., Waite, L.A., Leonard, D.M., Sefler, A.M., Kaltenbronn, J.S., Haleen, S.J., Walker, D.M., Flynn, M.A., Welch, K.M., Reynolds, E.E. and Doherty, A.M., Structure-activity relationships of the potent combined endothelin-A/endothelin-B receptor antagonist Ac-D-Dip[16]-Leu-Asp-Ile-Ile-Trp[21]: development of endothelin-B receptor selective antagonists, *J. Med. Chem.*, 38, 2809-2819, 1995b.

Coles, M., Munro, S.L.A. and Craik, D.J., The solution structure of a monocyclic analogue of endothelin [1,15 Aba]-ET-1, determined by ^{1}H NMR spectroscopy, *J. Med. Chem.*, 37, 656-664, 1994.

de Castiglione, R., Galantino, M., Giordano, P., Frigerio, R., Cristiani, C. and Vaghi, F., Endothelin synthetic analogs, in: *Peptides 1992*, Proc. Eur. Peptide Symp., Schneider, C.H. and Eberle A.N., Eds., ESCOM, Leiden, 1992, 681-682.

de Castiglione, R., Tam, J.P., Liu, W., Zhng, J.-W., Galantino, M., Bertolero, F. and Vaghi, F., Alanine scan of endothelin, in: *Peptides, Chemistry and Biology*, Proc. 12th Am. Peptide Symp., Smith, J.A. and Rivier, J.E., Eds., ESCOM, Leiden, 1992, 402-403.

Doherty, A.M., Cody, W.L., Leitz, N.L., DePue, P.L., Taylor, M.D., Rapundalo, S.T., Hingorani, G.P., Major, T.C., Panek, R.L. and Taylor, D.G., Structure-activity studies of the C-terminal region of the endothelins and the sarafotoxins, *J. Cardiovasc. Pharmacol.*, 17(S7), S59-S61, 1991.

Doherty, A.M., Endothelin: a new challenge, *J. Med. Chem.*, 35, 1493-1508, 1992.

Ducancel, F., Matre, V., Dupont, C., Lajeunesse, E., Wollberg, Z., Bdolah, A., Kochva, E., Boulain, J.C. and Menez, A., Cloning and sequence analysis of cDNAs encoding precursors of sarafotoxins. Evidence for an unusual rosary-type organization, *J. Biol. Chem.*, 268, 3052-3055, 1993.

Galantino, M., de Castiglione, R., Tam, J.P., Liu, W., Zhang, J.-W., Cristiani, C. and Vaghi, F., D-Amino acid scan of endothelin, *Proc. 12th American Peptide Symposium*, Smith, J.A. and Rivier, J.E., Eds., ESCOM, Leiden, 1992, 404-405.

Graur, D., Bdolah, A., Wollberg, Z. and Kochva, E., Homology between snake venom sarafotoxins and mammalian endothelins, *Isr. J. Zool.*, 35, 171-175, 1988.

Hasegawa, K., Hirata, M., Koshi, T., Ohshima, T., Miyashita, Y., Sasaki, S. and Okabe, T., Quantitative structure-activity relationships study of endothelin-1 analogs, *Bioorg. Med. Chem. Lett.*, 4, 1157-1160, 1994.

Hickey, K.A., Rubanyi, G., Paul, R.J. and Highsmith, R.F., Characterization of a coronary vasoconstrictor produced by cultured endothelial cells, *Am. J. Physiol.*, 248, C550-C556, 1985.

Hiley, C.R., Jones, C.R., Pelton, J.T. and Miller, R.C., Binding of [^{125}I]-endothelin-1 to rat cerebellar homogenates and its interactions with some analogs, *Br. J. Pharmacol.*, 1010, 319-324, 1990.

Holden, W.E. and McCall, E., Hypoxia-induced contractions of porcine pulmonary artery strips depend on intact endothelium, *Exp. Lung Res.*, 7, 101-112, 1983.

Huggins, J.P., Pelton, J.T. and Miller, R.C., The structure and specificity of endothelin receptors: their importance in physiology and medicine, *Pharmacol. Ther.*, 59, 55-123, 1993.

Hunt, J.T., Lee, V.G., Liu, E. C.-K., Moreland, S., McMullen, D., Webb, M.L. and Bolgar, M., Control of peptide disulfide regioisomer formation by mixed cystein-penicillamine bridges, *Int. J. Peptide Protein Res.*, 42, 249-258, 1993.

Hunt, J.T., Lee, V.G., Liu, E.C.-K., Moreland, S., McMullen, D., Webb, M.L. and Bolgar, M., Control of peptide disulfide regioisomer formation by mixed cysteine-penicillamine bridges, in: *Peptides 1992*, Proc. Eur. Peptide Symp., Schneider, C.H. and Eberle A.N., Eds., ESCOM, Leiden, 1992, 399-400.

Hunt, J.T., Lee, V.G., Stein, P.D., Hedberg, A., Liu, E. C.-K., McMullen, D. and Moreland, S., Structure-activity relationships of monocyclic endothelin analogs, *Bioorg. Med. Chem.*, 1, 33-38, 1991.

Inoue, A., Yanagisawa, M., Kimura, S., Kasuya, Y., Miyauchi, T., Goto, K. and Masaki, T., The human endothelin family: three structurally and pharmacologically distinct isopeptides predicted by three separate genes, *Proc. Natl. Acad. Sci. U.S.A.*, 86, 2863-2867, 1989.

Ishida, N., Tsujioka, K., Tomoi, M., Saida, K. and Mitsui, Y., Differential activities of two distinct endothelin family peptides on ileum and coronary artery, *FEBS Lett.*, 247, 337-340, 1989.

Jarvis, M.F., Assal, A.A. and Gessner, G., Pharmacological characterization of the rat cerebellar endothelin$_B$ (ET_B) receptor using the novel agonist radioligand [^{125}I]BQ3020, *Brain Res.*, 665, 33-38, 1994.

Johansen, N.L., Lundt, B.F., Madsen, K., Olsen, U.B., Suzdak, P.D., Phogersen, H. and Weis, J.U., Structure-activity relationships of endothelin analogs, in: *Peptides 1990*, Proc. 21st Eur. Pep. Symp., Giralt, E. and Andreu, D., Eds., ESCOM, Leiden, 1990, 680-681.

Kimura, S., Kasuya, Y., Sawamura, T., Shinmi, O., Sugita, Y., Yanagisawa, M., Goto, K. and Masaki, T., Structure-activity relationships of endothelin: importance of the C-terminal moiety, *Biochem. Biophys. Res. Commun.*, 156, 1182-1186, 1988.

Kitazumi, K., Shiba, T., Nishiki, K., Furukawa, Y., Takasaki, C. and Tasaka, K., Structure-activity relationship in vasoconstrictor effects of sarafotoxins and endothelin-1, *FEBS Lett.*, 260, 269-272, 1990.

Kloog Y. and Sokolovsky, M., Similarities in mode and sites of action of sarafotoxins and endothelins, *TIBS*, 10, 212-214, 1989.

Koshi, T., Suzuki, C., Arai, K., Mizoguchi, T., Torii, T., Hirata, M., Ohkuchi, M. and Okabe, T., Syntheses and biological activities of endothelin-1 analogs, *Chem. Pharm. Bull.*, 39, 3061-3063, 1991.

Koshi, T., Suzuki, C., Ohshima, T., Hirata, M. and Yokota, K., Structure-activity relationships on the C-terminal region of endothelin-1, *Annu. Rep. Tohoku Coll. Pharm.*, 39 201-208, 1992.

Kumagaye, I., Kuroda, H., Nakajima, K., Watanabe, T.X., Kimura, T., Masaki, T. and Sakakibara, S., Synthesis and disulfide structure determination of porcine endothelin: an endothelium-derived vasoconstricting peptide, *Int. J. Peptide Protein Res.*, 32, 519-526, 1988.

Lamthanh, H., Bdolah, A., Creminon, C., Grassi, J., Menez, A., Wollberg, Z. and Kochva, E., Biological activities of [Thr^{2}Sarafotoxin-b], a synthetic analogue of sarafotoxin-b, *Toxicon*, 32, 1105-1114, 1994.

Leonard, D.M., Reily, M.D., Dunbar, J.B., Holub, K.E., Cody, W.L., Hill, K.E., Welch, K.M., Glynn, M.A., Reynolds, E.E. and Doherty, A.M., Structure-activity and biophysical studies of the C-terminal hexapeptide of endothelin, *Bioorg. Med. Chem. Lett.*, 5, 967-972, 1995.

Magazine, H.I., Malik, A.B., Goligorsky, M.S., Bruner, C.A. and Andersen, T.T., Site specific biotinylation of endothelin-1. Potential use for characterization of endothelin-receptor populations, in: *Peptide, Chemistry and Biology*, Proc. 12th Am. Peptide Society, Smith, J. A. and Rivier, J. E. Eds., 1991, ESCOM, Leiden 1992, 427-428.

Maggi, C.A., Giuliani, S., Patacchini, R., Santicioli, P., Rovero, P., Giachetti, A. and Meli, A., The C-terminal hexapeptide, endothelin-(16-21) discriminates between different endothelin receptors, *Eur. J. Pharmacol.*, 166, 121-122, 1989.

Miller, R.C., Pelton, J.T. and Huggins, J.P., Endothelin — from receptors to medicine, *TiPS*, 14, 54-60, 1993

Morita, A., Nomizu, M., Okitsu, M., Horie, K., Yokogoshi, H. and Roller, P.P., D-Val^{22} containing human big endothelin-1 analog, [D-Val^{22}]Big ET-1[16-38], inhibits the endothelin converting enzyme, *FEBS Lett.*, 353, 84-88, 1994.

Nagase, T., Mase, T., Fukami, T., Hayama, T., Fujita, K., Niiyama, K., Takahashi, H., Kumagai, U., Urakawa, Y., Nagasawa, Y., Ihara, M., Nishikibe, M. and Ishikawa, K., Linear peptide ET_A antagonists: rational design and practical derivatization of N-terminal amino- and imino-carbonylated tripeptide derivatives, *Bioorg. Med. Chem. Lett.*, 5, 1395-1400, 1995.

Nakajima, K., Kubo, S., Kumagaye, S.-I., Nishio, H., Tsunemi, M., Inui, T., Kuroda, H., Chino, N., Watanabe, T.X., Kimura, T. and Sakakibara, S., Structure-activity relationship of endothelins: importance of charged groups, *Biochem. Biophys. Res. Commun.*, 163, 424-429, 1989a.

Nakajima, K., Kumagaye, S., Nishio, H., Kuroda, H., Watanabe, T.X., Kobayashi, Y., Tamaoki, H., Kimura, T. and Sakakibara, S., Synthesis of endothelin-1 analogues, endothelin-3, and sarafotoxin STXb: structure-activity relationships, *J. Cardiovasc. Pharmacol.*, 13 (S5), S8-S12, 1989b.

Nakamichi, K., Ihara, M., Kobayashi, M., Saeki, T., Ishikawa, K. and Yano, M., Different distribution of endothelin receptor subtypes in pulmonary tissues revealed by the novel selective ligands BQ-123 and [Ala^{1,3,11,15}]ET-1, *Biochem. Biophys. Res. Commun.*, 182, 144-150, 1992.

Panek, R.L., Major, T.C., Hingorani, G.P., Doherty, A.M., Taylor, D.G. and Rapundalo, S.T., Endothelin and structurally related analogs distinguish between endothelin receptor subtypes, *Biochem. Biophys. Res. Commun.*, 183, 566-571, 1992a.

Panek, R.L., Major, T.C., Taylor, D.G., Hingorani, G.P., Dunbar, J.B., Doherty, A.M. and Rapundalo, S.T., Importance of secondary structure for endothelin binding and functional activity, *Biochem. Biophys. Res. Commun.*, 183, 572-576, 1992b.

Pelton, J.T., Jones, R., Saudek, V. and Miller, R., Receptor binding and biophysical studies of monocyclic analogs of endothelin, in: *Proc. 11th American Peptide Symposium*, Rivier, J.E. and Marshall, G.R., Eds., ESCOM, Leiden, 1991, 274.

Rivero, R.A., Greenlee, W.J., Williams, P.D., Kieczykowski, G.R. and Williams, D.L., Discovery of substituted 8,9-dicarboxyldibenzo [2,3:5,6]bicyclo[5.2.0]nonan-4-ones with moderate binding affinity to the endothelin ET_A and ET_B receptors, *Bioorg. Med. Chem. Lett.*, 5, 1401-1404, 1995.

Ramalingam, K. and Snyder, G.H., Selective disulfide formation in truncated apamin and sarafotoxin, *Biochemistry*, 32, 11144-11161, 1993.

Rovero, P., Patacchini, R. and Maggi, C.A., Structure-activity studies on endothelin (16-21), the C-terminal hexapeptide of the endothelins, in the guinea-pig bronchus, *Br. J. Pharmacol.*, 101, 232-234, 1990.

Saeki, T., Ihara, M., Fukuroda, T., Yamagiwa, M. and Yano, M., [Ala^{1,3,11,15}]Endothelin-1 analogs with ET_B agonist activity, *Biochem. Biophys. Res. Commun.*, 179, 286-292, 1991.

Saida, K., Mitsui, Y. and Ishida, N., A novel peptide, vasoactive intestinal contractor, of a new (endothelin) peptide family, *J. Biol. Chem.*, 264, 14613-14616, 1989.

Sokolovsky, M., Endothelins and sarafotoxins: physiological regulation, receptor subtypes and transmembrane signaling, *TIBS*, 16, 261-264, 1991.

Sokolovsky, M., Galron, R., Kloog, Y., Bdolah, A., Indig, F.E., Blumberg, S. and Fleminger, G., Endothelins are more sensitive than sarafotoxins to neutral endopeptidase: possible physiological significance, *Proc. Natl. Acad. Sci. U.S.A.*, 87, 4702-4706, 1990.

Sokolovsky, M., Structure-function relationships of endothelins, sarafotoxins, and their receptor subtypes, *J. Neurochem.*, 59, 809-821, 1992.

Spellmeyer, D.C., Brown, S., and Stauber, G.B., Endothelin receptor ligands, replacement net approach to SAR determination of potent hexapeptides, *Bioorg. Med. Chem. Lett.*, 3, 519-524, 1993a.

Spellmeyer, D.C., Brown, S., Stauber, G.B., Geysen, H.M. and Valerio, R., Endothelin receptor ligands. Multiple D-amino acid replacement net approach, *Bioorg. Med. Chem. Lett.*, 3, 1253-1256, 1993b.

Spinella, M.J., Malik, A.B., Everitt, J. and Andersen, T.T., Design and synthesis of a specific endothelin 1 antagonist: effects on pulmonary vasoconstriction, *Proc. Natl. Acad. Sci. U.S.A.*, 88, 7443-7446, 1991.

Stephenson, K., Gandhi, C.R. and Olson, M.S., Biological actions of endothelin, *Vitamins Hormones*, 48, 157-198, 1994.

Takai, M., Umemura, I., Yamasaki, K., Watakabe, T., Fujitani, Y., Oda, K., Urade, Y., Inui, T., Yamamura, T. and Okada, T., A potent and specific agonist, Suc-[Glu^9, Ala^{11,15}]-endothelin-1(8-21), IRL 1620, for the ET_B receptor, *Biochem. Biophys. Res. Commun.*, 184, 953-959, 1992.

Takasaki, C., Aimoto, S., Kitazumi, K., Tasaka, K., Shiba, T., Nishiki, K., Furukawa, Y., Takayanagi, R., Ohnaka, K. and Nawata, H., Structure-activity relationships of sarafotoxins: chemical syntheses of chimera peptides of sarafotoxins STXb and STXc, *Eur. J. Pharmacol.*, 198, 165-169, 1991.

Takasaki, C., Aimoto, S., Takayanagi, R., Ohashi, M. and Nawata, H., Structure-receptor binding relationships of sarafotoxin and endothelin in porcine cardiovascular tissues, *Biochem. Int.*, 21, 1059-1064, 1990.

Takasaki, C., Tamiya, N., Bdolah, A., Wollberg, Z. and Kochva, E., Sarafotoxins STX: several isotoxins from *Atractaspis engaddensis* (burrowing asp) venom that affect the heart, *Toxicon*, 26, 543-548, 1988.

Takayanagi, R., Hashiguchi, T., Ohashi, M. and Nawata, H., Regional distribution of endothelin receptor in porcine cardiovascular tissues, *Regul. Peptides*, 27, 247-255, 1990.

Tam., J.P., Liu, W., Zhang, J.-W., Glantino, M., Bertolero, F., Cristiani, C., Vaghi, F. and de Castiglione, R., Alanine scan of endothelin: importance of aromatic residues, *Peptides*, 15, 703-708, 1994.

Topouzis, S., Huggins, J.P., Pelton, J.T. and Miller, R.C., Modulation by endothelium of the responses induced by endothelin-1 and by some of its analogs in rat isolated aorta, *Br. J. Pharmacol.*, 102, 545-549, 1991.

Topouzis, S., Pelton, J.T. and Miller, R.C., Effects of calcium entry blockers on contractions evoked by endothelin-1, [Ala^{3,11}]endothelin-1 and [Ala^{1,15}]endothelin-1 in rat isolated aorta, *Br. J. Pharmacol.*, 98, 669-677, 1989.

Tschirhart, E.J., Drijfhout, J.W., Pelton, J.T., Miller, R.C. and Jones, C.R., Endothelins: functional and autoradiographical studies in guinea pig trachea, *J. Pharmacol. Exp. Ther.*, 258, 381-387, 1991.

Urade, Y., Fujitani, Y., Oda, K., Watakabe, T., Umemura, I., Takai, M., Okada, T., Sakata, K. and Karak, H., An endothelin B receptor-selective antagonist: IRL 1038, [Cys^{11}-Cys^{15}]-endothelin-1(11-21), *FEBS Lett.*, 311, 12-16, 1992.

Watakabe, T., Urade, Y., Takai, M., Umemura, I. and Okada, T., A reversible radioligand specific for the ET_B receptor: [^{125}I]Tyr^{13}-Suc-[Glu^9, $Ala^{11,15}$]-endothelin-1(8-21), [^{125}I]IRL 1620, *Biochem. Biophys. Res. Commun.*, 867-873, 1992.

Watanabe, T.X., Itahara, Y., Nakajima, K., Kumagaye, S., Kimura, T. and Sakakibara, S., The biological activity of endothelin-1 analogues in three different assay systems, *J. Cardiovasc. Pharmacol.*, 17(Suppl. 7), S5-S9, 1991.

Weiser, E., Wollberg, Z., Kochva, E. and Lee, S.Y., Cardiotoxic effects of the venom of the burrowing asp, *Atractaspis engaddensis* (Atractaspididae, ophidia), *Toxicon*, 22, 767-774, 1984.

Williams, D.L., Jones, K.L., Pettibone, D.J., Lis, E.V. and Clineschmidt, B.V., Sarafotoxin STXc: an agonist which distinguishes between endothelin receptor subtypes, *Biochem. Biophys. Res. Commun.*, 175, 556-561, 1991.

Wollberg, Z., Bdolah, A. and Kochva, E., Vasoconstrictor effects of sarafotoxins in rabbit aorta: structure-function relationships, *Biochem. Biophys. Res. Commun.*, 162, 371-376, 1989.

Yanagisawa, M., Inoue, A., Ishikawa, T., Kasuya, Y., Kimura, S., Kumagaye, S., Nakajima, K., Watanabe, T.X., Sakakibara, S., Goto, K. and Masaki, T., Primary structure, synthesis, and biological activity of rat endothelin, an endothelium-derived vasoconstrictor peptide, *Proc. Natl. Acad. Sci. U.S.A.*, 85, 6946-6967, 1988.

Yanagisawa, M., Kurihara, H., Kimura, S., Tomobe, Y., Kobayashi, M., Mitsui, Y., Yazaki, Y., Goto, K. and Masaki, T., A novel potent vasoconstrictor peptide produced by vascular endothelial cells, *Nature*, 332, 411-415, 1988.

Chapter 16

Development of Endothelin Antagonists

Annette M. Doherty and William C. Patt

CONTENTS

I. PEPTIDES AND PEPTIDIC ENDOTHELIN ANTAGONISTS

Development of highly specific agonists and/or antagonists to the endothelin receptor subtypes will enable the evaluation of the role of endothelin isopeptides in physiology and pathophysiology.

Numerous SAR studies have been carried out using synthetic analogues and fragments of ET, with each portion of ET-1 being examined for relative importance to binding and functional activities.

Systematic replacement of each amino acid residue in ET-1 with Ala has revealed the relative importance of each side-chain functionality to receptor recognition (de Castiglione et al., 1992). It was shown that residues Phe^{14}, Leu^{17}, and Trp^{21} were essential for binding in human vascular smooth muscle (VSM) cells and agonist activity in contraction of rabbit vena cava. Substitution of Asp^{8} by Ala^{8} did not change binding affinity but greatly diminished agonist activity. In fact, ET-1[Ala^{8}] appears to be a noncompetitive antagonist. Ala substitution in the C-terminal region was generally much more sensitive than in the N-terminal region and ET-1[Ala^{21}] had no biological activity. Interestingly, ET-1[Ala^{16}] and ET-1[Ala^{19}] are more potent agonists than ET-1 itself. Ser^{2}, Ser^{4}, Leu^{6}, Met^{7}, Lys^{9}, and Glu^{10} can all be replaced with alanine while still retaining reasonable binding and agonist activity.

0-8493-6975-4/97/$0.00+$.50

TABLE 1
SAR of the Monocyclic Analogues of ET-1

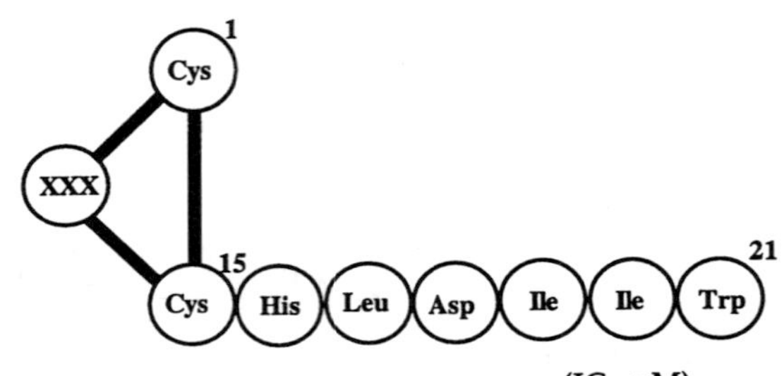

Compound XXX =	Binding[a] (IC_{50} μM) RaH	ET_A	ET_B	Functional[b] IP_3
-Ser-Ser-Leu-Met-Asp-Lys-Glu	49	>10	>10	[c]
-Ser-Apa-Val-Tyr-Phe-[d]	1.8	2.5	0.35	>10
-Ser-Aha-Val-Tyr-Phe-[d]	2.5	[c]	[c]	8.0
-Ser-Aoa-Val-Tyr-Phe-[d]	1.6	1.8	0.24	3.0
-Val-Tyr-Phe-	6.5	2.6	1.7	2.2
-Ser-Val-Tyr-Phe-	2.8	3.8	1.7	[c]
-Ser-Ser-Val-Tyr-Phe-	1.8	2.0	1.4	[c]
-Ser-Ser-Ser-Val-Tyr-Phe-	1.2	2.5	0.06	5.0
-Ser-Asp-Lys-Glu-Val-Tyr-Phe-	3.5	7.1	1.1	>10

[a] Binding data: rat heart ventricle (RaH); rabbit renal artery vascular smooth muscle cells (ET_A); rat cerebellar membranes (ET_B).
[b] Functional data (IP_3): inhibition of ET-1 induced accumulation of inositol phosphates in rat skin fibroblasts.
[c] Not determined.
[d] Apa = 5-aminopentanoic acid; Aha = 7-aminoheptanoic acid; Aoa = 8-aminooctanoic acid.

Since the full-length monocyclic ET analog ET-1[Ala^{3,11}, Nle^{7}] elicits about 10% of the vasoconstrictor activity of ET-1, Hunt et al. reported an SAR study of ET-1 by carrying out an Ala scan of ET-1[Ala^{3,11}] (Miasiro et al., 1993; Hunt et al., 1991). Their results support the observations of the ET-1 Ala scan, showing that Ser^{2}, Val^{12}, His^{16}, and Ile^{19} are the tolerant sites, while Asp^{8}, Tyr^{13}, Ile^{20}, and Trp^{21} replacements with Ala caused >10^3-fold loss in binding affinity. Substitution of Glu^{10}, Phe^{14}, Leu^{17}, and Asp^{18} with Ala produced analogues that showed appreciable affinity to the receptor at concentrations below those at which agonism was observed, making these candidate sites for further exploration of receptor antagonism.

Various monocyclic fragments of ET-1 were synthesized and evaluated for binding affinity in rabbit aorta (ET_A) and pulmonary artery (ET_B like) (Cody et al., 1991; Reynolds and Mok, 1991). It was shown that the loop region ET[3-11] does not bind to either receptor subtype up to concentrations of 100 μM. A series of monocyclic analogues without the [3-11] region binds to both receptors with μM affinity (Table 1). Several of these compounds also exhibited antagonist activity by inhibiting ET-1 induced inositol phosphate accumulation (Muldoon et al., 1989) in rat skin fibroblasts (IP_3) and ET-1 induced arachidonic acid release (AAR) (Cody et al., 1994) in rabbit renal artery smooth muscle cells (ET_A) (Table 1).

One of the compounds in Table 1, Cyclo-(Cys-Val-Tyr-Phe-Cys)-His-Leu-Asp-Ile-Ile-Trp, was reported independently by another group to be an ET_B antagonist (IRL 1038) inhibiting ET_B mediated contractions of the guinea-pig ileal and tracheal smooth muscle (Urade et al., 1992). It was also claimed to selectively inhibit the

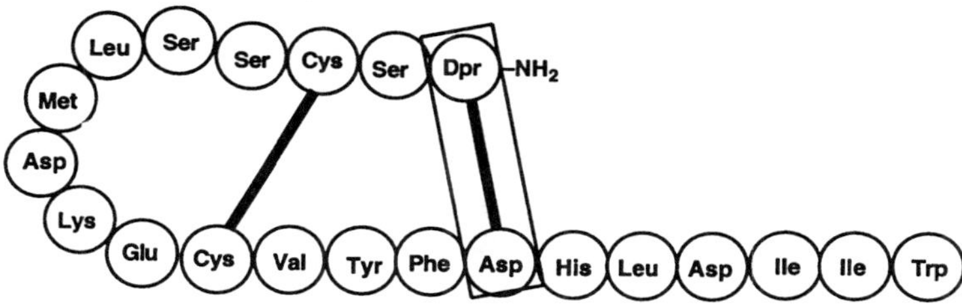

FIGURE 1
ET-1[Dpr[1]-Asp[15]].

ET-induced endothelium-dependent vascular relaxation (Karaki et al., 1993; Urade et al., 1992; Urade et al., 1994). Our results indicate that this compound is a weak nonselective ligand to both receptor subtypes (μM) exhibiting antagonist activity in secondary messenger assays such as inhibition of inositol phosphate accumulation or arachidonic acid release with only micromolar activity (Reynolds and Mok, 1991).

Spinella and co-workers reported a synthetic bicyclic ET-1 analog, ET-1[Dpr[1]-Asp[15]] (Figure 1) to be an endothelin antagonist (Spinella et al., 1991; Werber et al., 1992). This compound has a binding affinity of IC_{50} 2 nM in rat pulmonary smooth muscle cells. In isolated perfused guinea-pig lung, pretreatment with ET-1[Dpr[1]-Asp[15]] at 1×10^{-7} M substantially decreased the vasoconstrictor response induced by ET-1 but had no effect on ET-3 induced vasoconstriction, showing ET_A receptor selectivity. In conscious sheep, inhalation of ET-1 and, to a lesser extent, ET-3 causes concentration-dependent bronchoconstriction which was significantly inhibited by pretreatment with ET-1[Dpr[1]-Asp[15]] (25 breaths of 5×10^{-8} M) (Abraham et al., 1993).

ET-1[Pen[1,11],Nle[7], Ala[18]] and ET-1[Pen[1,11],Nle[7], Asn[18]] were shown to have ET_A antagonist activity in the rabbit carotid artery and partial agonist activity at an alternate non-ET_A receptor subtype which is present in the guinea-pig trachea and in the rabbit saphenous vein (Hunt et al., 1993).

Other ET-1 analogues such as ET-1[Thr[18], Leu[19]] and ET-1[Thr[18], γ-Me-Leu[19]] were reported to be ET_A/ET_B nonselective antagonists and ET-1[Asn[8], Glu[10]] was recently reported as an ET_A selective antagonist (Wakimasu et al., 1993; Akiyama et al., 1992).

Although being a poor ligand itself at both receptors, the hydrophobic C-terminal region (residues 16-21) of ET-1 is crucial for receptor recognition. Several groups have focused on this region of the molecule in efforts to develop ET antagonists (Doherty, 1992; Wakimasu et al., 1993). A mono-D-amino acid scan in the C-terminal hexapeptide of natural sequence (Figure 2) has demonstrated that Asp[18] and Ile[20] are tolerant to *D*-configuration to produce slightly more potent compounds. The most significant increase in binding activity at both ET_A and ET_B receptors was observed at position 16 where Ac-D-His-Leu-Asp-Ile-Ile-Trp was an ET antagonist with micromolar binding affinity (Doherty et al., 1993a).

Thus ET[D-His[16], 16-21] and ET[D-Phe[16], 16-21] are ET_A/ET_B nonselective antagonists. Selectivity is tunable by changing other residues that can also be modified by this region of ET. For example, Leu[17] was found to be reasonably tolerant to many changes. As shown in Table 2, more than 30-fold ET_B selectivity was obtained by replacing Leu[17] with D-Phe or Asp[18] with Phe. A moderate ET_A selectivity was observed with the Orn[17] analog. An extensive SAR study of Ac-D-Phe-Orn-Asp-Ile-Ile-Trp was reported recently using multipin peptide synthesis technology (Spellmeyer et al., 1993a; Spellmeyer et al., 1993b). Modifications in the Ile-Ile-Trp region generally cause a drop in activity, especially at the ET_A receptor (Table 2). Further truncations of the hexapeptide also lead to some loss in binding affinity (Doherty et al., 1993b).

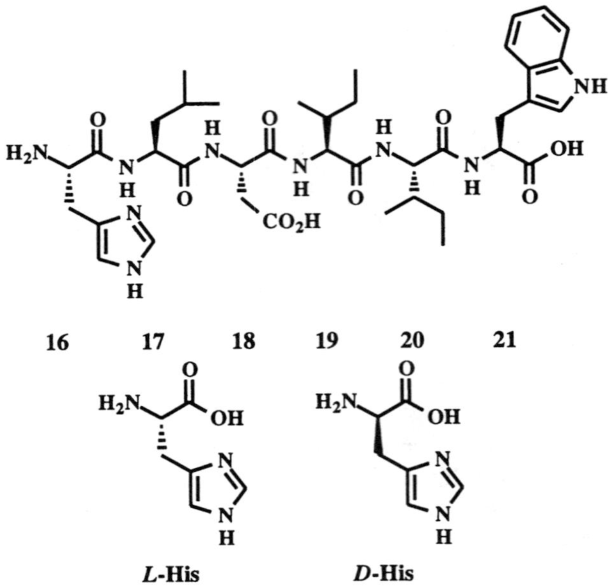

FIGURE 2
The C-terminus of ET-1.

TABLE 2
SAR of ET-1[D-Phe, 16-21]

	IC_{50} (μM)		
	Binding		Functional
Compound	ET_A[a]	ET_B[b]	AAR[f]
Ac-D-His-Leu-Asp-Ile-Ile-Trp	8.9[c]	9.1[c]	3.2
Ac-D-Phe-Leu-Asp-Ile-Ile-Trp	1.8[c]	2.0[c]	3.1
Ac-Phe-Leu-Asp-Ile-Ile-Trp	18.4	18.5	[e]
Phe-Leu-Asp-Ile-Ile-Trp	>100	2.6	[e]
Ac-D-Phe-Orn-Asp-Ile-Ile-Trp	0.58[c]	2.45[c]	2.0
Ac-D-Phe-Ala-Asp-Ile-Ile-Trp	0.42[c]	0.29[c]	0.33
Ac-D-Phe-Phe-Asp-Ile-Ile-Trp	0.20	0.06	2.6
Ac-D-Phe-D-Phe-Asp-Ile-Ile-Trp	0.91[d]	0.02[d]	2.2
Ac-D-Phe-Leu-Phe-Ile-Ile-Trp	1.4	0.036[d]	4.5
Ac-D-Phe-Leu-Lys-Ile-Ile-Trp	>10	>10	[e]
Ac-D-Phe-Leu-Asp-Ala-Ile-Trp	3.6	0.45	1.5
Ac-D-Phe-Leu-Asp-Glu-Ile-Trp	6.7	>10	>10
Ac-D-Phe-Leu-Asp-Lys-Ile-Trp	>10	>10	[e]
Ac-D-Phe-Leu-Asp-Ile-Ala-Trp	8.0	>10	>10
Ac-D-Phe-Leu-Asp-Ile-Ile-Tyr	>10	0.15c	[e]

[a] Rabbit renal artery vascular smooth muscle cells.
[b] Rat cerebellum.
[c] n = 2 IC_{50} determinations.
[d] n = 3 IC_{50} determinations. All other values represent 1 IC_{50} determination. IC_{50} values were derived from single competition experiments in which data points were measured in triplicate. Binding data were computer-analyzed by nonlinear least squares analysis giving the best fit for a one-site model.
[e] Not tested.
[f] Inhibition of ET-1 stimulated arachidonic acid release in rabbit renal artery vascular smooth muscle cells.

TABLE 3

SAR of 16 Position of the C-terminus of ET-1

	IC_{50} (μM)		
	Binding		Functional
Compound	ET_A[a]	ET_B[b]	AAR[a]
Ac-D-Phe-Leu-Asp-Ile-Ile-Trp	1.8	2.0	3.1
Ac-D-(2Nap)Ala-Leu-Asp-Ile-Ile-Trp	1.0	1.0	1.9
Ac-D-Tyr-Leu-Asp-Ile-Ile-Trp	0.40	7.0	0.25
Ac-D-Trp-Leu-Asp-Ile-Ile-Trp	0.13	1.8	0.45
Ac-D-Dip-Leu-Asp-Ile-Ile-Trp (PD 142893)	0.025	0.14	0.07
Ac-D-Bip-Leu-Asp-Ile-Ile-Trp	4.4	3.5	—

[a] Rabbit renal artery vascular smooth muscle cells.
[b] Rat cerebellum.
Dip = 3,3-diphenylalanine.
Bip = 4,4′-biphenylalanine.

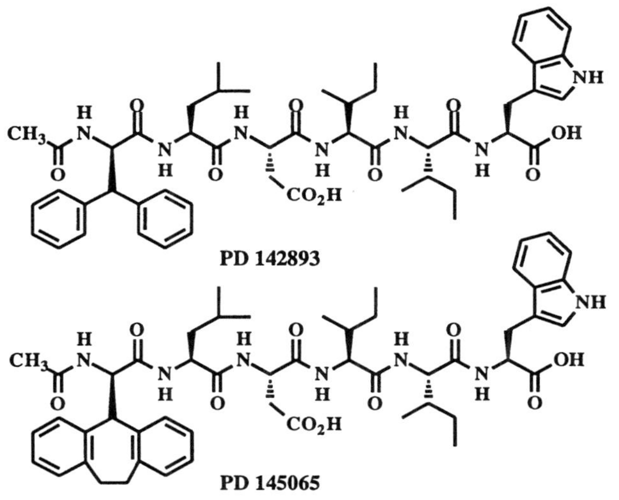

PD 142893

PD 145065

FIGURE 3
Hexapeptide ET antagonists.

Thorough studies around the 16-position were carried out by replacing D-Phe with various D-aromatic amino acids (Table 3) (Doherty et al., 1993b; Cody et al., 1992a; Cody et al., 1993, Cody et al., 1992b). Increasing the steric bulkiness of D-Phe by using 3,3-diphenylalanine (Dip) led to the discovery of PD 142893 (Figure 3). PD 142893 exhibits high affinity both to ET_A and ET_B receptors with a slight ET_A preference. PD 142893 inhibits arachidonic acid release induced by endothelin in the rabbit renal artery (ET_A) and in the rat cerebellum (ET_B). It also antagonizes ET-1-induced constriction in the rabbit femoral (ET_A) and pulmonary (ET_B) arteries with pA_2 values of 5.95 and 6.31, respectively (Hingorani et al., 1992; LaDouceur et al., 1992; Warner et al., 1993). Further exploration at this position by substitution with tricyclic amino acids provided a group of more potent compounds (Table 4). In particular, PD 145065 (Figure 3) possesses low nanomolar binding affinity at each receptor subtype and is a potent functional antagonist of vasoconstriction in tissues specifically expressing each receptor subtype (pA_2 = 6.8 to 7.2) (Cody et al., 1993; Doherty et al., 1993c).

TABLE 4
SAR of PD 142893

Compound	IC_{50} (μM) Binding ET_A[a]	Binding ET_B[b]	Functional AAR[c]
Ac-D-Dip-Leu-Asp-Ile-Ile-Trp · Na (PD 142893)	0.04	0.06	0.07
Ac-D-Bhg-Leu-Asp-Ile-Ile-Trp (PD 145065)	0.003	0.015	0.005
D-Dip-Leu-Asp-Ile-Ile-Trp	2.1	1.0	1.92
Ac-D-Dip-Orn-Asp-Ile-Ile-Trp	0.013	0.15	0.03
Ac-D-Dip-Glu-Asp-Ile-Ile-Trp	0.025	0.052	0.13
Ac-D-Dip-Leu-Phe-Ile-Ile-Trp	0.58	0.06	1.3
Ac-D-Dip-Leu-Asp-Val-Ile-Trp	0.017	0.05	0.029
Ac-D-Dip-Leu-Asp-Ile-Val-Trp	6.0	6.0	2.60
Ac-D-Dip-Leu-Asp-Ile-Ile-Tyr	2.7	6.1	4.3
H O CH_3 N O Leu-Asp-Ile-Ile-Trp	0.005	0.019	0.003

[a] Rabbit renal artery vascular smooth muscle cells.
[b] Rat cerebellum.
[c] Inhibition of ET-1-stimulated arachidonic acid release in rabbit renal artery vascular smooth muscle cells.

FIGURE 4
TTA-386.

Based on the C-terminal tail region of ET-1, it was reported recently that Fmoc-ET-1[17-21] (Fmoc-Leu-Asp-Ile-Ile-Trp) dose-dependently antagonized ET-3 but not ET-1-induced contractile activity in electrically stimulated rat vas deferens (Pegoraro et al., 1993).

A recently reported ET_A selective antagonist, TTA-386 (Figure 4), is also derived from the C-terminal tail region of ET-1 (Kito et al., 1993).

The first of the small cyclic peptide endothelin (ET) antagonists was reported by Banyu (Ihara et al., 1991). The ET_A-selective antagonists BE-18257A and B were isolated from a fermentation broth of *Streptomyces misakiensis* (Kojiri et al., 1991). These analogues have ET_A receptor (porcine smooth muscle membranes) binding of 3.0 μM (IC_{50}) and 1.4 μM (IC_{50}), respectively, while ET_B receptor binding affinity (porcine cerebellum membranes) is over 100 μM (IC_{50}) (Ishikawa et al., 1991). BE-18257B antagonized the ET-1-induced contraction of rabbit iliac arteries (an ET_A tissue) with a pA_2 value of 5.9. This compound had no effect upon vasoconstriction induced by histamine, potassium chloride, or acetylcholine, indicating that it is an ET receptor-specific antagonist. The *in vivo* antagonist effect of BE-18257B to an ET-1 challenge was evaluated in Wistar Kyoto rats (Ihara et al., 1992). The i.p. administration of BE-18257B 1 h prior to an i.v. bolus injection of ET-1 produced no effect on the initial depressor response (an ET_B-mediated event), however, BE-18257B did attenuate the sustained pressor response (an ET_A-mediated event) for almost 1 h post ET-1 challenge.

TABLE 5

SAR of Cyclic Pentapeptides

		Binding IC_{50}(μM)	
		ET_A[a]	ET_B[b]
BE-18257A	cyclo-(-D-Trp-D-Glu-Ala-D-Val-Leu-)	3.0	>100
BE-18257B	cyclo-(-D-Trp-D-Glu-Ala-D-alloIle-Leu-)	1.4	>100
	cyclo-(-D-Trp-D-Asp-Ala-D-Val-Leu-)	0.11	>100
	cyclo-(-D-Trp-D-Glu-Pro-D-Val-Leu-)	0.41	>100
	cyclo-(-D-Trp-D-Asp-Ala-D-Val-Pro-)	8.5	>100
BQ-123	cyclo-(-D-Trp-D-Asp-Pro-D-Val-Leu-)	0.022	18
	cyclo-(-D-Trp-D-Asp-Sar-D-Val-Leu-)	0.032	30
BQ-153	cyclo-(-D-Trp-D-Sal-Pro-D-Val-Leu-)	0.021	54

[a] Porcine aortic smooth muscle membranes.
[b] Porcine cerebellum membranes.

Extensive SAR of these cyclic pentapeptides has been carried out (Table 5) and has led to potent ET_A-selective analogues, (Ihara et al., 1992; Ishikawa et al., 1992a). Substitution of the D-Glu of BE-18257A with D-Asp resulted in a 30-fold increase in ET_A receptor affinity. A 10-fold increase in affinity was obtained after replacement of the Ala of BE-18257A with Pro. Substitution of the Leu with Pro led to essentially no enhancement in affinity. Combining the two successful substitutions gave compound BQ-123, with a 100-fold increase in ET_A receptor affinity (compared with BE-18257A) and a measurable effect at ET_B receptors. BQ-123 has an ET_A affinity of 22 nM (IC_{50}) and ET_B affinity of 23 μM (IC_{50}). Replacement of the D-Asp of BQ-123 with D-sulfoalanine (D-Sal) gave BQ-153 which is also a potent ET_A-selective agent. Additional substitutions of amino acids produced compounds with good receptor affinity but no significant improvement over BQ-123.

BQ-123 has a pA_2 value of 7.4 against ET-1-induced contraction of porcine coronary arteries (ET_A tissue). BQ-123 has demonstrated activity in preventing sudden death via a lethal dose of ET-1 in mice at doses of 0.5 and 1.0 mg/kg (i.v.) (Ihara et al., 1992). The main mechanism for ET-1-induced sudden death is thought to be from coronary vasoconstriction followed by ischemia. This suggests that ET_A-selective agents may be useful in the treatment of ischemic heart disease. BQ-123 has also been implicated in the prevention of cerebral vasospasm following subarachnoid hemorrhage (SAH) (Clozel and Watanabe, 1993). BQ-123 has been demonstrated to be efficacious in the treatment of hypoxic pulmonary hypertension (Bonvallet et al., 1993), malignant hypertension in stroke-prone spontaneously hypertensive rats (SHRSP), and elicited a protective effect in ischemic acute renal failure (Mino et al., 1992). Interestingly, BQ-123 had no effect in spontaneously hypertensive rats (SHR) (Nishikibe et al., 1993).

Fujisawa has reported a very closely related series of cyclic pentapeptides as ET_A-selective antagonists (Hasimoto et al., 1990).

Another series of cyclic peptide antagonists are represented by the depsipeptides cochinmicins I, II, III, IV, and V (Table 6) These agents, discovered by Merck researchers, are natural products produced by *Microbiospora* sp. ATCC 55140 and were found to be nonselective micromolar antagonists with IC_{50}s of 0.2 to 90 μM at either ET receptor subtype (Zink et al., 1992; Lam et al., 1992a,b). The most potent analog in this series is I with IC_{50}s of 0.24 μM in rat aorta (ET_A) and 1.5 μM in rat hippocampus (ET_B). Compound I also blocked ET-1 stimulated phosphatidyl inositol turnover in a dose-dependent fashion in rat atria (EC_{50} = 10.8 μM). Little SAR has been reported on these analogues, although it is known that the S configuration at the starred (*) site and chloro substitution on the pyrrole ring are both detrimental to receptor binding.

TABLE 6

Cochinmicin SAR

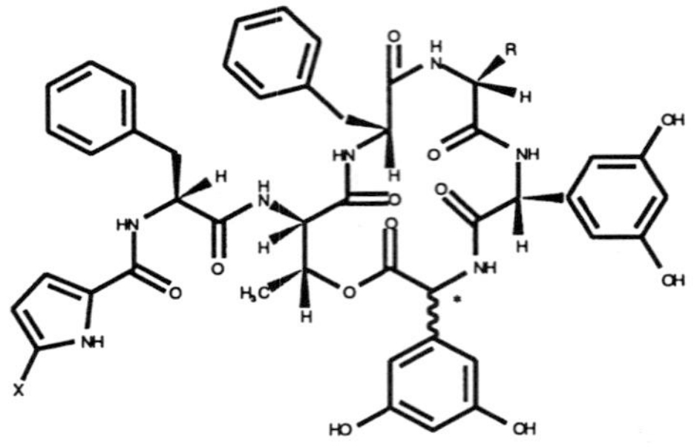

Cochinmicin	X	R	*	Binding IC_{50}(μM) ET_A	Binding IC_{50}(μM) ET_B
I	H	CH_3	R	0.24[a]	1.5[c]
II	Cl	CH_3	S	6.2[a]	22[c]
III	Cl	CH_3	R	0.42[a]	3.7[c]
IV	Cl	CH_2OH	S	3[b]	2[c]
V	H	CH_3	S	90[b]	25[c]

[a] Rat aorta.
[b] Bovine aorta.
[c] Rat hippocampus.

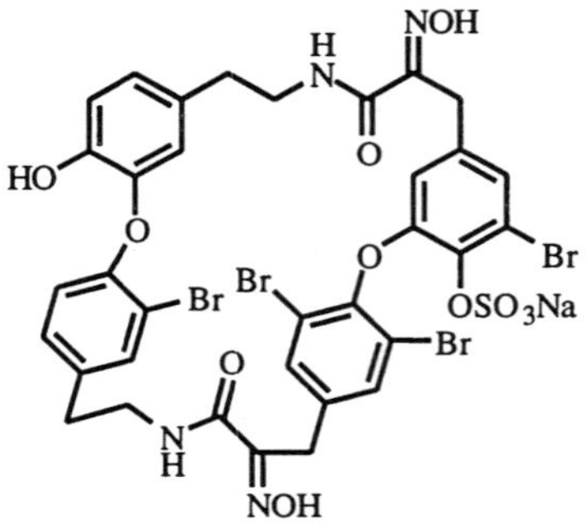

FIGURE 5
34-Sulfatobastadin-13.

The Ciba-Geigy group has reported a cyclic natural product structure, 34-Sulfatobastadin-13 (Figure 5), which is a weak ET_A-selective agent (Gulavita et al., 1993). This compound showed an IC_{50} = 39 μM at the ET_A receptor (porcine thoracic aorta) and only 27% inhibition at 70 μM at an ET_B receptor (rat cerebellum). The only published SAR is a desulfonylated derivative which had no measurable activity at either receptor.

Further size reduction of ET antagonists was accomplished by the Fujisawa group with the discovery of a series of tripeptide analogues (Table 7) (Hemmi et al., 1991). These derivatives were rationally designed mimics of a fermentation broth lead (isolated from a strain of *Streptomyces*). This series of compounds is thought to be ET_A selective but little ET_B binding data in this series have been published. These antagonists are all derivatives of D-(α-Me)Trp. The best known compound in this series is FR 139317. Structurally, all have a rather bulky and lipophilic N-terminus

TABLE 7

Tripeptide Endothelin Antagonists

	Binding IC_{50} (nM)	
	ET_A	ET_B
Cycloheptyl-NCO-Leu-D-Trp(Me)-D-Pya-OH (FR139317)	2.5[a]	4600[b]
Cyclohexyl-HNCO-Leu-D-Trp(Me)-D-Pya-OH.HCl	2.3[a]	nr
Cycloheptyl-NCO-Leu-D-Trp(Me)-D-Pya-Sar-ONa	7.6[a]	nr
Cycloheptyl-NCO-Leu-D-Trp(Me)-D-Pya-Phe-ONa	21[a]	nr
Cycloheptyl-NCOCH$_2$CH(S)(CH$_2$CH(Me)$_2$)CO-D-Trp(Me)-D-Pya-NHMe	7.6[a]	nr
[(2-Chlorophenyl)acetyl]Leu-D-Trp(Me)-D-Leu-OH	32[a]	nr
BOC-Leu-D-Trp(N^i-CO$_2$CH$_3$)-D-Gly(n-Butyl)-OH	nr	90%[c,d]
BOC-Leu-D-Trp(N^i-CO$_2$CH$_3$)-D-Gly(n-Propyl)-OH	nr	84%[c,d]
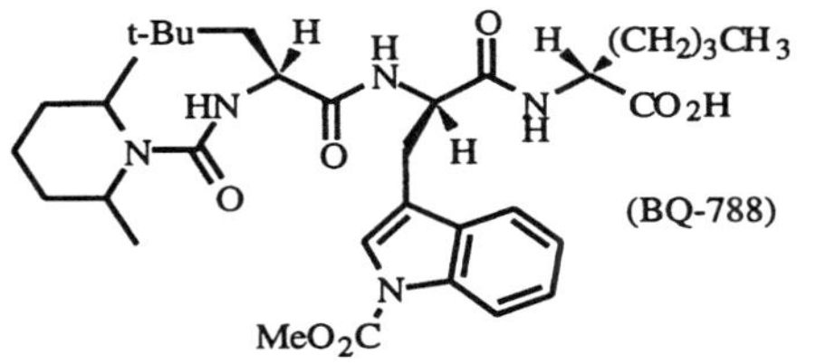 (BQ-788)	nr	99%[c,d]
Cycloheptyl-NCO-Leu-D-Trp-D-Trp (BQ-485)	3.4[e]	26[b]
Cycloheptyl-NCO-Leu-D-Trp(CHO)-D-Trp (BQ-610)	nr	nr

[a] Human aorta.
[b] Porcine brain.
[c] Rat cerebellum.
[d] At 10^{-6} M.
[e] Porcine aortic smooth muscle.
nr = not reported.

with an acidic C-terminal functionality. The SAR seems to indicate that a C-terminal amino acid that is capable of forming an acid addition salt improves activity (e.g., pyridylalanine (Pya) or Sar are better than Leu or Phe). The K_i values for FR 139317 are 2.5 nM at ET_A (human aorta) and 4.6 μM at ET_B (porcine brain). In the isolated rabbit aorta, FR 139317 shifted the ET-1-induced contraction response curve to the right in a dose-dependent fashion giving a pA_2 value of 7.2. As was the case with the other ET_A-selective compounds, an i.v. bolus dose of FR 139317 inhibited the pressor response due to an ET-1 challenge *in vivo*, but had no effect upon the initial depressor response as expected (Sogabe et al., 1992).

A very similar series of tripeptide ET antagonists, based upon D-Trp rather than D-(α-Me)Trp, were discovered by the Banyu group (Table 7, center section) (Ishikawa et al., 1992b). However, these tripeptides are reported to be ET_B ligands. Whether these agents are selective for the ET_B receptor is not known due to the fact that no binding data were given with an ET_A-containing tissue. These compounds also employ a bulky N-terminus and an acidic C-terminus, similar to the FR 139317 series. These compounds all have good pA_2 values (pA_2s = 6.7 to 7.1) for inhibition of ET-3-induced contraction in rabbit pulmonary artery (ET_B).

Banyu has also reported another set of tripeptides (Table 7, last section) which, like the Fujisawa derivatives, are ET_A-selective receptor antagonists. A continuous systematic infusion of BQ-485 has been shown effective in reversing the constriction of canine basilar arteries due to an injection of autologous blood in the cisterna magna, a model for SAH (Itoh et al., 1993; Cosentino and Katusic, 1994). In the same study intracisternal injection of FR 139317 on days 0, 2, and 4 showed similar activity.

TABLE 8

SAR of WS009 Series

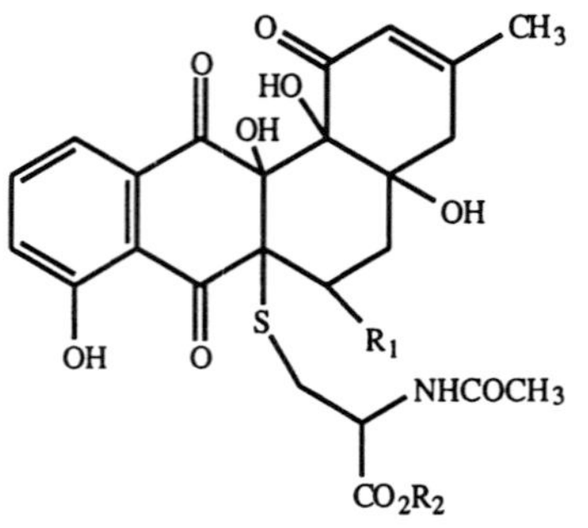

	R_1	R_2	Binding IC_{50} (μM) ET_A	ET_B[b]
WS009A	H	H	5.8[a]	ia
WS009Aester	H	CH_3	nr	ia
WS009B	OH	H	0.67[a]	ia

[a] Porcine aorta.

[b] Various tissues; nr = not reported; ia = inactive.

II. SMALL MOLECULE ENDOTHELIN ANTAGONISTS

The first true nonpeptide ET antagonists were reported in 1992. These antagonists, once again, were isolated as natural products produced by a strain of *Streptomyces* (Miyata et al., 1992a, b). These anthraquinones are reported to be ET_A-selective antagonists (Table 8). The two acids have micromolar affinity for the ET_A receptor (porcine aorta) and no activity against a variety of ET_B tissues. Esterification of the acid group eliminated all ET_A receptor affinity (WS009A ester). The two acids are antagonists blocking the ET-1-induced increase of inositol phosphates in rat aortic rings (ET_A). Furthermore, WS009A (10 mg/kg, i.v.) reduces the pressor effect of an i.v. bolus injection of ET-1 (3.3 mg/kg) in SHR without any effect on the depressor response.

Another early series of nonpeptide ET antagonists were the asterric acid derivatives (Table 9) (Ohashi et al., 1992). Asterric acid (R^1 = COOH, R^2 = OH, R^3 = CO_2CH_3, R^4 = OH) exhibits ET_A receptor binding in A10 cells with an IC_{50} of 10 μM. Conversion of the acidic groups of asterric acid to either an ester or a hydrazide gave inactive derivatives. Perhaps surprisingly, hydrolysis of the ester group of asterric acid to a second acid group, also abolished activity.

Another nonpeptide ET antagonist reported is myriceron caffeoyl ester (Figure 6) (Mihara and Fujimoto, 1993). This compound was isolated from *Myrica cerifera* (bayberry) and is a steroid-like structure that is ET_A selective. The ET_A receptor binding is 78 nM (IC_{50}) in rat cardiac membranes, with no measurable inhibition in a variety of ET_B tissues. This antagonist also inhibited the contractile response due to ET-1 in isolated rat thoracic aorta (ET_A tissue) giving a pA_2 value of 6.65 (Fujimoto et al., 1992).

Hoffmann-LaRoche disclosed a series of tri- and/or tetrasubstituted pyrimidines which are balanced ET_A/ET_B antagonists discovered by SAR of leads from compound library screening (Table 10) (Burri et al., 1992; Breu et al., 1993; Clozel et al., 1993a,b). Ro 46-2005 has ET_A receptor binding of IC_{50} = 216 nM in human smooth muscle cells and ET_B receptor binding of IC_{50} = 221 nM in porcine cerebellum.

TABLE 9

Asterric Acid Derivative

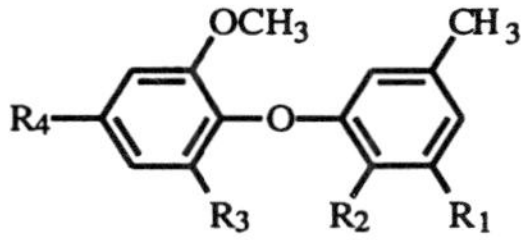

R^1	R^2	R^3	R^4	Binding IC_{50} (μM) ET_A	ET_B
COOH	OH	$COOCH_3$	OH	10[a]	ia
$COOCH_3$	OCH_3	$COOCH_3$	OCH_3	nr	nr
COOH	OH	COOH	OH	nr	nr
$CONHNH_2$	OH	$COOCH_3$	OH	nr	nr

[a] A10 cells; nr = not reported.

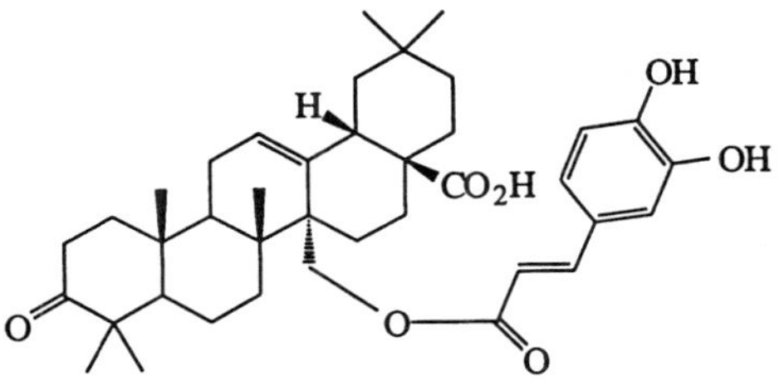

FIGURE 6
Myriceron caffeoyl ester.

TABLE 10

Ro 46-2005 Analogues

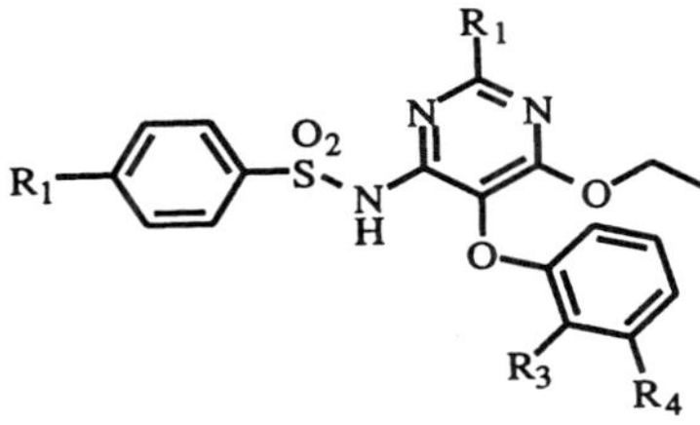

R_1	R_2	R_3	R_4	Binding IC_{50} (μM) ET_A	ET_B
H (Ro 46-2005)	tBu	H	OCH_3	0.216[a]	0.221[c]
H	iPr	OCH_3	H	0.073[a]	nr
CF_3 (Ro 46-203)	i-Pr	OCH_3	H	0.05[a]	nr
3-Pyrimidyl (Ro 47-0203)	t-Bu	OCH_3	H	0.020[b]	0.150[b]

[a] Human placental membranes.
[b] Human CHO-expressed ET_A or ET_B receptors.
[c] Porcine cerebellum membranes.

Ro 46-2005 inhibited the ET-1-induced contraction in rat aortic rings (ET_A) with a pA_2 of 6.5 and the sarafotoxin-6C (STX-c)-induced contraction in mesenteric arteries (ET_B) with a pA_2 of 6.5. This agent is a selective ET receptor antagonist as it does not antagonize contractions due to potassium, serotonin, AII, prostaglandin $F_{2\alpha}$ (PGF2), TXA_2 analog, or acetylcholine.

The *in vivo* efficacy of this antagonist has been reported in several animal models. Ro 46-2005 prevented postischemic renal vasoconstriction in rats. A 3 mg/kg i.v. dose of Ro 46-2005 effected a 66% increase in renal blood flow upon reperfusion vs. control. The decrease in cerebral blood flow due to subarachnoid hemorrhage in rats was also alleviated by Ro 46-2005. In this model, a 3 mg/kg dose of Ro 46-2005 dramatically increased blood flow vs. control when rats were subjected to subarachnoid hemorrhage by injection of autologous blood in the cisterna magna. Furthermore, although no significant effect on mean arterial blood pressure (MABP) was seen in the rat renal or subarachnoid hemorrhage models, Ro 46-2005 (doses of 10 to 100 mg/kg) exhibited a dose-dependent reduction in MABP in sodium-depleted squirrel monkeys.

By comparison of the oral dosing and i.v. injection of Ro 46-2005, an oral bioavailabilty of 30% has been determined in rats, with peak blood levels achieved 15 min post i.v. dosing and approximately 4 h post oral dosing.

There have been few SAR studies reported for this series although some conclusions can be noted. The addition of a fourth substituent at the 2 position of the pyrimidine ring seems to impart about a tenfold increase in ET_A potency. In addition, the electronic properties of substitution on the phenoxy group also seem to have an impact upon receptor affinity.

A recent compound from this series is Ro 47-0203 (Bosentan) (Clozel et al., 1993b). This derivative has improved *in vitro* binding affinity at both receptor subtypes: ET_A = 20 nM and ET_B = 150 nM (human). Ro 47-0203 inhibits contraction in isolated rat aorta (ET_A) pA_2 = 7.4 and in rat tracheal contraction (ET_B) pA_2 = 6.8. In a rabbit model of subarachnoid hemorrhage a 30 mg/kg dose reversed (rather than prevented) vasoconstriction. The same dose lowered MABP in SHRSP as well as in salt-depleted models but not in normotensive rats. An infusion of BQ-123 produced a 20-mmHg drop in MABP in SHR while an additional infusion of Bosentan lowered the MABP by 40 mmHg. This potent derivative has been selected for clinical studies and is reported to be under investigation in SAH, hypertension and congestive heart failure.

The Bristol-Myers Squibb group also reported a series of nonpeptide ET_A-selective antagonists (Stein et al., 1994). This series was developed from a compound library screening hit, sulfathiazole, *1*. These aryl sulfonamides (1 to 8 in Table 11) have IC_{50}s of from 150 nM to 69 μM at the ET_A receptor (A10 cells) with reportedly no binding affinity at the ET_B receptors. The sulfonamide hydrogen is critical for receptor binding affinity as the N-methyl derivative was devoid of activity. The two methyl groups of the isoxazole are also important for binding, the 4-methyl group being critical for binding. Modification of the 3-methyl group to larger alkyls and aryls led to a loss of binding affinity. Substitution on the phenyl group gave several potent compounds, but only those containing an alkyl- or arylalkyl-amino group exhibited functional activity (8, K_b = 100 μM vs. 7, $K_b \gg$ 100 μM).

Due to the increased functional activity with an increase in lipophilicity on the phenyl group, a series of naphthalenesulfonamides were synthesized. A large number of positional isomers were synthesized and led to the conclusion that a 1,5-disubstitution pattern was preferred for receptor affinity (5 to 8 Table 11). One of these substituted naphthalene compounds, (8, BMS 182874) has ET_A binding of IC_{50} = 150 nM (A10 cells). Functional antagonism was displayed by Compound 8, giving an IC_{50} of 570 nM as an antagonist of the ET-1-induced increase in intracellular Ca^{2+} in A10 cells and a K_b value of 520 nM in rabbit coronary artery rings. Oral activity

TABLE 11

Sulfathiazole Endothelin Antagonists

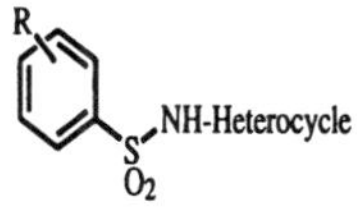

#	Heterocycle	R	Binding IC_{50} (μM) ET$_A$[a]	ET$_B$[b]
1	2-Thiazolyl	4-NH_2	69	ia
2	3,4-Dimethyl-5-isoxazolyl	4-NH_2	0.78	ia
3	3,4-Dimethyl-5-isoxazolyl	4-OH	9.2	ia
4	3,4-Dimethyl-5-isoxazolyl 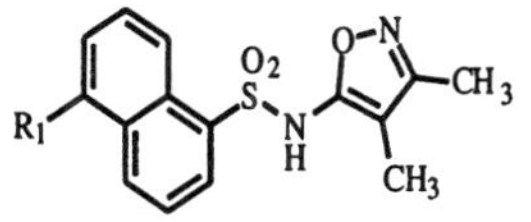	4-$NHCH_2C_6H_5$	9.8	ia
5		H	20	ia
6		OH	7.8	ia
7		NH_2	4.0	ia
8		$N(CH_3)_2$	0.15	>200[b]

[a] A10 cells.
[b] Rat cerebellum.

TABLE 12

CGS 27830 Analogues

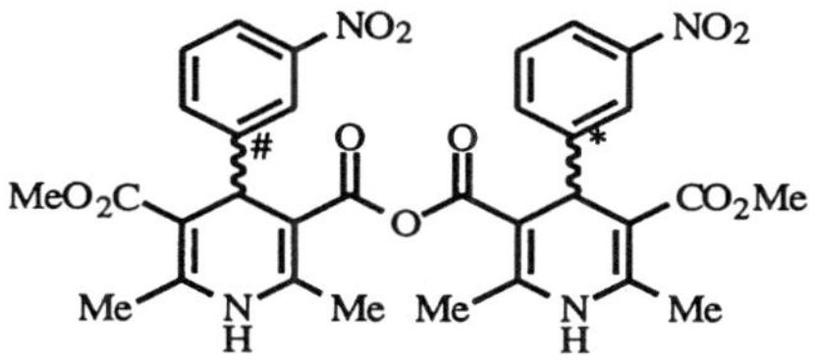

	#	*	Binding IC_{50} (μM) ET$_A$	ET$_B$
CGS 27830	S	R	0.0159[a]	0.295[b]
	S	S	ia	ia
	R	R	0.422[a]	2.7[b]

[a] Porcine thoracic aorta.
[b] Rat cerebellum.
ia = Inactive.

was also demonstrated for BMS 182874 in DOCA-salt hypertensive rats. A dose of 100 μmol/kg, either i.v. or p.o., provided a maximal drop in blood pressure of 32 to 45% with a sustained drop of 25 to 12% from 12 to 24 h post dosing.

Ciba-Giegy has also published a nonpeptide ET antagonist (CGS 27830) (Table 12), which is an asymmetrical dihydropyridine anhydride (Mugrage et al., 1993). This compound evolved from the calcium ion modulators of similar structure, presumably by compound library screening. The anhydride was inactive as a calcium channel antagonist but showed ET$_A$ (porcine thoracic aorta membranes) and ET$_B$ (rat cerebellum membranes) receptor binding affinity with IC_{50}s of 15.9 and 295 nM,

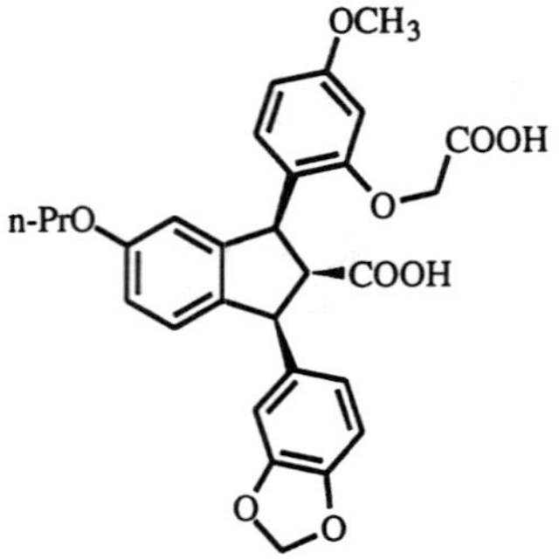

FIGURE 7
SKB 209670.

respectively. CGS 27830 also produced a dose-dependent inhibition of ET-1-induced contraction in isolated rabbit aorta, with a maximum attenuation of 77% at 10 μM. The compound was ineffective at altering the effects of KCl or phenylephrine, which suggests that CGS 27830 is a specific antagonist of ET receptors. A 5-min pretreatment (10 mg/kg, i.v.) in a conscious rat abolished the pressor response due to an ET-1 challenge and attenuated the depressor response. The half life of this antagonist, however, was found to be rather short ($T_{1/2}$ <60 min). The two optical centers in the anhydride appear to be important as the SR compound, CGS 27830, was the most potent isomer. The SS derivative was inactive at both receptors and the RR derivative was much less active (ET_A, IC_{50} = 422 nM and ET_B, IC_{50} = 2.7 μM). No additional ring substitutions have been published.

A series of trisubstituted indane carboxylic acids have also been recently discovered by researchers at Smith Kline Beecham (SKB) (Cousins et al., 1992). (+)SKB 209670 (Figure 7) is representative of this series (Ohlstein et al., 1994). This compound was rationally designed using conformational models of ET-1. (+)SB 209670 has potency at both ET_A and ET_B receptors with K_is = 0.2 nM and 18 nM, respectively (human cloned). SKB 209670 did not affect the basal hemodynamic parameters, continuous i.v., or bolus i.v. administration in male Sprague-Dawley rats. However, a 5-min pretreatment of SB 209670 (1 mg/kg) selectively attenuated the initial depressor response due to an i.v. bolus injection of ET-1 (0.3 nM/kg). The secondary pressor response was unaffected by this dose (Douglas et al., 1994). At a dose of 10 mg/kg SB 209670 abolished both responses due to ET-1. SKB 209670 has further been shown to be effective at reversing the effects of ischemia-induced acute renal failure in the rat (Gellai et al., 1994). A 30 mg/kg/min dose of SKB 209670 exhibited a significant change in several renal functions as well as a 75% increase in preventing mortality. SKB 209670 was also shown to be an effective antihypertensive agent in SHR (Elliott et al., 1994). MABP was reduced at doses of 10 mg/kg/min infusion and 10 mg/kg, i.d. Cardiac output was unaffected, as was heart rate. The compound was ineffective in Wistar-Kyoto rats.

Recently, a second series of ET antagonists from SKB has been disclosed. These derivatives are the tri- or tetrasubstituted propanoic (propenoic) acids of which Figure 8 is representative (Bryan and Elliot, 1993). No biological data have been reported on these analogues.

Immunopharmaceutics has also patented a series of nonpeptides as ET antagonists which can be represented by compounds in Table 15 (Chan and Balaji, 1993). These derivatives appear to be ET_A selective; however, only minimal data have been reported to date.

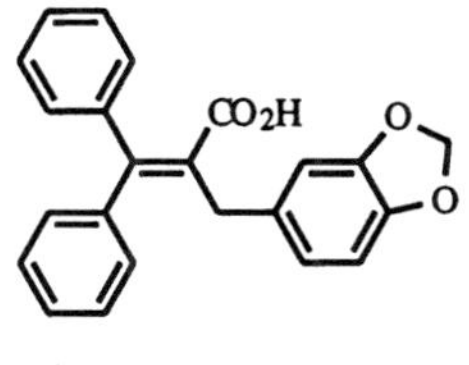

FIGURE 8
Tri- or Tetrasubstituted propanoic (propenoic) acids.

TABLE 13
Immunopharmaceutics Endothelin Antagonists

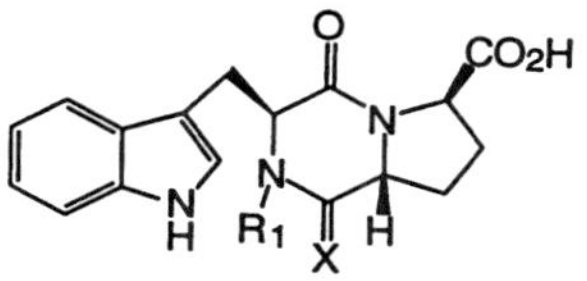

R_1	X	Binding IC_{50} (μM)[a] ETA	ETB
		ET_A	ET_B
H	O	22.5	nr
BOC	H_2	31.1	nr

[a] TE671 cells.

III. RECENTLY DISCLOSED ENDOTHELIN ANTAGONISTS

BQ-788, which was discussed briefly earlier in this chapter (Table 7), has indeed been shown to be an ET_B-selective antagonist. BQ-788 has only micromolar binding affinity at the ET_A receptor while it exhibits nanomolar binding affinity at the ET_B receptor (Table 14; Ishikawa et al.,1994). BQ-788 may be a useful tool in elucidating the importance of the ET_B receptor.

One of the most recently disclosed cyclic pentapeptides, designed by Takeda, is the nonselective antagonist TAK-044 (Figure 9) (Watanabe et al., 1995). It has nanomolar activity at the ET_A receptor but only 130 nM activity at the ET_B receptor (Table 14). Furthermore, TAK-044 has recently been shown to exhibit vasodilator effects in humans (Haynes et al., 1995). TAK-044 was well tolerated in all human volunteers and displayed dose-dependent decreases in systemic vascular resistance and increases in heart rate, cardiac index and circulating levels of plasma endothelin.

Parke-Davis has disclosed a series of nonpeptide ET_A-selective antagonists of which PD155080 and PD156707 (Figure 9) are representative (Doherty et al., 1995). PD155080 exhibits 400-fold selectivity for the human ET_A receptor, with IC_{50} = 8 nM (Table 14). PD156707 is approximately 20-fold more potent at the ET_A receptor (IC_{50} = 0.3 nM) and, most interestingly, is 1400-fold selective for the ET_A receptor. With this selectivity PD156707 is one of the most potent ET_A-selective nonpeptide antagonists reported. PD156707 exhibits a pA_2 value of 7.5 in rabbit femoral artery (ET_A) while displaying poor ET_B functional activity (pA_2 = 5.4, rabbit pulmonary artery). PD156707 is efficacious in inhibiting ischemic damage due to middle cerebral artery occlusion in a cat model of stroke (Patel et al., 1995).

More recently, this series of antagonists has yielded agents with balanced affinity for the human cloned ET receptors (Doherty et al., 1995). PD160874 is representative

TABLE 14

Receptor Binding of Recently Disclosed Endothelin Antagonists

	Binding IC_{50}(nM)	
Agent	ET_A	ET_B
BQ-788	1300[a]	1.2[b]
TAK-044	3.8[c]	130[d]
PD155080	8.0[e]	3510[f]
PD156707	0.3[e]	420[f]
PD160874	3.5[e]	8.9[f]
L-749,329	0.80[e]	16[f]
A-127722	0.15[e]	166[f]

[a] Human (Girardi heart).
[b] Human (SK-N-MC cells).
[c] Rat ventricle.
[d] Rat cerebellum.
[e] Human cloned.
[f] Human cloned.

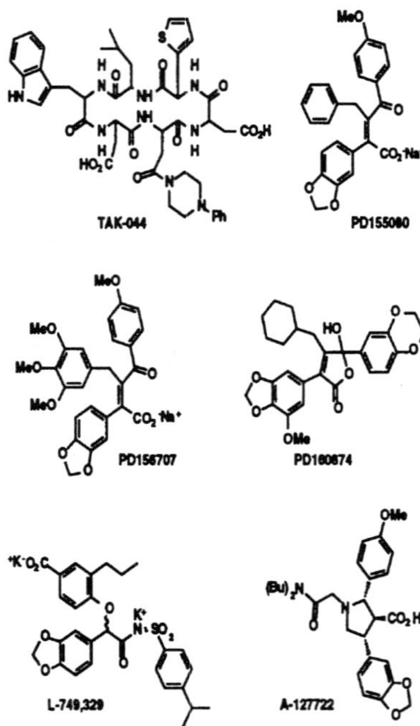

FIGURE 9
Recently disclosed antagonist structures.

of these modifications (Figure 9). The combination of these potent ET_A-selective and nonselective agents within a very closely related series may prove useful in defining the significance of endothelin in various human diseases.

Merck has also disclosed a nonpeptide antagonist L-749,329 (Figure 9) (Walsh et al., 1994; Greenlee et al., 1994). This antagonist is nonselective for human cloned receptors exhibiting subnanomolar affinity for the ET_A receptor and 16 nM affinity for the ET_B receptor (Table 14).

Abbott Laboratories has also very recently disclosed an ET_A-selective series of antagonists (Tasker et al., 1995,; Winn et al., 1995). A representative example (A-127722, Figure 9), was initially developed using SB 209670 as a model, replacing the indane ring with a pyrrolidine. It exhibits ET_A binding affinity of 0.15 nM and ET_B binding affinity of 166 nM in human cloned receptors (Table 14).

Additionally, several other patents for small-molecule ET antagonists have been disclosed although no structure-activity relationships have been published (Furuya and Ohtaki, 1994; Furuya et al., 1995; Sato et al., 1994; Ogawa et al., 1994; Ogawa et al., 1995).

REFERENCES

Abraham, W.M., Ahmed, A., Cortes, A., Spinella, M.J., Malik, A.B. and Andersen, T.T., A specific endothelin antagonist blocks inhaled endothelin-1-induced bronchoconstriction in sheep. *J. Appl. Physiol.*, 74, 2537, 1993.

Akiyama, H., Inagaki, Y., Kashiwabara, T., Ohta, H., Fushima, H. and Nishikori, K., *Jpn. J. Pharmacol.*, 58 (S1), 336, 1992.

Bonvallet, S., Morris, K., Zamora, M., Yano, M., McMurtry, I., Stelzner, T., A selective ET_A receptor antagonist (BQ123) attenuates the development of hypoxic pulmonary hypertension *in vivo*. *Annu. Rev. Respir. Dis.*, 147, A493, 1993.

Breu, V., Löffler, B.-M., and Clozel, M., *In vitro* characterization of Ro 46-2005, a novel synthetic nonpeptide endothelin antagonist of ET_A and ET_B receptors. *FEBS Lett.*, 334, 210-214, 1993.

Bryan, D. and Elliot, J., Endothelin Receptor Antagonists. WO 94/02474, July 15, 1993.

Burri, K., Clozel, M., Fischli, W., Hirth, G., Loffler, B.-M. and Ramuz, H., Phenyl-sulfonamido Pyrimidine(s), as Endothelin Receptor Inhibitors Used in Circulatory Disorders, Hypertension, Ischemia, Vasospasm and Angina. AU-A-18121/92, June 6, 1992.

Chan, M. and Balaji, V., Compounds That Modulate Endothelin Activity. WO 93/23404, May 17, 1993.

Clozel, M., Breu, V., Burri, K., Cassal, J-M., Fischli, W., Gray, G., Hirth, G., Löffler, B.-M., Müller, M., Neidhart, W., and Ramuz, H., Pathophysiological role of endothelin revealed by the first orally active endothelin receptor antagonist. *Nature*, 365, 759-761, 1993a.

Clozel, M., Clozel, S.-P., and Hess, P., Endothelin receptor antagonism: a new therapeutic approach in experimental hypertension. *Circulation*, 88, Abstr. 1689, 1993b.

Clozel, M. and Watanabe, H., BQ-123, a peptidic endothelin ET_A receptor antagonist, prevents the early cerebral vasospasm following subarachnoid hemorrhage after intracisternal but not intravenous injection. *Life Sci., 52*, 825-834, 1993.

Cody, W.L., Doherty, A.M., He, J.X., DePue, P.L., Rapundalo, S.T., Hingorani, G.A., Major, T.C., Panek, R.L., Dudley, D., Haleen, S.J., LaDouceur, D., Hill, K.E., Flynn, M.A. and Reynolds, E.E., Design of a functional hexapeptide antagonist of endothelin. *J. Med. Chem.*, 35, 3301, 1992a.

Cody, W.L., Doherty, A.M., He, J.X., DePue, P.L., Waite, L.A., Topliss, J.G., Haleen, S.J., LaDouceur, D., Flynn, M.A., Hill, K.E. and Reynolds, E.E., The rational design of a highly potent combined ET_A and ET_B receptor antagonist (PD 145065) and related analogues. *Med. Chem. Res.*, 3, 154, 1993.

Cody, W.L., Doherty, A.M., He, J.X., Topliss, J.G., Haleen, S.J., LaDouceur, D., Flynn, M.A., Hill, K.E. and Reynolds, E.E., Structure-activity relationships in the C-terminus of endothelin-1 (ET-1). The discovery of potent antagonists. *Peptides*, 687, 1992b.

Cody, W.L., He, J.X., DePue, P.L., Rapundalo, S.T., Hingorani, G.P., Dudley, D.T., Hill, K.E., Reynolds, E.E. and Doherty, A.M., Structure-activity relationships in a series of monocyclic endothelin analogs. *Bioorg. Med. Chem. Lett.*, 4, 567, 1994.

Cody, W.L., Doherty, A.M., He, X., Rapundalo, S.T., Hingorani, G.P., Panek, R.L. and Major, T.C., Monocyclic endothelins: examination of the individual disulfide rings. *J. Cardiovasc. Pharmacol.*, 17 (Suppl.7), S62, 1991.

Cosentino, F. and Katusic, Z., Does endothelin-1 play a role in the pathogenesis of cerebral vasospasm?, *Stroke*, 25, 904-908, 1994.

Cousins, R., Elliot, J., Lago, M., Leber, J. and Peishoff, C., Endothelin Receptor Antagonists. WO 93/08799, Oct. 29, 1992.

de Castiglione, R., Tam, J.P., Liu, W., Zhang, J.-W., Galantino, M., Bertolero, F. and Vaghi, F., Alanine scan of endothelin. *Peptides: Chemistry and Biology;* Proc. 12th Am. Peptide Symp., Smith, J.A. and Rivier, J.E., Eds., ESCOM, Leiden, Netherlands, 1992, 402.

Doherty, A., Patt, W., Edmunds, J., Berryman, K., Reisdorph, B., Plummer, M., Shahripour, A., Lee, C., Cheng, X., Walker, D., Haleen, S., Keiser, J., Flynn, M., Welch, K., Hallak, H., Taylor, D. and Reynolds, E., Discovery of a series of orally active nonpeptide endothelin (ET_A) receptor-selective antagonists. *J. Med. Chem.*, 38, 1259-1263, 1995.

Doherty, A., Patt, W., Repine, J., Edmunds, J., Berryman, K., Reisdorph, B., Walker, D., Haleen, S., Keiser, J., Flynn, M., Welch, K., Hallak, H. and Reynolds, E., Structure-activity relationships (SAR) of a novel series of orally active nonpeptide ET_A receptor selective antagonists. Fourth Int. Conf. Endothelin, London, April 23-26,1995; Abst. C40.

Doherty, A.M., Endothelin: a new challenge. *J. Med. Chem.*, 35, 1493, 1992.

Doherty, A.M., Cody, W.L., DePue, P.L., He, J.X., Waite, L.A., Leonard, D.M., Leitz, N.L., Dudley, D., Rapundalo, S.T., Hingorani, G.A., Haleen, S.J., LaDouceur, D., Hill, K.E., Flynn, M.A. and Reynolds, E.E., Structure-activity relationships of C-terminal endothelin hexapeptide antagonists. *J. Med. Chem.*, 36, 2585, 1993b.

Doherty, A.M., Cody, W.L., He, J.X., DePue, P.L., Cheng, X.-M., Welch, K.M., Flynn, M.A., Reynolds, E.E., LaDouceur, D.M., Davis, L.S., Keiser, J.A. and Haleen, S.J., *In vitro* and *in vivo* studies with a series of hexapeptide endothelin antagonists. *J. Cardiovasc. Pharmacol.*, 22 (Suppl. 8), 1993c.

Doherty, A.M., Cody, W.L., He, J.X., DePue, P.L., Leonard, D.M., Dunbar, J.B., Hill, K.E., Flynn, M.A. and Reynolds, E.E., Design of C-terminal peptide antagonists of endothelin. Structure-activity relationships of ET-1[16-21, D-His^{16}]. *Bioorg. Med. Chem. Lett.*, 4, 497, 1993a.

Douglas, S., Elliott, J. and Ohlstein, E., SB 209670 inhibits the hemodynamic effects of endothelin-1 (ET-1) in the anesthetized rat. *FASEB J. Abstr. (Exp. Biol. 94)*, April 24-28, 1994, Abstr. 246.

Elliott, J., Ezekiel, M., Gellai, M., Douglas, S. and Ohlstein, E., Antihypertensive effect of the endothelin receptor antagonist SB 209670 in conscious spontaneously hypertensive rats (SHR). *FASEB J. Abstr. (Exp. Biol. 94)*, April 24-28, 1994, Abstr. 40.

Fujimoto, M., Mihara, S-I., Nakajima, S., Ueda, M., Nakamura, M. and Sakurai, K.-S., A novel nonpeptide endothelin antagonist isolated from bayberry, *Myrica cerifera*. *FEBS Lett.*, 305, 41-44, 1992.

Fukuroda, T., Fujikawa, T., Ozaki, S., Ishikawa, K., Yano, M. and Nishikibe, M., Clearance of circulating endothelin-1 by ET_B receptors in rats. *Biochem. Biophys. Res. Commun.*, 199, 1461-1465, 1994.

Furuya, S., Choh, N., Ohtaki, T. and Watanabe, T. Thienopyrimidine Derivatives, Their Production, and Use as Endothelin Antagonists. EP 640606 A1, March 1, 1995.

Furuya, S. and Ohtaki, T., Pyrido[2,3-d]pyrimidines and Their Use as Endothelin Antagonists. EP 608565 A1, August 3, 1994.

Gellai, M., Jugas, M., Fletcher, T., Nambi, P., Brooks, D., Ohlstein, E., Elliott, J., Gleason, J. and Rutiolo, R., The endothelin receptor antagonist, (+)SB 209670, reverses ischemia-induced acute renal failure (ARF) in the rat. *FASEB J. Abstr. (Exp. Biol. 94)*, April 24-28, 1994, Abstr. 1506.

Greenlee, W., Walsh, T., Pettibone, D., Tata, J., Rivero, R., Williams, D., Bagley, S., Dhanoa, D., Chakravarty, P., et al. Phenoxyphenylacetic Acid Derivatives Useful as Endothelin Antagonists. EP-617001 A1, September 28, 1994.

Gulavita, N., Wright, A., McCarthy, P., Pomponi, S., Kelley-Borges, M., Chin, M. and Sills, M., Isolation and structure elucidation of 34-Sulfatobastadin 13, an inhibitor of the endothelin A receptor, from a marine sponge on the genus *Ianthella*. *J. Nat. Prod.*, 56, 1613-1617, 1993.

Hasimoto, M., Nishikawa, M., Esaki, M., Kiyoto, S., Okuhara, M. Takase, S., Henmi, K., Neya, M., Fukami, N. and Hashimoto, M., New Peptide Prepared by Culturing *Streptomyces* sp. Used as Endothelin Antagonist. JO 3130-299-A, filed August 8, 1990.

Haynes, W., Skwarski, K., Wyld, P. and Ripke, H., Vasodilator effects of the ET A/B receptor antagonist, TAK-044, in man. Fourth Int. Conf. Endothelin, April 23-26, 1995; Abstr. C24.

Hemmi, K., Neva, M. Fukami, N. Hashimoto, M., Tanaka, H. and Kayakiri, N., Peptides Having Endothelin Antagonist Activity, A Process for Preparation Thereof and Pharmaceutical Compositions Comprising the Same. EPA 0457 195 A2, May 9, 1991.

Hingorani, G.A., Major, T.C., Panek, R.L., Reynolds, E.E., He, J.X., Cody, W.L., Doherty, A.M. and Rapundalo, S.T., *In vitro* pharmacology of a nonselective ET_A/ET_B endothelin receptor antagonist, PD 142893 (Ac-(b-phenyl)D-Phe-L-Leu-L-Asp-L-Ile-L-Ile-L-Trp trifluoracetate). *FASEB J.*, 6 (Part 1, No.4), 392, 1992.

Hunt, J.T., Lee, V.G., McMullen, D., Liu, E.C.-K., Bolgar, M., Delaney, C.L., Festin, S.M., Floyd, D.M., Hedberg, A., Natarajan, S., Serafino, R., Stein, P.D., Webb, M.L., Zhang, R. and Moreland, S., Structure-activity studies of endothelin leading to novel peptide ET_A antagonists. *Bioorg. Med. Chem.*, 1, 59, 1993.

Hunt, J.T., Lee, V.G., Stein, P.D., Hedberg, A., Liu, E.C., McMullen, D. and Moreland, S., Structure-activity relationships of monocyclic endothelin analogues. *Bioorg. Med. Chem. Lett.*, 1, 33, 1991.

Ihara, M., Fukuroda, T., Saeki, T., Nishikibe, M., Kojiri, K., Suda, H. and Yano, M., An endothelin receptor (ET_A) antagonist isolated from *Streptomyces misakiensis*. *Biochem. Biophys. Res. Commun.*, 178, 132-137, 1991.

Ihara, M., Noguchi, K., Saeki, T., Fukuroda, T., Tsuchida, S., Kimura, S., Fukami, T., Ishikawa, K., Nishikibe, M. and Yano, M., Biological profiles of highly potent novel endothelin antagonists selective for the ET_A receptor. *Life Sci.*, 50, 247-255, 1992.

Ishikawa, K., Fukami, T., Hayama, T., Niiyama, K., Nagase, T., Mase, T., Fujita, K., Kumagai, U., Urakawa, Y., Ihara, M., Kimura, S. and Yano, M., Twelfth Am. Peptide Symp., Cambridge, MA, June 16-21, 1991: 506.

Ishikawa, K., Fukami, T., Nagase, T., Fujita, K., Hayama, T., Niiyama, K., Mase, T., Ihara, M. and Yano, M., Cyclic pentapeptide endothelin antagonists with high ET_A selectivity. Potency and solubility enhancing modifications. *J. Med. Chem.*, 35, 2139-2142, 1992a.

Ishikawa, K., Fukami, T., Nagase, T., Mase, T., Ihara, M., Yano, M. and Nishikibe, M., Peptides With Endothelin Antagonist Activity Useful as Bronchodilators and Vasodilators. EPA 0555-537-A, Nov. 1992b.

Ishikawa, K., Ihara, M., Noguchi, K., Mase, T., Mino, N., Saeki, T., Fukuroda, T., Fukami, T., Ozaki, S., et al., Biochemical and pharmacological profile of a potent and selective endothelin B-receptor antagonist, BQ-788. *Proc. Natl. Acad. Sci. U.S.A.*, 91(11), 4892-4896, 1994.

Itoh, S., Sasaki, T; Ide, K., Isikawa, K., Nishikibe, M. and Yano, M., A novel endothelin ET_A receptor antagonist, BQ-485, and its preventative effect on experimental cerebral vasospasm in dogs. *Biochem. Biophys. Res. Commun.*, 195, 969-975, 1993.

Karaki, H., Sudjarwo, S.A., Hori, M., Sakata, K., Urade, Y., Takai, M. and Okada, T., ET_B receptor antagonist, IRL 1038, selectively inhibits the endothelin induced endothelium-dependent vascular relaxation. *Eur. J. Pharmacol.*, 231, 371, 1993.

Kitada, C., Ohtaki, T., Masuda, V., Masuo, V., Nomura, H., Asami, T., Matsumoto, Y., Satou, M., and Fujino, M., Design and synthesis of ETA receptor antagonists and study of ETA receptor distribution, *J. Cardiovasc. Phamacol.*, 22 (Suppl. 8) S128-S131.

Kojiri, K., Ihara, M., Nakajima, S., Kawamura, K., Funaishi, K., Yano, M. and Suda, H. Endothelin-binding inhibitors, BE-18257A and BE-18257B. I. Taxonomy, fermentation, isolation and characterization. *J. Antibiot.*, 44, 1342-1347, 1991.

LaDouceur, D., Davis, L.S., Keiser, J.A., Doherty, A.M., Cody, W.L., He, J.X. and Haleen, S.J., Effects of the endothelin receptor antagonist PD 142893 (Ac-(b-phenyl)D-Phe-L-Leu-L-Asp-L-Ile-L-Ile-L-Trp trifluoro-acetate) on endothelin-1 (ET-1) induced vasodilation and vasoconstrictor in regional arterial beds of the anesthetized rat. *FASEB J.*, 6 (Part 1, No. 4), 390, 1992.

Lam, Y., Williams, D., Jr., Sigmond, J., Sanchez, M., Genilloud, O., Kong, Y., Stevens-Miles, S., Huang, L. and Garrity, G., Cochinmicins, novel and potent cyclodepsipeptide endothelin antagonists from a *Microbispora* sp. I. Production, isolation, and characterization. *J. Antibiot.*, 45, 1709-1716, 1992a.

Lam, Y., Zink, D., Williams, D., Jr. and Burgess, B., Additional cochinmicins from a *Microbispora* sp. *J. Antibiot.*, 45, 1792-1794, 1992b.

Miasiro, N., Nakaie, C.R. and Paiva, A.C.M., Endothelin (16-21): biphasic effect and no desensitization on the guinea-pig isolated ileum. *Br. J. Pharmacol.*, 109, 68, 1993.

Mihara, S-I. and Fujimoto, M., The endothelin ET_A receptor-specific effect of 50-235, a nonpeptide endothelin antagonist. *Eur. J. Pharm. (Mol. Pharm. Sec.)*, 246, 33-38, 1993.

Mino, N., Kobayashi, M., Nakajima, A., Amano, H., Shimamoto, K., Ishikawa, K., Watanabe, K., Nishikibe, M., Yano, M. and Ikemoto, F., Protective effect of a selective endothelin receptor antagonist, BQ-123, in ischemic acute renal failure in rats. *Eur. J. Pharmacol.*, 221, 77-83, 1992.

Miyata, S., Hashimoto, M., Fujie, K., Shouho, M., Sogabe, K., Kiyoto, S., Okuhara, M. and Kohsaka, M. WS009 A and B, new endothelin receptor antagonists isolated from *Streptomyces* sp. no. 89009. II. Biological characterization and pharmacological characterization of WS009 A and B. *J. Antibiot.*, 45, 1041-1046, 1992b.

Miyata, S., Ohhata, N., Murai, H., Masui, Y., Ezaki, M., Takase, S., Nishikawa, M., Kiyoto, S., Okuhara, M. and Kohsaka, M., WS009 A and B, new endothelin receptor antagonists isolated from *Streptomyces* Sp. no. 89009. I. Taxonomy, fermentation, isolation, physio-chemical properties and biological activities. *J. Antibiot.*, 45, 1029-1040, 1992a.

Mugrage, B., Moliterni, J., Robinson, L., Webb, R., Shetty, S., Lipson, K., Chin, M., Neale, R. and Cioffe, C., CGS 27830, A potent nonpeptide endothelin receptor antagonist. *Bioorg. Med. Chem. Lett.*, 10, 2099-2104, 1993.

Muldoon, L.L., Rodland, K.D., Forsythe, M.I. and Magnn, B.E., Stimulation of phosphatidylinositol hydrolysis, diacyl glycerol release and gene expression in response to endothelin, a potent new agonist for fibroblasts and smooth muscle cells. *J. Biol. Chem.*, 264, 8529, 1989.

Nishikibe, M., Tsuchida, S., Okada, M., Fukuroda, T., Shimamoto, K., Yano, M., Ishikawa, K. and Ikemoto, F., Antihypertensive effect of a newly synthesized endothelin antagonist, BQ-123, in a genetic hypertensive model. *Life Sci.*, *52*, 717-724, 1993.

Ogawa, T., Tanaka, T., Tsukoda, E., Kakita, S., Koda, M., Ochiai, K., Ando, K. and Matsuda, Y., Benz[a]anthracene-1,7,12 (2H)-Triones as Endothelin Antagonists. JP 07002664 A2, January 6, 1995.

Ogawa, T., Tanaka, T., Tsukuda, E., Uosaki, Y., Yoshida, M., Ando, K., Iwasaki, T., Kita, K. and Matsuda, Y. Endothelin Antagonist RES-1149-1 Manufacture With Aspergillus and Preparation of RES-1149-1 Derivatives. WO 9422806 A1, October 13, 1994.

Ohashi, H., Akiyama, H., Nishikori, K. and Mochizuki, J.-I., Asterric acid, a new endothelin binding inhibitor. *J. Antibiot.*, 45, 1684-1685, 1992.

Ohlstein, E., Nambi, P., Douglas, S., Beck, G. and Gleason, J. *In vitro* characterization of the nonpeptide endothelin receptor antagonist SB 209670. *FASEB J. Abstr. (Exp. Biol. 94)*, April 24-28, 1994, Abstr. 4655.

Patel, T., Galbraith, S., McAuley, M., Doherty, A. and McCulloch, J., Therapeutic potential of endothelin receptor antagonists in experimental stroke. Fourth Int. Conf. Endothelin, April 23-26, 1995; Abstr. C63.

Pegoraro, S., Rovero, P., Sedo, A., Revoltella, R.P., Mizrahi, J., Telemaque, S. and D'Orleans-Juste, P., Structure-activity relationship studies of the C-terminus of endothelin-1, ET(16-21). 13th Am. Peptide Symp. Edmonton, Alberta, 1993, Abstr. 362.

Reynolds, E.E. and Mok, L., ET-stimulated arachidonic acid release in vascular smooth muscle cells. *FASEB J.*, 5, A1066, 1991.

Rovero, P., Pegoraro, S., Sedo, A. and Rvoltella, R.P., A new endothelin C-terminal analogue IBDP 064 antagonizes endothelin-3-induced rat glioma cell proliferation. *Biomed. Pharmacother.*, 47, 249, 1993.

Sato, A., Watanabe, T., Iijima, Y. and Haruyama, H., Nahocols and/or Isonahocols as Endothelin-1 Antagonists. JP 06256261 A2, September 13, 1994.

Sogabe, K., Nirea, H., Shoubo, M., Nomoto, A., Henmi, K., Notsu, Y., and Ono, T., J. Vasc. Res. 2nd Int. Symp. Endothelium-Derived Vasoactive Factors., 1992, p. 201, Abstr. 367.

Spellmeyer, D.L., Brown, S., Stanber, G.B., Geysen, H.M. and Valerio, R., Endothelin receptor ligands. Replacement net approach to SAR determination of potent hexapeptides. *Bioorg. Med. Chem. Lett.*, 3, 519, 1993a.

Spellmeyer, D.L., Brown, S., Stanber, G.B., Geysen, H.M. and Valerio, R., Endothelin receptor ligands. Multiple D-amino acid replacement net approach to SAR determination of potent hexapeptides. *Bioorg. Med. Chem. Lett.*, 3, 1253, 1993b.

Spinella, M.J., Malik, A.B., Everitt, J. and Anderson, T.T. Design and synthesis of a specific endothelin 1 antagonist. Effects on pulmonary vasoconstriction. *Proc. Natl. Acad. Sci. U.S.A.*, 88, 7443, 1991.

Stein, P., Hunt, J., Floyd, D., Moreland, S., Dickenson, K., Mitchell, C., Liu, E., Webb, M., Murugesan, N., Dickey, J., McMullen, D., Zhang, R., Lee, V., Serafino, R., Delany, C., Schaeffer, T. and Kozlowski, M., The discovery of sulfonamide endothelin antagonists and the development of the orally active ET_A antagonist 5-(dimethylamino)-N(3,4-dimethyl-5-isoxazolyl)-1-naphthalenesulfonamide. *J. Med. Chem.*, 37, 329-331, 1994.

Tasker, A., Sorenson, B., Jae, H., von Geldern, T., Dixon, D., Chiou, W., Dayton, B., Calzadila, S., Hernandez, L., Marsh, K., WuWong, J. and Opgenorth, T., Potent and selective non benzodioxole-containing endothelin-A receptor antagonists. Book of Abstracts, 210th National A. C. S. Meeting, Chicago, IL, 1995. Abstract MEDI-34.

Urade, Y., Fujitani, Y., Oda, K., Watakabe, T., Umemura, I., Takai, M., Okada, T., Sakata, K. and Karaki, H., An endothelin B receptor-selective antagonist: IRL 1038, [Cys^{11}-Cys^{15}]-endothelin-1(11-21). *FEBS Lett.*, 311, 12, 1992.

Urade, Y., Fujitani, Y., Oda, K., Watakabe, T., Umemura, I., Takai, M., Okada, T., Sakata, K. and Karaki, H., Retraction concerning an endothelin B receptor-selective antagonist: Urade, Y., Fujitani, Y., Oda, K., Watakabe, T., Umemura, I., Takai, M., Okada, T., Sakata, K., and Karaki, H. (1992), *FEBS Lett., 311, 12-16; FEBS Lett.*, 342, 103, 1994.

Wakimasu, M., Kikuchi, T., Kubo, K., Asami, T., Ohtaki, T. and Fujino, M., Studies on endothelin antagonists. *Perspectives in Medical Chemistry,* Barnard, T., Ed., Verlag Helvetica Chim. Acta, Basel, 1993, 165.

Walsh, T., Fitch, K., Chakravarty, P., Williams, D., Murphy, K., Nolan, N., O'Brien, J., Lis, E., Pettibone, D., Kivlighn, S., Gabel, R., Zingaro, G., Krause, S., Siegl, P., Clineschmidt, B. and Greenlee, W., The discovery of L-749,329, a highly potent, orally active antagonist of endothelin receptors. Book of Abstracts, 208th National A. C. S. Meeting, Washington, D.C. 1994. Abstr. MEDI-145.

Warner, T.D., Allcock, G.H. and Vane, J.R., The endothelin receptor antagonist PD 142893 inhibits endothelium-dependent vasodilatations induced by endothelin sarafotoxin peptides. *Br. J. Pharmacol.*, 109, 56, 1993.

Watakabe, T., Urade, Y., Takai, M. Umemura, I and Okada, T., A reversible radioligand specific for the ET_B receptor. *Biochem. Biophys. Res. Commun.* 185, 867-73, 1992.

Watanabe, T., Awane, Y., Ikeda, S., Fujiwara, S., Kubo, K., Kikuchi, T., Kusumoto, K., Wakismasu, M. and Fujino, M. Pharmacology of a nonselective ET_A and ET_B receptor antagonist, TAK-044 and the inhibition of myocardial infarct size in rats. *Br. J. Pharmacol.*, 114(5), 949-954, 1995.

Werber, A.W., Spinella, M.J. and Anderson, T.T., [DPR-1 Asp-15]endothelin-1 does not antagonize endothelin-1 in the superior cerebellar artery of rats. *FASEB J.*, 6, A1005, 1992.

Winn, M., Tasker, A., Boyd, S., Jae, H., von Geldern, T., Bal, R., Mantei, R., Wu-Wong, R., Chiou, W., Dixon, D. and Opgenorth, T., Endothelin antagonists (I)- S.A.R. Of 2,4-diaryl-3-carboxy-pyrrolidine-1-acetamides. Book of Abstracts, 210th National A. C. S. Meeting, Chicago, IL, 1995, Abstr. MEDI-30.

Zink, D., Hensens, O., Lam, Y., Reaner, R. and Liesch, J., Cochinmicins, novel and potent cyclodepsipeptide endothelin antagonists from a *Microbispora* sp. II. Structure determination. *J. Antibiot.*, 45, 1717-1722, 1992.

Part V

Clinical Studies

Chapter **17**

Clinical Pharmacology and Pathophysiology of the Endothelins

William G. Haynes and David J. Webb

CONTENTS

0-8493-6975-4/97/$0.00+$.50

I. INTRODUCTION

The widespread expression of mRNA for members of the ET family, and the distribution of their receptors, suggests that these peptides play an important role in regulation of the cardiovascular system. ET-1 is probably of most importance in regulation of vascular tone, given its predominant expression in the endothelium (Howard et al., 1992). ET-1 appears to be primarily a locally acting substance because, under physiological circumstances, circulating plasma concentrations are usually not sufficient to elicit direct vasoconstriction (Webb and Cockcroft, 1989). However, concentrations at the interface between endothelial cells and vascular smooth muscle cells are probably higher, because more ET-1 is secreted abluminally than luminally (Yoshimoto et al., 1991).

Studies in animals and *in vitro* have shown that ET-1 is a potent vasoconstrictor, has inotropic and mitogenic properties, influences salt and water homeostasis, and stimulates generation of renin, angiotensin II, aldosterone, and epinephrine. In addition, centrally and peripherally administered ET-1 enhances sympathetic activity. Together, these actions promote vasoconstriction and increase blood pressure. Although there is a wealth of data from such studies in animals, the known interspecies heterogeneity of endothelium-dependent vascular responses limits the value of extrapolating from animals to humans (Lüscher and Vanhoutte, 1988). This chapter, therefore, addresses the cardiovascular and renal pharmacology of ET as demonstrated *in vivo* in humans. The potential physiological significance of ET in cardiovascular regulation in humans is also covered. In addition, we discuss the evidence suggesting a pathophysiological role for ET in cardiovascular and renal diseases.

II. HUMAN VASCULAR PHARMACOLOGY OF ENDOTHELIN

A. Vascular Actions

The most valuable technique for elucidating the direct vascular effects of ET-1 in humans has been the assessment of the effects on forearm blood flow of brachial artery administration of nonsystemic doses of the peptide. This technique does not require systemically active doses of drugs that may obscure any direct vascular action by direct effects on other organs such as the heart and kidney, or activate reflex mechanisms due to changes in blood pressure. In addition, responses in resistance vessels of this circulation appear to parallel those in other beds (Collier et al., 1978). Brachial artery infusion of ET-1 causes a characteristically slow-onset forearm

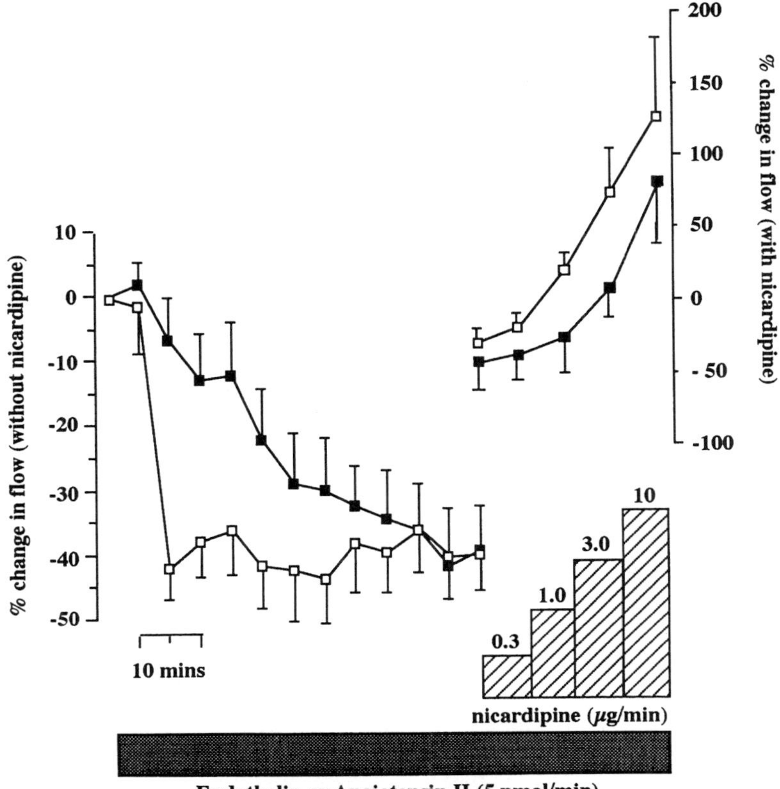

FIGURE 1
Effect of brachial artery administration of ET-1 (■; n = 9) and angiotensin II (□; n = 8) for 60 min on forearm blood flow and the effect of incremental doses of nicardipine on these responses (n = 4 each agonist). The left-hand scale refers to ET-1 and angiotensin II alone, and the right-hand scale refers to the coinfusion of nicardipine. ET-1 and angiotensin II cause similar vasoconstriction, but the response to ET-1 is characteristically slower in onset, taking 60 min to reach maximal. Coinfusion of nicardipine (0.3 to 10 μg/min) after development of vasoconstriction to ET-1 or angiotensin II results in a similar vasodilatation, implying that ET-1 does not have a specific action at dihydropyridine-sensitive Ca^{2+} channels. Values are mean ± SEM. (From Haynes, W. G., Clarke, J., Cockcroft, J. R., and Webb, D. J., *J. Cardiovasc. Pharmacol.*, 17 (Suppl. 7), S284, 1991. With permission.)

vasoconstriction, taking up to 60 min to reach a steady-state response (Figure 1), and with vasoconstriction sustained for more than 2 h after halting the infusion (Clarke et al., 1989; Hughes et al., 1989; Dahlöf et al., 1990; Pernow et al., 1991, Cockcroft et al., 1991; Kiowski et al., 1991; Kiowski and Linder, 1992; Andrawis et al., 1992; Haynes and Webb, 1994, 1995a). In contrast to the slow onset of action for ET-1, forearm vasoconstriction to angiotensin II, given intra-arterially at the same dose, is rapid in onset with a maximal effect apparent after only 5 min (Figure 1) (Clarke et al., 1989). Intradermal injection of nonsystemic doses of ET-1 into the skin of the forearm causes a local decrease in skin microcirculatory blood flow, but with an increase in flow to the surrounding area (Crossman et al., 1991; Wenzel et al., 1994). Local intravenous administration of ET-1 causes constriction of human dorsal hand veins *in vivo* (Clarke et al., 1989; Haynes and Webb, 1993a, b; Haefeli et al., 1993; Haynes et al., 1995a). As in arteries, this constriction is characteristically slow in onset and sustained for several hours (Haynes et al., 1991). In contrast to the *in vitro* data obtained using animal vessels (Cocks et al., 1989), human veins appear to be less sensitive than human arteries to ET-1.

These studies thus demonstrate that ET-1 is capable of constricting human resistance, nutritive, and capacitance vessels *in vivo.*

B. Endothelin Converting Enzyme

ET-1 is generated through the action of an ET converting enzyme (ECE) on a 38-amino-acid precursor, big ET-1 (see Chapter 6). In humans, brachial artery administration of big ET-1 causes a dose-dependent forearm vasoconstriction that is completely blocked by phosphoramidon (Figures 2A and 2B), suggesting that the effects of the precursor are mediated through conversion to the mature peptide by ECE (Haynes and Webb, 1994). Blockade of vasoconstriction to big ET-1 by phosphoramidon is unlikely to be due to inhibition of ET receptor binding, given that vasoconstriction to ET-1 is unaffected by phosphoramidon (Figure 2C). Because circulating blood does not exhibit ECE activity (Watanabe et al., 1991a), conversion of big ET-1 in the forearm probably occurs via vascular ECE situated within the forearm blood vessels. Big ET- 1 is approximately tenfold less potent then ET-1, suggesting that local ECE converts about 10% of luminally presented big ET-1 to ET-1. Interestingly, big ET-1 does not constrict dorsal hand veins *in vivo*, even at supramaximal doses, implying that ECE is poorly expressed in cutaneous capacitance vessels (Haynes et al., 1995b).

Further evidence that generation of mature ET-1 is responsible for forearm vasoconstriction to big ET-1 through the action of a phosphoramidon-sensitive ECE is derived from measurement of the concentrations of ET peptides in blood drawn from antecubital veins draining infused and noninfused forearms during these local infusion studies. Plasma immunoreactive ET-1, big ET-1, and the inactive C-terminal fragment (CTF) of ET-1 that is formed during the cleavage of big ET-1 by ECE, have been measured (Plumpton et al., 1995). Local infusion of big ET-1 causes a large increase in big ET-1 in plasma from the infused arm and a slight increase in big ET-1 in plasma from the control arm. Concentrations of ET- 1 and CTF in venous plasma from the infused arm, but not the control arm, also increase significantly during infusion of big ET-1, indicating local cleavage of big ET-1. The ratio of CTF to big ET-1 is about ~0.1, indicating 10% conversion of the precursor by ECE, in keeping with the functional results. Interestingly, the increase in ET-1 in the infused arm is relatively more modest, with a ratio of ET-1 to big ET-1 of ~0.02. This probably reflects rapid binding of newly formed ET-1 to receptors near the site of conversion, before ET- 1 can pass into blood. There is no significant increase in plasma ET-1 or CTF when big ET-1 is coinfused with phosphoramidon. In addition to confirming biochemically the *in vivo* activity of ECE demonstrated functionally, these results suggest that circulating concentrations of CTF are more likely than circulating ET-1 concentrations to accurately reflect generation of ET-1. Thus, assay of CTF and big ET-1 may prove a valuable tool in future investigations of ET generation in health and disease.

Xu and colleagues have reported the isolation, structure, and characterization of ECE-1, a membrane-bound neutral metalloprotease that is expressed abundantly in endothelial cells *in vivo*, and that is structurally related to neutral endopeptidase 24.11 (Xu et al., 1994). Interestingly, compared with 50 to 90% conversion of endogenous big ET-1 in cells transfected with cDNA for ECE-1, the conversion of exogenous big ET-1 by such cells is much less efficient (~10%) and similar to that found in the functional and biochemical studies in humans reviewed above.

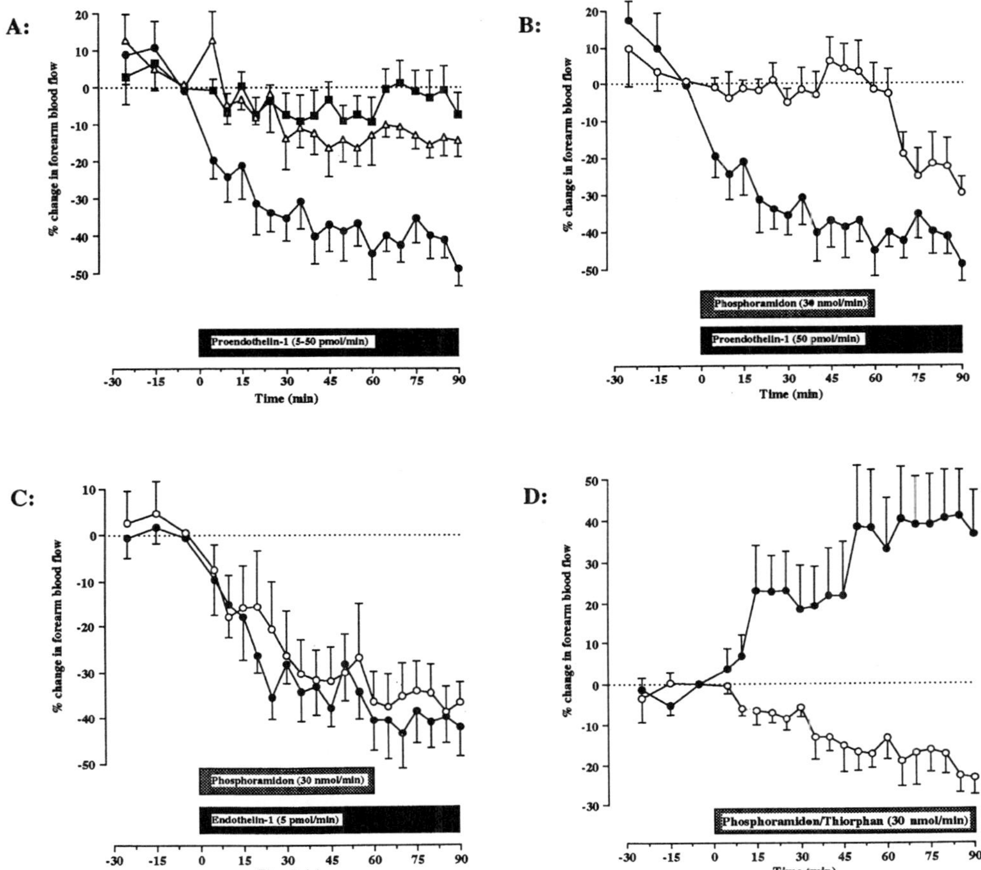

FIGURE 2
A: Brachial artery infusion of local doses of big ET-1 at 5 (■), 15 (Δ), and 50 pmol/min (●) causes slow-onset forearm vasoconstriction in healthy men, demonstrating the presence of ECE in human blood vessels *in vivo*. B: Forearm vasoconstriction to intraarterial big ET-1 (●; 50 pmol/min) is abolished by coinfusion of the ECE and neutral endopeptidase inhibitor phosphoramidon (○; 30 nmol/min). C: Forearm vasoconstriction to intraarterial ET-1 (●; 5 pmol/min) is unaffected by coinfusion of the ECE inhibitor phosphoramidon (○; 30 nmol/min), demonstrating that the effect of phosphoramidon on big ET-1 responses is not due to inhibition of responses to the mature peptide. D: Brachial artery infusion of the ECE and neutral endopeptidase inhibitor phosphoramidon (●; 30 nmol/min) results in slow-onset forearm vasodilatation. In contrast, the selective neutral endopeptidase inhibitor thiorphan (○; 30 nmol/min) causes modest forearm vasoconstriction. Thus, ECE inhibition with phosphoramidon causes forearm vasodilatation, offset to some degree by vasoconstriction caused by neutral endopeptidase inhibition. These results suggest a role for generation of ET-1 in the maintenance of vascular tone. Values are mean ± SEM (n = 6 for each group). (Adapted from Haynes, W. G. and Webb, D. J., *Lancet*, 344, 852, 1994. With permission.)

C. Receptors

Although molecular biology techniques suggest expression of both ET_A- and ET_B-receptors in vascular smooth muscle (Davenport et al., 1993), the functional significance of such vascular smooth muscle ET_B-receptors in humans has been unclear. *In vitro* studies using human blood vessels have shown that ET_B-receptors make either a minimal (Riezebos et al., 1994; Fukuroda et al., 1994) or, at most, a moderate contribution (Seo et al., 1994; White et al., 1994) to vasoconstriction to ET-1. *In vivo* in humans, selective ET_B-receptor agonists, such as ET-3 and STXc, constrict human

A:

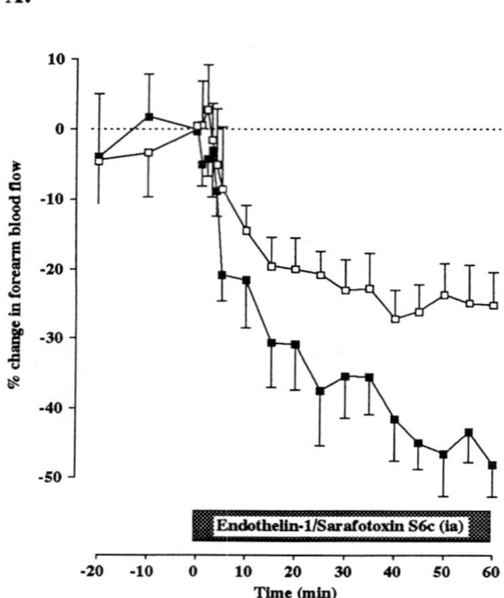

B:

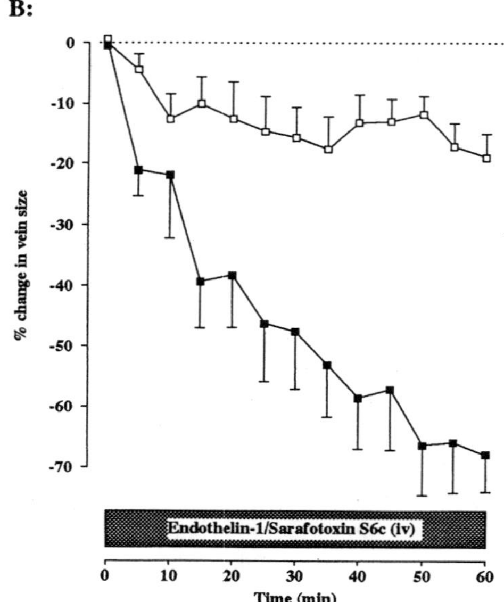

FIGURE 3
A: Effect of brachial artery infusion of the nonselective $ET_{A/B}$-receptor agonist, ET-1 (■; 5 pmol/min), and on a separate occasion, STXc (❏; 5 pmol/min), each for 60 min. B: Effect of dorsal hand vein infusion of ET-1 (■; 5 pmol/min) and on a separate occasion, STXc (❏; 5 pmol/min), each for 60 min. Both peptides cause significant constriction of forearm resistance and cutaneous capacitance vessels, with the effect of STXc less than that of ET-1 in both resistance and capacitance vessels. These results suggest that a substantial proportion of vasoconstriction to ET-1 is mediated by ET_B-receptors; ia = intraarterial; iv = intravenous. Values are mean ± SEM (n = 6 for each group). (From Haynes, W. G., Strachan, F. E., and Webb, D. J., *Circulation*, 92, 357, 1995. With permission.)

resistance and capacitance vessels (Figure 3) (Haynes et al., 1995a). The magnitude of this effect is such that ET_B-receptors may account for up to 50% of vasoconstrictor ET receptors in human subjects.

In contrast, human skin microcirculatory flow is reduced by intradermal injection of ET-1, but not by ET-3, suggesting a predominance of ET_A-receptors in skin nutritive microvessels (Wenzel et al., 1994). A relative absence of vasoconstrictor ET_B-receptors in these vessels is confirmed by the almost complete blockade of local vasoconstriction to ET-1 when the selective ET_A antagonist PD 147,953 is coinjected. ET-3, and to a lesser extent ET-1, cause transient forearm vasodilatation when administered as a bolus via the brachial artery, consistent with the presence of an endothelial ET_B-receptor that mediates release of endothelium-dependent dilators (Kiowski et al., 1991; Haynes et al., 1995a; see below). However, venodilatation to the ETs has not been shown in human skin capacitance vessels (Haefeli et al., 1993).

D. Second Messengers

In view of the characteristically sustained vasoconstrictor effects of ET-1, much interest has focused on the mechanisms underlying the sustained rise in free intracellular Ca^{2+} caused by ET-1. It was initially proposed that ET-1 might be an endogenous agonist of the dihydropyridine-sensitive voltage-operated Ca^{2+} channel (Yanagisawa et al., 1988; Goto et al., 1989). In humans, however, although infusion of dihydropyridine Ca^{2+} channel antagonists blocks forearm vasoconstriction to ET-1 (Clarke et al., 1989; Kiowski et al., 1991; Kiowski and Linder, 1992; Andrawis et al., 1992), this blockade is not specific to ET-1 because vasoconstriction to angiotensin II is blocked to an equal or greater degree (Figure 1) (Clarke et al., 1989). Thus, at least some of the effects of Ca^{2+} channel antagonists are likely to be mediated against basal arteriolar tone, rather than specifically against vasoconstriction to ET-1. However, there is increasing evidence that ET-1 interacts closely with membrane K^+ channels. The ATP-sensitive K^+ channel openers, cromakalim and pinacidil, and the K^- ionophore, valinomycin, antagonize ET-1-induced contractions of animal vascular tissue *in vitro*, while the Ca^{2+} channel antagonists, verapamil and nicardipine, are less effective (Kim et al., 1989; O'Donnell et al., 1990). There is also evidence that ET-1 closes ATP-sensitive K^+ channels in porcine coronary smooth muscle cells (Miyoshi et al., 1992), and that the activity of a cloned cardiac K^+ channel can be suppressed by ET-1 (Ishii et al., 1992). In humans, dorsal hand vein constriction to ET-1 is blocked more effectively by an opener of ATP-sensitive K^+ channels than a Ca^{2+} antagonist (Haynes and Webb, 1993a), suggesting that ET-1 responses in human veins depend only in part on Ca^{2+} entry through dihydropyridine-sensitive Ca^{2+} channels. The greater efficacy of K^+ channel-opening agents is consistent with ET-1 acting to close ATP-sensitive K^+ channels, thus leading to plasma membrane depolarization and vasoconstriction by mechanisms additional to opening of voltage-operated Ca^{2+} channels.

E. Other Mediators of Vascular Tone

As discussed above, the ETs cause transient arterial vasodilatation *in vivo* in humans (Kiowski et al., 1991; Haynes et al., 1995a), before the development of sustained vasoconstriction. *In vitro* evidence suggests that this is due to generation of endothelium-derived dilators. Inhibition of PG, but not NO, generation potentiates venoconstriction to ET-1 and STXc *in vivo* in humans (Haynes and Webb, 1993b; Strachan et al., 1995). However, as noted earlier, ET-1 does not cause venodilatation when administered to preconstricted hand veins (Haefeli et al., 1993). Thus, the venous endothelium may generate vasodilator substances in response to ET-1, but the vasodilator effects of such substances appear to be masked by the simultaneous direct venoconstriction caused by the peptide, and serve only to modulate venoconstriction. Vasodilatation to ET-1 in human skin appears to be mediated by NO and histamine, probably through an axon reflex (Crossman et al., 1991).

There is *in vitro* and whole animal evidence to suggest that ET-1 has central and peripheral actions on the nervous system to potentiate sympathetic nerve activity (see Chapters 11 and 12). However, in humans, forearm resistance vessel constriction to lower-body negative pressure and norepinephrine is not potentiated by low doses of ET-1 administered via the brachial artery (Cockcroft et al., 1991). In addition, sympathetically mediated venoconstriction of human dorsal hand veins *in vivo* is unaltered by local infusion of ET-1 at doses that cause local venoconstriction (Haynes et al., 1995b). Thus, under physiological circumstances, there is no evidence *in vivo* in humans for an interaction between ET-1 and the peripheral autonomic nervous system.

III. SYSTEMIC ACTIONS OF ENDOTHELIN IN HUMANS

A. Blood Pressure

Intravenous infusion of ET-1 at doses of 0.4 to 4 pmol/kg/min, sufficient to increase plasma immunoreactive ET concentrations by 2- to 50-fold, has been shown to elevate blood pressure in humans by 5 to 10% (Figure 4) (Vierhapper et al., 1990, 1992; Weitzberg et al., 1991; Wagner et al., 1992; Gasic et al., 1992; Sørensen et al., 1994). As with forearm vasoconstriction, this pressor effect is characteristically slow in onset, taking at least 60 min to reach maximum. The similarity between venous plasma ET concentrations achieved in these studies and those observed in cardiovascular disease has led some investigators to suggest a pathophysiological role for ET-1. However, a number of factors may confound this interpretation. It is likely that arterial ET concentrations will be substantially higher than venous concentrations in these studies, and arterial rather than venous concentrations are more relevant to the exposure of resistance vessels to ET-1. This may lead to an overestimation of the importance of ET-1 in cardiovascular regulation. On the other hand, the preferential secretion of ET-1 abluminally implies that circulating plasma concentrations are likely to be less than concentrations at the endothelial cell-vascular smooth muscle interface. In this case, hemodynamic responses at a given plasma concentration following systemic infusion may underestimate the physiological or pathophysiological importance of ET-1. Thus, any role that ET-1 may play in health or disease is probably best investigated using blockade of its generation or action. The pressor response to ET-1 is not affected by pretreatment with cyclosporin, the Ca^{2+} antagonist nifedipine, or cyclooxygenase inhibition with indomethacin (Vierhapper et al., 1992). Unlike in animals, depressor responses to ET-1 have not been demonstrated in humans probably because, for reasons of safety, ET-1 has been administered by intravenous infusion rather than as bolus doses.

B. Cardiac

Infusion of systemic doses of ET-1 to human subjects tends to decrease cardiac output (Wagner et al., 1992), probably through a baroreceptor-mediated decrease in heart rate (Figure 4), although an increase in afterload may also contribute. Coronary vasoconstriction has not been observed in clinical studies, but does occur in humans

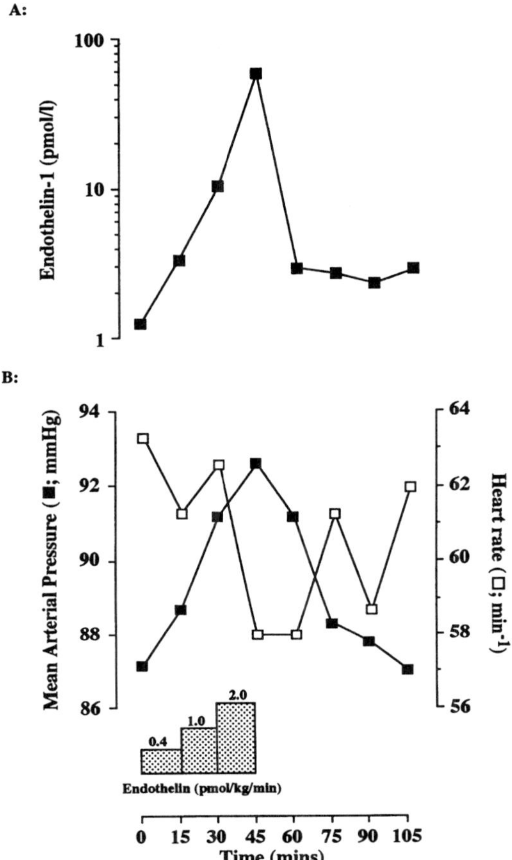

FIGURE 4
Effect of intravenous infusion of ascending systemic doses of ET-1 on venous plasma immunoreactive ET concentrations (■; panel A), and blood pressure (■; panel B), and heart rate (□; panel B) in six subjects. ET-1 markedly increases circulating ET concentrations, accompanied by a modest increase in blood pressure and a decrease in heart rate. (Adapted from Vierhapper, H., Wagner, O., Nowotny, P., and Waldhausl., W., *Circulation*, 81, 1415, 1990. With permission.)

who are bitten by the burrowing asp *Actractaspis engaddensis* (Weiser et al., 1984), whose venom contains sarafotoxins that act as ET receptor agonists.

C. Renal

The renal vascular bed appears to be particularly sensitive to the vasoconstrictor effects of ET-1. Systemic infusion of ET-1 causes renal vasoconstriction *in vivo* in humans, with renal vascular resistance increasing by 33 to 66% (Weitzberg et al., 1991; Gasic et al., 1992; Sørensen et al., 1994; Rabelink et al., 1994). Infusion of ET-1 decreases renal sodium excretion (Sørensen et al., 1994) even at doses insufficient to cause renal vasoconstriction (Rabelink et al., 1994). Thus there is no *in vivo* evidence in humans for a renal tubular diuretic action of ET-1, in contrast to the situation observed *in vitro* and in animals *in vivo*. It is possible that any such tubular effect is obscured by the potent renal vasoconstrictor effects of ET-1 and is most likely to be detected using selective ET receptor antagonists.

D. Pulmonary

Given that ~50% of ET-1 is cleared from the blood when passing through the pulmonary circulation (Wagner et al., 1992), one might expect this bed to be more responsive than the systemic vasculature following intravenous administration of the peptide. However, in one study, intravenous infusion of a relatively low dose of ET-1 (0.4 pmol/kg/min) did not increase pulmonary vascular resistance, although the dose was sufficient to increase blood pressure by 5% and splanchnic and total systemic vascular resistance by about 20% (Wagner et al., 1992). A more recent study, using a higher intravenous dose of ET-1 (3 pmol/kg/min), did show an increase in pulmonary vascular resistance which was greater than that observed for systemic vascular resistance (Kiely et al., 1996). Studies using ECE inhibitors or ET receptor antagonists will help to elucidate whether the pulmonary circulation is more or less responsive to ET-1 than the systemic circulation.

E. Other Beds

Intravenous infusion of ET at a pressor dose (0.4 pmol/kg/min) causes splanchnic vasoconstriction of ~20% (Weitzberg et al., 1991; Wagner et al., 1992), a little less than that observed for the renal circulation. In contrast, no significant change in leg blood flow occurs at this dose of ET-1 (Gasic et al., 1992), suggesting that the splanchnic and renal circulations are more sensitive than skeletal muscle to the vasoconstrictor effects of ET-1.

IV. PHYSIOLOGICAL ROLE OF ENDOTHELINS

There has been much interest in whether endogenous generation of ETs plays a physiological role in the maintenance of basal vascular tone and blood pressure. However, results of animal studies using ECE inhibitors and ET receptor antagonists have been contradictory. Some have shown no apparent effect of anti-ET therapy on blood pressure in normotensive animals (Gardiner et al., 1991, 1992, 1994; Douglas et al., 1992; Ihara et al., 1992; Cirino et al., 1994; Filep et al., 1994). However, most of these studies had not been primarily designed to test this hypothesis, with the result that they may have lacked statistical power to confidently exclude a hypotensive effect. In addition, in some, blood pressure was not measured for sufficient time after dosing to detect the expected slow-onset hypotensive effect of anti-ET therapy. Other studies have shown that ECE inhibitors and ET-receptor antagonists do decrease blood pressure in normotensive animals (McMahon et al., 1991; Bigaud and Pelton, 1992; Pollock and Opgenorth, 1993; Véniant et al., 1994; Clozel et al., 1993; Hoffman et al., 1994). These positive studies have usually examined haemodynamic responses for several hours after drug administration, thereby taking into account the known slow reversal of ET-1-induced vasoconstriction by anti-ET therapy (Warner et al., 1994). To date, there have been a few studies that have specifically examined this issue in humans; these are discussed below.

A. Local Inhibition of Endothelin Generation

As noted earlier, forearm resistance vessels appear to contain a phosphoramidon-sensitive ECE, able to convert exogenous big ET-1 to the mature peptide (Haynes and Webb, 1994). Inhibition of endogenous generation of ET-1 by brachial artery infusion of phosphoramidon alone causes progressive vasodilatation of forearm resistance vessels, suggesting that generation of ET-1 by ECE provides an important contribution to the maintenance of basal vascular tone in humans (Figure 2D) (Haynes and Webb, 1994). However, phosphoramidon also inhibits neutral endopeptidase, an enzyme that degrades a number of peptide hormones including the natriuretic peptides, angiotensins, and ETs (Erdos and Skidgell, 1989; Abassi et al., 1993). Vasodilatation to phosphoramidon could therefore be due to inhibition of forearm neutral endopeptidase causing local accumulation of dilator natriuretic peptides. This possibility appears highly unlikely because brachial artery infusion of the selective neutral endopeptidase inhibitor, thiorphan, causes forearm vasoconstriction rather than vasodilatation (Figure 2D). Thus, the forearm vasodilator effects of phosphoramidon through ECE inhibition may be offset to some degree by vasoconstriction through inhibition of neutral endopeptidase causing impaired breakdown of constrictor peptides, such as angiotensin II and ET-1. Such vasoconstriction may explain why systemic inhibition of neutral endopeptidase fails to decrease blood pressure despite increasing sodium excretion (O'Connell et al., 1993).

There appear to be two cellular sites for ECE, with low-efficiency cell surface conversion of exogenous big ET-1 and high-efficiency processing of endogenous big ET-1 in membrane-associated organelles (Xu et al., 1994). Given that phosphoramidon is less potent as an inhibitor of ECE-1-mediated cleavage of endogenous big ET-1, it is possible that the vasodilatation observed with phosphoramidon infusion was not maximal, even though the dose was sufficient to block conversion of exogenous big ET- 1. This contention is supported by the results of studies using ET receptor blockade.

B. Local Endothelin Receptor Antagonism

Brachial artery infusion of the specific and selective ET_A-receptor antagonist BQ 123 prevents vasoconstriction to locally administered ET-1, even 30 min after halting BQ 123 (Figure 5). Infusion of BQ 123 alone causes progressive and substantial forearm vasodilatation, greater than that observed with phosphoramidon. These results suggest that endogenous generation of ET-1 maintains vascular tone in humans through activation of ET_A-receptors. However, a role for ET_B-receptors in mediating vasoconstriction to endogenously generated ET-1 cannot presently be excluded, given that exogenous ET-1 can produce vasoconstriction through an action on ET_B-receptors (Haynes et al., 1995a). Exact delineation of the physiological role of ET_B-receptors awaits the results of similar studies with selective ET_B-receptor antagonists.

C. Systemic Endothelin Receptor Antagonism

We have reported, in preliminary form, the effects of systemic ET receptor antagonism in healthy normotensive subjects, using a potent peptide $ET_{A/B}$-receptor antagonist, TAK 044 (Haynes et al., 1995c). Intravenous administration of TAK 044 decreases blood pressure by ~15% and systemic vascular resistance by ~25%, with these effects sustained for ~8 h. In addition, TAK 044 increases plasma immunoreactive ET concentrations by up to tenfold, probably through inhibition of ET_B-receptor-mediated clearance of ET-1. Furthermore, TAK 044, in doses as low as 30 mg, completely blocks forearm vasoconstriction to locally administered ET-1 for several hours. These results support the findings in the forearm resistance bed with phosphoramidon and

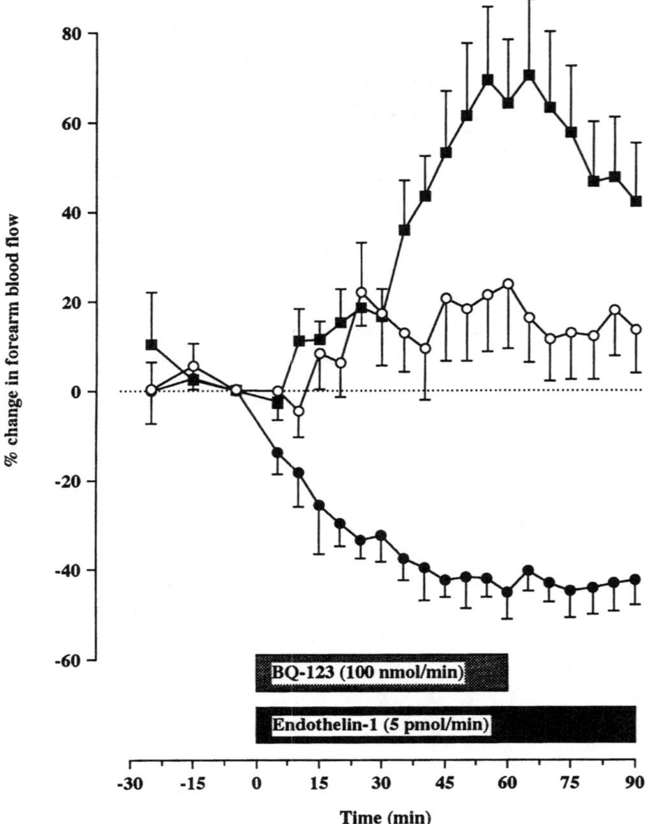

FIGURE 5
Brachial artery infusion of ET-1 to healthy human subjects causes significant forearm vasoconstriction (●) that can be abolished by coinfusion of the ET_A-receptor antagonist BQ 123 (○). Infusion of BQ 123 alone causes progressive and substantial forearm vasodilatation (■). These results support a physiological role for endogenous generation of ET-1, acting through ET_A-receptors, in the maintenance of basal vascular tone in healthy humans. Values are mean ± SEM (n = 6 for each group). (Reproduced from Haynes, W. G. and Webb, D. J., *Lancet*, 344, 852, 1994. With permission.)

BQ 123, and confirm an important physiological role for endogenously generated ET-1 in maintenance of peripheral resistance and blood pressure.

The only other factors, besides ET-1, that have been shown to have a similar fundamental physiological role in maintenance of basal vascular tone are the sympathetic nervous system and NO (Vallance et al., 1989). Sustained vasoconstriction elicited by ET-1 may act in concert with the short-lived effects of the sympathetic nervous system and NO to stabilize vasomotor tone, while preserving flexibility in its dynamic control. These results in humans suggest that orally active ECE inhibitors and ET receptor antagonists may have potential therapeutic uses as novel vasodilator agents.

V. PATHOPHYSIOLOGICAL SIGNIFICANCE OF ENDOTHELIN

In the absence of human studies using ECE inhibitors or ET receptor antagonists, much of the work on the role of ET in human pathophysiology has been based on changes in circulating plasma concentrations of immunoreactive ETs. It is worth remembering that these are dependent not only on endogenous generation, but also on renal and receptor-mediated clearance and enzyme-mediated metabolism of the peptide. In some studies sensitivity to ET-1 has been examined, but because receptor number can be downregulated by increased ET-1 concentrations (Hirata et al., 1988), these studies may also prove difficult to interpret.

A. Vascular Disease

1. *Systemic Hypertension*

a. *Endothelin Generation in Hypertension*

In animal models of hypertension, immunoreactive ET concentrations are not raised unless accelerated hypertension is present, where they are positively correlated with plasma creatinine (Kohno et al., 1991). There appear to be strain-related differences, because although ET immunoreactivity is increased in blood vessels from deoxycorticosterone acetate (DOCA)-salt rats, it is decreased in blood vessels from SHR as compared to normotensive WKY rats (Larivière et al., 1993). A lack of a role for altered ET-1 generation in the pathophysiology of experimental hypertension is supported by the fact that polymorphisms of the preproET-1 gene do not cosegregate with blood pressure or cardiac weight in inbred Dahl rats (Cicila et al., 1994). Interestingly, there is linkage between a locus near the preproET-3 gene and blood pressure in these rats.

In human essential hypertension, several investigators have used elevated concentrations of circulating immunoreactive ET to suggest increased production (Kohno et al., 1990; Saito et al., 1990). However, because clearance of ET-1 depends on normal renal function (Kohno et al., 1989; Koyama et al., 1989), the elevated ET concentrations found in severe and accelerated phase hypertension are probably secondary to impaired renal clearance. Studies in hypertensive patients with normal renal function have shown similar concentrations of ET to those in normotensives (Davenport et al., 1990; Schiffrin and Thibault, 1991; Haynes et al., 1994). Indeed, in one study a negative correlation between blood pressure and plasma ET was observed in the hypertensive group (Davenport et al., 1990), making a global increase in generation of ET-1 unlikely as a cause of essential hypertension.

In pre-eclampsia, plasma concentrations of ET-1 are elevated in the presence of normal renal function and ET-1 may therefore play a role in the pathophysiology of this hypertensive condition of pregnancy (Florijn et al., 1991). Increased production of ET-1 also occurs in one secondary form of hypertension, albeit rare. Yokokawa and colleagues have described two cases of the skin tumor, hemangio-endothelioma, in which hypertension was associated with increased ET-1 plasma concentrations (Figure 6) (Yokokawa et al., 1991). Biopsies of tumor cells displayed increased expression of mRNA for preproET-1, and strong immunohistochemical staining for the peptide. Blood pressure and plasma ET concentrations returned to normal in both cases following surgical resection of the tumors and, in one patient, recurrence of the tumor led to another increase in both blood pressure and plasma ET-1 (Figure 6).

b. *Sensitivity to Endothelin in Hypertension*

The results of studies examining vascular sensitivity to ET- 1 in hypertension are difficult to interpret, due partly to the confounding potentiating effects of vascular hypertrophy. In animal studies comparing WKY and SHR, both conduit (renal artery and aorta) and mesenteric resistance vessels from the hypertensive rat have been shown to be more sensitive to the effects of ET-1 (Tomobe et al., 1988; Criscione et al.,

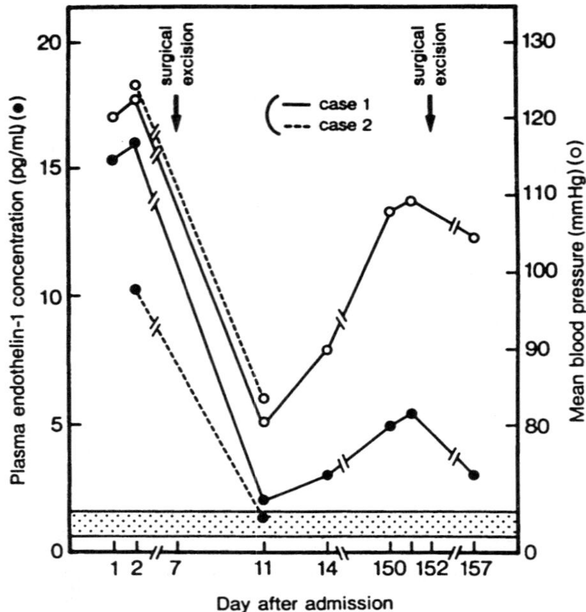

FIGURE 6
Sequential plasma immunoreactive ET concentrations (●) and mean blood pressure (○) in two patients with malignant hemangioendothelioma. In both cases, baseline blood pressure and circulating ET concentrations are elevated, and fall with surgical resection of the tumor. In case 1, a recurrence of the tumor is associated with increased blood pressure and plasma ET. The shaded area indicates the normal range of plasma immunoreactive ET concentrations in humans with this assay. (Reproduced from Yokokawa, K., Tahara, H., Kohno, M., et al., *Ann. Intern. Med.*, 114, 213, 1991. With permission.)

1990). In the Dahl salt-sensitive rat, vascular responsiveness to ET-1 is enhanced prior to, but not after, the development of hypertension (Goligorsky et al., 1991). However, other investigators have reported decreased sensitivity to ET-1 in the aorta and mesenteric resistance arteries from SHR (Dohi and Lüscher, 1991), DOCA-salt rats (Deng and Schiffrin, 1992) and renovascular hypertensive animals (Dohi et al., 1991; Roberts-Thomson et al., 1994). It is possible that decreased sensitivity to ET-1 may be related to downregulation of ET receptors secondary either to increased local generation of ET-1 or raised blood pressure (Larivière et al., 1993). Downregulation of receptors is suggested by the decreased Ca^{2+} response to ET-1 of isolated vascular smooth muscle cells from SHR (Touyz et al., 1994). Systemic doses of ET-1 have greater pressor effects in SHR than WKY rats (Miyauchi et al., 1989a) and in renovascular hypertensive rabbits (Roberts-Thomson et al., 1994). The number of binding sites for ET-1 in aortic smooth muscle (Clozel, 1989) and heart (Gu et al., 1990) are lower in SHR. There is, however, a relative increase in the number of binding sites for ET-1 in the brain of SHR, as compared to WKY rats (Gu et al., 1990), so there may be increased central nervous system sensitivity to the peptide in SHR.

Overall, the results of animal studies examining sensitivity to ET-1 in hypertension suggest that sensitivity is reduced in resistance vessels *in vitro* but may be increased in conduit vessels, particularly renal arteries. Increased sensitivity in these latter vessels may contribute to the hypertensive process. Some of the differences in responsiveness to ET-1 that have been observed may be related to the level of local ET-1 generation and to the relative proportions of ET_A- and ET_B-receptor subtypes in different vessels. However, it should be remembered that studies of vascular responses in hypertension may be confounded by the presence of vascular hypertrophy

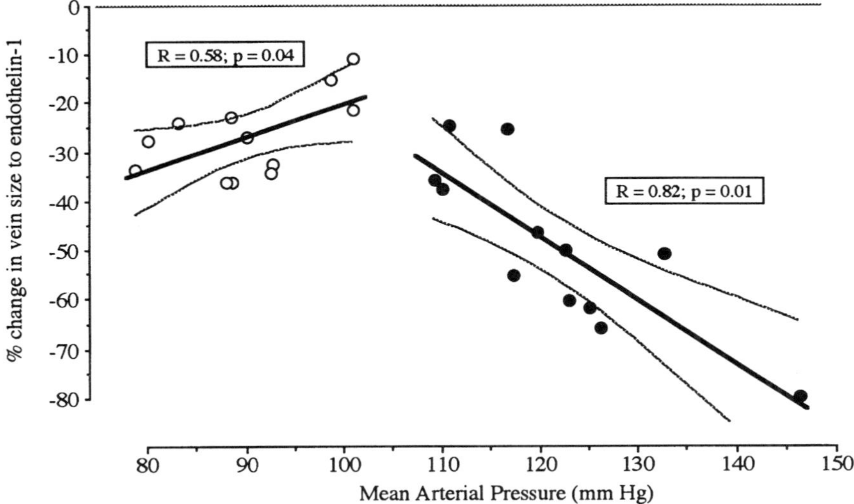

FIGURE 7
Dorsal hand vein constriction to locally administered ET-1 in untreated normotensive (○) and hypertensive (●) subjects and its relationship with resting blood pressure. Venoconstriction to ET-1 is greater in the hypertensive than in the control subjects. There is also a negative correlation between blood pressure and responsiveness to ET-1 in normotensive subjects but a positive correlation in hypertensive subjects. (Reproduced from Haynes, W. G., Hand, F. M., Johnstone, H. A., et al., *J. Clin. Invest.*, 94, 1359, 1994. With permission.)

in resistance vessels. In patients with essential hypertension, *in vitro* efficacy of ET-1 in subcutaneous resistance arteries appears to be reduced (Schiffrin et al., 1992). In untreated patients with essential hypertension, responsiveness to ET-1, but not norepinephrine, is increased in cutaneous hand veins, vessels in which vascular hypertrophy does not occur (Haynes et al., 1994). In addition, there is a positive correlation between venoconstriction to ET-1 and blood pressure in hypertensive subjects (Figure 7). Furthermore, ET-1 appears to potentiate sympathetically mediated vasoconstriction in hypertensive but not normotensive subjects. These findings suggest that ET-1 may contribute to the elevated preload observed in the early stages of essential hypertension. In summary, the results of studies examining sensitivity to ET-1 in hypertension suggest that sensitivity is reduced in resistance vessels *in vitro* but may be increased in capacitance and conduit vessels, particularly veins and renal arteries.

Some investigators (Panza et al., 1990; Linder et al., 1990) have shown impaired resistance vessel responses to endothelium-dependent dilators in essential hypertension, although this finding is by no means universal (Cockcroft et al., 1994). In addition, decreased basal generation of NO has been demonstrated in untreated essential hypertension (Calver et al., 1992). Alternatively, there may be greater production of, or sensitivity to, endothelium-derived vasoconstrictors such as ET-1. Even if impaired endothelial dilator function in hypertension is solely due to decreased production of NO, the balance between endothelium-derived dilator and constrictor factors will be altered to favor vasoconstriction to ET-1 or other hormones.

c. *Renal Effects of Endothelin in Hypertension*

There appear to be two separate ET "systems" mediating opposing actions in the kidney. The first is vasoconstriction of both afferent and efferent glomerular arterioles, ultimately leading to sodium retention (López-Farré et al., 1989), probably mediated through activation of ET_A-receptors (Maguire et al., 1994). Because the renal artery of the SHR is more sensitive to ET-1 (Tomobe et al., 1988), increased renal vascular generation of ET-1 may contribute to the pathophysiology of hypertension. The second major site of ET-1 generation in the kidney is the renal tubule, where it mediates salt and water excretion (Oishi et al., 1991; Kohan and Padilla, 1993), probably through activation of ET_B-receptors (Terada et al., 1992). Generation of ET-1 in the renal medulla, particularly the collecting duct, is decreased in SHR compared to WKY rats (Kitamura et al., 1989; Hughes et al., 1992). Because tubular generation of ET-1 increases urinary sodium and water excretion, any deficiency in renal tubule ET may cause sodium retention and thus hypertension in this strain of rats. Patients with essential hypertension have been shown to have less urinary immunoreactive ET excretion than normotensive controls (Hoffman et al., 1990), suggesting that the findings in SHR may be relevant to humans.

d. ECE Inhibitors and Endothelin Antagonists in Hypertension

Systemic administration of ECE inhibitors and ET receptor antagonists lowers blood pressure in rats, including hypertensive models. Although some studies have reported a decrease in blood pressure only in hypertensive rats (Bazil et al., 1992; Douglas et al., 1994), others have observed proportional decreases in blood pressure in normotensive and hypertensive strains (McMahon et al., 1991; Nishikibe et al., 1993). Studies with sufficient power have not yet been performed, so it is still unclear whether anti-ET therapy causes a proportionately greater fall in blood pressure in hypertensive than in normotensive animal models.

In subjects with essential hypertension, brachial artery adminstration of BQ 123 causes forearm vasodilatation of similar degree to that observed in normotensive subjects. A single oral dose (200 mg) of the nonpeptide $ET_{A/B}$-receptor antagonist bosentan has been shown to lower blood pressure by ~6 mmHg in patients with essential hypertension (Schmitt et al., 1995). The fact that $ET_{A/B}$-receptor antagonism also lowers blood pressure in normotensive human subjects (Haynes et al., 1995c) suggests that it is unlikely that anti-ET therapy will prove to have specific antihypertensive properties without effects on blood pressure in normotensive subjects.

e. Endothelin and Complications of Hypertension

The potent mitogenic effects (Komuro et al., 1988) of ET-1 might contribute to hypertension-induced hypertrophy of vascular smooth muscle, thus amplifying any vasoconstrictor influences (Folkow, 1978; Lever, 1986). In addition, ET-1 may play a role in promoting the development of left ventricular hypertrophy in hypertension, a factor that adversely affects prognosis (Levy et al., 1990). The growth-promoting properties of ET-1 may also contribute to the development of atherosclerosis. Atherosclerotic human blood vessels show increased mRNA expression and immunostaining for ET-1, and patients with atherosclerosis have raised plasma concentrations of the peptide, with the highest concentrations in patients with the largest number of affected vessels (Lerman et al., 1991; Winkles et al., 1993). Thus, although ET receptor blockade may not be selectively "antihypertensive", such therapy may be antimitogenic, with potential advantages in patients with disordered cardiac and vascular growth.

2. *Pulmonary Hypertension*

Pulmonary hypertension causes progressive shortness of breath and, ultimately, symptoms and signs of right ventricular failure. Pulmonary hypertension is usually secondary to conditions such as chronic obstructive airway disease (cor pulmonale) or chronic heart failure. A rare primary form afflicts mainly young people. Whatever the cause, pulmonary hypertension has poor prognosis and is difficult to treat. Pulmonary hypertension is characterized pathophysiologically by endothelial injury, vascular smooth muscle proliferation, and vasoconstriction of pulmonary resistance vessels. The vasoconstrictor and mitogenic effects of ET-1 therefore make it a plausible contributor to the pathophysiology of this disease.

In healthy subjects, 30 min of hypoxemia causes about a twofold increase in circulating immunoreactive ET-1, similar to that observed in patients with cor pulmonale (Cargill et al., 1995). Thus, it is possible that ET- 1 is involved in this form of secondary pulmonary hypertension. Indeed, plasma concentrations of ET-1 in venous blood are elevated in patients with both primary and secondary pulmonary hypertension (Stewart et al., 1991a). Also, the pulmonary circulation seems to generate more ET-1 than it clears in primary pulmonary hypertension, because the ratio of arterial to venous concentrations of ET-1 is significantly greater than unity (~2.2). This is not the case for healthy controls who have an arterial to venous ratio substantially less than unity (~0.6). Patients with primary and secondary pulmonary hypertension exhibit increased immunoreactivity and mRNA expression for ET-1 in the endothelial cells of hypertrophied pulmonary vessels, with the degree of expression proportional to pulmonary vascular resistance (Giaid et al., 1993). These findings are consistent with a pathophysiological role for ET-1 in the progression of pulmonary hypertension.

3. *Vasospastic Disorders*

Raynaud's disease and Prinzmetal's angina are characterized by instability of tone in small arteries and are associated with vasospasm in digital and coronary vessels, respectively. Patients with the latter have an increased incidence of Raynaud's phenomenon, migraine, and ocular vasospasm (Miller et al., 1981), suggesting that such instability of vascular tone may be a more widespread phenomenon, perhaps due to an imbalance between endothelium-derived dilator and constrictor factors. Indeed, circulating ET concentrations are elevated in patients with Raynaud's disease and coronary vasospasm, and between episodes of spasm, consistent with a generalized endothelial abnormality associated with increased ET-1 production (Zamora et al., 1990; Toyo-oka et al., 1991). In Raynaud's disease, circulating ET concentrations increase further following a cold challenge to a hand (Zamora et al., 1990), and ET concentrations are higher still in venous blood draining the cold-challenged hand. By contrast, in coronary vasospasm, ET concentrations do not increase further during ischemic episodes (Toyo-oka et al., 1991), suggesting that another factor, such as decreased NO production, may be responsible for initiating the attack. Because there is an association between coronary vasospasm, Raynaud's phenomenon, and migraine, it is also possible that ET-1 plays a role in the pathophysiology of migraine. Indeed, plasma concentrations of ET are elevated during acute migraine attacks (Farkilla et al., 1992).

The major complication of subarachnoid hemorrhage and the cause of considerable morbidity and mortality is later vasospasm of other cerebral arteries, leading to cerebral ischemia and infarction. Several lines of evidence support a role for ET-1 in the pathophysiology of this condition. First, exogenous application of ET-1 causes profound and sustained constriction of the cerebral vessels that undergo chronic constriction after subarachnoid hemorrhage (Asano et al., 1989; Kobayashi et al., 1991). Second, plasma and cerebrospinal fluid ET concentrations are elevated in patients with subarachnoid hemorrhage, and are higher in those who later develop cerebral vasospasm, with elevations in cerebrospinal concentrations temporally related to the occurrence of vasospasm (Suzuki et al., 1992; Ehrenreich et al., 1992). Third, cerebral vasospasm can be both prevented and reversed by the ECE inhibitor phosphoramidon (Matsumura et al., 1991) and by ET_A- (Clozel and Watanabe, 1993; Nirei et al., 1993; Foley et al., 1994) and $ET_{A/B}$-receptor antagonists (Clozel et al., 1993). These results suggest that cerebral vasospasm is associated with increased local generation of ET-1 that causes vasospasm predominantly through activation of ET_A-receptors.

B. Cardiac Disease

1. *Angina and Myocardial Infarction*

Plasma concentrations of ET are normal in patients with stable, and mild to moderate unstable, angina (Yasuda et al., 1990; Stewart et al., 1991b), although they are increased in patients with severe unstable angina, in whom they are associated with an increased risk of an adverse outcome (Weiczorek et al., 1994).

Cardiac tissue immunoreactive ET-1 increases substantially (about sixfold) after experimental myocardial infarction in rats (Watanabe et al., 1991b), with this tissue increase lasting longer than that in plasma (Watanabe et al., 1991b). In addition, the number of cardiac binding sites for ET-1 increases markedly following either ischemia alone, or ischemia with reperfusion (Liu et al., 1990). Furthermore, pretreatment with anti-ET gamma globulin or ET receptor antagonists in a rat model of myocardial infarction causes a 40% decrease in infarct size (Watanabe et al., 1991b, 1995; Grover et al., 1993). Thus, a combination of increased cardiac ET-1 content and additional binding sites for the peptide may contribute to extension of infarct size by compromising blood flow in reperfused areas. However, reduction of myocardial infarction size has not been confirmed in other animal studies using ET receptor antagonists (McMurdo et al., 1994; Richard et al., 1994).

In humans, circulating immunoreactive ET concentrations rise rapidly in the hours following myocardial infarction (Ray et al., 1992), with concentrations raised 2- to 3-fold on the day of infarction; and only returning to basal over the next 7 to 10 days (Miyauchi et al., 1989b). The duration and degree of elevation is proportional to the severity of infarction, as judged by Killip class, with the highest concentrations found in cardiogenic shock (Yasuda et al., 1990). These findings are consistent with the elevated ET-1 concentrations described in other forms of shock (Weitzberg et al., 1991b; Sanai et al., 1995), and may reflect a homeostatic response to maintain arterial pressure. High plasma ET concentrations 3 days after myocardial infarction are strongly predictive of death in the subsequent 12 months (Omland et al., 1994).

2. *Chronic Heart Failure*

Chronic heart failure is characterized by major left ventricular dysfunction at rest, breathlessness or fatigue, and evidence of sodium retention (The Task Force on Heart Failure of the European Society of Cardiology, 1995). Neurohumoral activation occurs in chronic heart failure, with increases in sympathetic nerve activity and circulating concentrations of epinephrine, norepinephrine, renin, angiotensin II, aldosterone, and vasopressin. It has been hypothesized that these neurohumoral changes, by causing peripheral vasoconstriction and sodium retention, contribute to the pathophysiology of chronic heart failure. The proven mortality benefits of angiotensin converting enzyme (ACE) inhibitor therapy in heart failure support the neurohumoral hypothesis (Dargie and McMurray, 1994). However, despite the benefits of ACE inhibitor therapy, morbidity and mortality are still markedly increased in patients with chronic heart failure.

As discussed earlier, systemic infusion of ET-1 increases blood pressure and peripheral vascular resistance, decreases cardiac output, and causes renal vasoconstriction and sodium retention. These properties suggest that ET-1 may also play a role in the neurohumoral maladaption to chronic heart failure and, like the renin-angiotensin system, contribute to the morbidity and mortality associated with this condition. Immunoreactive ET concentrations are elevated in patients with chronic heart failure (Hiroe et al., 1991; McMurray et al., 1992), correlate closely with the degree of haemodynamic and functional impairment (Cody et al., 1992; Pacher et al., 1993; Wei et al., 1994), and are associated with increased mortality or need for cardiac transplantation (Pacher et al., 1993). Big ET-1 represents a substantial component of total circulating immunoreactive ET in chronic heart failure (Pacher et al., 1993; Wei et al., 1994), perhaps reflecting increased generation of ET, rather than decreased clearance, in this condition.

Local administration of the ECE inhibitor phosphoramidon, and the ET_A-receptor antagonist BQ 123, cause forearm vasodilatation in patients with chronic heart failure already receiving maximal therapy with ACE inhibitors and diuretics (Love et al., 1994). Interestingly, in comparison with healthy subjects, vasodilatation to BQ 123 appears to be reduced and that to phosphoramidon increased in these patients with chronic heart failure, consistent with upregulation of ET_B-mediated vasoconstriction. Similar forearm studies using the receptor agonists ET-1 and STXc support the hypothesis that vascular smooth muscle ET_B-receptors may be of greater functional significance in chronic heart failure than in health (Love et al., 1995). Early phase clinical studies with the $ET_{A/B}$-receptor antagonist, bosentan, in patients with severe chronic heart failure have shown sustained peripheral, pulmonary, and venous vasodilatation and improved cardiac performance without reflex tachycardia (Kiowski et al., 1995). These hemodynamic effects are similar to those seen with ACE inhibitors and thus justify the continued development of anti-ET therapies for chronic heart failure.

C. Renal Disease

1. Acute Renal Failure

Acute renal failure due to postischemic acute tubular necrosis is characterized by intense renal vasoconstriction and impairment of renal function that may necessitate hemodialysis. This may be mediated by ET-1, which is a potent renal vasoconstrictor, because production of and sensitivity to the peptide is increased by hypoxia (Rakugi et al., 1990a; Liu et al., 1990). Renal ischemia and reperfusion increase gene expression and production of ET-1 (Firth and Ratcliffe, 1992; Shibouta et al., 1990), and also renal ET-1 binding affinity (Clozel et al., 1991), and ET_A- and ET_B-receptor number (Nambi et al., 1993; Roubert et al., 1994). Circulating plasma concentrations of ET are increased in patients with postischemic renal failure (Tomita et al., 1989; Moore et al., 1992).

In rats, renal vasoconstriction following ischemia is substantially ameliorated by administration of antibodies to ET, given either before (Shibouta et al., 1990; López-Farré et al., 1991) or after the insult (Kon et al., 1989). Similar findings have been reported for BQ 123, the selective ET_A-receptor antagonist (Gellai et al., 1994; Chan et al., 1994) and Ro 462,005, the combined ET_A/ET_B-receptor antagonist (Clozel et al., 1993). However, in dogs BQ 123 does not prevent renal vasoconstriction after ischemia (Stingo et al., 1993; Brooks et al., 1994), even though ET_A-receptors appear to predominate in the canine renal vasculature (Karet and Davenport, 1994). However, ischemia-induced renal dysfunction in the dog is prevented by the $ET_{A/B}$-receptor antagonist, SB 209,670 (Brooks et al., 1994), suggesting that ET-1 may contribute to the pathophysiology of acute renal failure through actions on nonvascular ET_B-receptors, including those situated on tubular epithelial cells.

Gram-negative septic shock is associated with a reduction in peripheral vascular resistance and blood pressure, which is thought to be due to endotoxin and other mediators stimulating NO production. Paradoxically, renal vasoconstriction occurs during septic shock, leading to acute renal failure. Because endotoxin stimulates ET-1 release, both *in vitro* and *in vivo* (Sugiura et al., 1989), and local administration of anti-ET gamma globulin completely reverses endotoxin-induced renal vasoconstriction (Kon and Badr, 1991), ET may contribute to this form of acute renal failure. Indeed, plasma ET concentrations are strikingly high in patients with septic shock (Weitzberg et al., 1991b; Sanai et al., 1996). Acute renal failure associated with radiocontrast media may also be mediated by ET-1 because these agents increase plasma and urine immunoreactive ET concentrations (Marguiles et al., 1991).

2. *Chronic Renal Failure*

Chronic renal failure, once established, tends to progress to end-stage renal failure requiring therapy with dialysis. In animal models of progressive chronic renal failure, there are increases in renal preproET-1 mRNA, cortical tissue immunoreactive ET-1, and urinary ET excretion (Brooks et al., 1991; Benigni et al., 1991; Orisio et al., 1993). In addition, renal ET-1 generation is positively correlated with urinary protein excretion and glomerulosclerosis in such animals (Brooks et al., 1991; Orisio et al., 1993), supporting a role for ET-1 in progression of chronic renal failure. Furthermore, glomerular expression of mRNA for ET-1, and for ET_A- and ET_B-receptors, is increased in experimental glomerulosclerosis (Nakamura et al., 1995). Finally, administration of an ET_A-receptor antagonist prevents the development of hypertension, glomerular damage, and renal insufficiency in rats (Benigni et al., 1993), suggesting that ET-1 contributes to the progression of chronic renal failure.

Plasma concentrations of immunoreactive big ET-1, ET-1, and ET-3 are increased in patients with chronic renal failure, and are raised two- to fourfold in those on hemodialysis (Koyama et al., 1989; Warrens et al., 1990). Although these increased concentrations may reflect impaired clearance, the data from animal models, and the fact that urinary excretion of immunoreactive ET is also increased in patients with chronic renal failure (Ohta et al., 1991), suggest excessive renal generation of ETs in this disease. Elevated plasma concentrations of ETs may also play a role in the development or maintenance of hypertension in chronic renal failure. In contrast to essential hypertension, ET-1 concentrations in hemodialysis patients are positively correlated with blood pressure (Miyauchi et al., 1991; Tsunoda et al., 1991). Interestingly, ET-3 concentrations, but not those of ET-1, rise further during hemodialysis (Miyauchi et al., 1991).

Responsiveness to ET-1 has been examined in cutaneous veins of normotensive and hypertensive patients with chronic renal failure who are not yet on dialysis (Hand et al., 1994). Although there is no difference in response between controls and normotensive patients with chronic renal failure, venoconstriction to ET-1 is attenuated in hypertensive patients. Because these patients have elevated immunoreactive ET concentrations, decreased responsiveness may reflect ET receptor downregulation secondary to increased generation or decreased clearance of the peptides. Further studies have demonstrated that patients with chronic renal failure have impaired forearm vasodilatation to brachial artery infusion of the ET_A-receptor antagonist BQ 123, suggesting that vascular generation of ET-1 is probably decreased in this disease (Hand et al., 1995). Nevertheless, ET-1 may still contribute to vascular disease in chronic renal failure through its mitogenic actions. Erythropoietin therapy is used to reverse the anemia of chronic renal failure, but is associated with clinically important hypertension in a substantial number of patients. Alterations in ET generation or receptors are unlikely to underlie the association of erythropoietin therapy with hypertension, because plasma ET is unchanged and forearm vasoconstriction to ET-1 decreased in patients receiving erythropoietin (Hand et al., 1995b).

3. *Immunosuppressive Therapy*

ET-1 may also contribute to the hypertension and renal impairment caused by cyclosporin. Production of ET-1 *in vitro* (Bunchman and Brookshire, 1991) and *in vivo* (Kon et al., 1990) is stimulated by cyclosporin, which also increases the number of renal ET-1 binding sites (Nambi et al., 1990). In addition, cyclosporin-induced renal vasoconstriction is substantially attenuated by anti-ET gamma-globulin (Kon et al., 1990). Furthermore, mesangial cell and renal afferent arteriolar contractile responses to cyclosporin are substantially attenuated by the ET_A-receptor antagonist, BQ 123 *in vitro* (Takeda et al., 1992; Lanese and Conger, 1993) and *in vivo* (Fogo et al., 1992). The novel immunosuppressant FK 506 (tacromilus), also appears to increase ET-1 secretion *in vitro* (Moutabarrik et al., 1991).

REFERENCES

Abassi, Z. A., Golomb, E., Bridenhaugh, R. and Keiser, H. R., Metabolism of endothelin-1 and big endothelin-1 by recombinant neutral endopeptidase EC.3.4.24.11. *Br. J. Pharmacol.*, 109, 1024-1028. 1993.

Andrawis, N. S., Gilligan, J. and Abernethy, D. R., Endothelin-1-mediated vasoconstriction: specific blockade by verapamil. *Clin. Pharmacol. Ther.*, 52, 583-589. 1992.

Asano, T., Ikegaki, I., Suzuki, Y., Satoh, S. and Shibuya, M., Endothelin and the production of cerebral vasospasm in dogs. *Biochem. Biophys. Res. Commun.*, 159, 365-372. 1989.

Bazil, M. K, Lappe, R. W. and Webb, R. L., Pharmacologic characterization of an endothelinA (ET_A) receptor antagonist in conscious rats. *J. Cardiovasc. Pharmacol.*, 20, 940-948. 1992.

Benigni, A., Perico, N., Gaspari, F., Zoja, C., Bellizi, L., Gabanelli, M. and Remuzzi, G. Increased renal endothelin production in rats with reduced renal mass. *Am. J. Physiol.*, 260, F331-F339. 1991.

Benigni, A., Zoja, C., Corna, D., Orisio, S., Longaretti, L., Bertani, T. and Remuzzi, G., A specific endothelin subtype A receptor antagonist protects against injury in renal disease progression. *Kidney Int.*, 44, 440-445. 1993.

Bigaud, M. and Pelton, J. T., Discrimination between ET_A- and ET_B-receptor-mediated effects of endothelin-1 and [$Ala^{1,3,11,15}$]endothelin-1 by BQ-123 in the anesthetized rat. *Br. J. Pharmacol.*, 107, 912-918. 1992.

Brooks, D. P., Contino, L. C., Storer, B. and Ohlstein, E. H., Increased endothelin excretion in rats with renal failure induced by partial nephrectomy. *Br. J. Pharmacol.*, 104, 987-989. 1991.

Brooks, D. P., dePalma, P. D., Gellai, M., Nambi, P., Ohlstein, E. H., Elliott, J. D., Gleason, J. G. and Ruffolo, R. R., Jr., Nonpeptide endothelin receptor antagonists. III. Effect of SB 209670 and BQ123 on acute renal failure in anesthetized dogs. *J. Pharmacol. Exp. Ther.*, 271, 769-75. 1994.

Bunchman, T. E. and Brookshire, C. A., Cyclosporine-induced synthesis of endothelin by cultured human endothelial cells. *J. Clin. Invest.*, 88, 310-314. 1991.

Calver, A., Collier, J., Moncada, S. and Vallance, P., Effect of intra-arterial N^G-monomethyl-L-arginine in patients with hypertension: the nitric oxide mechanism appears abnormal. *J. Hypertens.*, 10, 1025-1031. 1992.

Cargill, R. I., Kiely, D. G., Clarke, R. A. and Lipworth, B. J., Hypoxia and release of endothelin-1. *Thorax* 50, 1308-1310. 1995.

Chan, L., Chittinandana, A., Shapiro, J. I., Shanley, P. F. and Schrier, R. W., Effect of an endothelin-receptor antagonist on ischemic acute renal failure. *Am. J. Physiol.*, 266, F135-F138. 1994.

Cicila, G. T., Rapp, J. P., Bloch, K. D., Kurtz, T. W., Pravenec, M., Kren, V., Hong, C. C., Quertermous, T. and Ng, S. C., Cosegregation of the endothelin-3 locus with blood pressure and relative heart weight in inbred Dahl rats. *J. Hypertens.*, 12, 643-651. 1994.

Cirino, M., Battistini, B., Yano, M. and Rodger, I. W., Dual cardiovascular effects of endothelin-1 dissociated by BQ-153, a novel ET_A receptor antagonist. *J. Cardiovasc. Pharmacol.*, 24, 587-594. 1994.

Clarke, J. G., Benjamin, N., Larkin, S. W., Webb, D. J., Davies, G. J. and Maseri, A., Endothelin is a potent long-lasting vasoconstrictor in men. *Am. J. Physiol.*, 257, H2033-H2035. 1989.

Clozel, M., Endothelin sensitivity and receptor binding in the aorta of spontaneously hypertensive rats. *J. Hypertens.*, 7, 913-917. 1989.

Clozel, M., Löffler B. M. and Gloor, H., Relative preservation of the responsiveness to endothelin-1 during reperfusion following renal ischemia in the rat. *J. Cardiovasc. Pharmacol.*, 17 (Suppl. 7), S313-S315. 1991.

Clozel, M. and Watanabe, H., BQ-123, a peptidic endothelin ET_A receptor antagonist, prevents the early cerebral vasospasm following subarachnoid hemorrhage after intracisternal but not intravenous injection. *Life Sci.*, 52, 825-834. 1993.

Clozel, M., Breu, V., Burri, K., Cassal, J. M., Fischli, W., Gray, G. A., Hirth, G., Löffler, B. M., Müller, M., Neidhart, W. and Ramuz, H., Pathophysiological role of endothelin revealed by the first orally active endothelin receptor antagonist. *Nature* 365, 759-761. 1993.

Cockcroft, J. R., Chowiencyk, P. J., Benjamin, N. and Ritter, J. M., Preserved endothelium-dependent vasodilatation in patients with essential hypertension. *N. Engl. J. Med.*, 330, 1036-1040. 1994.

Cockcroft, J. R., Clarke, J. G. and Webb, D. J., The effect of intra-arterial endothelin on resting blood flow and sympathetically mediated vasoconstriction in the forearm of man. *Br. J. Clin. Pharmacol.* 31, 521-524. 1991.

Cocks, T. M., Broughton, A., Dib, M., Sudhir, K. and Angus, J. A., Endothelin is blood vessel selective: studies on a variety of human and dog vessels *in vitro* and on regional blood flow in the conscious rabbit. *Clin. Exp. Pharmacol. Physiol.*, 16, 243-246. 1989.

Cody, R. J., Haas, G. J., Binkley, P. F., Capers, Q. and Kelly, R., Plasma endothelin correlates with the extent of pulmonary hypertension in patients with chronic congestive heart failure. *Circulation*, 85, 504-509. 1992.

Collier, J. G., Lorge, R. E. and Robinson, B. F., Comparison of the effects of tolmesoxide (RX71107), diazoxide, hydrallazine, prazosin, glycerol trinitrate and sodium nitroprusside on forearm arteries and dorsal hand veins of man. *Br. J. Clin. Pharmacol.* 5, 33-44. 1978.

Criscione, L., Nellis, P., Riniker, B., Thomann, H. and Burdet, R., Reactivity and sensitivity of mesenteric vascular beds and aortic rings of spontaneously hypertensive rats to endothelin: effects of calcium entry blockers. *Br. J. Pharmacol.*, 100, 31-36. 1990.

Crossman, D. C., Brain, S. D. and Fuller, R. W., Potent vasoactive properties of endothelin in human skin. *Am. J. Physiol.*, 70, 260-266. 1991.

Dahlöf, B., Gustaffsson, D., Hedner, T., Jern, S. and Hansson, L., Regional haemodynamic effects of endothelin- 1 in rat and man: unexpected adverse reactions. *J. Hypertens.*, 8, 811-817. 1990.

Dargie, H. J. and McMurray, J. J. V., Diagnosis and management of heart failure. *Br. Med. J.*, 308, 321-328. 1994.

Davenport, A. P., Ashby, M. J., Easton, P., Ella, S., Bedford, J., Dickerson, C., Nunez, D. J., Capper, S. J. and Brown, M. J., A sensitive radioimmunoassay measuring endothelin-like immunoreactivity in human plasma: comparison of levels in patients with essential hypertension and normotensive control subjects. *Clin. Sci.*, 78, 261-264. 1990.

Davenport, A. P., O'Reilly, G., Molenaar, P., Maguire, J. J., Kuc, R. E., Sharkey, A., Bacon, C. R. and Ferro, A., Human endothelin receptors characterized using reverse transcriptase-polymerase chain reaction, in-situ hybridization, and subtype-selective ligands BQ123 and BQ3020: evidence for expression of ET_B receptors in human vascular smooth muscle. *J. Cardiovasc. Pharmacol.*, 22 (Suppl. 8), S22-S25. 1993.

Deng, L. Y. and Schiffrin, E. L., Effects of endothelin on resistance arteries of DOCA-salt hypertensive rats. *Am. J. Physiol.*, 262, H1782-H1787. 1992.

Dohi, Y. and Lüscher, T.F., Endothelin-1 in hypertensive resistance arteries, intraluminal and extraluminal dysfunction. *Hypertension*, 18, 543-549. 1991.

Dohi, Y., Criscione, L. and Lüscher, T.F., Renovascular hypertension impairs formation of endothelium-derived relaxing factors and sensitivity to endothelin-1 in resistance arteries. *Br. J. Pharmacol.*, 104, 349-354. 1991.

Douglas, S. A., Elliot, J. D. and Ohlstein, E. H., Regional vasodilation to endothelin-1 is mediated by a non-ET_A receptor subtype in the anaethetised rat: effect of BQ-123 on systemic haemodynamic responses. *Eur. J. Pharmacol.*, 221, 315-324. 1992.

Douglas, S. A., Gellai, M., Ezekial, M. and Ohlstein, E. H., BQ-123, a selective endothelin subtype A-receptor antagonist, lowers blood pressure in different rat models of hypertension. *J. Hypertens.*, 12, 561-567. 1994.

Ehrenreich, H., Lange, M., Near, K. A., Anneser, F., Schoeller, L. A. C., Schmid, R., Winkler, P. A., Kehrl, J. H., Schmiedek, P. and Goebel, F.-D., Long term monitoring of immunoreactive endothelin-1 and endothelin-3 in ventricular cerebrospinal fluid, plasma, and 24 hour urine of patients with subarachnoid hemorrhage. *Res. Exp. Med.*, 192, 257-268. 1992.

Erdos, E. G. and Skidgell, R. A., Neutral endopeptidase 24.11 (enkephalinase) and related regulators of peptide hormones. *FASEB J.*, 3, 145-151. 1989.

Farkilla, M., Falo, J., Saijonmaa, O. and Fyhrquist, F., Raised plasma endothelin during acute migraine attack. *Cephalalgia*, 12, 383-384. 1992.

Filep, J. G., Clozel, M., Fournier, A. and Földes-Filep, E., Characterization of receptors mediating vascular responses to endothelin-1 in the conscious rat. *Br. J. Pharmacol.*, 113, 845-852. 1994.

Firth, J. D. and Ratcliffe, P. J., Organ distribution of three rat endothelin messenger RNAs and the effects of ischemia on renal gene expression. *J. Clin. Invest.*, 90, 1023-1031. 1992.

Florijn, K. W., Derkx, F. H. M., Visser, W., Hofman, J. A., Rosmalen, F. M. A., Wallenberg, H. C. S. and Schalekamp, M. A. D. H., Plasma immunoreactive endothelin-1 in pregnant women with and without pre-eclampsia. *J. Cardiovasc. Pharmacol.*, 17 (Suppl. 7), S446-S448. 1991.

Fogo, A., Hellings, S. E., Inagami, T. and Kon, V., Endothelin receptor antagonism is protective in *in vivo* acute cyclosporin toxicity. *Kidney Int.*, 42, 770-774. 1992.

Foley, P. L., Caner, H. H., Kassell, N. F. and Lee, K. S., Reversal of subarachnoid hemorrhage-induced vasoconstriction with an endothelin receptor antagonist. *Neurosurgery*, 34, 108-113. 1994.

Folkow, B., Cardiovascular structural adaption: its role in the initiation and maintenance of primary hypertension. *Clin. Sci. Mol. Med.*, 55 3S-22S. 1978.

Fukuroda, T., Kobayashi, M., Ozaki, S., Yano, M., Miyauchi, T., Onizuka, M., Sugishita, Y., Goto, K. and Nishikibe, M., Endothelin receptor subtypes in human vs. rabbit pulmonary arteries, *J. Appl. Physiol.*, 76, 31976-31982. 1994.

Gardiner, S. M., Compton, A. M., Kemp, P. A. and Bennett, T., The effects of phosphoramidon on the regional haemodynamic responses to human proendothelin [1-38] in conscious rats. *Br. J. Pharmacol.*, 103, 2009-2015. 1991.

Gardiner, S. M., Kemp, P. A. and Bennett, T., Inhibition by phosphoramidon of the regional haemodynamic effects of proendothelin-2 and-3 in conscious rats. *Br. J. Pharmacol.*, 107, 584-590. 1992.

Gardiner, S. M., Kemp, P. A., March, J. E. and Bennett, T., Effects of bosentan (Ro 47-0203), an ET_A-, ET_B-receptor antagonist, on regional haemodynamic responses to endothelins in conscious rats. *Br. J. Pharmacol.*, 112, 823-830. 1994.

Gasic, S., Wagner, O. F., Vierhapper, H., Nowotny, P. and Waldhäusl, W., Regional haemodynamic effects and clearance of endothelin-1 in humans: renal and peripheral tissues may contribute to the overall disposal of the peptide. *J. Cardiovasc. Pharmacol.*, 19, 176-180. 1992.

Gellai, M., Jugus, M., Fletcher, T., DeWolf, R. and Nambi, P., Reversal of postischemic acute renal failure with a selective $endothelin_A$ receptor antagonist in the rat. *J. Clin. Invest.*, 93, 900-906. 1994.

Giaid, A., Yanagisawa, M., Langleben, D., Michel, R. P., Levy, R., Shenib, H., Kimura, S., Masaki, T., Duguid, W. P. and Stewart D. J., Expression of endothelin-1 in lungs of patients with pulmonary hypertension. *N. Engl. J. Med.*, 328, 1732-1739. 1993.

Goligorsky, M. S., Iijima, K., Morgan, M., Yanagisawa, M., Masaki, T., Lin, L., Nasjletti, A., Kaskel, F., Frazer, M. and Badr, K. F., Role of endothelin in the development of Dahl hypertension. *J. Cardiovasc. Pharmacol.*, 17 (Suppl. 7), S484-S491. 1991.

Goto, K., Kasuya, Y., Matsuki, N., Takuwa, Y., Kurihara, H., Ishikawa, T., Kimura, S., Yanagisawa, M. and Masaki, T., Endothelin activates the dihydropyridine-sensitive, voltage-dependent Ca^{2+} channel in vascular smooth muscle. *Proc. Natl. Acad. Sci. U.S.A.* 86, 3915-3918. 1989.

Grover, G. J., Dzwoncyk, S. and Parham, C. S., The endothelin-1 receptor antagonist BQ-123 reduced infarct size in a canine model of coronary occlusion and reperfusion. *Cardiovasc. Res.*, 27, 1613-1618. 1993.

Gu, X. H., Casley, D. J., Cincotta, M. and Nayler, W. G., ^{125}I-endothelin-1 binding to brain and cardiac membranes from normotensive and spontaneously hypertensive rats. *Eur. J. Pharmacol.*, 177, 205-209. 1990.

Haefeli, W. E., Linder, L., Kiowski, W. and Lüscher, T. F., *In vivo* properties of endothelin-1 and endothelin-3 in human hand veins and its reversal by bradykinin and verapamil. *Hypertension*, 22, 343 (Abstr.). 1993.

Hand, M. F., Haynes, W. G., Anderton, J. L. and Webb, D. J., Endothelin-dependent basal vascular tone is decreased in chronic renal failure. *Kidney Int.*, (abstract in press). 1996.

Hand, M. F., Haynes, W. G., Anderton, J. L., Winney, R. and Webb, D. J., Responsiveness of veins to endothelin- 1 in chronic renal failure. *Nephrol. Dial. Transplant.*, 9, 1349-1350. 1994.

Hand, M. F., Haynes, W. G., Johnstone, H. A., Anderton, J. L. and Webb, D. J., Enhanced vascular responsiveness to norepinephrine but not endothelin-1 during erythropoietin treatment in end-stage renal failure. *Kidney Int.*, 48, 806-813, 1995b.

Haynes, W. G. and Webb, D. J., Endothelium-dependent modulation of responses to endothelin-1 in human veins. *Clin. Sci.*, 84, 427-433. 1993a.

Haynes, W. G. and Webb, D. J., Venoconstriction to endothelin-1 in humans: the role of calcium and potassium channels. *Am. J. Physiol.*, 265, H1676-H1681. 1993b.

Haynes, W. G. and Webb, D. J., Contribution of endogenous generation of endothelin-1 to basal vascular tone. *Lancet*, 344, 852-854. 1994.

Haynes, W. G., Clarke, J., Cockcroft, J. R. and Webb, D. J., Pharmacology of endothelin-1 *in vivo* in man. *J. Cardiovasc. Pharmacol.*, 17 (Suppl. 7), S284-S286. 1991.

Haynes, W. G., Hand, M. F., Johnstone, H. A., Padfield, P. L. and Webb, D. J. Direct and sympathetically mediated venoconstriction in essential hypertension: enhanced responses to endothelin-1. *J. Clin. Invest.*, 94, 1359-1364. 1994.

Haynes, W. G., Moffat, S. and Webb, D. J., An investigation into the direct and indirect venoconstrictor effects of endothelin-1 and big endothelin-1 in man. *Br. J. Clin. Pharmacol.* 40, 307-311, 1995b.

Haynes, W. G., Skwarski, K. M., Wyld, P. J., Ripke, H. and Webb, D. J., Vasodilator effects of the $ET_{A/B}$ receptor antagonist, TAK-044, in man, presented at Fourth Int. Conf. Endothelin, London, April 23 to 26, 1995c.

Haynes, W. G., Strachan, F. E. and Webb, D. J., Endothelin ET_A and ET_B receptors mediate vasoconstriction of human resistance and capacitance vessels *in vivo*. *Circulation*, 92, 357-363. 1995a.

Hirata, Y., Yoshimi, H., Takachi, S., Yanagisawa, M. and Masaki, T., Binding and receptor downregulation of novel vasoconstrictor endothelin in cultured rat vascular smooth muscle cells. *FEBS Lett.*, 239, 13-17. 1988.

Hiroe, M., Hirata, Y., Fujita, N., Umezawa, S., Ito, H., Tsujino, M., Koike, A., Nogami, A., Takamoto, T. and Marumo, F., Plasma endothelin-1 levels in idiopathic dilated cardiomyopathy. *Am. J. Cardiol.*, 68, 1114-1115. 1991.

Hoffman, A., Grossman, E., Goldstein, D. S., Gill, J. R. and Keiser, H. R., Low urinary levels of endothelin-1 in patients with essential hypertension. *J. Am. Soc. Nephrol.*, 1, 417 (Abstr.). 1990.

Hoffman, A., Haramati, A., Dalal, I., Shuranyi, E. and Winaver, J., Diuretic-natriuretic actions and pressor effects of big endothelin (1-39) in phosphoramidon treated rats. *Proc. Soc. Exp. Med. Biol.*, 205, 168-173. 1994

Howard, P. G., Plumpton, C. and Davenport, A. P., Anatomical localization and pharmacological activity of mature endothelins and their precursors in human vascular tissue. *J. Hypertens.*, 10, 1379-1386. 1992.

Hughes, A. D., Thom, S. A. M., Woodall, N., Schachter, M., Hair, W. M., Martin, G. N. and Sever, P. S., Human vascular responses to endothelin-1: observations *in vivo* and *in vitro*. *J. Cardiovasc. Pharmacol.*, 13 (Suppl. 5), S225-S228. 1989.

Hughes, A. K., Cline, R. C. and Kohan, D. E., Alterations in renal endothelin-1 production in the spontaneously hypertensive rat. *Hypertension*, 20, 666-673. 1992.

Ihara, M., Noguchi, K., Saeki, T., Fukuroda, T., Tsuchida, S., Kimura, S., Fukami, T., Ishikawa, K., Nishikibe, M. and Yano, M., Biological profiles of highly potent novel endothelin antagonists selective for the ET_A receptor. *Life Sci.* 50, 247-255. 1992.

Ishii, K., Nunoki, K., Murakoshi, H. and Taira, N., Cloning and modulation by endothelin-1 of rat cardiac potassium channel. *Biochem. Biophys. Res. Commun.*, 184, 1484-1489. 1992.

Karet F. E. and Davenport, A. D., Endothelin and the human kidney: a potential target for new drugs. *Nephrol. Dial. Transplant.*, 9, 465-468. 1994.

Kiely, D. G., Cargill, R. I., Struthers, A. D. and Lipworth, B. J., Systemic and pulmonary haemodynamic effects of endothelin-1 in humans. *Clin. Sci.*, (in press). 1996.

Kim, S., Morimoto, S., Koh, E., Miyashita, Y. and Ogihara, T., Comparison of effects of a potassium channel opener BRL34915, a specific potassium ionophore valinomycin and calcium channel blockers on endothelin-induced vascular contraction. *Biochem. Biophys. Res. Commun.*, 164, 1003-1008. 1989.

Kiowski, W. and Linder, L., Reversal of endothelin-1-induced vasoconstriction by nifedipine in human resistance vessels *in vivo*. *Am. J. Cardiol.*, 69, 1063-1066. 1992.

Kiowski, W., Lüscher, T. F., Linder, L. and Bühler, F., Endothelin-1 induced vasoconstriction in humans: reversal by calcium channel blockade but not by nitrovasodilators or endothelium-derived relaxing factor. *Circulation*, 83, 468-475. 1991.

Kiowski, W., Sütsch, G., Hunziker, P., Müller, P., Kim, J., Oechslin, E., Schmitt, R., Jones, R. and Bertel, O., Evidence for endothelin-1 mediated vasoconstriction in severe chronic heart failure. *Lancet*, 346, 732-736. 1995.

Kitamura, K., Tanaka, T., Kato, J., Ogawa, T., Eto, T. and Tanaka, K., Immunoreactive endothelin in rat kidney inner medulla: marked decrease in spontaneously hypertensive rats. *Biochem. Biophys. Res. Commun.*, 162, 38-44. 1989.

Kobayashi, H., Hayashi, M., Kobayashi, S., Kabuto, M., Handa, Y., Kawano, H. and Ide, H., Cerebral vasospasm amd vasoconstriction caused by endothelin. *Neurosurgery*, 28, 673-679. 1991.

Kohan, D. E. and Padilla, E., Osmolar regulation of endothelin-1 production by rat inner medullary collecting duct. *J. Clin. Invest.*, 91, 1235-1240. 1993.

Kohno, M., Murakawa, K.-I., Horio, T., Yokokawa, K., Yasunari, K., Fukui, T. and Takeda, T. Plasma immunoreactive endothelin-1 in experimental malignant hypertension. *Hypertension*, 18, 93-100. 1991.

Kohno, M., Murakawa, K., Yasunari, K., Yokokawa, K., Horio, T., Kurihara, N. and Takeda, T., Prolonged blood pressure elevation after endothelin administration in bilaterally nephrectomised rats. *Metabolism*, 38, 712-713. 1989.

Kohno, M., Yasunari, K., Murakawa, K.-I., Yokokawa, K., Horio, T., Fukui, T. and Takeda, T., Plasma immunoreactive endothelin in essential hypertension. *Am. J. Med.*, 88, 614-618. 1990.

Komuro, I., Kurihara, H., Sugiyama, T., Yoshizumi, M., Takaku, F. and Yazaki, Y., Endothelin stimulates *c-fos* and *c-myc* expression and proliferation of vascular smooth muscle cells. *FEBS Lett.*, 238, 249-252. 1988.

Kon, V. and Badr, K. F., Biological actions and pathophysiologic significance of endothelin in the kidney. *Kidney Int.*, 40, 1-12. 1991.

Kon, V., Sugiura, M., Inagami, T., Harvie, B. R., Ichikawa, I. and Hoover, R. L., Role of endothelin in cyclosporine-induced glomerular dysfunction. *Kidney Int.*, 37, 1487-1491. 1990.

Kon, V., Yoshioka, T., Fogo, A. and Ichikawa, I., Glomerular actions of endothelin *in vivo*. *J. Clin. Invest.*, 83, 1762-1767. 1989.

Koyama, H., Tabata, T., Nishzawa, Y., Inoue, T., Morii, H. and Yamaji, T., Plasma endothelin levels in patients with uraemia. *Lancet*, 333, 991-992. 1989.

Lanese, D. M. and Conger, J. D., Effects of endothelin receptor antagonist on cyclosporin-induced vasoconstriction in isolated rat renal arterioles. *J. Clin. Invest.*, 91, 2144-2149. 1993.

Larivière, R., Thibault, G. and Schiffrin E. L., Increased endothelin-1 content in blood vessels of deoxycorticosterone acetate-salt hypertensive rats but not in spontaneously hypertensive rats. *Hypertension*, 21, 294-300. 1993.

Lerman, A., Edwards, B. S., Hallett, J. W., Heublein, D. M., Sandberg, S. M. and Burnett, J. C., Circulating and tissue endothelin immunoreactivity in advanced atherosclerosis. *N. Engl. J. Med.*, 325, 997-1001. 1991.

Lever, A.F., Slow pressor mechanisms in hypertension: a role for hypertrophy of resistance vessels?, *J. Hypertens.*, 4, 515-524. 1986.

Levy, D., Garrison, R. J., Savage, D. D., Kannel, W. B. and Castelli, W. P., Prognostic implications of echocardiographically determined left ventricular mass in the Framingham Heart Study. *N. Engl. J. Med.*, 322, 1561-1566. 1990.

Linder, L., Kiowski, W., Buhler, F. R. and Lüscher, T. F., Indirect evidence for release of endothelium-derived relaxing factor in human forearm circulation *in vivo*: blunted response in essential hypertension. *Circulation*, 81, 1762-1767. 1990.

Liu, J., Chen, R., Casley, D. J. and Nayler, W. N., Ischemia and reperfusion increase ^{125}I-labeled endothelin-1 binding in rat cardiac membranes. *Am. J. Physiol.*, 258, H829-H835. 1990.

López-Farré, A., Gómez-Garre, D., Bernabeu, F. and López-Nova, J. M., A role for endothelin in the maintenance of post-ischemic acute renal failure in the rat. *J. Physiol. (London)* 444, 513-522. 1991.

López-Farré, A., Montanes, I., Millas, I. and López-Nova, J. M., Effect of endothelin on renal function in rats. *Eur. J. Pharmacol.*, 163, 187-189. 1989.

Love, M. P., Haynes, W. G., Webb, D. J. and McMurray, J. J. V., Anti-endothelin therapy is of potential benefit in heart failure. *Circulation* 90, I-547 (Abstr.). 1994.

Love, M. P., Haynes, W. G., Webb, D. J. and McMurray, J. J. V., Which endothelin receptors are functionally important in chronic heart failure? *Circulation*, 92, I-331 (Abstract), 1995.

Lüscher, T. F. and Vanhoutte, P. M., Endothelium dependent responses in human blood vessels. *Trends Pharmacol. Sci.*, 9, 181-184. 1988.

Maguire, J. J., Kuc, R. E., O'Reilly, G. and Davenport, A. P., Vasoconstrictor endothelin receptors characterized in human renal artery and vein *in vitro*. *Br. J. Pharmacol.*, 113, 49-54. 1994.

Margulies, K. B., Hildebrand, F. L., Heublein, D. M. and Burnett, J. C., Jr. Radiocontrast increases plasma and urinary endothelin. *J. Am. Soc. Nephrol.*, 2, 1041-1045. 1991.

Matsumura, Y., Ikegawa, R., Suzuki, Y., Takaoka, M., Uchida, T., Kido, H., Shinyama, H., Hayashi, K., Watanabe, M. and Morimoto, S., Phosphoramidon prevents cerebral vasospasm following subarachnoid hemorrhage in dogs: the relationship to endothelin-1 levels in the cerebrospinal fluid. *Life Sci.*, 49, 841-848. 1991.

McMahon, E. G., Palomo, M. A. and Moore, W. M., Phosphoramidon blocks the pressor activity of big endothelin[1-39] and lowers blood pressure in spontaneously hypertensive rats. *J. Cardiovasc. Pharmacol.*, 17 (Suppl. 7), S29-S33. 1991.

McMurdo, L., Thiemermann, C. and Vane, J. R., The effects of the endothelin ET_A receptor antagonist FR 139317, on infarct size in a rabbit model of acute myocardial ischemia and reperfusion. *Br. J. Pharmacol.*, 113, 75-80. 1994.

McMurray, J. J. V., Ray, S. G., Abdullah, I., Dargie, H. J. and Morton, J. J. Plasma endothelin in chronic heart failure. *Circulation*, 85, 1374-1379. 1992.

Miller, D., Waters, D. D., Warnica, W., Szlachcic, J., Kreeft, J. and Thèroux, P., Is variant angina the generalised manifestation of a generalised vasospastic disorder?, *N. Engl. J. Med.*, 304, 763-766. 1981.

Miyauchi, T., Ishikawa, T., Tomobe, Y., Yanagisawa, M., Kimura, S., Sugishita, Y., Ito, I., Goto, K. and Masaki, T., Characteristics of pressor response to endothelin in spontaneously hypertensive and Wistar-Kyoto rats. *Hypertension*, 14, 427-434. 1989a.

Miyauchi, T., Suzuki, N., Kurihara, T., Yamaguchi, I., Sugishita, Y., Matsumoto, H., Goto, K. and Masaki, T., Endothelin-1 and endothelin-3 play different roles in acute and chronic alterations of blood pressure in patients with chronic hemodialysis. *Biochem. Biophys. Res. Commun.*, 178, 276-281. 1991.

Miyauchi, T., Yanagisawa, M., Tomizawa, T., Sugishita, Y., Suzuki, N., Fujino, M., Ajisaka, R., Goto, K. and Masaki, T., Increased plasma concentrations of endothelin-1 and big endothelin-1 in acute myocardial infarction. *Lancet*, 334, 53-54. 1989b.

Miyoshi, Y., Nakaya, Y. and Wakatsuki, T., Endothelin blocks ATP-sensitive K^+ channels and depolarises smooth muscle cells of porcine coronary artery. *Circ. Res.*, 70, 612-616. 1992.

Moore, K., Wendon, J., Frazer, M., Karani, J., Williams, R. and Badr, K., Plasma endothelin immunoreactivity in liver disease and the hepatorenal syndrome. *N. Engl. J. Med.*, 327, 1774-1778. 1992.

Moutabarrik, A., Ishibashi, M., Fukunaga, M., Kameoka, H., Takano, Y., Kokado, Y., Takahara, S., Jiang, H., Sonoda, T. and Okuyama, A., FK 506 mechanism of nephrotoxicity: stimulatory effect on endothelin secretion by cultured kidney cells and tubular cell toxicity *in vitro*. *Tranplant. Proc.*, 23, 3133-3136. 1991.

Nakamura, T., Fukui, M., Ebihara, I., Osada, S., Takahashi, T., Tomino, Y. and Koide, H., Effects of a low protein diet on glomerular endothelin family gene expression in experimental focal glomerular sclerosis. *Clin. Sci.*, 88, 29-37. 1995

Nambi, P., Pullen, M., Contino, L. C. and Brooks, D. P., Upregulation of renal endothelin receptors in rats with cyclosporin A-induced nephrotoxicity. *Eur. J. Pharmacol.*, 187, 113-116. 1990.

Nambi, P., Pullen, M., Jugus, M. and Gellai, M., Rat kidney endothelin receptors in ischemia-induced acute renal failure. *J. Pharmacol. Exp. Ther.*, 264, 345-348. 1993.

Nirei, H., Hamada, K., Shoubo, M., Sogabe, K., Notsu, Y. and Ono, T., An endothelin ET_A receptor antagonist, FR139317, ameliorates cerebral vasospasm in dogs. *Life Sci.*, 52, 1869-1874. 1993.

Nishikibe, M., Tsuchida, S., Okada, M., Fukuroda, T., Shimamoto, K., Yano, M., Ishikawa, K. and Ikemoto, F., Antihypertensive effect of a newly synthesised endothelin antagonist, BQ123, in a genetic hypertensive model. *Life Sci.*, 52, 717-724. 1993.

O'Connell, J. E., Jardine, A. G., Davidson, G., McQueen, J. and Connell, J. M. C. Renal and hormonal effects of chronic inhibition of neutral endopeptidase (EC 3.4.24.11) in normal man. *Clin. Sci.*, 85, 19-26. 1993.

O'Donnell, S. R., Wanstall, J. C. and Zeng, X. P., Pinacidil antagonism of endothelin-induced contractions of smooth muscle in the lungs: differences between tracheal and pulmonary artery preparations. *J. Pharmacol. Exp. Ther.*, 252, 1318-1323. 1990.

Ohta, T., Hirata, Y., Shichiri, M., Kano, K., Emori, T., Tomita, K. and Marumo, F., Urinary excretion of endothelin-1 in normal subjects and patients with renal disease. *Kidney Int.*, 39, 307-311. 1991.

Oishi, R., Monoguchi, H., Tomita, K. and Murumo, F., Endothelin-1 inhibits AVP stimulated osmotic water permeability in rat inner medullary collecting duct. *Am. J. Physiol.*, 261, F951-F956. 1991.

Omland, T., Lie, R. T., Aakvaag, A., Aarsland, T. and Dickstein, K., Plasma endothelin determination as a prognostic indicator of 1-year mortality after acute myocardial infarction. *Circulation*, 89, 1573-1579. 1994.

Orisio, S., Benigni, A., Bruzzi, I., Corna, D., Perico, N., Zoja, C., Benatti, L. and Remuzzi, G., Renal endothelin gene expression is increased in remnant kidney and correlates with disease progression. *Kidney Int.* 43, 354-358. 1993.

Pacher, R., Bergler-Klein, J., Globits, S., Teufelsbauer, H., Schuller, M., Krauter, A., Ogris, E., Rödler, S., Wutte, M. and Hartter, E., Plasma big endothelin-1 concentrations in congestive heart failure patients with or without systemic hypertension. *Am. J. Cardiol.*, 71, 1293-1299. 1993.

Panza, J. A., Quyyumi, A. A., Brush, J. E. and Epstein, S. E., Abnormal endothelium-dependent relaxation in patients with essential hypertension. *N. Engl. J. Med.*, 323, 22-27. 1990.

Pernow, J., Hemsén, A., Lundberg, J. M., Nowak, J. and Kaijser, L., Potent vasoconstrictor effects and clearance of endothelin in the human forearm. *Acta Physiol. Scand.*, 141, 319-324. 1991.

Plumpton, C., Haynes, W. G., Webb, D. J. and Davenport, A. P., Phosphoramidon inhibition of the *in vivo* conversion of big endothelin-1 to endothelin-1 in human forearm. *Br. J. Pharmacol.*, 115, 116, 1821-1828. 1995.

Pollock, D. M. and Opgenorth, T. J., Evidence for endothelin-induced renal vasoconstriction independent of ET_A receptor activation. *Am. J. Physiol.*, 264, R222-R226. 1993.

Rabelink, T. J., Kaasjager, K. A. H., Boer, P., Stroes, E. G., Braam, B. and Koomans, H. A., Effects of endothelin-1 on renal function in humans: implications for physiology and pathophysiology. *Kidney Int.*, 46, 376-381. 1994.

Rakugi, H., Tabuchi, Y., Nakamaru, M., Nagano, M., Higashimori, K., Mikami, H., Ogihara, T. and Suzuki, N., Evidence for endothelin-1 release from resistance vessels of rats in response to hypoxia. *Biochem. Biophys. Res. Commun.*, 169, 973-977. 1990.

Ray, S. G., McMurray, J. J., Morton, J. J. and Dargie, H. J., Circulating endothelin in acute ischaemic syndromes. *Br. Heart J.*, 67, 383-386. 1992.

Richard, V., Knaeffer, N., Hogie, M., Tron, C., Blanc, T. and Thuillez, C., Role of endogenous endothelin in myocardial and coronary endothelial injury after ischemia and reperfusion in rats: studies with bosentan, a mixed ET_A-ET_B antagonist. *Br. J. Pharmacol.*, 113, 869-876. 1994.

Riezebos, J., Watts, I. S. and Vallance, P. J. T., Endothelin receptors mediating functional responses in human small arteries and veins. *Br. J. Pharmacol.* 111, 609-615. 1994.

Roberts-Thomson, P., McRitchie, R. J. and Chalmers, J. P., Experimental hypertension produces diverse changes in the regional vascular responses to endothelin-1 in the rabbit and the rat. *J. Hypertens.*, 12, 1225-1234. 1994.

Roubert, P., Gillard-Roubert, V., Pourmarin, L., Cornet, S., Guilmard, C., Plas, P., Pirotzky, E., Chabrier, P. E. and Braquet, P. Endothelin receptor subtypes A and B are upregulated in an experimental model of acute renal failure. *Mol. Pharmacol.*, 45, 182-188. 1994.

Saito, Y., Nakao, K., Mukoyama, M. and Imura, H., Increased plasma endothelin level in patients with essential hypertension. *N. Engl. J. Med.*, 322, 205. 1990.

Sanai, L., Haynes, W. G., MacKenzie, A., Grant, I. S. and Webb, D. J., Endothelin production in sepsis and the adult respiratory distress syndrome. *Intensive Care Med.*, (in press). 1996.

Schiffrin, E. L. and Thibault, G., Plasma endothelin in human essential hypertension. *Am. J. Hypertens.*, 4, 303-308. 1991.

Schiffrin, E. L., Deng, L. Y. and Larochelle, P., Blunted effects of endothelin upon small subcutaneous resistance arteries of mild essential hypertensive patients. *J. Hypertens.*, 10, 437-444. 1992.

Schmitt, R., Belz, G. G., Fell, D., Lebmeier, R., Prager, C., Stahnke, P. L., Sittner, W. D., Karwoth, A. and Jones, C. R., Effects of the novel endothelin receptor antagonist bosentan in hypertensive patients. Presented at Seventh Eur. Meet. Hypertension, Milan, June 9 to 12, 1995.

Seo, B., Oemar, B. S., Siebermann, R., Segesser, L. and Lüscher, T. F., Both ET_A and ET_B receptors mediate contraction to endothelin-1 in human blood vessels. *Circulation*, 89, 1203-1208. 1994.

Shibouta, Y., Suzuki, N., Shino, A., Matsumoto, H., Terashita, Z., Kondo, K. and Nidshikawa, K., Pathophysiological role of endothelin in acute renal failure. *Life Sci.*, 46, 1611-1618. 1990.

Sørensen, S. S., Madsen, J. K. and Pedersen, E. B., Systemic and renal effect of intravenous infusion of endothelin-1 in healthy human volunteers. *Am. J. Physiol.*, 266, F411-F418. 1994.

Stewart, D. J., Levy, R. D., Cernacek, P. and Langleben, D., Increased plasma endothelin-1 in pulmonary hypertension: marker or mediator of disease?, *Ann. Intern. Med.*, 114, 464-469. 1991a.

Stewart, J. T., Nisbet, J. A. and Davies, M. J., Plasma endothelin in coronary venous blood from patients with either stable or unstable angina. *Br. Heart J.*, 66, 7-9. 1991b

Stingo, A. J., Clavell, A. L., Aarhus, L. L. and Burnett, J. C., Jr., Biological role for the endothelin-A receptor in aortic cross-clamping. *Hypertension*, 22, 62-66. 1993.

Strachan, F. E., Haynes, W. G., Storey, P. and Webb, D. J., Endothelin type A and B receptors mediate venoconstriction in humans. *Br. J. Pharmacol.*, (abstract in press). 1995.

Sugiura, M., Inagami, T. and Kon, V., Endotoxin stimulates endothelin-release *in vivo* and *in vitro* as determined by radioimmunoassay. *Biochem. Biophys. Res. Commun.*, 161, 1220-1227. 1989.

Suzuki, R., Masaoka, H., Hirata, Y., Marumo, F., Isotani, E. and Hirakawa, K., The role of endothelin-1 in the origin of cerebral vasospasm in patients with aneurysmal subarachnoid hemorrhage. *J. Neurosurg.*, 77, 96-100. 1992.

Takeda, M., Breyer, M. D., Noland, T. D., Homma, T., Hoover, R. L., Inagami, T. and Kon, V., Endothelin-1 receptor antagonist: effects on endothelin- and cyclosporin-treated mesangial cells. *Kidney Int.*, 42, 1713-1719. 1992.

Terada, Y., Tomita, K., Nonoguchi, H. and Marumo, F., Different localization of two types of endothelin receptor mRNA in microdissected rat nephron segments using reverse transcription and polymerase chain reaction assay. *J. Clin. Invest.*, 90, 107-112. 1992.

The Task Force on Heart Failure of the European Society of Cardiology. Guidelines for the diagnosis of heart failure. *Eur. Heart J.*, 16, 741-751. 1995.

Tomita, K., Ujiie, K., Nakanishi, T., Tomura, S., Matsuda, O., Ando, K., Scichibi, M., Hirata, Y. and Marumo, F., Plasma endothelin levels in patients with acute renal failure. *N. Engl. J. Med.*, 321, 1127-1127. 1989.

Tomobe, Y., Miyauchi, T., Saito, A., Yanagisawa, M., Kimura, S., Goto, K. and Masaki, T., Effects of endothelin on the renal artery from spontaneously hypertensive and Wistar Kyoto rats. *Eur. J. Pharmacol.*, 152, 373-374. 1988.

Touyz, R. M., Tolloczko, B. and Schiffrin, E. L., Mesenteric vascular smooth muscle cells from spontaneously hypertensive rats display increased calcium reponses to angiotensin II but not to endothelin-1. *J. Hypertens.*, 12, 663-673. 1994.

Toyo-oka, T., Aizawa, T., Suzuki, N., Hirata, Y., Miyauchi, T., Shin, W. S., Yanagisawa, M., Masaki, T. and Sugimoto, T., Increased plasma level of endothelin-1 and coronary spasm induction in patients with vasospastic angina pectoris. *Circulation*, 83, 476-483. 1991.

Tsunoda, K., Abe, K. and Yoshinaga, K., Endothelin in hemodialysis-resistant hypertension. *Nephron*, 59, 687-688. 1991.

Vallance, P., Collier, J. and Moncada, S., Effects of endothelium derived nitric oxide on peripheral arteriolar tone in man. *Lancet*, ii, 997-1000. 1989.

Véniant, M., Clozel, J. P., Hess, P. and Clozel, M., Endothelin plays a role in the maintenance of blood pressure in normotensive guinea pigs. *Life Sci.*, 55, 445-454. 1994.

Vierhapper, H., Wagner, O., Nowotny, P. and Waldhäusl, W., Effect of endothelin-1 in man. *Circulation* 81, 1415-1418. 1990.

Vierhapper, H., Wagner, O. F., Nowotny, P. and Waldhäusl, W., Effect of endothelin-1 in man: pretreatment with nifedipine, with indomethicin and with cyclosporine A. *Eur. J. Clin. Invest.*, 22, 55-59. 1992.

Wagner, O. F., Vierhapper, H., Gasic, S., Nowotny, P. and Waldhäusl, W., Regional effects and clearance of endothelin-1 across pulmonary and splanchnic circulation. *Eur. J. Clin. Invest.*, 22, 277-282. 1992.

Warner, T. D., Allcock, G. H. and Vane, J. R., Reversal of established responses to endothelin-1 *in vivo* and *in vitro* by the endothelin receptor antagonists, BQ-123 and PD 145065. *Br. J. Pharmacol.*, 112, 207-213. 1994.

Warrens, A. N., Cassidy, M. J. D., Takahashi, K., Ghatei, M. A. and Bloom, S. R., Endothelin in renal failure. *Nephrol. Dial. Transplant.*, 5, 418-422. 1990.

Watanabe, T., Awane, Y., Ikeda, S., Fujiwara, S., Kubo, K., Kikuchi, T., Kusumoto, K., Wakimasu, M. and Fujino, M., Pharmacology of a nonselective ET_A and ET_B receptor antagonist, TAK-044 and the inhibition of myocardial infarct size in rats. *Br. J. Pharmacol.*, 114, 949-954. 1995.

Watanabe, Y., Naruse, M., Monzen, C., Naruse, K., Ohsumi, K., Horiuchi, J., Yoshihara, I., Kato, Y., Nakamura, N., Kato, M., Sugino, N. and Demura, H., Is big endothelin converted to endothelin-1 in circulating blood?, *J. Cardiovasc. Pharmacol.*, 17 (Suppl. 7), S503-S505. 1991a.

Watanabe, T., Suzuki, N., Shimamoto, N., Fujino, M. and Imada, A., Contribution of endogenous endothelin to the extension of myocardial infarct size in rats. *Circ. Res.*, 69, 370-377. 1991b.

Webb, D. J. and Cockcroft, J. R., Plasma immunoreactive endothelin in uraemia. *Lancet*, ii, 1211. 1989.

Wei, C.-M., Lerman, A., Rodeheffer, R. J., McGregor, C. G. A., Brandt, R. R., Wright, S., Heublein, D. M., Kao, P. C., Edwards, W. D. and Burnett, J. C., Endothelin in human congestive heart failure. *Circulation*, 89, 1580-1586. 1994.

Weiczorek, I., Haynes, W. G., Ludlam, C., Fox, K. and Webb, D. J., Plasma endothelin predicts outcome in patients with unstable myocardial ischemic syndromes. *Br. Heart J.*, 72, 436-441. 1994.

Weiser, E., Wollberg, Z., Kochva, E. and Lee, S. Y., Cardiotoxic effects of the venom of the burrowing asp, *Actractaspis engaddensis* (Actractaspididae, Ophidia). *Toxicon*, 22, 767-774. 1984.

Weitzberg, E., Ahlborg, G. and Lundberg, J. M., Long-lasting vasoconstriction and efficient regional extraction of endothelin-1 in human splanchnic and renal tissues. *Biochem. Biophys. Res. Commun.*, 180, 1298-1303. 1991a.

Weitzberg, E., Lundberg, J. M. and Rudehill, A., Elevated plasma levels of endothelin in patients with sepsis syndrome. *Circ. Shock*, 33, 222-227. 1991b.

Wenzel, R. R., Noll, G. and Lüscher, T. F., Endothelin receptor antagonists inhibit endothelin in human skin microcirculation. *Hypertension*, 23, 581-586. 1994.

White, D. G., Garratt, H., Mundin, J. W., Sumner, M. J., Vallance, P. J. and Watts, I. S., Human saphenous vein contains both endothelin ET_A and ET_B receptors. *Eur. J. Pharmacol.*, 257, 307-310. 1994.

Winkles, J. A., Alberts, G. F., Brogi, E. and Libby, P., Endothelin-1 and endothelin receptor mRNA expression in normal and atherosclerotic human arteries. *Biochem. Biophys. Res. Commun.*, 191, 1081-1088. 1993.

Xu, D., Emoto, N., Giaid, A., Slaughter, C., Kaw, S., deWit, D. and Yanagisawa, M., ECE-1: A membrane bound metalloprotease that catalyses the proteolytic activation of big endothelin-1. *Cell*, 78, 473-485. 1994.

Yanagisawa, M., Kurihawa, H., Kimura, S., Tomobe, Y., Kobayashi, M., Mitsui, Y., Yazaki, Y., Goto, K. and Masaki, T., A novel potent vasoconstrictor peptide produced by vascular endothelial cells. *Nature*, 332, 411-415. 1988.

Yasuda, M., Kohno, M., Tahara, A., Itagani, H., Toda, I., Akioka, K., Teragaki, M., Oku, H., Takeuchi, K. and Takeda, T., Circulating immunoreactive endothelin in ischemic heart disease. *Am. Heart J.*, 119, 801-806. 1990.

Yokokawa, K., Tahara, H., Kohno, M., Murakawa, K., Yasunari, K., Nakagawa, K., Hamada, T., Otani, S., Yanagisawa, M. and Takeda, T., Hypertension associated with endothelin-secreting malignant haemangioendothelioma. *Ann. Intern. Med.*, 114, 213-215. 1991.

Yoshimoto, S., Ishizaki, Y., Sasaki, T. and Murota, S. I., Effect of carbon dioxide and oxygen on endothelin production by cultured porcine cerebral endothelial cells. *Stroke*, 22, 378-383. 1991.

Zamora, M. R., O'Brien, R. F., Rutherford, R. B. and Weil, J. V., Serum endothelin-1 concentrations and cold provocation in primary Raynaud's phenomenon. *Lancet*, 336, 1144-1147. 1990.

Index

A

B

C

D

F

G

H

I

Q

R

S

T

U

V